Directions for the Next Generation of MMIC Devices and Systems

Directions for the Next Generation of MMIC Devices and Systems

Edited by

Nirod K. Das

Weber Research Institute
Polytechnic University
Farmingdale, New York

and

Henry L. Bertoni

Weber Research Institute
Polytechnic University
Brooklyn, New York

Springer Science+Business Media, LLC

Library of Congress Cataloging-in-Publication Data

Directions for the next generation of MMIC devices and systems / edited by Nirod K. Das and Henry L. Bertoni.
p. cm.
Expanded selected papers from a symposium held September 11-13, 1996, Brooklyn, N.Y.
Includes bibliographical references and index.

1. Microwave integrated circuits. 2. Microwave communications systems. 3. Millimeter wave devices. 4. Microelectronic packaging. I. Das, Nirod K. II. Bertoni, Henry L.
TK7876.D57 1998
621.381'3--dc21 97-41794
CIP

DOI 10.1007/978-1-4899-1480-4

Cover illustration: Photo of a millimeter-wave/CMOS multichip module in a 24-pin ceramic package consisting of multiple functions—low-frequency CMOS processing electronics, a printed antenna array, and millimeter-wave detector circuit, operating at 94.6 GHz. (see Chapter 27).

Proceedings of an International Symposium on Directions for the Next Generation of MMIC Devices and Systems, held September 11 – 13, 1996, in Brooklyn, New York

Originally published by Plenum Press, New York in 1997.
MyCopy version of the original edition 1997
http://www.plenum.com

10 9 8 7 6 5 4 3 2 1

PREFACE

Microwave and millimeter-wave integrated circuits (MMICs) are of increasing importance in modern military and commercial wireless communication systems. Current trends are towards low-cost, high-density, multilevel, and multifunctional integration, covering millimeter and submillimeter wave regions. The integration of diverse subfunctions, such as light-wave devices, superconductor circuits, digital circuits and ferrite devices, together with conventional microwave or millimeter-wave devices, circuits and antennas, will allow implementation of large systems on a single chip. Research on advanced device concepts, 3-D interconnects, high-performance packaging methods, advanced CAD-tools, measurement and testing techniques, as well as material and fabrication technologies, are being directed to meet these new challenges.

Continuing on the series of symposia sponsored by the Weber Research Institute of Polytechnic University, an international symposium focusing on the current developments and new research initiatives for the next generation of microwave and millimeter wave integrated circuits and systems was held at Brooklyn, New York, during September 11–13, 1996. The symposium was organized as a 3-day event, running mostly in a single-session format of regular papers and panel discussions, It was co-sponsored by the Army Research Office, Research Triangle Park, NC, in cooperation with the IEEE Microwave Theory and Techniques Society, the IEEE Antennas and Propagation Society, and IEEE Long Island and New York Metropolitan Sections. The papers published in this volume are extended versions of selected papers presented at this symposium.

Nirod K. Das
Henry L. Bertoni
Co-Editors
May 1997

CONTENTS

INTRODUCTION

SYSTEMS AND INTEGRATION

NOVEL ANTENNAS AND DEVICE TECHNOLOGY

MODELING AND CAD

Directions for the Next Generation of MMIC Devices and Systems

INTRODUCTION

Scanning the Conference

N. K. Das and H. L. Bertoni

Weber Research Institute
Polytechnic University

Microwave and millimeter-wave integrated circuits (MMICs) are of increasing importance in modern military and commercial wireless communication systems. Current trends are towards low-cost, high-density, multilevel, and multifunctional integration, covering millimeter and submillimeter wave regions. The integration of diverse subfunctions, such as light-wave devices, superconductor circuits, digital circuits and ferrite devices, together with conventional microwave or millimeter-wave devices, circuits and antennas, will allow implementation of large systems on a single chip. Research on advanced device concepts, 3-D interconnects, high-performance packaging methods, advanced CAD-tools, measurement and testing techniques, as well as material and fabrication technologies, are being directed to meet these new challenges.

In response to this current interest and future technology needs, the Weber Reseach Institute of Polytechnic University hosted the International Symposium on Directions for the Next Generation of Microwave and Millimeter-wave Integrated Circuits and Systems. It was held at Brooklyn, New York, during September 11-13, 1996, and was organized as a 3-day symposium, running mostly in a single-session format of opening presentations, regular papers and panel discussions. All together 13 sessions were held, including the opening session, a panel session, and over 50 regular papers presented in 11 technical sessions. The participants included distinguished leaders in the field from industry, government and academia from United States and abroad.

The opening session of the conference included a talk on "Historical Perspectives on Microwave and Millimeter-Wave Integrated Circuits" by Prof. A. A. Oliner, Polytechnic University, a "Technical Introduction" by Dr. J. F. Harvey, Army Reseach Office, and the Keynote Address by Dr. Elliot Brown, DARPA MAFET Program. These presentations provided an outstanding survey of the field as well as an introduction to the conference.

Six panelists participated in the Panel Session entitled, "Future Directions for MMIC Research and Technology," moderated by Dr. J. F. Harvey, Army Research Office, and Prof. D. M. Bolle, Polytechnic University. Among the panelists, **(I.)** Dr. E. R. Brown,

Directions for the Next Generation of MMIC Devices and Systems
Edited by N. K. Das and H. L. Bertoni, Plenum Press, New York, 1997

DARPA MAFET Program, discussed on *"DARPA MAFET-3 Initiatives for MMIC Research and Development,"* **(II.)** Dr. B. S. Perlman, US Army Research Laboratory, EPSD, Fort Monmouth, NJ, discussed on *"Advances in MMIC Packaging and Interconnect Technology,"* **(III.)** Prof. G. M. Rebeiz, University of Michigan, Ann Arbor, discussed on *"Low-Cost GaAs/Si Front-End Electronics,"* **(IV.)** Dr. P. W. Staecker, M/A COM, Inc., Past President of IEEE Microwave Theory and Techniques Society, discussed on *"Future Directions in MMIC Research and Technology: The Wireless Pull,"* and **(V.)** Prof. R. J. Trew, Case Western Reserve University, Editor of IEEE Transactions on Microwave Theory and Techniques, discussed on *"Microwave Power Amplifiers: The Emerging High- and Low-Power Markets."*

As discussed in the Panel Session, the military needs for advanced MMICs would continue to drive some of the key research and technology areas. It is clear, however, that the commercial wireless market would be the strong driving force for low-cost, high-density, and multifunctional MMICs of the future. Research and technology development in advanced packaging and interconnect methods, GPS systems, power amplifiers, low-power electronics, and millimeter and sub-millimeter wave applications, are some of the areas of strategic needs. University research on EM-field theoretic and sophisticated numerical models, as well as advanced device and technology concepts, tied with curriculum development in the related subjects, would be critical in order to support the future research and technical manpower needs.

As classified in this volume, the topics covered in the conference can be grouped under three main subject areas: (1) Systems Integration, (2) Novel Antennas and Device Technology, and (3) Modeling and Computer Aided Design (CAD:)

I. The subject area of Systems Integration was broadly covered in four separate sessions: (i) Packaging and Interconnect Technology, (ii) Millimeter and Sub-millimeter wave systems, (iii) Advanced Integration Concepts, and (iv) Phased Array Systems. Advanced methods of interconnection and packaging are addressed as the areas of strategic importance to the next-generation of integrated circuits and systems. Low-cost, high- and ultra high-density integration, integration of diverse functions together in a single package, and applications in millimeter, sub-millimeter and terahertz frequency range, are a few of the advanced packaging needs of the future.

II. The subject area of Novel Antennas and Device Technology was broadly covered in three separate sessions: (i) Novel Microstrip Antennas, (ii) Si and SiGe MMIC Devices, and (iii) Microwave-Photonics devices. Innovations in integrated microstrip antennas and feeding techniques for sophisticated functions would be topics of continued research. Further, new device and fabrication technologies, and integration of MMICs with photonic and digital functions, would be in growing needs for future radar.communication systems.

III. The subject area of Modeling and CAD was broadly covered in four separate sessions: (i) New Guided Surface- and Leaky-Wave Effects in MMICs, (ii) Numerical/CAD methods, (iii) Next Generation of CAD, and (ii) Packaging and Interconnect Modeling. With the growing complexities of advanced devices, circuits and packages of the future, it is clear that sophisticated modeling and design tools would play a vital role in the future research and development. These include advanced field-theoretic as well as numerical models for device, circuit and system simulation, with increased emphasis on fast design iterations and "intelligence-based" algorithms.

HISTORICAL PERSPECTIVES ON MICROWAVE AND MILLIMETER-WAVE INTEGRATED CIRCUITS

Arthur A. Oliner

Department of Electrical Engineering
Polytechnic University
Brooklyn , NY 11201

INTRODUCTION

The principal highlights in the early history of electromagnetic guided waves are reviewed first, in order to show when and why hollow pipes were proposed and built as guiding structures for microwaves. The paper then develops the transition from that elemental form to modern-day microwave and millimeter-wave integrated circuits, including the competition between strip line and microstrip line, by combining the underlying physical principles with historical developments and anecdotes. The paper concludes by proposing that the development of microwave integrated circuits be viewed in terms of three stages, corresponding simultaneously to time periods and to the types of solutions required to characterize the circuit performance.

I. MAJOR DEVELOPMENTS LEADING TO THE USE OF HOLLOW PIPES AS WAVEGUIDES

A. The Early History (Before About 1910)

Although many individuals made important contributions to the early developments in electromagnetic waves, we select here only four outstanding individuals and their particular contributions to indicate the major steps in these developments. These four individuals are Maxwell, whose theory showed that electromagnetic waves were possible, Hertz, who verified Maxwell's theory experimentally, Rayleigh, who was the first to derive the properties of waveguides, and Marconi, whose successful demonstration that low-frequency waves could be propagated over long distances was a major element in postponing any further interest in microwaves for over two decades.

1. James Clerk Maxwell: Before Maxwell developed his well-known theory, much was already known about electromagnetics, from contributions made by many people with famous names like Ampere and Faraday. Maxwell's elegant and beautiful theory, which he

Directions for the Next Generation of MMIC Devices and Systems
Edited by N. K. Das and H. L. Bertoni, Plenum Press, New York, 1997

perfected over a long period of time around 1860 or so, tied all these separate bits of understanding into a unified whole, but he also introduced a new concept, the displacement current, to make electricity and magnetism more symmetrical. With this added term, his equations showed that electromagnetic ***waves*** were possible. When he solved for the velocity of these waves and then compared them with the measured velocity of light, he found excellent agreement.

His theory indeed demonstrated that light was an electromagnetic wave, but the more general, and more important, implication was that it should be possible to produce electromagnetic waves at ***any*** frequency. At that time, however, his theory was not well understood, and many of his contemporaries were skeptical. On the other hand, it was clear that his results would be very important, if true. Some German learned societies offered prizes to anyone who could prove Maxwell's theory experimentally, but there were no takers because there were no sources or detectors; they had to be invented. It was not until about 20 years later, some 10 years after Maxwell died, that the theory was experimentally proved.

The famous physicist, Richard P. Feynman, has stated the truly great importance of Maxwell's contribution in this way [1]: "From a long view of the history of mankind...there can be little doubt that the most significant event of the 19th Century will be judged as Maxwell's discovery of the laws of electromagnetics."

2. Heinrich Hertz: The experimental verification of Maxwell's theory was performed by Heinrich Hertz in a series of brilliant experiments that he began in 1886. Hertz created a spark generator with a resonant circuit attached, devised detectors, invented the dipole antenna, built parabolic-cylinder reflectors, showed that waves could be propagated in air and also along wires, and produced standing waves. He also did original theoretical work. His contributions are described in detail in a relatively recent short book [2].

Unfortunately, he died from an illness at an early age (36 years); he would undoubtedly have contributed much more had he lived longer. His experiments not only verified Maxwell's theory very clearly, but he showed that electromagnetic waves could be excited at various frequencies. In his own experiments he produced electromagnetic waves with wavelengths of 6 meters, 3 meters and 60 centimeters. After his landmark experiments, the field of electromagnetic-wave engineering moved rapidly in various directions, and in various countries.

For his important fundamental contributions, the unit of frequency, the Hertz, has been appropriately named after him. In addition, the IEEE has established the Hertz Medal as one of its major awards. Since this symposium is sponsored by the Polytechnic University, it is appropriate to mention that a Polytechnic University professor, Nathan Marcuvitz, was the first recipient of that award.

3. John William Strutt, Lord Rayleigh: Rayleigh, who succeeded Maxwell as Cavendish Professor at Cambridge, was a prolific contributor to all sorts of topics in classical physics. He seems to have been the first in nearly everything, including the resolving power of gratings, an explanation of why the sky is blue, a host of new results on the theory of sound, and the discovery of argon, for which he received the Nobel Prize.

In microwave theory, he was the first to derive and then discuss in detail (in 1897) the electromagnetic modes that can propagate through hollow metallic tubes [3]. This work actually contains the fundamental ideas of mode propagation and cutoff in waveguides. He also derived the scattering of electromagnetic waves by circular apertures and by ellipsoidal obstacles [4], laying the foundation for the very useful "small aperture" and "small obstacle" methods which were revived and developed further during World War II.

Interest in microwave wavelengths was strong from about 1890 to the very early 1900s, but it dropped significantly after Marconi's dramatic demonstration (see immediately below). Because of this, Lord Rayleigh's basic and pioneering work on waveguides became buried in the literature and needed to be rediscovered during the 1930s.

4. Guglielmo Marconi: Marconi was fascinated by the idea of wave propagation in air, and he conducted many simple early experiments in Italy before moving to England, where he founded the British Marconi Company and continued this work. In 1901 he was ready for his big experiment, the transmission of simple Morse-Code telegraphy across the Atlantic Ocean, between Cornwall, England and Newfoundland, Canada. His success created a sensation, and it marked the beginning of the era of wireless communication.

It also led to proposals of various mechanisms to explain why such transmission over long distances was possible. One mechanism was based on surface waves guided around the earth, which stimulated basic investigations of surface waves and involved such prominent scholars as J. Zenneck and A. Sommerfeld. The correct explanation, however, was proposed independently by A. E. Kennelly and O. Heaviside in 1902, and later confirmed experimentally by various researchers. They proposed that the radiated waves were successively reflected between the earth and an electrically conducting layer in the upper atmosphere, with the result that the waves would be guided around the curved earth.

Marconi's historic demonstration had an additional very important consequence, Many of the early studies of wave propagation were made at UHF and microwave frequencies, even though low-frequency sources were easier to build and yielded higher power. As an example of such interest, we recall Lord Rayleigh's basic theory of wave guidance by hollow pipes, and we may add that even Marconi's early experiments involved frequencies around 1 GHz. Marconi's long-distance transmission experiment used low-frequency waves, however, and its dramatic success (and the rapid appearance of further successes with long-wavelength radio) effectively put an end to further interest in the utilization of short wavelengths for about two decades.

There are many sources in the literature to which one may go to find further information about various contributions made in this early period, roughly before 1910. An excellent source is the Special Centennial Issue of the IEEE Transactions on Microwave Theory and Techniques, Vol. MTT-32, September 1984. Many of the Societies within the IEEE agreed to prepare special historical issues to commemorate the 100th anniversary of the IEEE. The special issue prepared by the MTT Society Transactions won an award for the best issue among all the Transactions, and its Guest Editor, Theodore S. Saad, received a special commendation. Although none of the papers in that issue is devoted specifically to this early period, the portions at the beginnings of several of the papers contain relevant information [5-8]. Of course, other portions of these papers and of various other papers in this issue are directly relevant to other sections of this historical paper, and they will be referenced in context. A notable paper [9] not in this special issue is devoted fully to this early period, and stresses the activity involving microwave frequencies.

B. Early Work on Hollow Waveguides (1930-1940)

Shorter waves were neglected until about 1930, and interest in such waves was revived only after ship-to-shore and transoceanic wireless transmission became commonplace. The two major figures in the development of hollow-pipe waveguides for these shorter wavelengths were George C. Southworth and Wilmer L. Barrow, who headed groups at the Bell Laboratories and at MIT, respectively. It is quite astonishing that each worked independently and without any knowledge of the accomplishments of the other, and yet the results of their investigations were remarkably similar.

We are very fortunate that Dr. Southworth has written a book [10] detailing his personalized history of this period; it is very revealing not only with respect to technical details but also the attitudes of the time. He was working for the Bell Laboratories, but he was curious about how to guide these higher-frequency waves without loss due to radiation, and he set out to conduct some experiments on his own. He concentrated on hollow circular metal guides, but he was unaware of Lord Rayleigh's early analysis [3] and had to rely loosely on a 1920 paper by Otto Schriever on dielectric rods. He did in fact perform some experiments with dielectric guides. In his early experiments the sources available to him were rather low in frequency, but he filled his waveguides with water ($\varepsilon_r \approx 80$) so that the guide diameters could be reduced by a factor of about nine. Later, with better sources, he worked with air-filled pipes.

The Bell Laboratories were skeptical of the value and even the validity of his experiments, and at one stage he was "ordered to be assigned to more constructive work." Later, he was transferred to the Research Department, and he was then able to work with some of the mathematicians there, including S. A. Schelkunoff. They discovered Lord Rayleigh's original paper and extended it to include loss, and they also discovered the TE_{01} mode, for which the attenuation ***decreased*** as the frequency is increased. After further successful work, Southworth wanted to publish his results but his superiors were reluctant because of a fear that the work was fallacious. Not until almost two years later was the approval for publication finally given.

The other major figure in these developments was Wilmer L. Barrow, a professor at the Massachusetts Institute of Technology (MIT). His initial motivation for a hollow metal tube to guide electromagnetic waves was to have it as a feed for a radiating horn. He was also unaware of Lord Rayleigh's earlier work, and the results of his initial experiments turned out to be different from what he had expected. As a result, he had to engage in a set of basic investigations that involved theoretical analysis as well as experiments. His efforts overlapped in many ways those of the people at the Bell Laboratories, but there were differences, primarily in his investigations of radiation from the open end of the guide.

Both Southworth and Barrow began their investigations in 1931, and both were ready to publish their results in 1936. Amazingly, an oral presentation was to be given in Washington, D.C., by Southworth on April 30, 1936, and a corresponding one, on a similar topic, was scheduled to be presented in the same city by Barrow on the very next day, May 1, 1936. To compound the coincidence, a colleague working at the Bell Laboratories who was a graduate of MIT visited his former school at the end of March, saw Barrow's laboratory, and then informed both Barrow and Southworth about the work of the other. Both were stunned by the news, and Southworth immediately wrote a letter to Barrow, enclosing his paper, which was about to be published [11], and a companion paper on the theory by his mathematics colleagues [12]. Barrow reciprocated by exchanging information, but his paper [13] was not published until October, several months later. Both the Bell Laboratories and the MIT groups continued to make further contributions and to publish additional papers.

Two of the papers [5,7] in the Special Centennial Issue of the MTT Transactions contain detailed treatments of the contributions made by Southworth and Barrow. A two-page summary (pp. 1024 and 1025) is given in paper [7], whereas essentially the whole of paper [5] (pp. 963-969) is devoted to these developments.

C. The World-War II Period

The microwave field made striking advances during the World-War II period as a result of the tremendous push to develop centimeter-wave radar systems. The development in the late 1930s in Great Britain of a magnetron capable of reliably producing centimeter wavelengths served as the motivation for this push because it provided the Allies with the promise of something unique -- centimeter-wave radar. It was known that Japan had developed UHF radar, but further progress there was hampered by internal rivalry between the Army and the Navy. Curiously, microwave activity in Germany was forbidden during World War II because it was believed that it was useless for electronic warfare.

The center of activity for microwave research and development in the USA was the Radiation Laboratory at the Massachusetts Institute of Technology (MIT), where many prominent and highly capable individuals were grouped together to produce an unusually stimulating and productive working environment. This laboratory had also established a variety of satellite laboratories at places like Harvard, Stanford, Brooklyn Polytechnic, and Columbia, and also worked closely with the Telecommunications Research Establishment in England and McGill University in Canada. After the war, the Radiation Laboratory coordinated the writing of a set of 28 volumes covering in great detail the many achievements developed during the five-year period of the Laboratory's existence. This Radiation Laboratory Series of books established the state of the art and served as an invaluable guide for the next generation of engineers in the microwave field. The enormous advances made during the war period, and the knowledge made available by the Radiation Laboratory Series of books, allowed the microwave field to develop very rapidly after the war and enabled the U.S. to establish a predominant position in this field.

The fundamental guiding structure on which most of the microwave circuitry was based was the hollow metal rectangular waveguide. An extensive number of clever and effective components were invented and developed during this period, and these in turn were employed in a variety of systems. The early designs of these components were hampered by the lack of information regarding the behavior of discontinuities in these structures. A systematic program was established, addressing both the development of new theoretical techniques and the precise measurements of various specific discontinuities. Nathan Marcuvitz had the responsibility for the measurements, and Julian S. Schwinger was the outstanding figure among the theorists.

By chance, they lived in the same rooming house and grew to be good friends. However, Schwinger worked during the night and slept all day. Marcuvitz would therefore wake him up at 7:30 PM, and they would go to dinner. After that they would often discuss their research problems until midnight, after which Marcuvitz would go to bed and Schwinger would begin his work. From these discussions Schwinger learned about engineering and Marcuvitz developed into an excellent theorist, for which he is known primarily. As is well known, theoretical and measured results for many discontinuity structures in rectangular waveguide were compiled by Marcuvitz and published as the Waveguide Handbook [14], which was Vol. 10 of the Radiation Laboratory Series, and was very widely used. Schwinger was later prevailed upon to get up earlier (in the afternoon) in order to present a series of lectures to his fellow theorists; notes on these lectures, which later became famous, were taken by David S. Saxon [15]. Further details regarding the characterization of waveguide discontinuities during the World-War II period may be found in pp. 1027-1030 of [7].

Using these theoretical expressions for the equivalent circuits of discontinuities in rectangular waveguide, derived by Schwinger and his colleagues, and also by others, it became easy to design components in rectangular waveguide quite accurately. The demand for guiding structures capable of greater bandwidth, however, caused people to look at alternative structures, and ultimately led to the design and widespread use of microwave integrated circuits. The highlights in the transition from hollow rectangular waveguides to the printed-circuit transmission lines used in microwave integrated circuits are summarized in the next section, together with remarks concerning the physical principles involved.

II. THE TRANSITION FROM HOLLOW PIPES TO PRINTED CIRCUITS

A. Rectangular Waveguides (One Conductor) and Coaxial Lines (Two Conductors)

The microwave industry blossomed rapidly after World War II, with applications to radar, communications, electronic warfare, and so on. Hollow rectangular waveguides were employed almost exclusively in these microwave systems, and the circuits that used them leaned heavily on the designs developed during World War II. Further improvements in performance were of course being made continually, but, after a few years, it was recognized that the type of guiding structure being used suffered from an inherent limitation with respect to bandwidth.

Rectangular waveguide was a member of the class of guiding structures called "uniconductor waveguides," in which the power was transmitted along the inside of a hollow metallic pipe in order to prevent loss by radiation. The basic properties of guidance by such hollow pipes was understood very early on, beginning with Lord Rayleigh's analysis in 1897 [3] and then rediscovered [11-13] during the 1930s. The major physical feature was the phenomenon of cutoff frequency, below which the mode could not propagate. The cross section of the rectangular waveguide was designed to have a width that was equal to or slightly greater than twice that of its height, to insure that the cutoff frequency of the first higher-order mode was not less than twice that of the dominant mode.The dominant mode could then propagate alone between its cutoff frequency and that of the first higher-order mode, so that, in principle, the guide could support single-mode propagation over a two-to-one frequency range. Because of increased loss due to the finite conductivity of the metal in the neighborhood of the cutoff for each mode, however, the frequency range for single-mode propagation was significantly less than two-to-one in practice.

Much creative effort was expended in attempting to increase the frequency range for single-mode transmission. One direction involved modifying the rectangular waveguide cross section, but retaining the uniconductor-guide approach. A notable example was ***ridge waveguide***, in which the height of the center portion of the cross section was depressed, thereby reducing the cutoff frequency of the dominant mode strongly while changing that for the first higher-order mode only slightly. Sketches of rectangular waveguide and of one type of ridge guide are shown in Fig. 1. The improvement with ridge guide was significant but not sufficient, and it was finally appreciated that the uniconductor approach had to be given up altogether.

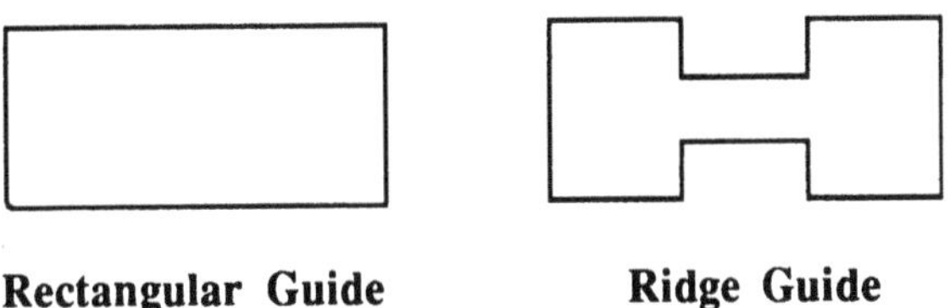

Rectangular Guide **Ridge Guide**

Fig. 1 Cross sections of two examples of uniconductor (hollow-pipe) waveguides: rectangular guide and one form of ridge guide. Ridge guide serves to increase the bandwidth of the dominant mode.

The next step, a very big one, was to recognize that ***two conductors*** were necessary. Examples of such guiding structures are coaxial line, two-wire line, parallel-plate guide, etc., which were certainly not new to electromagnetics, but they were now being looked at in a new light.

Two conductors allow the presence of an additional mode, the TEM mode, which can propagate down to zero frequency but cannot exist in uniconductor waveguides. All of the TE and TM modes that are possible in uniconductor guides are also present in two-conductor guides, as higher-order modes. These TE and TM modes all have cutoff frequencies, however, whereas the TEM mode exists at all frequencies. The bandwidth for single-mode propagation therefore extends from zero frequency to the cutoff frequency of the first higher-order mode. If the cross section of the two-conductor guiding structure is made smaller, furthermore, the cutoff frequency of the first higher-order mode is increased, thereby extending even further the range of single-mode propagation. Two-conductor guides therefore offer these two huge advantages:

(a) ***Extremely wide-band single-mode performance***, as discussed above, and

(b) The possibility of ***miniaturization***, because the TEM mode can propagate no matter how small the cross section is made. For uniconductor guides, the cross section must maintain a certain size which depends on the operating frequency.

The first two-conductor structure that was examined seriously in this new light was ***coaxial line***. However, two problems presented themselves. The first was the fact that it was expensive and even difficult to build components in coaxial line; the second was the absence of a longitudinal component of field in the TEM mode, so that designs for many components in rectangular waveguide which made use of the longitudinal component could not be carried over to coaxial line.

In order to minimize the fabrication difficulties and expense, ***flattened versions*** of coaxial line were introduced. The design of new components using the TEM mode required ingenuity, however. In Fig. 2 we see the cross sections of the usual coaxial line and of one of the flattened versions. The flattened versions were fitted with input and output coaxial connectors.

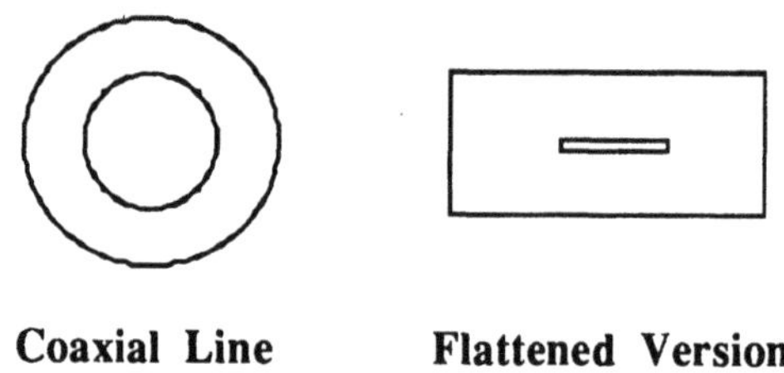

Fig. 2 Cross sections of two examples of early two-conductor transmission lines: the usual coaxial line and a flattened version of it. Two-conductor lines permit very wide-band single-mode performance, and miniaturization.

Soon afterwards these flattened versions were supplanted by an even more flattened version, the printed-circuit strip line, and later by its competitor, microstrip line. These lines are discussed in the next section, but it is important to note that they use two conductors. The use of rectangular waveguide continued strongly for applications not requiring wide-band performance, but the extent of its use gradually diminished with time. Rectangular waveguides are still employed today, but usually for specialized applications such as high-Q filters and high power.

B. The Competition Between Strip Line and Microstrip Line

The concept that led to the flattened versions of coaxial line was extended in a much bolder way by Robert M. Barrett in 1952 [16]. He envisionised a transmission line that would not only be a flattened coaxial line but which could also be ***printed,*** and fabricated by etching or by a silk-screen process using a metallic ink, for example. His structure was essentially that shown in Fig. 3(a), and became known as ***strip transmission line,*** or ***strip line***. He actually foresaw complete "microwave printed circuits" using this transmission line as the basis.

Barrett was the prime mover in the development of this concept, and, in his role at the Air Force Cambridge Research Center (now called Rome Laboratory at Hanscom Field, MA), he furnished contract money and coordinated the exchange of information among the organizations he supported in these studies. These organizations included the Airborne Instruments Laboratory (AIL), Tufts College, The Polytechnic Institute of Brooklyn, and Sanders Associates. Barrett wrote a popular article [17] early on, in 1952, to encourage interest in this new type of waveguide. He stressed the simplicity of the structure, its printed-circuit nature, and other of its virtues. In recognition of his fundamental contribution, the IEEE Microwave Theory and Techniques Society presented him its Pioneer Award in 1992.

Shortly after the appearance of Barrett's article [17], a group of engineers from the Federal Telecommunications Laboratories of ITT presented a series of three papers [18-20] on ***microstrip line.*** Developed completely independently, the structure of microstrip line, shown in Fig. 3(b), was conceived as a modification of a (bisected) two-wire transmission line rather than coaxial line. In their papers, the authors presented the concept, an approximate theory, and various components.

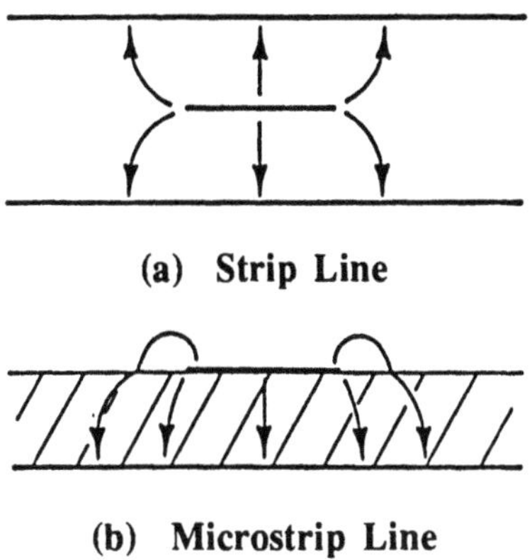

Fig. 3 The earliest two examples of printed-circuit transmission lines: (a) strip line, and (b) microstrip line. The cross sections are shown, together with the electric field lines.

Progress on strip line and microstrip line proceeded so rapidly that a full-scale symposium on Microwave Strip Circuits was held in October, 1954 at Tufts College under the sponsorship of AFCRC. At this symposium, the Proceedings of which were published as a special issue of the IRE Transactions on Microwave Theory and Techniques in March 1955 [21], the extensive developments on microstrip line were summarized in a paper by M. Arditi [22], but it became clear from the rest of the presentations that only ITT was working on microstrip and that everyone else was using strip line.

There were good and valid technical reasons for the preference for ***strip line.*** There were two basic structural versions of strip line available commercially. The one made by AIL, called "Stripline," supported the center strip by using a very thin dielectric sheet on which the center strip was printed on both sides and registered together. The dielectric sheet is located in a region of minimum field, so that the structure is basically air-filled. The second version, made by Sanders Associates and called "Tri-plate," placed the strip centrally between two identical layers of dielectric material; the structure is therefore completely dielectric-filled. Because, in either version, the region between the two outer metal plates of strip line contains only a single medium, the phase velocity and the characteristic impedance of the dominant (purely TEM) mode do not vary with frequency. Furthermore, because of the symmetry of the structure, all discontinuity elements in the plane of the center strip are purely reactive.

In the case of ***microstrip line,*** the strip must be supported by placing it on a dielectric layer. As a result, the transmission-line cross section contains two different media, so that the dominant mode becomes hybrid, not TEM. Therefore the phase velocity, characteristic impedance, and field variation in the guide cross section all become mildly frequency dependent. Furthermore, because of the symmetry unbalance, all discontinuity elements possess some resistive content and therefore radiate to some extent in the form of surface waves on the surrounding dielectric layer and into space waves if there is no top cover.

Practioners in the field quickly compared the advantages and disadvantages of strip line with those of microstrip line, and saw the following: For strip line, the dominant mode was a ***TEM mode*** with a simple field configuration, the mode is ***nondispersive***, and discontinuities in the strip plane were purely ***reactive***. For microstrip line, the dominant mode was a ***hybrid mode***, with all six field components, the mode was ***dispersive*** so that frequency dependences had to be taken into account in the designs, and discontinuities in the strip plane were ***resistive*** as well as reactive, resulting in some ***radiation*** into surface waves and space waves. We must also recall that at that time each component was packaged separately, with coaxial input and output connectors; there was therefore little attempt to reduce the size of the line's cross section, so that the performance differences mentioned above were actually not negligible.

AIL stressed the radiation feature of microstrip discontinuities by calling AIL's strip line "High-Q Stripline." In an oral technical presentation, Eugene G. Fubini of AIL jokingly suggested that one could actually use a microstrip dipole element as an antenna. That remark led to a confrontation with ITT, and Fubini withdrew the comment (but not the name High-Q Stripline). It is ironic that some 15 years later microstrip patch antennas were proposed and turned out to be highly successful.

With all of these negative features arrayed against microstrip, it is not surprising that within a relatively short time the symmetrical form of strip line was the clear winner, and microstrip stayed in the background for another decade.

Some time around the mid-1960s, the concept of ***microwave integrated circuits*** began to take hold. It was realized that it was not necessary to package each component separately and then connect them using separate connectors. Since the strip circuits were basically printed anyway, why not combine several components into a single package and connect them directly to each other within the same package? An example of a portion of a microwave integrated circuit is sketched in Fig. 4, which consists of, from left to right, a directional coupler, a simple filter, a mitered right-angle bend, and a tee junction, all connected directly. Because several components were contained within the same package, there was a natural tendency to reduce the cross sections of the printed-circuit transmission lines to prevent the package from appearing too large. This tendency toward miniaturization was enhanced by the growing availability of improved substrate materials, particularly alumina with its high value of dielectric constant. The trend toward miniaturization accelerated rapidly, and the contrast between the integrated-circuit packages and those they were designed to replace, which used rectangular waveguide, were dramatic with respect to the reduction in size and weight.

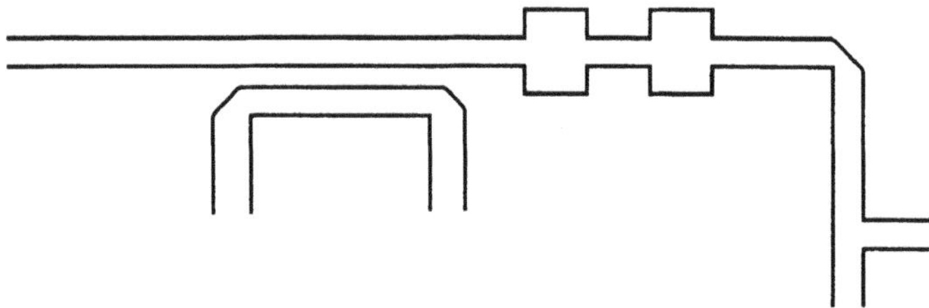

Fig. 4 A sketch of a portion of a microwave integrated circuit, where the individual components are connected directly together.

Both strip line and microstrip line (one ***or*** the other) were used in these miniaturized microwave integrated circuits. As a result of the miniaturization process, which substantially reduced the size of the transmission-line cross section, both the reactive and the resistive contents of the microstrip discontinuity elements were greatly reduced, thereby removing one of the prime objections to the original microstrip circuits. In addition, the effect of dispersion on the phase velocity and the characteristic impedance became greatly reduced. The electrical performance properties of microstrip line therefore became essentially equivalent to those of strip line at the UHF and low microwave frequencies used then.

When active circuit elements were employed with semiconductor substrates, it was more convenient to use microstrip line because it was no longer necessary to maintain symmetry in the cross section. The first big push in this direction involved the development by Texas Instruments [23] of a sophisticated transmit-receive module for a phased array program named MERA (Molecular Electronics for Radar Applications). Microstrip line gradually caught the imagination of microwave designers, and, after only a few short years, microstrip line effectively replaced strip line as the dominant transmission line used in microwave integrated circuits. The symmetrical form of strip line is still essential when extremely wide-band performance is required, for power dividers or directional couplers, for example, because strip line is dispersionless. It is interesting, if not ironic, that the roles of strip line and microstrip line had ***reversed*** so completely as a result of miniaturization, which removed the performance limitations possessed earlier by microstrip line, and the feeling that microstrip line was easier to use.

Further details regarding the theoretical research efforts on strip line may be found on pp. 1032-1034 of [7]; an overview of microwave integrated circuits is given in [24]; a detailed description of strip-line circuits and circuit designs appeared in [25]; and many details regarding the MERA program and early developments in monolithic microwave integrated circuits more generally may be found in [23].

C. Assessment of Developments Since the 1960s

During the late 1960s new types of printed-circuit transmission lines were proposed, although they did not become popular until later. These were the ***slot line*** [26] and the ***coplanar waveguide*** [27], due to Seymour B. Cohn and C. P. Wen, respectively, both in 1969. They both involved printing on only the top of the dielectric layer, but they possessed opposite electrical bisection symmetries. Modifications in these lines were introduced as time passed. The side conducting plates in these lines extended laterally to the edges of the package, to transverse "infinity." One type of modification was therefore to make these side plates finite in width, effectively changing them from plates to strips. For slot line, the modified structure was called "coplanar strips." A second modification involved the introduction of a ground plane under the dielectric layer; the lines were then designated as "conductor-backed." Conductor backing produced several advantages, such as greater mechanical strength, a heat sink, and a better value of characteristic impedance. When the side plates were "infinite" transversely, however, these conductor-backed lines become leaky at all frequencies [28], requiring some modification to make them practical [29,30]. Cross sections of these various lines are illustrated in Fig. 5. In some circuits, transmission lines of opposite bisection symmetry are combined together; sometimes, when conductor backing is not used, these lines are called "uni-planar" lines.

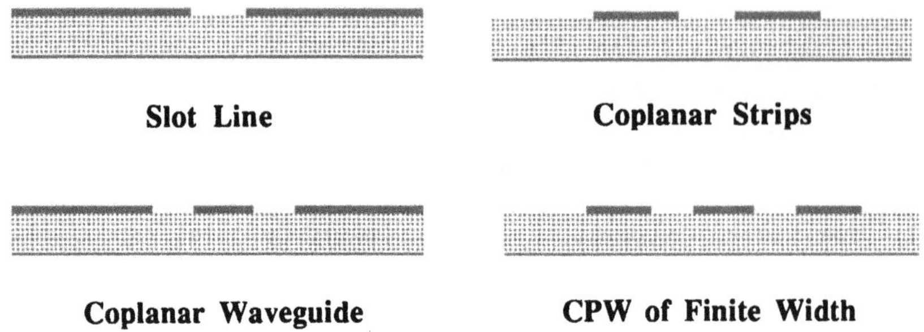

Fig. 5 Other printed-circuit transmission lines: slot line and coplanar waveguide of infinite width and finite width. Slot line of finite width is called coplanar strips. Conductor-backed versions of these lines involve placing a ground plane underneath the dielectric layer.

As time passed, and circuit sophistication improved, more accurate characterizations of the line discontinuities were required, and higher frequencies were employed. In response to these needs, remarkable progress was made in the development of new and improved ***numerical techniques*** and their combination with analytical methods. Such progress is actively continuing today.

With respect to the frequencies employed, the important consideration is the substrate height h relative to the wavelength λ_0. If one operates at a higher frequency, but scales all dimensions accordingly, maintaining the ratio of h/λ_0, then the calculation method used at the lower frequency will yield similar accuracy at the higher frequency. (Of course, the value of ε_r (or μ_r) is not expected to be scalable.) As the millimeter-wave range is approached, however, it becomes difficult to scale the substrate height because the substrate cannot be made too thin. As a result, the ratio h/λ_0 must be increased, so that the dispersion behavior becomes more pronounced and the reactive and resistive contents of the line discontinuities increase significantly. Dynamic, or full-wave, solutions for the transmission-line behavior then become necessary.

When the value of h/λ_0 increases even further, it is found that all kinds of ***new physical effects*** occur. Among these effects are leakage from dominant modes, which were initially assumed to be purely bound at all frequencies, the behavior of higher-order modes, particularly in the vicinity of cutoff, where they become leaky over a narrow frequency range, interactions with modes of the package, and loss effects, which may also be unusual at lower frequencies. When these effects occur they are not simply interesting, but they can cause crosstalk, power loss, and package resonances, which can seriously affect circuit performance. Most of these effects either do not arise at lower frequencies or their impact is less significant.

I have found it conceptually convenient to view the development of microwave integrated circuits as occurring in three stages, corresponding simultaneously to time periods and to the types of solutions required to characterize the circuit performance. These ***three stages of microwave integrated circuits*** are summarized in Table 1, and they follow the discussion above.

Table 1. THE THREE STAGES OF MICROWAVE INTEGRATED CIRCUITS

1. QUASI-STATIC SOLUTIONS (Early to late 1970s)

Valid for:	**thin substrates** **cross sections small compared to λ_0**
Yielded:	**low-frequency solutions** **reactive discontinuities (L and C)**

2. DYNAMIC (FULL-WAVE) SOLUTIONS (Late 1970s, 1980s)

Valid for:	**operation at higher frequencies**
Yielded:	**dispersion behavior** **resistive, as well as reactive, discontinuities**

3. NEW PHYSICAL EFFECTS AT HIGHER FREQUENCIES (Late 1980s, 1990s)

Behavior of higher-order modes (leaky near cutoff)
Power leakage from dominant modes
Interactions with modes of the package
Anomalous loss effects

An additional remark should be made in connection with the ***monolithic*** microwave integrated circuits (MMICs) made on a ***chip***, which have within the past decade become much more widely used and more complex and sophisticated in character. Because the fabrication process permits these chips to become so small, the regime on the chip itself corresponds to stage 1, for which quasi-static solutions are valid, even if the operating frequency is very high. On the chip, therefore, the discontinuities are characterized in low-frequency terms. However, we must remember that the MMIC chip must connect to other portions of the overall circuit, so that, depending on the operating frequency, those other portions may need to be viewed as being in stage 2 or stage 3.

REFERENCES

1. R. P. Feynman, R. B. Leighton, and M. L. Sands, *The Feynman Lectures on Physics, Vol. II*, p. 1-11, Addison-Wesley, Reading, MA (1964).
2. J. H. Bryant, *Heinrich Hertz, The Beginning of Microwaves*, Institute of Electrical and Electronics Engineers, New York (1988).
3. Lord Rayleigh, On the passage of electric waves through tubes, or the vibration of dielectric cylinders, *Phil. Mag.* 43:125 (1897).
4. Lord Rayleigh, On the incidence of aerial and electric waves on small obstacles in the form of elliptic cylinders; on the passage of electric waves through a circular aperture in a conducting screen, *Phil. Mag.* 44:28 (1897).
5. K. S. Packard, The origin of waveguides: a case of multiple rediscovery, *IEEE Trans. Microwave Theory Tech.*, Special Centennial Issue 32:961 (1984).
6. J. H. Bryant, Coaxial transmission lines, related two-conductor transmission lines, connectors, and components: a U.S. historical perspective, *IEEE Trans. Microwave Theory Tech.*,

Special Centennial Issue 32:970 (1984).
7. A. A. Oliner, Historical perspectives on microwave field theory, *IEEE Trans. Microwave Theory Tech.*, Special Centennial Issue 32:1022 (1984).
8. H. Sobol, Microwave communications -- an historical perspective, *IEEE Trans. Microwave Theory Tech.*, Special Centennial Issue 32:1170 (1984).
9. J. F. Ramsay, Microwave antenna and waveguide techniques before 1900, *Proc. IRE* 46:405 (1958).
10. G. C. Southworth, *Forty Years of Radio Research*, Gordon and Breach, New York (1962).
11. G. C. Southworth, Hyper-frequency wave guides -- general considerations and experimental results, *Bell Syst. Tech. J.* 15:284 (1936).
12. J. R. Carson, S. P. Mead, and S. A. Schelkunoff, Hyper-frequency wave guides -- mathematical theory, *Bell Syst. Tech. J.* 15:310 (1936).
13. W. L. Barrow, Transmission of electromagnetic waves in hollow tubes of metal, *Proc. IRE* 24:1298 (1936).
14. N. Marcuvitz, *Waveguide Handbook*, MIT Radiation Laboratory Series vol. 10, McGraw-Hill, New York (1951).
15. D. S. Saxon, Notes on lectures by Julian Schwinger: Discontinuities on waveguides, (1945).
16. R. M. Barrett and M. H. Barnes, Microwave printed circuits, *IRE National Conf. on Airborne Electronics*, Dayton, OH (1951).
17. R. M. Barrett, Etched sheets serve as microwave components, *Electronics* 25:114 (1952).
18. D. D. Grieg and H. F. Engelman, Microstrip -- a new transmission technique for the kilomegacycle range, *Proc. IRE* 40:1644 (1952).
19. F. Assadourian and E. Rimai, Simplified theory of microstrip transmission systems, *Proc. IRE* 40:1651 (1952).
20. J. A. Kostriza, Microstrip components, *Proc. IRE* 40:1658 (1952).
21. *Proceedings of Symposium on Microwave Strip Circuits*, Tufts College, 1954, Special issue of *IRE Trans. Microwave Theory Tech.* vol. 3 (1955).
22. M. Arditi, Characteristics and applications of microstrip for microwave wiring, *IRE Trans. Microwave Theory Tech.*, Special issue: Symposium on Microwave Strip Circuits 3:31 (1955).
23. D. N. McQuiddy, Jr., J. W. Wassel, J. B. LaGrange, and W. R. Wisseman, Monolithic microwave integrated circuits: an historical perspective, *IEEE Trans. Microwave Theory Tech.*, Special Centennial Issue 32:997 (1984).
24. H. Howe, Jr., Microwave integrated circuits -- an historical perspective, *IEEE Trans. Microwave Theory Tech.*, Special Centennial Issue 32:991 (1984).
25. H. Howe, Jr., *Stripline Circuit Design*, Artech House, Dedham, MA (1974).
26. S. B. Cohn, Slot line on a dielectric substrate, *IEEE Trans. Microwave Theory Tech.* 17:768 (1969).
27. C. P. Wen, Coplanar waveguide: a surface strip transmission line suitable for nonreciprocal gyromagnetic device applications, *IEEE Trans. Microwave Theory Tech.* 17:1087 (1969).
28. H. Shigesawa, M. Tsuji, and A. A. Oliner, Conductor-backed slot line and coplanar waveguide: dangers and full-wave analyses, *IEEE MTT-S Int. Microwave Sympos. Digest* p. 199 (1988).
29. N. Das, Characteristics of modified slotline configurations, *IEEE Int. Microwave Sympos. Digest* p. 777 (1991). Also, N. Das, Methods of suppression or avoidance of parallel-plate power leakage from conductor-backed transmisssion lines, *IEEE Trans. Microwave Theory Tech.* 44:169 (1996).
30. Y. Liu and T. Itoh, Leakage phenomena in multilayered conductor-backed coplanar waveguide, *IEEE Microwave and Guided Wave Lett.*39:426 (1993).

MAFET THRUST 3: A REVOLUTIONARY PROGRAM FOR SOLID-STATE MICROWAVE AND MILLIMETER-WAVE POWER

E. R. Brown

Defense Advance Research Projects Agency
Electronics Technology Office
3701 N. Fairfax Dr., Room 850
Arlington, VA 22203

INTRODUCTION

The United States is embarking on a period of great opportunity in the field of monolithic microwave integrated circuits (MMICs) created by two phenomena of historical significance. The first was the large U.S. Investment in MMIC technology driven largely by the military need for small, lightweight, and conformable electronically steered active antennas (i.e., phased arrays). Much of this investment was made through the DARPA MIMIC (Microwave and Millimeter-Wave Monolithic Integrated Circuit) and MAFET (Microwave and Analog Front-End Technology) Programs. It has created a design and manufacturing base in MMICs and MMIC-based assemblies and modules that spans across the industrial, military, and academic circles.

The second phenomenon has been the recent explosive growth of wireless communications at lower microwave frequencies. The wireless phenomenon has been driven to some extent by the "information revolution" and the associated spectacular success of the Internet (a DARPA by-product from the 1970s). Wireless communications will support many features of the Internet but with the additional benefits of operation in a mobile or remote environment and communication with much greater instantaneous bandwidth. The latter feature is noteworthy because it will enable the transmission of video in real time. For example, the FCC has allocated bandwidth in the "supernet band" between 5.15 and 5.35 GHz for wireless communications, which represents roughly 2000 times the data-rate capacity as available on the Internet (assuming standard 128 KB ISDN connection).

Along with this opportunity comes a challenge for the MMIC community. With the diminishing investment in MMIC research and development by the U.S. Department of Defense and the strict cost/size/weight requirements on components in commercial applications, MMICs must make a quantum leap of improvement in performance-to-cost (P-C) ratio to stay ahead of the competition: high-speed Si-based ICs (e.g., silicon-on-insulator and SiGe-on-Si) at the lower microwave frequencies and lower power levels, and miniaturized vacuum tubes (e.g., microwave power modules, or MPMs) at higher

Directions for the Next Generation of MMIC Devices and Systems
Edited by N. K. Das and H. L. Bertoni, Plenum Press, New York, 1997

frequencies and power levels. Given this situation, it is prudent for any investor in MMIC technology to adopt a plan whereby limited investment time and funds can result in significant advantages over the competition or, better yet, in unique capabilities that meet critical technology needs within the growing microwave and millimeter-wave systems arena.

The purpose of this overview is to document the technical plan of the most recent Phase (Thrust-3) of the DARPA Microwave and Analog Front-End Technology (MAFET) Program. It begins with a discussion of program strategy and then briefly summarizes the technical areas that the program is divided into. Following this summary, several near-term system applications are mentioned and a few long-term applications are postulated.

PROGRAM STRATEGY

In active microwave and millimeter-wave systems, there is probably no greater enabling factor than low-cost solid-state coherent power. Transmit power directly improves the signal-to-noise ratio in communications and radar systems under ideal (i.e., receiver-noise-limited) conditions. And when transmit-power-dependent noise occurs, such as clutter in radar or multipath in communications, excess power is enabling because it can be traded off in favor other system functionality (e.g., polarization diversification) that helps ascertain the nature of the noise and leads to its reduction by further alteration in system conditions or by signal processing.

Low-cost solid-state power is also enabling from a system-economy standpoint. Because the cost of the power components usually exceeds the cost of the other electronics greatly, a substantial reduction in these components translates directly into a significant reduction in overall system cost. Given this advantage, the system designer can invest the savings in more power components, greater system functionality, signal processing, or other system improvements.

In the MAFET Thrust-3 program, radical technological approaches are targeted for solid-state generator and transmitter technology that alleviate the difficulty and expense of conventional MMICs and MMIC-based modules in generating and controlling the large transmit power levels (i.e., > 100 W at microwave frequencies; > 10 W at millimeter wave) of most military radar and communications systems. Two of these approaches pertain to power generation specifically: (1) high-power transistor technology, primarily for microwave frequencies, and (2) quasi-optical power combining, primarily for millimeter-wave frequencies. Two of the other technical approaches pertain to the control of large power levels: (1) microelectromechanical-(MEM)-switch phase shifters for microwave and millimeter-wave frequencies, and (2) novel electromagnetic materials and micromachining for millimeter-wave frequencies.

In defining these technical approaches, it was decided to limit the microwave band of interest from 10 to 30 GHz, and limit the millimeter-wave band from 30 to 100 GHz. In the lower band, the microwave frequencies below 10 GHz are viewed with somewhat less interest in the MAFET-3 Program because of the rather satisfactory performance of existing solid-state power generation and control (e.g., transmit/receive-module) technology in this region and because of the already-large R&D activity by other sectors within the Department of Defense. This is not to imply that the cost of this technology is low enough for existing applications, particularly those in the commercial market. However, cost alone is not considered to be an adequate driving factor for a Program as far-reaching as MAFET.

In the millimeter-wave band, frequencies above 100 GHz are viewed with somewhat less interest because of the lack of military system applications in this region. This is not to imply that technical arguments for military systems are lacking or to deny the rather large scientific (e.g., ultrafast, chemistry, astrophysics, materials science) activity in this region. For example, it has long been agreed that operating at the higher frequencies will yield improved spatial resolution per unit area of aperture, so that the system size and weight can be reduced substantially. The problem is that these advantages tend to get offset by the well-known opacity of the atmosphere, and the unavoidable inferior performance and high cost of electronic devices that operate at these frequencies.

In all of the MAFET-3 technical approaches, monolithic integration is a paramount requirement for two reasons. First, integration will reduce the development time by taking timely advantage of the widespread MMIC and MCA design and manufacturing capability established by the DARPA MIMIC Program and the first two Thrusts of the MAFET Program. Second, integration should lead to lower cost than made possible by bulk or hybrid fabrication given high enough process yield and an adequate throughput. This reasoning has spawned a MAFET-3 programmatic goal of a 10x increase in the P-C ratio of solid-state power-generation and control MMICs. The logical performance metric for this program is transmitter power radiated into free space since that necessarily includes the performance of power-control components and circuits that occur just before or after the power-generation stage. And, of course, a substantial improvement in the watts-per-dollar of solid-state transmitters is fully consistent with the compelling DoD need to make active microwave systems more affordable.

POWER-GENERATION TECHNOLOGY

Within the MMIC context, the problem of power generation follows largely from the fact that individual semiconductor transistors can not produce cw power levels close to those required for most active systems. Consequently, power combining must be practiced, which is greatly limited in efficiency by losses in the circuits used for the combining. For example, in the standard binary-type corporate combiner shown in Fig. 1, the output power can be derived as $P_{OUT} = P_0 2^K E^K$, where P_{OUT} is the total output power, P_O is the output power that each amplifier would produce acting alone, K is the number of stages of combining, E is the efficiency per stage, and $N=2^K$ is the total number of devices.[1] This leads to a combining efficiency of $\eta = E^K$. As an example, with a combiner loss of 1.0 dB (E=79%) per stage, the combining efficiency for three stages (16 transistors) is limited to 50%. The more stages that are required to achieve sufficient power, the more that small unavoidable losses can hinder performance, not to mention the limitations that can be imposed on the instantaneous bandwidth of the overall circuit.

Microwave Frequencies

A straightforward solution to the power-combining problem at microwave frequencies is to reduce the number of stages by virtue of increasing the output power of each

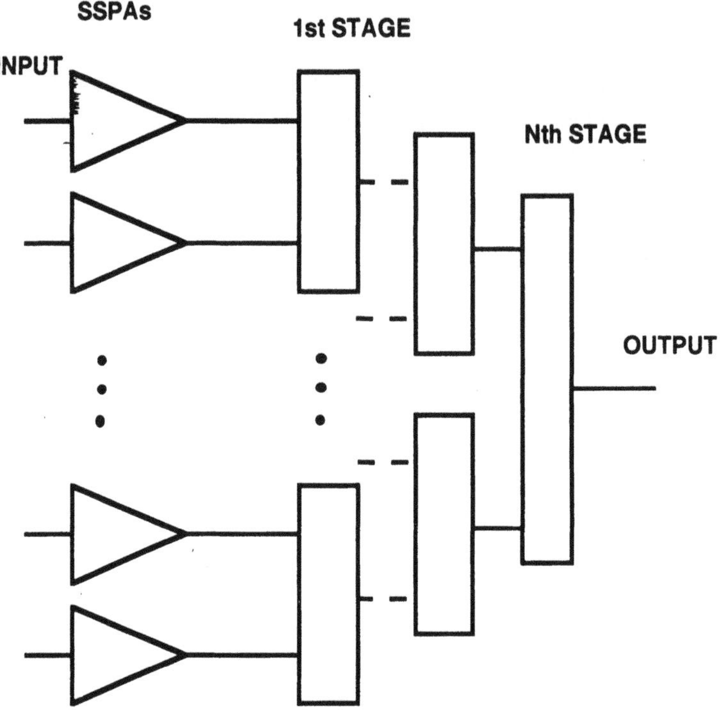

Figure 1. Schematic diagram of binary power combining circuit, commonly used in planar microwave integrated circuits (after Ref. 1).

individual amplifier. This is the strategy that has been adopted in the MAFET-3 through investment in transistor amplifiers made from the wide-band-gap materials GaN and SiC. Transistors made from these materials have three advantages over the conventional GaAs and InP. First, the electric breakdown field is much higher, for example, 5×10^6 V/cm in GaN compared to 5×10^5 V/cm in GaAs. This means that the transistor can sustain much larger bias voltage, roughly one order of magnitude, in a transistor of a given type and geometry. Second, the thermal conductivity is higher, 1.3 W/cm-K in GaN compared to 0.45 W/cm-K in GaAs. This is very important because the dc power dissipated by these transistors will necessarily increase compared to conventional transistors, not because they are any less efficient, but because they are operating with so much more bias power. Third, the electron drift velocity is much higher at the large electric fields that necessarily occur in power amplifiers. In fact, the electronic drift velocity in GaN is about 2.5×10^7 cm/s at an electric field of 1.5×10^5 V/cm. This is to be compared to a peak velocity in GaAs of 2.0×10^7 cm/s at an electric field of 1.0×10^5 V/cm. The benefit of the higher drift velocity is that these transistors will operate with greatest speed and bandwidth at a high field.

An impressive demonstration of the high-field advantage has been presented recently by the group at U.C. Santa Barbara.[2] Shown in Fig. 2 is are common-source current-voltage curves of their device, a GaN/AlGaN MODFET having a 1.0-µm gate length, a 2.0-µm

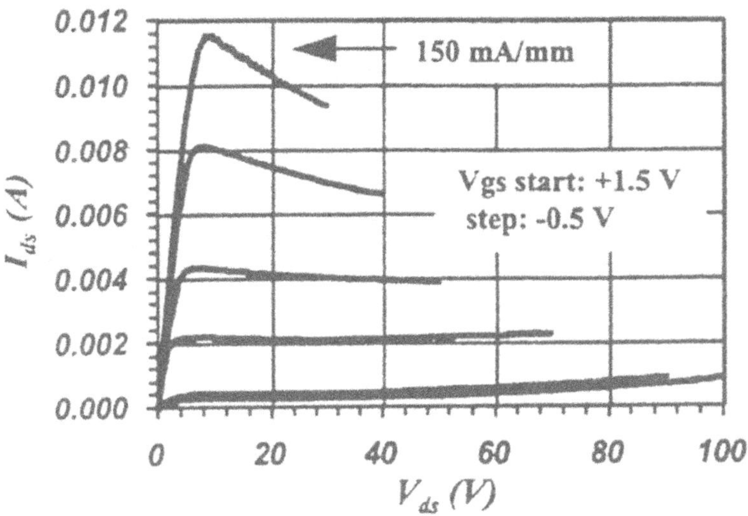

Figure 2. Common-source room-temperature current-voltage characteristics of GaN/AlGaN MODFET fabricated at the University of California, Santa Barbara (after Ref. 2).

gate-drain separation, a 4.0-μm drain-source separation, and a 75-μm gate width. The active layer of this device was grown by organometallic chemical vapor deposition on a sapphire substrate. Note that the drain-source voltage is plotted up to 100 V, proving that the GaN device can withstand very high internal electric fields compared to conventional microwave transistors. In a follow-on experiment, the maximum drain-source voltage, V_{DS}^{max}, on this same device was increased up to 220 V.[3] In both experiments, the maximum specific drain-source current, I_{DS}^{max}, is approximately 150 A/mm of gate width. In contrast, a GaAs/AlGaAs MODFET having practically the same geometry as the above GaN device would likely undergo electrical breakdown at a drain-source voltage of roughly 15 V. The greater current density of the GaAs device follows largely from its higher mobility at low bias fields.

From general power amplifier theory, MODFETs can deliver a specific power of $P_{OUT} = (1/8)V_{DS}^{max} \cdot I_{DS}^{max}$ when operated as a class-A amplifier (quiescent point in the center of the active drain-source bias region). Taking these parameters directly from the I-V curves yields $P_{OUT} \approx 2$ W per mm of gate width from the UCSB GaN MODFET. This is already higher than most, if not all, microwave GaAs-based MODFETs fabricated to date. A reduction in gate length to 0.5 micron or less, as commonly done in GaAs MODFETs, would likely increase this power density substantially simply through the associated increase in I_{DS}^{max}. A specific power in the range of 5-to 10 W/mm is quite conceivable. GaN is able to reach this high power density in a way similar to vacuum-tube amplifiers (e.g., traveling-wave tubes): much higher bias voltages than in conventional semiconductor power transistors at lower current density. Note, however, that the attainment of just 2 W/mm in GaN transistors will require excellent thermal management - difficult to achieve in the present state of GaN fabricated on a sapphire substrate. To address this problem, thermal management will be pursued vigorously in all of the MAFET-3 projects on high-power transistors.

Millimeter-Wave Frequencies

At millimeter-wave frequencies (> 30 GHz), the strategy given above becomes problematic simply because of the present-day difficulty in making the wide-band-gap semiconductor transistors operate efficiently at these frequencies. On the other hand, devices that are fast enough to operate at these frequencies, such as MODFETs made from InP-based materials, produce substantially lower levels of power at millimeter-waves than they do at microwave frequencies, exacerbating the problem of power combining. One possible solution to this problem, and the strategy adopted in MAFET-3 at millimeter-wave frequencies, is quasi-optical power combining.

Quasi-optics is a branch of millimeter-wave technology whereby optical methods are used to perform at least one function within a component or sub-system. By definition, the optical methods involve the interaction of radiation with components that are many wavelengths in spatial extent, so that diffraction effects can be largely ignored. Such is the case in the plane-wave type quasi-optical amplifiers, such as that shown schematically in Fig. 3(a). The amplifier consists of a two-dimensional grid of transistor gain elements connected to nearest neighbors by metal lines. The spacing between nearest neighbors is necessarily much less than $\lambda/2$, so that the entire grid can be thought of as an active layer, in some cases representable electrically by a negative sheet resistance. In this way, the grid behaves in a similar way to an optical laser, each gain element being much smaller than a wavelength and contributing to the passing electromagnetic wave in a minuscule way. But, as in a laser, the collective action of all elements together leads to a large net gain in the passing wave.

One important distinction between electronic plane-wave amplifiers and lasers is that the electronic gain elements, be they transistors or diodes, have a much broader gain bandwidth, generally extending from dc to millimeter-wave frequencies. In contrast, atomic gain elements usually have a very narrow gain bandwidth associated with an atomic or molecular quantum-state transition. Hence it is quite possible for the electronic elements to oscillate at frequencies far away from the plane-wave input frequency through spurious modes of the grid circuit or by regenerative feedback between the output of the grid and the input. To suppress the former effect, loss has to be introduced in the grid bias circuit, which causes consumes dc and rf power and causes inefficiency in the amplification process. To suppress the latter effect, the gain elements are connected on the grid so that the output plane-wave polarization is orthogonal to the input polarization. The polarizers shown in Fig. 3(a) impose this orthogonality.

One group that has been very active in the development of quasi-optical devices is at the California Institute of Technology. Their pioneering work was with quasi-optical grid oscillators, which culminated in 1994 with the demonstration of a 10-W-total-output single-frequency hybrid-MESFET oscillator near 10 GHz.[4] Remarkably, the dc-to-microwave generation efficiency of this oscillator was 23%. In more recent work, they have concentrated on grid amplifiers under the notion that amplifiers are simpler to design and fabricate, and generally more useful than oscillators. Several monolithic grid amplifiers have been fabricated in collaboration with the Rockwell Science Center using both HBT and HEMT transistors. For example, an HBT grid amplifier containing 36 elements displayed an output power of 0.67 W at 40 GHz with a "plane-wave" gain of 5 dB, a 3-dB instantaneous bandwidth of 1.8 GHz, and a power-added efficiency of 4%.[5] In addition, a 36-element

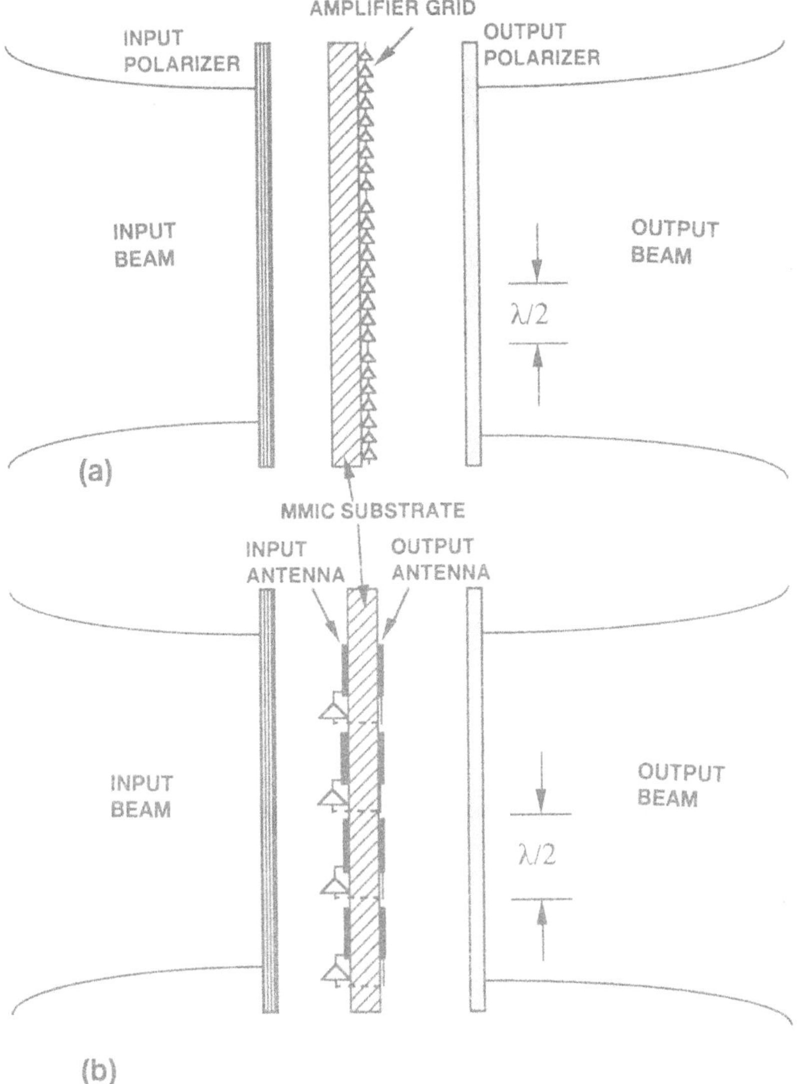

Figure 3. Schematic diagram of power combining strategy in quasi-optical power amplifiers. (a) Plane-wave type, in which the triangles represent gain elements (e.g, transistor amplifiers) that are connected on a two dimensional grid. Although not visible here, the input connection to the amplifier and the output connections are generally perpendicular so that the input and output beams are coupled to orthogonal polariations. This is important in isolating the input and output beams. (b) Active array type, in which the triangles represent gain elements that are connected directly to a planar (e.g., dipole) antenna. The output of the gain element is fed through the substrate to a second (output) antenna. The output antenna is designed to radiate with an orthogonal polarization to the input, again, for input-output isolation.

pHEMT grid displayed a peak gain of 6.5 dB at 44 GHz with an instantaneous bandwidth of approximately 5%.[6] The output power of this grid has not been reported yet; however, the gain bandwidth has been varied from 44-to-60 GHz by mechanically changing the positions of the external polarizer and a tuning slab. This degree of tuning is very difficult to achieve in conventional MMIC power amplifiers.

Note that the shift of emphasis from grid oscillators to amplifiers is very similar to the trend that has occurred over the past decade with powerful semiconductor sources at optical wavelengths. In the 1980s, researchers investigated diode-laser arrays with the hope of achieving coherent optical combining and single frequency operation. Great difficulties ensued, which lead to the simpler and more successful approach of a diode-laser power amplifier fed by a diode-laser master oscillator. This approach is now producing Watt-level single-frequency sources at wavelengths around 1 micron.

An alternative geometry to the grid that still produces beam amplification and, hence, power combining in free space is the active array approach, shown schematically in Fig. 3(b). In this case, the elements of the amplifier are separated by approximately $\lambda/2$ and are connected to on-chip planar antennas to increase the electromagnetic fill factor. In many ways, this approach is a monolithic version of the phased arrays often used in electronically steered active antennas. The majority of contemporary phased arrays use a T/R module for each element, or rows of modules to build up the entire array. Fully monolithic arrays could greatly reduce the cost per radiated watt, particularly at millimeter-wave frequencies. The potential of the active-array architecture has been demonstrated by a hybrid approach has been demonstrated recently by an active array consisting of low-cost GaAs-MESFET MMICs mounted at each element of the array.[7] The measured output power was approximately 3 W near 29 GHz, which represents the highest power measured from any quasi-optical amplifier at mm-wave frequencies.[8]

POWER-CONTROL TECHNOLOGY

Another important factor in generating large power levels with solid-state circuits is the efficiency of power-control devices and circuits. One of the most ubiquitous power control functions at microwave and millimeter-wave frequencies is phase shifting since it is intrinsic to the operation of beam steering in phased array active antennas. For the reason of integration (and hence, low cost), the phase shifting has often been carried out with the same device used in the power amplifiers - most commonly a microwave FET. Unfortunately, an FET, by its nature, is not a particularly good device for phase shifting. This is because the FET is used as a resistive (on/off) switch in this role, and the on resistance of FETs is very difficult to make negligible. Typically, the finite on-resistance leads to an insertion loss of 1 dB (i.e., 21%). And because two or more FET switches are required for each bit of phase shift (in a binary phase shifter), more than half of the transmit power can easily be lost to these phase shifters, which is costly in several ways.

Two ubiquitous power-control circuits are combiners (e.g., 3-dB Lange coupler) and planar antennas. In both circuits, performance is often limited by the problem of losses and parasitics in planar transmission lines when fabricated on substrates, such as GaAs or Si, having a large dielectric constant. While progress has been made in solving this problem by recent MCA and high-density microwave packaging (HDMP) development, these techniques

techniques lack the monolithic integration advantage so that packaging costs can dominate the overall cost of the transmitter electronics. An alternative approach, addressed below, is to maintain the MMIC approach but eliminate the source of the parasitics by removing the substrate (or superstrate) material in the region immediately below or above the circuit.

MEM-Switch Phase Shifters

A much more natural device for phase shifting is the microelectromechanical (MEM) switch. In essence, this is a miniaturized version of the venerable toggle switch so familiar in electronic components. One particularly interesting version is the diaphragm switch shown in Fig. 4. It consists of a thin film of metal suspended over a bottom metal contact. The bottom contact has a thin layer of insulator on top of it to prevent metal-to-metal contact when

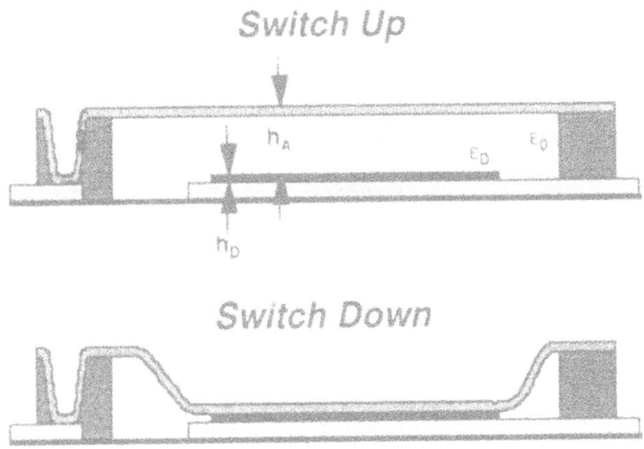

Fig. 4. Cross-sectional view of MEM diaphragm switch in the off (diaphragm up) and on (diaphragm down) states (after Ref. 9)

the suspended film is lowered. This lowering action is actuated by an electrostatic field applied between the top and bottom contacts. The advantage of this type of MEM switch is the ratio of on-capacitance to off-capacitance. In the off state (with the diaphragm suspended), the capacitance is given roughly by the expression $C_{OFF} = [h_A / \varepsilon_0 A + h_D / \varepsilon_D A]^{-1}$, where h_A is the thickness of the air gap, h_D is the thickness of the insulating layer, ε_D is the dielectric constant of the insulating layer, and A is the effective area of the capacitor. In the on state (with the diaphragm lowered), the capacitance is given simply by $C_{ON} = \varepsilon_D A/h_D$. Hence, the ratio of on-capacitance to off-capacitance is given by $C_{ON}/C_{OFF} = 1 + \varepsilon_D h_A / \varepsilon_0 h_D$, independent of area. As an example, the air gap in a typical diaphragm switch is approximately 2 µm, the insulator thickness is approximately 0.1 micron, and its dielectric constant is typically 7.5. This yields an on-to-off ratio of 151.

This large on-to-off ratio allows the MEM switch to have a low insertion loss in the on state and a high isolation in the off state. In data recently taken by the group at Texas Instruments, an insertion loss of 0.2 dB has been measured in a MEM diaphragm switch at 10 GHz.[9] The same switch demonstrated an isolation of 15 dB. This insertion loss is competitive with the best phase-shifter switches presently being used. The isolation requires improvement, and will be investigated rigorously in the MAFET-3 Program.

In the quasi-optical area, MEM switches can also be beneficial as phase shifters, but in a different implementation than the planar case. Since quasi-optical planar arrays operate on the entire wave-front, the MEM phase shifters can be added to each cell to introduce the proper amount of phase shift required to steer the beam. In many ways, this function is similar to that carried out in optics by a prism with the additional benefit that the difference between the entry and exit angles of the beam is under electronic control.

Other Passive Components

To reduce the losses and parasitics that commonly occur in MMIC combining and antenna circuits, the MAFET-3 Program is investigating the following two fabrication techniques: (1) removal of the vast majority of the substrate or superstrate in the vicinity of the circuit by micromachining, and (2) patterning of the substrate into a photonic crystal - an artificial dielectric designed to have a strong polarization-dependent stop band at specified frequencies. The limited space of this overview allows for discussion only of the micromachining, which offers the following three benefits. First, by only removing material from the vicinity of the specified circuit as shown in the cross-sectional views of a patch antenna in Fig. 5, the structural integrity and planarity of the MMIC is maintained. Hence, micromachined circuits and components can be integrated on the same chips as high-speed transistors and other active devices. Second, micromachining tends to become easier as the frequency increases. This is because the circuits and passive devices become smaller with frequency and, hence, less material has to be removed. Third, micromachining tends to eliminate parasitic effects, such as surface mode generation, that are difficult to model and impede the MMIC design process. The performance that results from micromachining can be close to that of the same circuit or component suspended in free space.[10]

One of the pioneering efforts in micromachining has taken place at the Univ. of Michigan. In work conducted primarily on silicon substrates, near-ideal performance has been demonstrated a directional coupler,[11] and similar performance is pending in filters[12] and interconnects.[10] As a results of this success, the MAFET-3 Program has invested in this technology to construct a mm-wave solid-state source in which micromachining enables highly-efficient three-dimensional combining of the power from InP-based MMICs.

APPLICATIONS

MMIC technology is currently planned for insertion into a variety of existing or planned military systems. Among these are the Longbow missile-guidance system for the Army, the F-22 radar system for the Air Force, and the Aegis fire-control radar for the Navy. These are large systems in which the cost of the transmitter electronics is a significant fraction of the overall system cost. Hence, a large improvement in the P-C ratio of power MMICs should expedite and, hopefully, accelerate the insertion process.

Realization of the MAFET-3 Program goal of a ten-fold improvement in the P-C ratio could greatly augment the insertion opportunities by qualifying for several commercial applications. Two examples are wireless radio for personal communications and collision-avoidance radar for automobiles. The communications application is particularly timely following the recent FCC auctions of bandwidth for wireless communications well into the millimeter-wave region. As discussed above, these bands are broad enough to support video transmission. The radar application is well known, but has long been waiting for FCC frequency allocation. In the likely event that a frequency of 77 GHz is adopted in accordance with the European standards, automobile-scale radar systems will be required in great abundance, and low-cost high-power MMICs could offer a very effective solution.

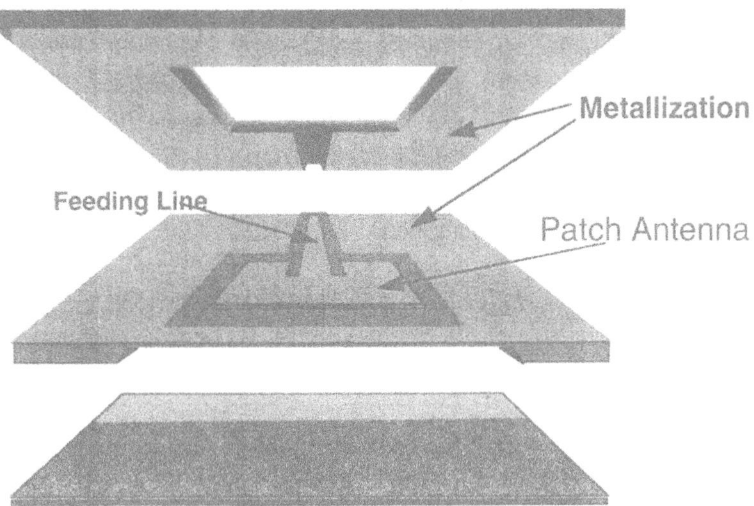

Figure 5. Cross-sectional view through a micromachined planar patch antenna showing shielding in the Si superstrate above the coplanar feedline, and significant thinning of the Si substrate below the antenna (after Ref. 14).

In the long term, low-cost high-performance MMICs could drive a class of applications associated with what could be coined the "popularization" of communications and radar. One example in communications would revolve around the recent FCC allocation of 1.1 GHz of bandwidth in the 27.5-to-30 GHz ("28 GHz") band for Local Multipoint Distribution Services. This format is intended for a wide variety of new consumer-based broadband, two-way interactive video, voice, and data services aimed at two-way communications, particularly satellite communications.[13] Similar systems would be useful in constructing wireless indoor local-area networks that would obviate many of the difficulties in communicating with base stations and satellites from inside buildings. Another example would be compact, perhaps hand-held, millimeter-wave radar for applications in law enforcement, construction, motion detection, etc. In this way, radar would acquire a personal embodiment analogous to what it has occurred in wireless communications.

ACKNOWLEDGEMENTS

Support and guidance for the MAFET-3 Program have been provided by Drs. K. Gabriel and L. Glasser. Technical contributions to this manuscript came from Dr. J. Harvey of the U.S. Army Research Office, Mr. B. Norvell of Texas Instruments, Dr. J. Hubert of Lockheed-Martin, Dr. R. Kihm of Hughes Aircraft, Dr. C. Nguyen of Hughes Research Labs, Dr. A. Higgins of Rockwell, Prof. D. Rutledge of Caltech, Profs. U. Mishra and R. York of the U.C.S.B., and Prof. L. Katehi of the U. of Michigan.

REFERENCES

1. R. A. York, "Quasi-Optical Power Combining Techniques," in *Millimeter and Microwave Engineering for Communications and Radar, ed. by J. C. Wiltse* (Critical Reviews of Optical Science and Technology, 1994), vol. CR54, pp. 63-97.
2. Y-F. Wu, B. P. Keller, S. Keller, D. Kapolnek, S. P. DenBaars, and U. K. Mishra, "GaN HFETs and MODFETs with Very High Breakdown Voltage and Large Transconductance" in *Device Research Conference Digest,* (IEEE, New York, 1996), p. 60.
3. Prof. U. K. Mishra, private communication.
4. J. B. Hacker et al., "A 10-W X-Band Grid Oscillator," 1994 IEEE MTT-S Int. Microwave Symposium Digest (IEEE, New York, 1994), pp. 823-826.
5. C-M. Liu, E. O. Sovero, W-J. Ho, J. A. Higgins, M. P. DeLisio, and D. B. Rutledge, "Monolithic 40-GHz 670-mW Grid Amplifier," in *IEEE MTT-S Digest* (IEEE, New York, 1996), p. 1123.
6. M. P. DeLisio, S. W. Duncan, D-W. Tu, S. Weinreb, C-H. Liu, and D. B. Rutledge, "A 44-60 GHz Monolithic pHEMT Grid Amplifier," *in IEEE MTT-S Digest (*IEEE, New York, 1996), p. 1127.
7. J. Hubert, J. Schoenberg, and Z. B. Popovic, "High-Power Hybrid Quasi-Optical Ka-Band Amplifier Design," *in IEEE MTT-S Digest* (IEEE, New York, 1995), p. 585.
8. Dr. John Hubert, 9 Sept. 1996, private communication.
9. C. Goldsmith, J. Randall, S. Eshelman, T.H. Lin, D. Denniston, S. Chen, and B. Norvell, "Characteristics of Micromachined Switches at Microwave Frequencies," in *IEEE MTT-S Digest* (IEEE, New York, 1996), p. 1141.
10. L. P. B. Katehi and G. M. Rebeiz, "Novel Micromachined Approaches to MMICs Using Low-Parasitic, High-Performance Transmission Media and Environments," *in 1996 IEEE MTT-S Digest* (IEEE, New York, 1996), p. 1145.
11. S. V. Robertson, L. P. B. Katehi, and G. M. Rebeiz, "A 10-50 GHz Micromachined Directional Coupler," *in 1996 IEEE MTT-S Digest* (IEEE, New York, 1996), p. 797.
12. T. M. Weller and L. P. B. Katehi, "A Millimeter-Wave Micromachined Lowpass Filter Using Lumped Elements," *in 1996 IEEE MTT-S Digest* (IEEE, New York, 1996), p. 631.
13. David L. Lyon, "Trends in Digital Wireless Communications," in *Device Research Conference Digest* (IEEE, New York, 1996), p. 4.
14. L. P. B. Katehi, private correspondence.

Future Directions in MMIC Research and Technology: The Wireless *Pull*

Peter Staecker
M/A-COM, Inc. Corporate R&D Center
100 Chelmsford St.
Lowell, MA 01853
508 656 2607 (V) 508 656 2777 (F)
p.staecker@ieee.org

Abstract:

MMIC solutions are a probable/eventual method of driving costs of commercial wireless products to the needed goals, and the continuous reduction in cost will be facilitated by increasing volume of product. The bill of materials for these RF front ends which drives the cost of the end product, however, includes much more than the MMIC: resonators, filters, crystals, and other passive components account for a major part of the material cost. With low cost the driver, system integration will be more and more important; "MMIC" will take on an increasingly broader meaning: integration in three dimensions, in antennas with active elements, and in new materials which allow inexpensive integration of high-performance circuit functions.

Introduction:

The market drivers for the commercial wireless activity in the future separate neatly into frequency regions: low cost 1-5 GHz transceiver circuits for cellular communications, 25-60 GHz power amplifiers and receivers for point-to-(multi)point radio, and 77 GHz transceivers for automotive radar. Each area promises extremely high volumes if the unit cost is right. In wireless telephony size and weight tracks cost, and its relation to volume is shown in Fig. 1 and Fig. 2.

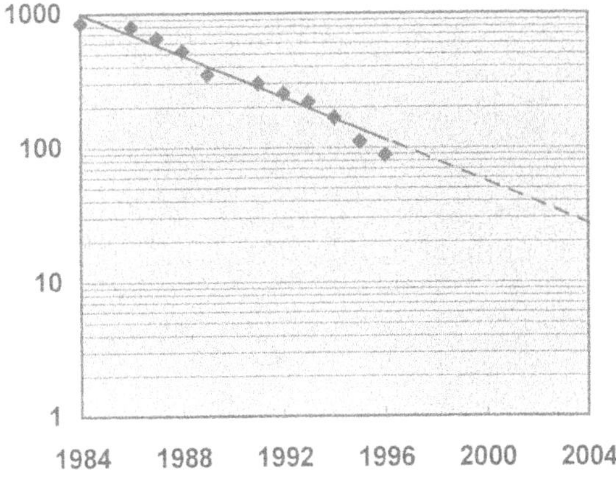

Fig. 1 Weight (in grams) of cellular telephones[1]

[1] data: H. Shosteck Associates, Ltd.

Directions for the Next Generation of MMIC Devices and Systems
Edited by N. K. Das and H. L. Bertoni, Plenum Press, New York, 1997

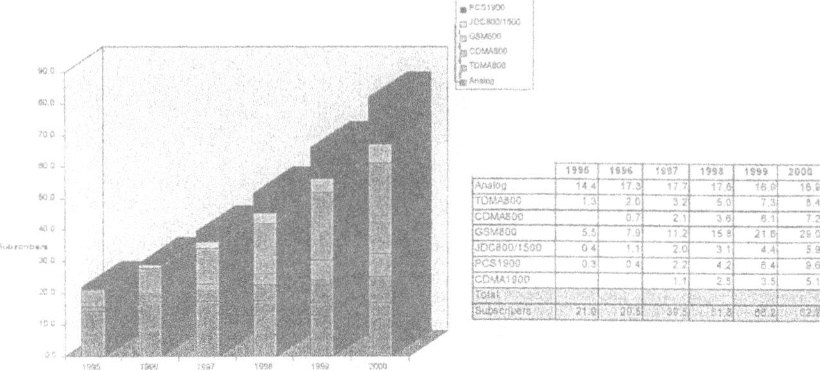

	1995	1996	1997	1998	1999	2000
Analog	14.4	17.3	17.7	17.6	16.9	16.9
TDMA800	1.3	2.0	3.2	5.0	7.3	8.4
CDMA800		0.7	2.1	3.6	6.1	7.2
GSM800	5.5	7.9	11.2	15.8	21.6	29.0
JDC800/1500	0.4	1.1	2.0	3.1	4.4	5.9
PCS1900	0.3	0.4	2.2	4.2	6.4	9.6
CDMA1900			1.1	2.5	3.5	5.1
Total						
Subscribers	21.9	29.5	39.5	51.8	66.2	82.2

Fig. 2 Annual New Subscribers by Technology (Worldwide): 1995-2000[2]

Microwave Semiconductor Materials:

Methods of cost reduction start with the material cost. Gallium arsenide is currently available in 100 mm diameter (Fig. 3) and development has begun on 150 mm diameter crystals. The maturity of GaAs crystal growth is still far behind silicon, however, which is available in 200 mm ingots, with research proceeding on 300 mm. RF silicon has a distinct cost advantage over GaAs, and is successfully competing at frequencies up to 2 GHz; GaAs has slight performance advantages in switch, LNA and power amplifier applications at L-band, with increasing leverage at higher frequencies. Recent advances in Si:Ge technology, however, will extend the silicon cost advantage to frequencies approaching 10 GHz, at least for medium power (0.1-5W) applications. GaAs currently is the material of choice for analog functions in the millimeter wave frequency region. InP, a material which offers higher frequency performance, will challenge GaAs for these applications as its manufacturing technology matures.

Fig. 3: Standard 6 kg GaAs ingot (foreground) with 22 kg boules of improved production process (rear)

[7] Strategies Unlimited

For infrastructure (basestation) markets, there will be needs for higher power solid state devices to replace TWTs, and these will be chosen from silicon carbide and perhaps gallium nitride, both high bandgap materials which are capable of operating at very high temperatures and impedance levels.

These material and device developments were strongly supported from 1987 through early 1995 by government infrastructure programs in the United States, particularly the MIMIC Program. This level of support allowed military RF requirements to be met, in a cost-effective and reliable manner, and was largely responsible for the US assuming its present position of leadership in the worldwide commercial wireless market. While it is important to continue these efforts, a strong and concerted effort in integration technology is absolutely mandatory to fashion these developments into affordable end products.

Integration Methods:

Single-chip solutions to high-volume commercial microwave functions remain the *holy grail* of the industry. While progress in that direction continues, a great deal of attention is being directed to integration methods which allow low-cost assembly while preserving or even improving performance of current assembly techniques. As part of advanced chip-on-board developments, for example, MMICs are now being developed with bumped 3-dimensional bond pads so that the entire circuit can be mounted flip-chip fashion to reduce reactive parasitics, improve heat flow when properly designed, and speed the assembly cycle (Fig. 4)

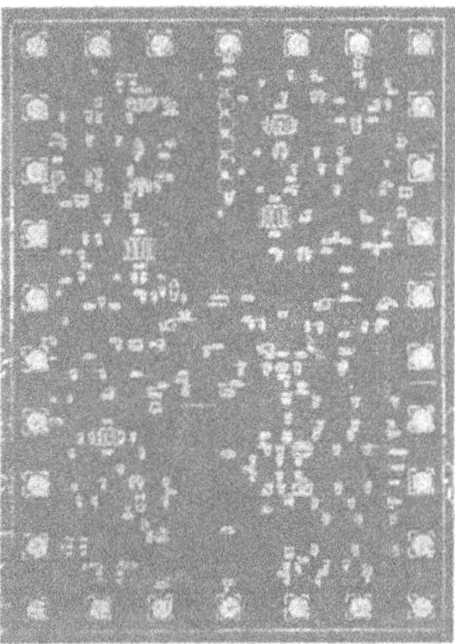

Fig. 4: GaAs X-band MMIC (3 x 5 mm^2) building block for TR module application with bumped perimeter for flip-chip mounting

For low frequency application, multi-layer PC boards can be used for integration of multi-function assemblies. Fig. 5 shows a 4 x 6 switch matrix used in 900 MHz infrastructure applications which integrates individual chip/functions implemented in glass, GaAs and silicon.

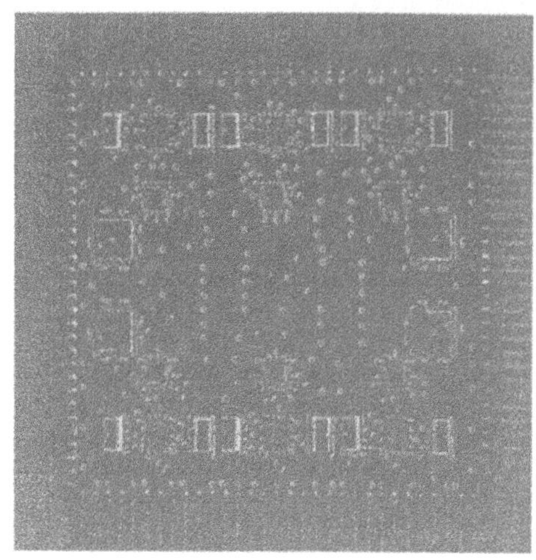

Fig. 5: 900 MHz 4x6 switch matrix (50 x 50 mm^2)

For higher frequencies, glass has been used to integrate MMICs and discrete devices in demonstrations of high performance TR module front ends[3] (Fig. 6). Glass has been tested to 80 GHz and shows promise for application at millimeter-wave frequencies as an integration medium.

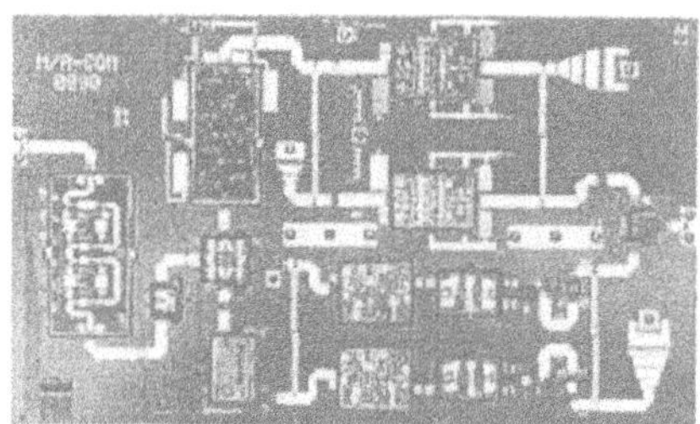

Fig. 6: 2W, 5 dB NF X-band TR module (15 x 25 mm^2)

Conclusion:

Material and device development for MMIC applications have been strongly supported for the past 15 years; demonstration of high performance functionality has been impressive, and low-volume applications have benefited from this support. Only by including a vigorous program of integration and manufacturing technology in the present infrastructure, however, will low-cost techniques be developed which make microwave and millimeter-wave wireless markets accessible.

[3] B.A. Ziegner and T. Murphy, "Radar Module on Glass Substrate," *Applied Microwave*, Winter 91/92.

SYSTEMS AND INTEGRATION

HIGH DENSITY MICROWAVE PACKAGING TECHNOLOGY DEVELOPMENT FOR DEPARTMENT OF DEFENSE APPLICATIONS

Franklin Lamb, Bradley Paul, and Christopher Lesniak

Avionics Directorate
WL/AADI Bldg 620
2241 Avionics Circle
Wright-Patterson AFB, OH 45433-7322

ABSTRACT

Future microwave system applications will require greater packaging density and increased functionality at the multi-chip module level. The DARPA/Tri-service "High Density Microwave Packaging (HDMP) For Next Generation Aircraft and Space Based Phased Array Radars" program is developing ways to increase packaging density. The primary military application for the HDMP technology is thin profile conformal active arrays and in particular 3-D or "tile" Transmit/Receive (T/R) modules. This program is in the third year of a four year effort and now in the prototype development phase. The program has three primary contracts, each having different approaches to develop 3-D tile modules. In this paper an overview of the program will be discussed. A summary of the different technical approaches, special technical challenges/issues, lessons learned and accomplishments on the program will be provided. In addition a brief summary will be given of other current and future Air Force microwave packaging programs. Future microwave packaging applications need to address and make advancements in the areas of advanced substrates, key materials and material properties, Computer Aided Design (CAD), Manufacturing (CAM), and Test (CAT) tools, and vertical interconnects. Because affordability is becoming a critical factor in the development of current and future DoD systems, the HDMP program and other current and future programs must develop packaging technologies that can leverage on both military and commercial applications to support dual-use systems to improve performance and reduce cost.

Directions for the Next Generation of MMIC Devices and Systems
Edited by N. K. Das and H. L. Bertoni, Plenum Press, New York, 1997

INTRODUCTION

The DARPA funded program titled "High Density Microwave Packaging (HDMP) for Next Generation Aircraft and Space-Based Radars" was started in September 1993 to develop three dimensional "tile" architecture for T/R modules. The motivation was to significantly reduce size, weight, and cost of next generation phased array radar systems. This program requires technology advancements in the areas of architecture, interconnects, substrates and module cost. The HDMP vision (figure 1) is to have a highly reliable, very thin profile, light weight active phased array system that can be used conformally These systems must also have efficient assembly and test methods.

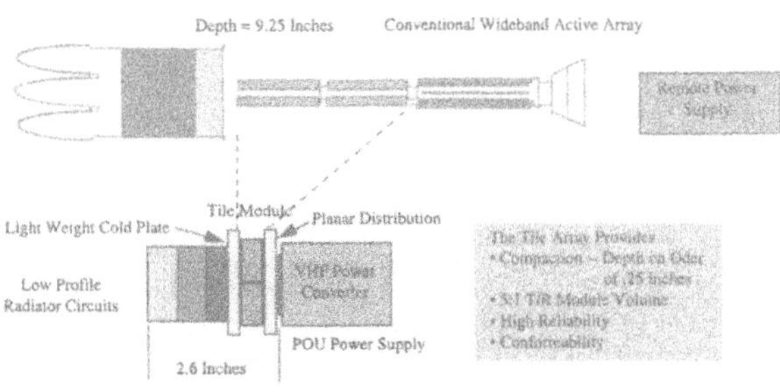

Figure 1: HDMP Vision

Advancements are needed in microwave and millimeter wave module integration technology to meet current and future applications. Future electronic systems require low cost, high density multi-chip assemblies (MCAs) and assemblies of MCA's. Thin profile arrays are needed for conformal and conventional installations with decreased weight, volume, and depth. The development of microwave high density packaging will provide opportunity for system performance enhancements for defense applications such as radar, electronic warfare, communications, missile systems and smart munitions. Potential commercial applications are in the area of wireless communications, direct broadcast satellites, surveillance and identification, intelligent vehicle highway systems and aviation electronics. Projected requirement for high density and high performance military microwave MCAs is in excess of eight million over the next twenty years.

The three dimensional "tile" module architecture has several advantages and significant improvements over traditional two dimensional "brick" architecture used today. This new architecture will result in module depths on the order of 0.25 inches compared to 4.5 inches with current architectures. In addition, low cost batch fabrication and increased test capability are realized. Significant improvements are listed in table1.

Table 1. Potential Improvements

Array Level		Module Level	
Volume Reduction	5:1	Volume Reduction	4:1
Weight Reduction	5:1	Weight Reduction	2.5:1
Cost Reduction	2:1	Cost Reduction	2:1

The disadvantages to the tile approach are 1) thermal management becomes more of an issue, 2) assembly alignment becomes critical, and 3) sophisticated CAD tools are required for modeling and simulation.

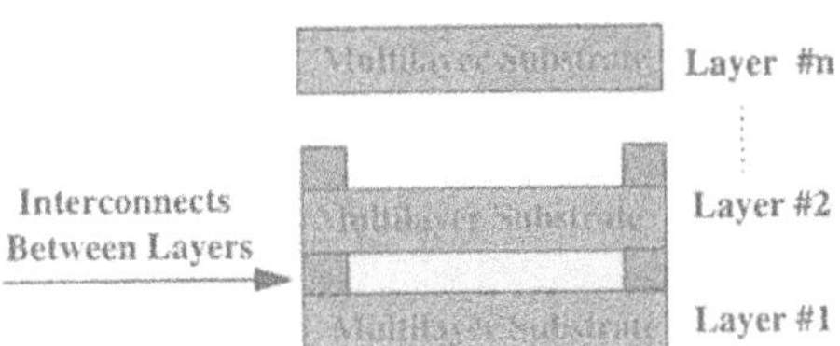

Figure 2: Tile Module Concept

The tile module concept, shown in figure 2, required technical development in several areas. Advancements were achieved in multilayer substrates, RF, digital and analog vertical interconnects, wire bond-less chip to chip and chip to substrate interconnects, solder free interconnects between layers, compact highly integrated MMICs and the use of mixed mode circuits.

One of the biggest technical challenges to the tile approach is signal transmission between layers via vertical interconnects. Normally RF signals do not readily make ninety degree turns. Other challenges associated with vertical interconnects are RF performance and heat conduction. Several vertical RF feedthroughs and solder free interconnects were investigated on this program such as fuzz buttons, elastomers, filled adhesives and filled elastomers. Another requirement for the HDMP program was to eliminate the need for wirebonds. One way to eliminate wire bonds is to use the High Density Interconnects (HDI) process developed by GE for routing lines in polyimide structures. Another way to eliminate wirebonds is a flip-chip approach for microwave and digital ICs and/or to flip-chip co-planar MMICs.

The HDMP approach addresses the affordability issue. The approach reduces cost because multiple modules can be fabricated "semiconductor style" on 6 inch substrates, touch labor is significantly reduced, and wirebonds are eliminated.

PROGRAM ORGANIZATION

The HDMP program has three prime contractors. They are: 1) the Texas Instruments/ Lockheed Martin/ GE CR&D Team, 2) Northrop Grumman with subcontracts with IBM, MCNC and TRW, and 3) Hughes Aircraft Company. In addition to the three primary programs there are four support programs to develop advancements in their area of expertise to help design and fabricate the modules. M/A-COM is developing hermetic like coatings at the MMIC chip level. Georgia Tech Research Institute (GTRI) is developing a packaging materials database to be used by the microwave designer. The database contains critical material properties and vendor information which can be easily accessed using a searchable application. Compact Software and HP-EEsof are developing models and new EM capability to model interconnects and structures used in 3D packaging. The original cost of this four year technical effort was $26.3M for the three prime contracts and an additional $5.4M for the support programs for a total of $31.7M.

TECHNICAL APPROACHES

Each of the prime contractors have a different approach to developing the HDMP T/R module. The differences are primarily in materials and module configuration. Each T/R module however, is X-Band with output power of 8-10 watts per channel, and configured in a 2X2 tile (4 elements per module) module package. The dimension of these 2x2 modules is approximately 1.1" x 1.1" x 0.27".

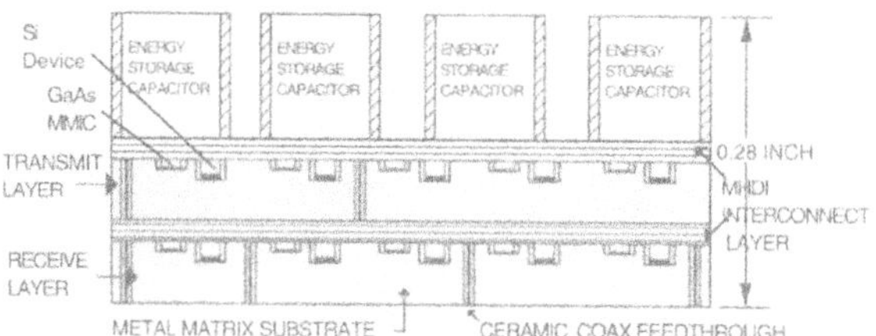

Figure 3: Texas Instruments Approach

The Texas Instruments (TI) approach (figure 3) uses the General Electric/Martin Marietta developed Microwave High Density Interconnect (MHDI) process. This approach uses two active layers, a transmit layer and a control layer. The MMICs are mounted face up in wells formed in an aluminum-silicon carbide substrate. Aluminum-silicon carbide was selected for its good CTE (coefficient of thermal expansion) match to the gallium arsenide MMIC chips. In addition TI is developing a capping process for putting a thick dielectric cap on top of the MMIC for isolation and protection of the MMIC. A key technical challenge to the TI approach is increasing the electrical performance and reliability as well as heat conduction of the vertical interconnects. Texas Instruments will demonstrate a working 2x2 T/R module and the end of 1996.

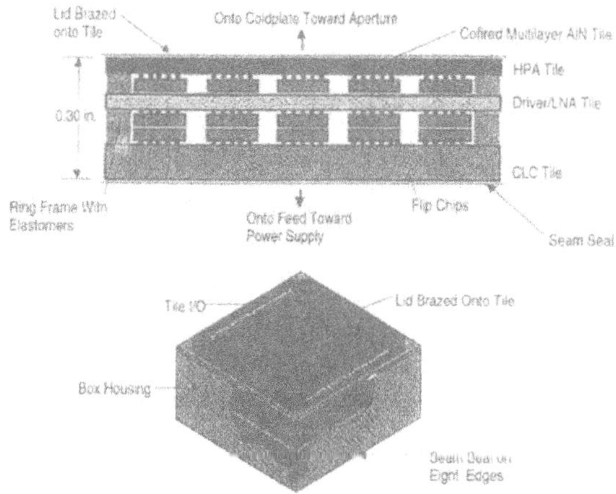

Figure 4: Hughes approach

The Hughes approach (figure 4) is a 3 active layers (tile) approach consisting of a High Power Amplifier tile, a Driver/Low Noise Amplifier tile and a Control Logic Circuit

(CLC) tile. This approach uses 9 MMICs per channel. Aluminum nitride was chosen for the substrates because of its superior thermal properties, low cost, and its coefficient of thermal expansion (CTE), which is a good match to both silicon and gallium arsenide. Hughes is using the "flip-chip" attachment for all silicon and MMIC chips. Hughes has already assembled a module and is now in testing. Hughes will demonstrate a 56 element array in June 1997.

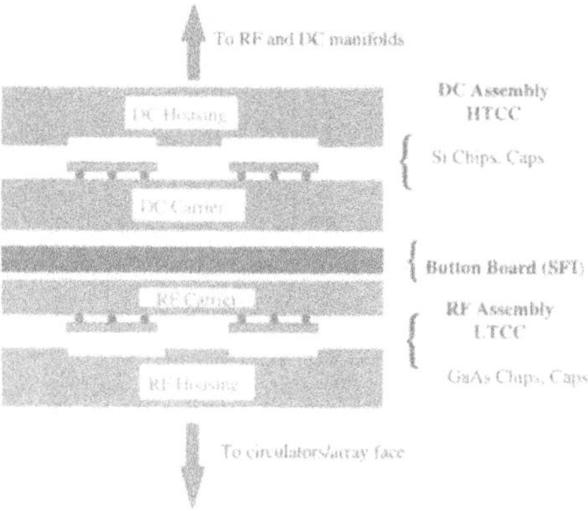

Figure 5: Northrop Grumman approach

The Northrop Grumman approach consists of an RF subassembly and a DC subassembly. Figure 5 provides a cross section view of the Northrop Grumman 3-D T/R module. High Temperature Co-fired Ceramic (HTCC) is used for the DC and low frequency section and Low Temperature Co-fired Ceramic (LTCC) material is used for the RF section. The Northrop Grumman approach utilizes flip chip technology and C4 (Controlled Collapsed Chip Connection - developed by IBM) interconnects for DC and a modified C4 like bump process being developed by Microelectronics Center of North Carolina (MCNC) for RF interconnects. The two sets of layers are placed together using a "button board" interconnect. Northrop Grumman will demonstrate working 2x2 T/R modules at the end of 1996.

ACCOMPLISHMENTS

There have been several technical accomplishments made over the first three years of the HDMP program. The Texas Instruments team has demonstrated extended MHDI features including a new polymer material and a new layer stack up (6 layers) with buried passive functions. They have demonstrated near-net shape AlSiC substrate fabrication with hermetic in situ feedthrough capture. The largest technical challenge in the Texas Instrument approach is the vertical interconnects. The vertical interconnects must not only perform well electrically they must also conduct heat. Although fuzz buttons and elastomers work well electrically they do not work well for heat conduction. A material being developed under the DARPA/Tri-Service MAFET program that has the necessary thermal resistance (45 W/m-^{0}K) is being developed. The original TI approach used "Capped MMICs". The intent was to have a 10 mil thick dielectric cap using Cyanide Ester over the MMIC for protection and isolation from the MHDI interconnect layers above. Much effort was spent in developing the solid cap approach which presented several tough issues that still need to be overcome; such as, capping material cracking and problems with metal adhesion to the deep vias. There were many lessons learned but in order to complete the program TI decided to use the "air cap approach" developed by GE (Lockheed Martin). In this approach there is a 2.5 mil air gap above the active device areas on the MMICs. Another accomplishment on the program was the integrated demo structure that was built to test the RF signal routes needed in this approach. The RF signal passed through the manifold, vertical interconnect, housing, 6 layers of MHDI down to the MMIC and then out through the MHDI layers, housing, vertical interconnect and back through the manifold. The results were excellent with small losses reported. TI performed a cost analysis for this approach, assuming a reasonable production order they are projecting a per element cost of less than $300 at the module level and close to $600 at the array level (including module cost). This represents a considerable reduction in cost compared to today's technology. Another array program using brick modules is estimating a module cost greater than $550 per element in large production volume.

Hughes has done extensive studies on transmission media and interconnect structures used in their HDMP MCA assemblies. Stripline, coplanar wave guide (CPW), coax line, dielectric filled slabline and 3-wire lines have been investigated. Initial vertical alignment tests of the fuzz button interconnects show tolerances of at least 10 mil misalignment being acceptable. Hughes has developed a new Wilkinson Combiner and fabricated it with excellent test results. Dummy aluminum nitride modules were manufactured for fuzz button vibration tests over a large area which proved the fuzz button interconnects superior reliability over elastomer interconnects. They have demonstrated multilayer AlN substrates and developed a test strategy to characterize each substrate prior to assembly allowing each stage to be troubleshot and reworked if necessary. Sort of a Known Good Die or in this case a Known Good Substrate approach to module build. Hughes has compared the brick style architecture to the tile architecture and shown that a significant weight, volume and cost reduction can be achieved using the tile modules. Hughes has demonstrated the first working 3-layer tile T/R module. They are currently testing the module with only a single channel turned on and will soon begin testing with all four channels activated.

Northrop Grumman has shown a very successful mechanical tile module demonstration. The XYZ tolerances for their approach are well characterized and under control for robust processing. Their co-fired solderable metalization meets leach resistance and wetting requirements. The DC assemblies are electrically continuous and hermetic to 10E-8 cc/sec. Northrop has had zero failures on AVIP tests of phase 1 hardware (1800 cycles completed). Accurate thermal analyses show junction temperatures well below standard derating criteria. The IBM gallium arsenide bumping process is under control, with over 8000 bumps processed and a total yield in excess of 94 percent. Northrop Grumman has developed the RF test hardware used to test the Solder Free Interconnects (SFI). The SFI test demon-strated good microwave performance. Misalignment experiments show up to a 20 mil misalignment without significant degradation. The SFI, or button board, has been demonstrated for DC and digital interconnects. Northrop has successfully developed the capability to flip chip circuits and will use flip chip technology in their design. Their approach uses standard microstrip MMICs. Gold bumps are placed on the MMIC and solder balls on the LTCC. The MMICs are flipped and joined by reflowing the solder bumps.

As mentioned previously there are four support contracts on the HDMP program that are to help fill in some of the needed technical development. The GTRI contract was funded to collect microwave packaging materials data and create a database that can be used by HDMP designers. The database has been successfully developed and demonstrated. GTRI has measured some critical material properties on limited samples of high interest materials to the primary contractors. GTRI hosted a packaging workshop in February 1996 to disseminate information and foster cooperation among those working to reduce microwave packaging costs. This was a very successful workshop and a very good technology discussion on materials used in HDMP structures. The M/A-COM contract is responsible for developing "hermetic like" coatings for MMICs. M/A-COM developed a two layer coating process (fist layer - BCB, second layer - SiC). This material coating program is a success. M/A-COM has transferred this technology to their commercial products. On the HDMP program HP-EEsof has completed the HDMP 2.5D EM simulator with new capabilities, improved quality, performance and accuracy. The 3D EM simulator development is in progress. Compact Software has demonstrated Microwave Explorer's (2.5D) capability to analyze complex HDMP structures. Microwave Conqueror (3D arbitrary) simulator is in development.

FUTURE PACKAGING NEEDS

The Department of Defense requires future RF sensor systems to be low profile, lightweight, low cost, and perform in the microwave and millimeter wave frequency ranges. Advanced packaging technologies are critical to meeting these requirements. In addition advances in millimeter wave power sources and low power consumption electronics will enable the DoD to meet current and future military threats. Future systems must be easy to retrofit and have easy installation capability. They must be manufacturable and above all affordable.

The HDMP program has been successful in developing some of the technologies needed to make the HDMP vision a reality. Over the next year the technology will be demonstrated. Some technical issues still require further development. In the area of multilayer substrates issues still remaining are metal fatigue, CTE mismatch, thermal management, testing, hermeticity RWoH and buried passive structures. For the vertical interconnects and solderless interconnects improvements are needed in reliability, noise performance, increased thermal and electrical conductivity, manufacturability and reliable mating connections. Mixed mode operations in high density packages will be demonstrated on this program. The modules have been modeled and simulated; however, all four channels have not been turned on simultaneously so module resonances, the effects of crosstalk or what amount of electrical coupling will be seen have not been characterized yet. For the microwave packaging materials database, material properties have to continually be collected, characterized, validated and added to the database. Another area that needs to be accomplished is to make the database easily connected to the individual CAD, CAM and CAT tools.

There are some other packaging needs that must be addressed for future DoD systems. Future systems will require increased density, increased power, increased manufacturability and decreased cost. There are currently other programs, in addition to the HDMP program, that are addressing these issues. Probably the largest program currently is the DARPA funded, Tri-Service managed Microwave and Analog Front End Technology (MAFET) program. The MAFET program is funded to provide additional computer aided design resources and to extend the development, assembly, and test of advanced MCAs to other frequency ranges and other applications.

The HDMP objectives are important and we look forward to seeing the technology demonstrated during the next year. We have to keep in mind that this program was developed at microwave frequencies and that the general concepts and designs are applicable to millimeter wave, but other considerations will have to be taken into account at millimeter wave frequencies. There are new Air Force programs planned to address the need to develop some new and innovative technologies such as optical interconnects, more compact and combined MMICs, MCA protective coatings, multilayer substrates, diamond based substrates and the use of plastics to further enhance the performance and reduce the cost of future phased array radar systems.

CONCLUSION

In summary, The HDMP technology offers a significant reduction in size and weight of active arrays and is a potential low cost approach. The success of this program is important in demonstrating the capability to package microwave modules in a very small 3 dimensional profile stacked architecture. Although we have come a long way, there are areas that require further development for future systems. There is a high probability of transition of various packaging elements developed by the HDMP program into military and commercial applications. The Texas Instrument team and Northrop Grumman will demonstrate working T/R modules at the end of 1996. Hughes will have a demonstration subarray using their 3 layer tile approach in June of 1997. This has been a very exciting program and we are looking forward to seeing the technologies developed, demonstrated and applied to future DoD and commercial systems.

FLIP CHIP POWER MMICS PACKAGING TECHNOLOGY

Terry C. Cisco

Hughes Aircraft Company
El Segundo, CA 90245

WHY FLIP CHIP?

Conventional microwave design uses thin (4-mil thick) MMIC chips epoxied or soldered down to a metal housing or a dielectric substrate. Thermocompression wire bonds are used to connect the chips to other die or microwave circuits. Flip chip technology uses unthinned (25-mil thick) MMIC chips with bumps on its surface, which, when inverted, are soldered or epoxied directly to a microwave circuit substrate to provide simultaneous die attach and circuit interconnection. Figure 1 compares the two mounting configurations. The flip chip process is lower in cost than the conventional mounting technique, because it eliminates the high cost and low process yields of wafer thinning and via formation while simultaneously improving the visual inspection yield by reducing wear and tear on the frontside of the chip caused by backside processing. Furthermore, an entire manufacturing process (wire bonding) is eliminated, reducing manufacturing capital and support costs as well as wire bond yield losses. The flip chip die attach process has been thoroughly demonstrated in the commercial automotive industry, which has shown that the flip chip process yield exceeds the combined die attach and wire bond yield that it replaces. For microwave devices, the flip chip process provides a very low inductance chip

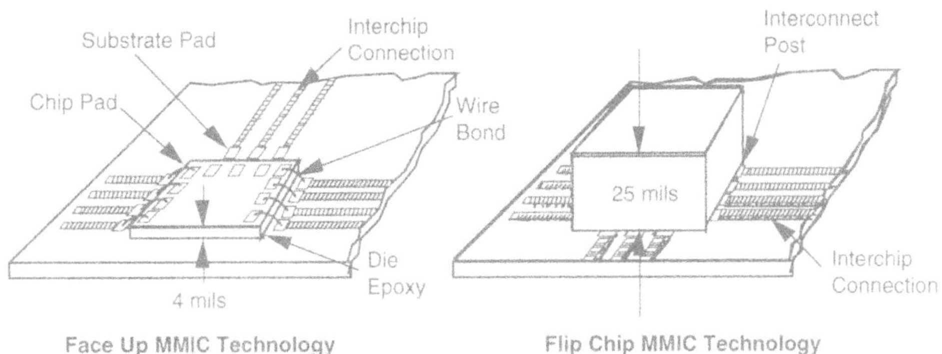

Figure 1. Flip chip technology eliminates wire bonds and enhances chip and assembly yields.

interconnection suitable for millimeter wave use and for microwave power devices. A properly designed custom bump can reduce thermal resistance to half that of face-up mounting. Examples of solid metal and thermal bumps are shown in Figure 2, together with a mounted flip chip assembly.

Figure 2. Microwave bump technology is based on the commercial automotive silicon flip chip process, which uses plated bumps.

HISTORY AND EVOLUTION

In 1960, the Delco division of General Motors introduced the flip chip process into its commercial manufacturing process for automotive electronics components in Kokomo, Indiana. The result was improved manufacturing yields, improved reliability, and a more competitive price, which, in turn, led to a steady growth in world-wide market share. In 1989, Hughes translated the Delco silver bump flip process for silicon integrated circuit chips into a silver bump MMIC chip process and successfully demonstrated the first flipped GaAs MMIC chip LNA that uses coplanar waveguide (CPW) microwave transmission lines to eliminate the grounded microstrip transmission lines and vias in conventional MMIC chip designs. Currently, an entire family of wideband MMIC chip designs have been developed to provide low cost T/R module circuit functions for the growing active array, ground and airborne, radar market. These wideband MMIC functions at X-band include LNAs with 1.7 dB noise figures (see Figure 3), 6-bit phase shifters, 5-bit variable gain amplifiers, 12-watt limiters, 12-watt transfer switches, 1-watt driver amplifiers, and 5-watt high power amplifiers.

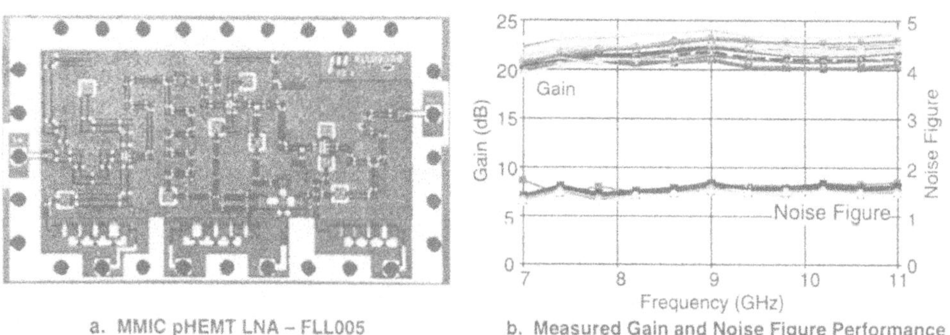

Figure 3. The measured gain and noise figure performance for a flipped pHEMT MMIC low noise amplifier designed for a ground based RADAR T/R module demonstrates 1.5 dB noise figures with an associated gain of 22 dB. The device is 5.1 mm × 3.0 mm in size and is currently being compacted to a 3.8 mm × 2.5 mm size.

PLATED VERSUS SOLDER BUMPS

Both commercial automotive (Delco) and computer (IBM) flip products have evolved into a solder bump process from a plated bump process to further reduce cost and to improve market share. Commercial wafer bumping services and companies have come into being to support the growing flip chip market; yet microwave products for the most part continue to use the solid metal bump. Metal bumps provide for a well-defined microwave structure and spacing between the MMIC chip and the substrate microwave interconnect circuitry. Metal bumps can be taller than solder bumps to prevent circuit interactions with the host substrate. Solder bumps, shown in Figure 4 for an InP HBT power cell, are under study for chips with bumps only for input/output connections. However, applications requiring good thermal conductivity, such as high power amplifiers, continue to use metal bumps or thermal shunts. Metal bumps can be configured into custom shapes and sizes, suitable for thermal bumps, while solder bumps tend to be restricted to spherical shapes and poor thermal resistance.

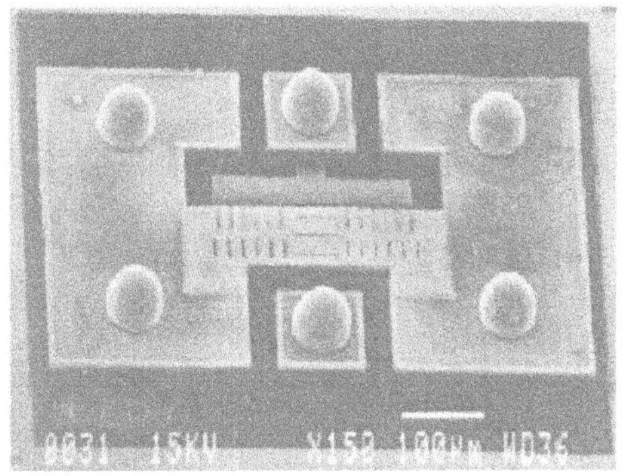

a. 12x2x30 μm² DHBT Power Cell

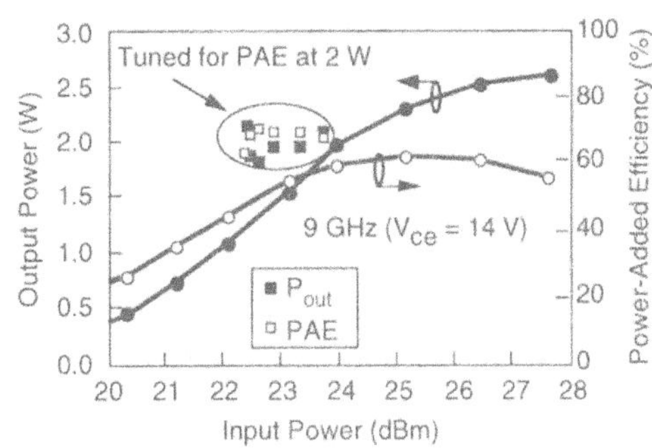

b. X-Band Power Performance

Figure 4. This flipped InP HBT power cell uses solder bumps for input and output connections to deliver 2 W of output power. A plated thermal shunt equalizes the temperature of the emitter fingers, preventing thermal runaway.

MMIC pHEMT POWER CHIP EXAMPLES

Recently, Hughes demonstrated the world's first flipped pHEMT MMIC high-power amplifier and driver chips. The two-stage, high-power amplifier shown in Figure 5 was designed for active array applications and achieves more than 35% power-added efficiency, with an associated output power of 5 W. The output gate periphery is 10 mm and is driven by an input periphery of 2.5 mm. The three-stage driver amplifier chip shown in Figure 6 has an output power of 1 W with an associated gain of 21 dB. Its output periphery is 2 mm driven by 0.35 and 0.5 mm peripheries.

Both chips were fabricated by a 0.25 μm gate length, double-recess, double-sided doped pHEMT process reported earlier.[1,2] The input/output (I/O) silver bumps are approximately 0.003-in. high by 0.006-in. in diameter. The thermal bumps are also silver and are grown directly on the source interconnect bridge of the power FETs and are also 0.003-in. high. Bump formation is accomplished by a laminated dry-film photo resist process and silver plating process.

Each pHEMT power chip is complete with integral input and output power matching circuits and integral bias and decoupling capacitors. The basic circuit design is accomplished using coplanar waveguide (CPW) transmission lines because there is no

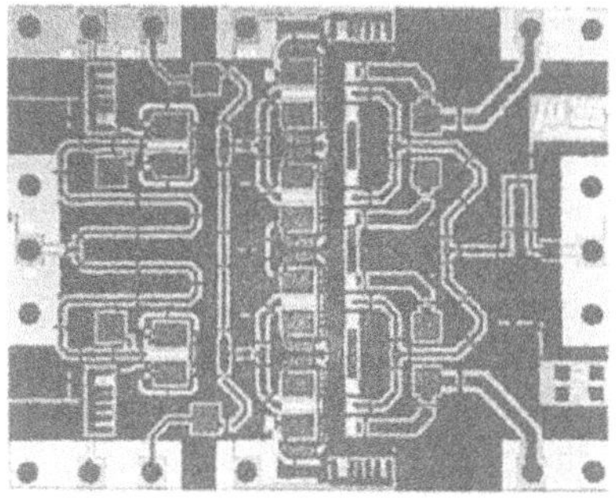

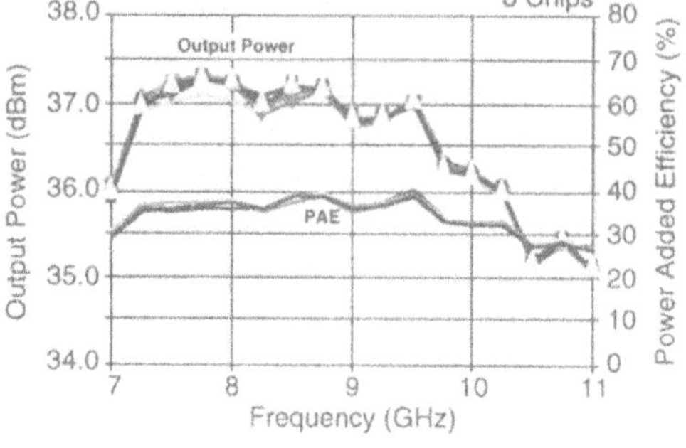

a. Flipped MMIC PHEMT HPA **b. Measured Power and Efficiency Performance**

Figure 5. Measured large signal performance for eight flipped pHEMT HPA chips from a multidesign wafer achieves peak efficiencies of 40% with associated output powers of over 5 W. The compact chip is 4.6 mm by 3.64 mm, and the pulsed measurements were performed with a 21 dBm input power level, a 10 μs pulse width, and a 10% duty cycle.

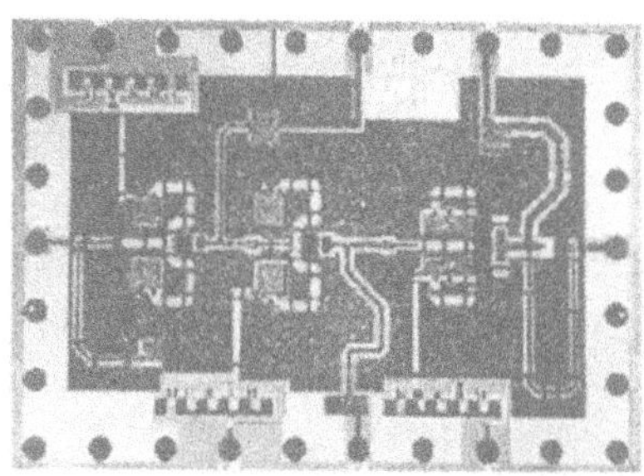

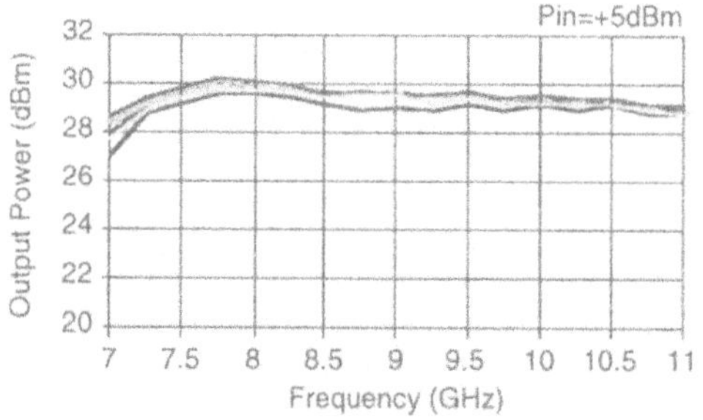

a. Flipped MMIC Driver Amplifier **b. Measured Output Power**

Figure 6. Measured large signal performance for the flipped pHEMT driver chip from the same wafer achieves a saturated output power of 1 W with an associated gain of 21 dB and associated efficiency of 30%. The compact chip is 4.6 mm by 3.2 mm, and the measurements were performed with a 5 dBm input power level in a CW mode.

substrate thinning and no backside metallization in a flip chip for microstrip matching circuits. Electromagnetic simulation tools and measured passive element data are used to model the CPW discontinuities and transitions, as well as for the gate manifold structures of the active devices. Loadpull impedance data and S-parameter device data are used to develop circuit models for MMIC chip circuit designs.

COST ADVANTAGES

The greatest advantages of the MMIC flip chip approach are low cost and low thermal resistance. Flip chip fabrication cost is reduced by eliminating the backside processes entirely and substituting the relatively low cost, high yield, plated bump process. While the flip chip CPW design process produces chips equal to or slightly larger than a face up microstrip design, our cost models indicate an approximately 50% lower cost for the flip chip designs when compared to comparable face-up designs.

In addition, the assembly costs associated with flip chip designs are reduced. The die attach and wire bond processes are reduced to a single process that simultaneously accomplishes both functions. The solder reflow process provides for self-alignment and low interconnect inductance, producing uniform repeatable microwave performance at a high process yield. In a recent manufacturing run of 400, all-flipped, active-array receive modules, a 99% yield was associated with the MMIC die attach and interconnect process. At millimeterwave frequencies, the flip chip approach provides the additional benefits of very precise chip alignment and low interconnect inductance. In an alignment experiment at Hughes, the chips were self-aligned by solder surface tension, resulting in placement accuracies of a few microns.

Finally, the flipped chip I/O bumps overlap the associated substrate pad locations, reducing the overall substrate size when compared with face-up mounting. Face-up I/O pads are located on both the chip and host substrate, but with the host pads taking up more substrate space by being external to the chip pads to allow for wirebonding room.

THERMAL ADVANTAGES

Flip chip technology, when applied to high power amplifiers, yields low thermal resistances. The thermal resistance of a power device is computed by dividing the temperature rise of the active device channel over its base plate temperature by the device dissipation power. Because the useful life of a power device is directly related to its junction or channel temperature, the thermal resistance of a device is an important parameter for estimating the useful operating life at a rated power output.

Besides simulations, there are three measurement techniques to estimate the thermal resistance of a power MMIC device: direct IR measurements, liquid crystal measurements, and diode IV measurements. The most convenient method uses a calibrated scanning IR sensor to develop a thermal image of a power device. For the Barnes Computherm instrument, the smallest imaging spot resolution is 15 microns. However, the region of peak temperature rise is the gate finger, which is many times smaller in size, causing the measurement to be an average over the measurement spot. Conventional IR measurements are made by viewing the top surface of a power device. For flip chip mounted devices this is not possible, but, because GaAs is reasonably transparent at IR frequencies, IR measurements can be made by viewing from the backside, with some loss of accuracy as a result of internal IR scattering and emissivity calibration.

The liquid crystal method uses a thin coating of temperature sensitive nematic crystals on the surface of the power device to measure channel temperatures by observing changes in the polarization of light reflected from the device under test. While the 2-micron resolution of the liquid crystal method is much superior to the IR scanning resolution, the method is time consuming and difficult to apply under actual device operating conditions, especially for flip chip assemblies.

The third measurement method[3] uses the power device Schottky diode IV data to compute the channel temperature rise. The current in a diode is exponentially dependent on the ratio of its temperature to its voltage. A simple temperature calibration procedure establishes the linear relationship of the change in channel temperature to the change in the device diode voltage, for a fixed current level. In a pulsed operating environment, the device can be biased for normal operating conditions and then rapidly switched into a constant gate current condition to measure the gate diode voltage, which is proportional to the gate temperature. This method averages the temperature of the entire gate periphery of the device and thus does not find localized hot spots. Nevertheless, the method is well suited for flip chip assemblies and can be correlated to liquid crystal measurements.

The pulsed IV method was used to measure,[4] for thermal resistance, a total of six face-up 4 W HPAs and five flipped 4 W HPAs of equal size and power dissipation. Figure 7 shows the detailed thermal stack-up of the two test configurations being compared, including material properties and sizes. The measured average thermal resistance of the face-up design was 21.45°C/W compared with an average of 11.59°C/W

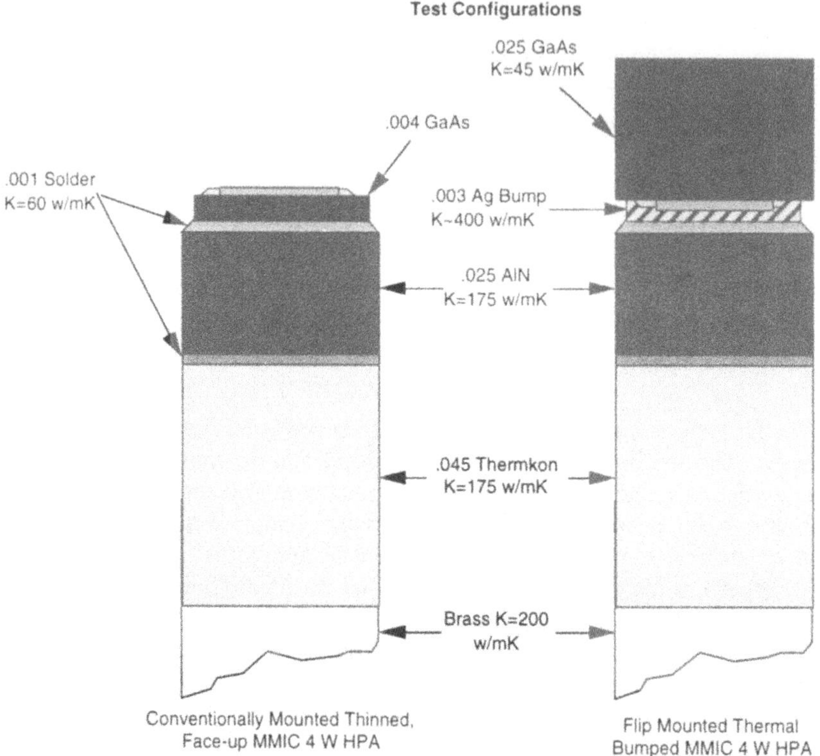

Figure 7. Two thermal environments, illustrated here and typical of T/R module assemblies, were used to compare the thermal performance of flip mounted chips to face-up mounted chips.

for the flipped configuration. While the absolute of a thermal resistance measured with the pulsed IV method may not be as accurate as a liquid crystal measurement, the relative thermal resistance comparison of the flipped chip to face-up chip is accurate and compelling. The flipped chip assembly is nearly twice as good thermally as the face-up assembly. The standard deviations for the two test results were 2.14°C/W and 1.44°C/W, respectively.

For a face-up mounted chip HPA, the heat generated in the gate region of the FET must travel through the 0.004–inch thick GaAs substrate, which has a relatively high thermal resistance compared with silicon or metal. In the flipped case, the heat has to travel only from the gate region to a neighboring source pad where it is conducted away by a thermal bump. Simulated and measured data indicates that the thermal resistance of the flipped configuration is only 60% of the, thermal resistance of the face-up configuration. Figure 8 illustrates the test configurations' thermal simulation plots. The simulations also predict the measured superiority of the flip chip configuration while illustrating both the peak and average improvement in thermal resistance as a function of spatial position within the power device. CINDA software was used to generate detailed computer models for the device and assembly shown in Figure 8. The FET assembly was simulated in two stages: a fine model showing the detail at the gate finger and a coarse model that includes the entire assembly stackup. The fine model had more than 35,000 nodes, and the coarse model had 7000 nodes.

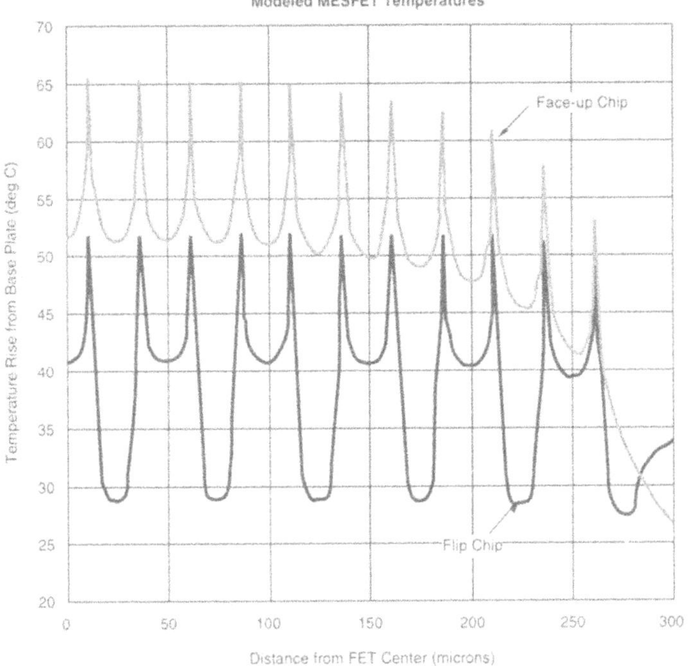

Figure 8. Simulations of the two configurations illustrated here show the temperature rise above baseplate along a slice through the middle of a power FET. The FET being modeled had 3.0 mm of gate periphery, was a MESFET, and assumed a 3V drain bias and a drain current of 450 mA. Assuming 700 mW/mm of dissipation, the flipped chip is cooler than the face up chip, suggesting a more reliable device.

FUTURE

Because flip chip technology reduces the cost of an in-place MMIC chip and supports planar chip mounting, it is suitable either for multisubstrate tile array module applications or single substrate array module applications. In either case, flip chip technology will be the driving force that moves a module design towards affordability with little or no loss of functionality or performance. Flip chip based T/R modules are being designed and tested for active array Radars, and commercial point-to-point and cellular telephone units are moving toward flip chip integrated circuit designs for low cost device and assembly designs to meet the challenge of a highly competitive market. A two-to-one price advantage and a two-to-one thermal resistance advantage will be hard to beat.

ACKNOWLEDGMENTS

The flip chips illustrated in this article were made at the Gallium Arsenide Operations of Hughes Aircraft with the help of Peter Chu, Tom Midford, and Danny Wong and designed by a team consisting of Paul Cameron, Mike Cole, Curtis Hanz, Rick Nicklaus, Wes Pan, and Doug Tonomura. The thermal studies were provided by Leah Felton and Jennifer Snopkowski.

REFERENCES

1. Y. C. Chen, et al., High PAE Pseudomorphic InGaAs/AlGaAs HEMT X-band Power Amplifiers, *1995 IEEE GaAs IC Symposium Digest*, September 1995, pp. 281-283.
2. R. Wang, et al., A 55% Efficiency 5 W pHEMT X-band High Power Amplifier, *1996 IEEE GaAs IC Symposium Digest*, September 1996, pp. 111-114.
3. D. L. Blackburn, An Electrical Technique for the Measurement of the Peak Junction Temperature of Power Transistors, Transactions on Electron Devices, Vol ED-23, no. 8, August 1976, pp 831-338.
4. L Felton, Flip Chip Thermal Study, internal report, Hughes Aircraft Company, January 1995.

MULTILEVEL VERTICAL INTERCONNECTION TECHNOLOGY

Francois Y. Colomb

Raytheon Electronics
362 Lowell Street
Andover, MA 01810

INTRODUCTION

Multilevel vertical interconnections are playing an increasingly important role in radar and communication circuits. In advanced systems, front ends with high circuit density and thus small footprint area are obtained using multilevel circuits. This approach potentially lends itself to a high degree of component integration and reduced cost. Additionally, the small overall depth of the front end allows minimal cost of installation and attractive esthetics. These factors could enable a multitude of new applications particularly in the commercial market. An example of this technology is a flat panel phased array which is manufactured by stacking thin panels for the distribution circuit, the T/R modules, and the patch antenna elements on top of each other.

The success of the multilevel approach relies critically on the ability to design dependable vertical interconnections for DC, digital, and RF signals. In this paper, several categories of vertical interconnections are identified on the basis of simple criteria including performance and manufacturing. Examples of practical implementations are then discussed for each category.

SELECTION CRITERIA

In determining which type of interconnection is best suited to a particular situation, one approach is to make a list of evaluation criteria for the size, performance, and manufacturing requirements. For the majority of cases, designers can significantly narrow down the options by considering:

- Z-axis interconnection distance
- Electrical performance: frequency range, insertion loss, dispersion, ...
- Manufacturing tolerances and cost
- Operation environment and reliability: thermal stress, vibrations, ...

In this paper, the selection approach is illustrated with a set of examples representing a progression of z-axis distance from very short to arbitrary long. For a specific distance, there may exist several candidate implementations. In practice, the approach would proceed by considering all the requirements and selecting the best comprise.

INTERCONNECTION APPROACHES

For very small distances, such as for thin multilayer flex circuits, simple integrated metallized via connections are very compact and result in very small electrical discontinuities.

Directions for the Next Generation of MMIC Devices and Systems
Edited by N. K. Das and H. L. Bertoni, Plenum Press, New York, 1997

One can add tuning elements such as a shunt capacitance to tune the equivalent series inductance of the via for enhanced performance. At frequencies below 20 GHz, the vertical section can be implemented as a separate component. An example is the elastomer interconnection. In both cases the interconnection length is a small fraction of the wavelength and these approaches form a category of electrically short via interconnections. Two examples will be shown: 1) an integrated tri-line; 2) an elastomer multistrip interconnection.

For interlayers ranging from very thin to approximately a tenth of a free-space wavelength, electromagnetic coupling by means of an aperture in a common ground plane of two transmission lines is an effective way to provide transition from layer to layer . The vias at the end of each transmission line can optionally be replaced by quarter wave open stubs thereby reducing fabrication cost at the expense of bandwidth and a larger circuit area. Two examples are discussed: 1) an aperture coupled stripline to microstrip transition; and 2) an aperture coupled patch antenna with thick ground plane.

Longer vertical interconnections are needed when a heatsink interlayer is required for removing heat dissipated by power amplifiers. Heatsinks are typically much thicker than the dielectric layers and so manufacturing long and narrow isolated metallized vias becomes impractical. One approach is to create a single circular or rectangular waveguide in the heatsink and design appropriate coupling structures between the planar transmission lines and the waveguide. In some cases there can be a large mismatch between the planar transmission line and the waveguide mode characteristic impedances so this approach is not appropriate. However, over a limited frequency bandwidth, it is possible to remedy to this situation by creating half wave sections that will produce some cancellation of the reflections emanating from the two waveguide to planar transmission line transitions. This approach is briefly discussed conceptually.

Electrically Short Via Interconnections

Integrated Tri-line Interconnection. A tri-line via interconnection is shown in Figure 1. Two 4 mil GaAs substrates with 50 ohm microstrip lines are interconnected through a 16 mil diamond substrate that serves as a heatsink. The pair of outer vias is used to connect the ground planes. The dimensions of the tri-line were initially designed for a 50 ohm characteristic impedance. As the design evolved, the position of the vias and the size of the apertures in the ground planes were adjusted for best performance. The S parameters were simulated using Sonnet including metal and dielectric losses. Good performance is obtained up to 40 GHz. Increased ripple and attenuation occur at higher frequencies as the interconnection becomes longer electrically . The total vertical section is 0.56 λ_g at 100 GHz.

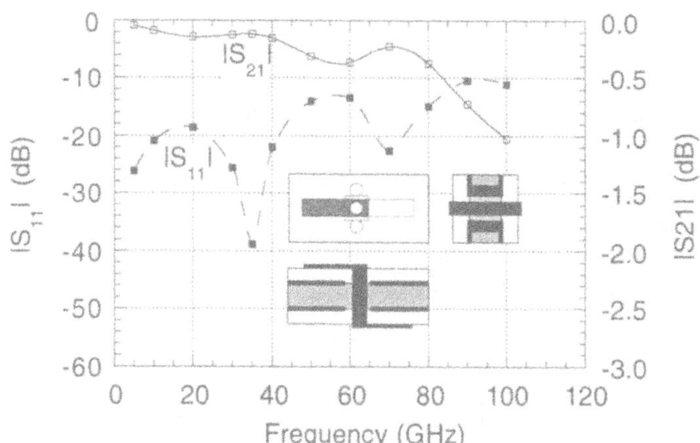

Figure 1. S-parameters of tri-line via interconnection simulated using Sonnet.

Elastomer Interconnection. Elastomer vertical interconnections were tested as part of an ongoing demonstration of a multilevel X band transmit module for a tile phased array

Figure 2. X band tile module prior to final assembly.

antenna. The module is shown in Figure 2 prior to final assembly. All DC, digital, and microwave level to level connections are implemented using elastomers inserted in slots in the perimeter of the conductive spacers.

A simplified exploded view of the interconnect is shown in Figure 3. The level to level connection is established by rows of vertical gold strips embedded in a block of silicone. Ground signal ground (GSG) pads on the substrates define the transverse electrical cross-section of this multiconductor transmission line. Each pad contacts the end of several strips so that bundles of strips are excited against other bundles. This offers some robustness of performance with respect to misalignment of the substrates.

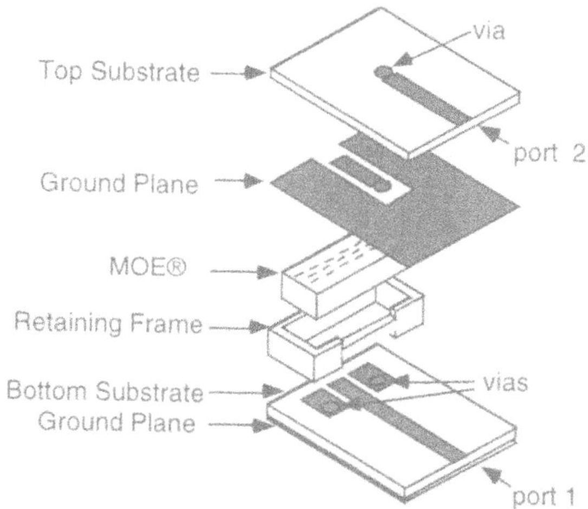

Figure 3. Elastomer vertical interconnect.

A two-tier design approach has been adopted in this study. First the vertical strips are considered alone and analyzed as a multiconductor transmission line using the Galerkin method. Second, the complete interconnection is modeled in 3D with finite elements (HFSS). The Galerkin method allows one to determine the charge distribution on the strips. The static approximation is made. Inverse Maxwell distribution basis functions with even and odd symmetry are used for representing the charge on the strips. The characteristic impedance Z_0 is then determined from the total charge after solving the integral equation using the method of moments (MoM). Figure 4 shows the variation of Z_0 associated with the relative position of the strips for two strip configurations. The plain line is the solution of the (MoM) for a balanced boxed stripline. For this simple geometry, Z_0 values can be calculated using the formula. These values are shown in Figure 4 as small squares and are in very good agreement with the MoM solution. The dotted line is the MoM solution for a balanced stripline that has a third conductor placed at mid-distance and floating electrically. The potential of this strip is zero due to the symmetry of this case. The effect of the third strip is particularly noticeable when the distance between the strips is small (small D/a values) as expected. For D/a=0.2 which corresponds to gaps between the edges of the strips equal to the width of the strips, a 3 percent decrease of Z_0 is predicted. The final geometry of the GSG pads and the sensitivity to misalignment are then determined by 3D modeling using HFSS.

Measurements were taken using the fixture shown in Figure 5. The substrates are supported by two fixture halves which compress the elastomer and can be laterally displaced to study the effects of misalignment. Verniers are printed directly on the face of the fixture halves for accurate positioning. Measured S parameters are shown in Figure 6. The S parameters include the loss of the 2 inch long feedlines of each port and the K connectors. The insertion loss of the interconnection after removing this loss is shown in Table 1. There is a marked roll-off in performance above 12 GHz which was predicted by HFSS. Recent simulations indicate that the bandwidth can be increased to 18 GHz by modifying the layout of the GSG pads.

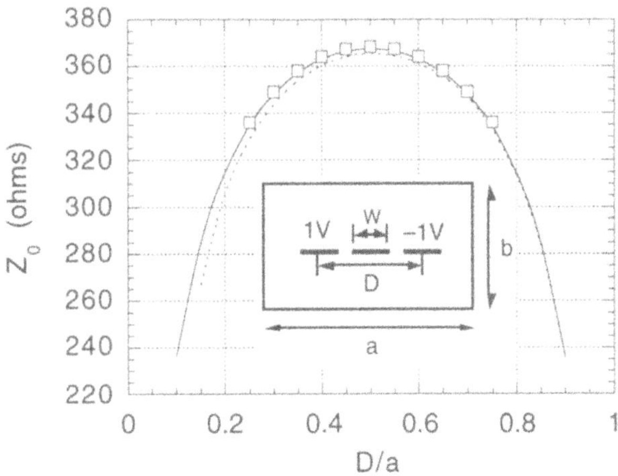

Figure 4. Simulated characteristic impedance of boxed stripline; solid line: balanced stripline (MoM); squares: balanced stripline (closed form); dotted line: balanced stripline with floating center strip (MoM). Parameters: a=2b; W=a/20; ε_r=1.

A valuable feature of the elastomer interconnect is the tolerance to misalignments. Figure 7 shows the insertion loss when the substrates are misaligned laterally in the direction perpendicular to the feedlines. A relatively flat insertion loss over a 13 mil misalignment range and a sharp increase on either side are observed. This is a favorable behavior for design purposes since there is little change in performance as long as the misalignment is in the permissible window.

Aperture Coupled Lines

Aperture Coupled Microstrip to Stripline Transition. Microwave energy can travel from one level of a multilevel circuit to another by means of electromagnetic coupling. One of the simplest examples is the microstrip to slotline transition. Several implementations with broadband performance have been demonstrated in the literature.

In the following, an aperture coupled microstrip to stripline transition is described. This structure has been designed for testing the X band module shown in Figure 2. The output of the power amplifier of the module is coupled to a stripline using a small aperture in the ground plane metallization common to the module and the stripline. For enhancing the transmission in the operating band, a stripline matching network consisting of two symmetric shunt stubs has been added.

The stripline consists of two 0.030 inch Duroid 5880 (ε_r=2.20) layers. The microstrip substrate in the module is 0.020 inch BeO. Figure 8 shows the magnitude of the S parameters of the transition modeled using full wave analysis (HP Momentum). A 2:1 VSWR bandwidth of 20 percent with center frequency of 9.85 GHz is obtained.

Aperture Coupled Patch With Thick Groundplane. In the previous case, the thickness of the conductors is small and except for loss considerations, the metallization can be modeled as infinitely thin without comprising accuracy. In some applications, the coupling aperture may need to be incorporated in a thick plate such as a heatsink. This small aperture acts as a short piece of waveguide below cutoff and produces significant attenuation. Appropriate modeling is required to account for this effect and restore adequate coupling.

In the example of Figure 9, a microstrip line is coupled to a patch antenna through an aperture in a 1/16 inch thick Aluminum ground plane (0.05 λ_0). The patch substrate is 0.030 inch Duroid 5880. A measured 2:1 VSWR bandwidth of 3.4 percent is obtained. This value is identical to the bandwidth obtained by simulating the case of a zero thickness ground plane and neglecting losses.

Figure 5. Test fixture for elastomer vertical interconnection.

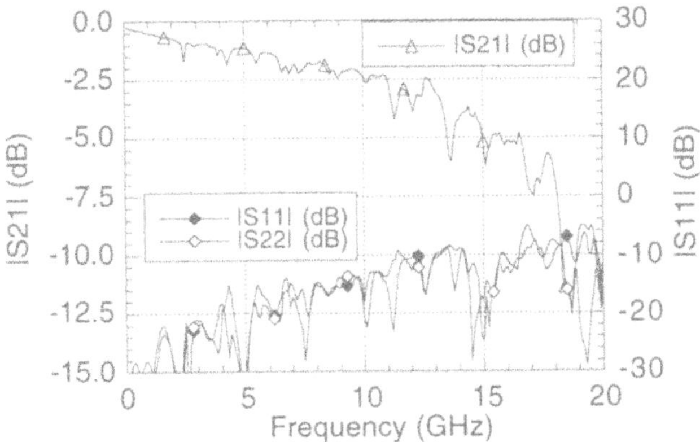

Figure 6. Measured $|S_{11}|$ and $|S_{21}|$ of elastomer vertical interconnection. Reference planes at the K-connectors.

Table 1. Measured insertion loss of elastomer vertical interconnection after removing losses due to feedlines and connectors.

	2 GHz	6 GHz	10 GHz	14 GHz	18 GHz
Insertion Loss	0.1 dB	0.4 dB	0.9 dB	2.2 dB	4.2 dB

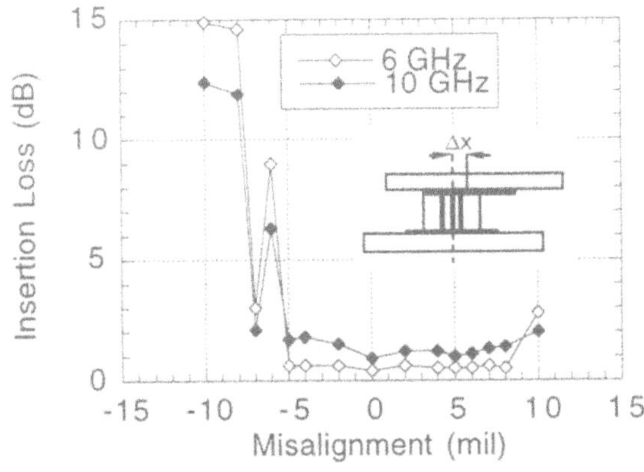

Figure 7. Measured sensitivity of $|S_{21}|$ of elastomer vertical interconnection to lateral misalignment of the substrates in the direction perpendicular to the feedlines.

Electrically Long Interconnections

High aspect ratio vias are difficult to fabricate and thus widely separated layer connections require a very different approach. One possibility consists in replacing stacked multiple vias by a single waveguide of circular or rectangular cross-section. If a good transition between the planar transmission line and the waveguide can be designed, the interlayer thickness becomes arbitrary.

It may however be difficult to avoid some mismatch if the characteristic impedance of the transmission line and that of the waveguide mode are greatly different. In this case, designing the waveguide such that the electrical length between the transitions at both ends is a multiple of one half waveguide wavelength will result in cancellation of the reflections and result in good transmission over some bandwidth.

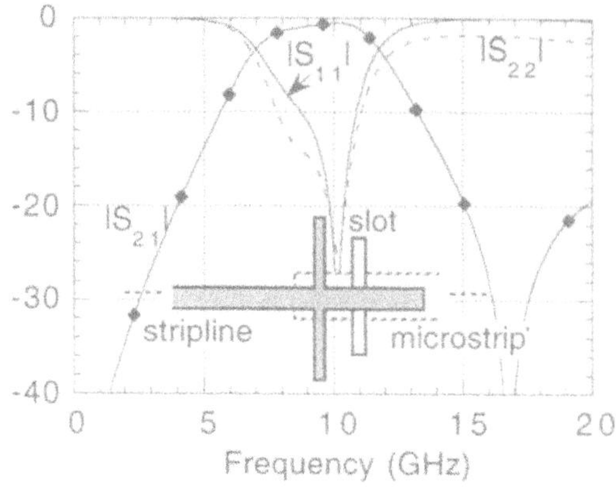

Figure 8. Simulated magnitude of S-parameters of microstrip to stripline transition (HP Momentum).

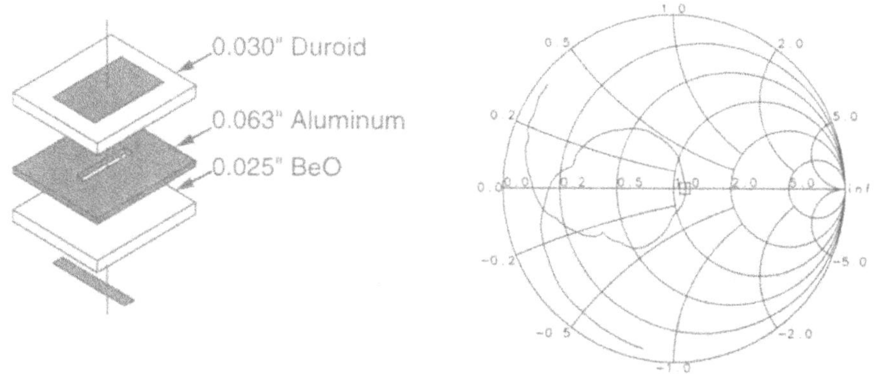

Figure 9. Measured reflection coefficient of aperture coupled patch with thick ground plane. Reference plane through the center of the aperture; sweep: 8 to 12 Ghz; square marker at 9.740 GHz.

SUMMARY AND PERSPECTIVE

Selecting the right interconnection implementation for a particular application depends in practice on a large number of criteria. This paper classifies examples by interlayer thickness ranging from very thin to arbitrary thick. For thin layers, via connections offer a compact solution with broadband performance. Strips and slots electromagnetically coupled with small apertures are an alternative approach and may also be used for moderate thickness up to approximately $\lambda_0/10$. For transitions through thick interlayers such as heatsinks, simple interconnection cross-sections such as waveguides are essential since the fabrication of high aspect ratio through holes is difficult.

It is anticipated that the role of multilevel circuits in the next generation of MMIC systems will be primordial. The driving factor is the reduced cost which is enabled by high density circuits resulting in reduced die size, fast and accurate CAD tools resulting in shorter design cycles, and novel low cost fabrication processes such as multilayer polymer materials.

ACKNOWLEDGMENTS

The author would like to thank his colleagues for their many contributions to this work, particularly J. Roman, K. Eastman, and E. Tarnuzzer and the remaining team members of the X band tile module program for their commitment and their relentless efforts. Many thanks also to M. Adlerstein and K. O'Shea for innovative concepts and practical suggestions.

MULTILEVEL PACKAGING FOR LOW COST MICROWAVE FUNCTIONS

G. Brehm, R. Peterson, H. Tserng, D. Purinton, A. Ketterson, and
B. Ables, Texas Instruments, Dallas, TX

Today's hybrid Microwave Integrated Circuit technology, utilizing active devices (transistors, diodes, and integrated circuits) integrated with passive devices by microstrip transmission lines, provides a high level of performance in a small module size at a reasonable cost. For the future, however, a step function improvement in size, weight, and cost of microwave circuit functions is needed. At Texas Instruments, to increase the density and reduce the cost of integrated microwave functions, we are developing a range of techniques for multilevel horizontal and vertical interconnects and an embedded transmission line MMIC especially well suited for wire-bond-free packaging.

HIGH DENSITY MICROWAVE CIRCUITS

To increase the density of integrated microwave functions, technologies allowing the stacking of layers of microwave integrated circuits are under development[1]. This involves the development of "tile" modules in which MMIC chips and other basic components are integrated using multilevel interconnect circuits into layers. Each layer is, in turn, stacked upon another and interconnected in the vertical dimension. This technique provides unprecedented density of microwave functions and makes possible the construction of a full T/R aperture in a layer as thin as one inch. Such a thin antenna can be designed to readily conform to the skin of an aircraft. Technologies necessary to realize such three-dimensional stacked modules for a tile array are: wire-bond-free interconnection of microwave and digital chips using multilayer circuits, "z-axis" films for high performance vertical interconnections between circuit layers, and low-expansion-coefficient, compatible housing techniques.

Figure 1 shows two layers of a three-layer stacked 3-D tile module assembly under development at Texas Instruments. This assembly is designed to perform the functions of four 10-W peak power X-band solid-state T/R modules in a little over one square inch (30 mm by 27 mm) of surface area. In this photograph the bottom layer is shown connected to a multilayer manifold and cooling structure. High power dissipation active devices, such as the power amplifier, are housed in this first layer; low power active devices, the second layer. The third layer, not shown, houses energy storage capacitors. Size and weight for the four-channel assembly in a two-by-two configuration are:

---

[1] J. Reddick, R. Peterson, M. Lang, W. Kritzler, P. Piancente, W. Kornrumpf, 1995 IEEE MTT Symposium Digest, p. 173

Directions for the Next Generation of MMIC Devices and Systems
Edited by N. K. Das and H. L. Bertoni, Plenum Press, New York, 1997

Figure 1. Tile Module -- a Stacked 2 X 2 Assembly of T/R Functions

	Without Energy Storage	With Energy Storage
Module footprint	30 mm X 27 mm	30 mm X 27 mm
Module thickness	3.5 mm	6.9 mm
Module weight	6.2 g	15.8 g

In each of the two active layers, capped (shielded) MMIC devices are mounted face up in chip wells and are interconnected to other MMICs or digital devices using polymer/metal interconnecting lines. This avoids wire bonds and the large air cavity commonly found in microwave modules. Interconnection between the layers of this module and between the bottom layer and the manifold are made using z-axis interconnect material.

Metal matrix composite substrates support each layer of the tile module. These substrates are fabricated from AlSiC (63% SiC) metal matrix composite material, a composition selected to closely match the thermal expansion coefficient of the GaAs MMIC chips. Chip wells of the proper size for each chip and coaxial feedthroughs with alumina insulators are formed in situ during the substrate fabrication process. Nickel and gold plating are used for conductivity and environmental protection.

Manifolds interconnecting tile modules of the type shown in Figure 1 use polymer/ metal multilayer circuits. In this approach layers of copper conductors alternate with Kapton E dielectric films. The exposed copper regions are TiW/Au metalized for good contact resistance and environmental robustness. By using these materials, we are able to form a compact, flexible board having extremely low-loss transmission characteristics for DC, digital, or microwave signals. It is lightweight and durable and batch fabricated to minimize cost. Figure 2 shows such a manifold.

Key to construction of the 3-D tile module is a means to form the many small, blind electrical interconnects between layers. These vertical connections are essential for increased routing density. For this purpose we use a sheet of insulating material having imbedded metal oriented such that electrical conduction is only possible through the material along the z-axis. This so-called "z-axis" material is sandwiched between two circuit layers. Contact pads on both layers are connected together by the vertically oriented conductors within the material. The portion of the z-axis material not between metal pads does not conduct current even though vertical conductors are present. Z-axis material allows repeatable electrical contact without critical alignment of the material itself.

Figure 2. RF/DC Interconnect Manifold

The tile module shown in Figure 1 uses a commercially-available z-axis material of the elastomeric gasket type. This silicone rubber material is available from 125- to 250-µm thick and is filled with spherical nickel conductors. It was used by clamping between circuit layers. Its compressibility provides conformance to surface features. In fact, one type of material was fabricated with through holes to provide increased compliance. This type of material, however, is limited in many applications by its relatively poor thermal conductivity and coarse feature size, as determined by the size and shape of the metal fill.

Micro-Z-Axis Material

We are developing techniques to overcome the limitations of the thick z-axis material discussed above. Thermal and electrical conductivity is poor for such z-axis materials because they are thick and provide very limited contact area. Large spherical conductors contact only a small fraction of the pad for smooth surfaces and less yet for surfaces with normal roughness. Our micro-z-axis concept utilizes a thin polymer layer with many vertically-oriented, high-aspect-ratio fibers or "wires" of high electrical and thermal conductivity. Using fine, high-aspect-ratio wires will allow very high heat and electrical conduction because they are compliant and form many tiny contact points even for pads having normal surface roughness. In addition, the polymer material used will have a thermally-activated adhesive property so that clamping will not be needed. Figure 3 illustrates, in four cases of decreasing z-axis metal size and increasing vertical aspect ratio, how fine wires provide increased contact area and therefore reduced contact resistance. Each of these cases has the same 20% metal fill of the z-axis insulator, but the finer wires have much better contact resistance and therefore better total resistance.

Initial trials have resulted in developmental micro-z-axis materials with gold wires as small as 1 µm in diameter at densities greater than 10^5 cm^{-2}. In one specific case, a sheet of 15-µm-thick polycarbonate material was produced with $4x10^5$ cm^{-2} 5-µm diameter Au-filled holes; in another, 15-µm-thick polycarbonate material with 10^6 cm^{-2} 1-µm diameter Au-filled holes. The manufacturing process used for this micro-z-axis material is shown schematically in Figure 4. Heavy ions such as gold or uranium accelerated to energies near 10^9 eV are used to bombard sheets of the polymer material. The resulting damage tracks are essentially straight and parallel through materials up to 100 µm thick. The tracks are then etched in

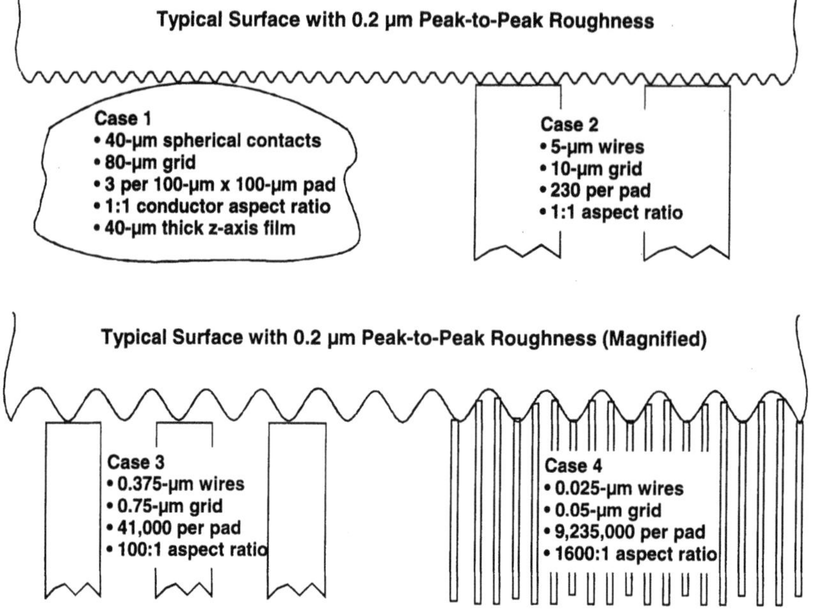

Figure 3. Small Z-Axis Wires Provide Superior Contact Resistance

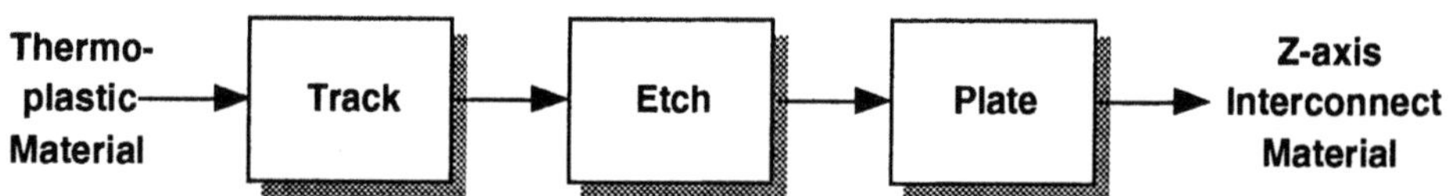

Figure 4. Manufacturing Process for Micro-Z-Axis Material

chemicals that do not dissolve the base polymer, and the holes are plated with gold or other suitable metal.

A number of candidate polymer materials have been evaluated to date by tracking, etching, and plating. Table 1 gives data on several such materials. In order to provide adequate compliance to accommodate expected surface topography and to provide good thermal and electrical resistance, we have defined the following goals for a z-axis die-attach system with micron-sized conductors through a thermoplastic material:

Thickness	50 - 75 µm
Conductor size	0.1 - 5 µm
Metal fill	15%
Thermal conductivity	30 W/m-°K

ULTRA-LOW-COST MICROWAVE CIRCUITS

In today's microwave industry cost reduction is of greater importance than circuit size and performance improvement. To dramatically reduce the cost of microwave functions, we have relied heavily on the experience of the mainstream electronics industry where electronic

Table 1. Status of Candidate Z-Axis Materials

Property	Units	AT&T ECPI (demonstrated)	Micro Z-Axis (demonstrated)	Micro Z-Axis (projected)
Dielectric Material		Silicone Rubber	Polycarbonates, Polyesters	Polycarbonates, Polyesters, Alpha Metals Staystik
Conductor Material		Nickel Plated Steel Spheres	Gold	Gold
Thickness	μm	150 to 250	10	50 to 75
Conductor Spacing	μm	250	1.8	0.25 to 15
Conductor Size	μm	25	0.08	0.1 to 5
Metal Content	volume percent	0.8	0.15	10 to 15
Electrical Resistivity	Ω-cm	0.33	N/A	$< 10^{-6}$
Thermal Conductivity	W/m°K	0.39	0.27	30 to 45

circuits are fabricated by soldering ICs in ever-shrinking surface-mount packages onto multilayer polymer/metal circuit boards whose design rules for lines and spacing are shrinking to keep pace. Low-frequency, narrow-band RF and microwave functions are being fabricated using this same circuit board and packaging technology, and there is great demand now for GaAs monolithic microwave integrated circuits (MMICs) in surface mount packages. The cost per function for assemblies using this technology is orders of magnitude lower than that using conventional microwave integrated circuit techniques.

To provide similar cost improvement for higher frequency, greater bandwidth, and higher power microwave assemblies, at Texas Instruments we are exploring techniques to extend the performance of circuits based on multilayer polymer/metal circuit boards. These performance requirements, for the most part, preclude the use of packaged devices because of the lead inductance of even the smallest surface mount packages. The resulting "chip-on-board" technology must have very fine linewidth and control, a viable vertical interconnect technique, and means of thermal management. To allow mounting of bare GaAs MMICs directly to multilayer interconnect circuits, we have developed a class of MMICs having the active circuitry completely covered by a metal top-side ground plane, the Embedded Transmission Line MMIC. In this approach, the microwave circuitry is completely isolated from electrical or environmental effects of the packaging techniques. RF I/O contacts are provides by pads isolated from the ground plane. These I/O pads are connected to corresponding pads on the multilayer interconnect circuits by an adhesive z-axis material layer or by solder bumps.

The Embedded Transmission Line (ETL) MMIC

To be compatible with low-cost flip-chip circuit technology as well as with the 3-dimensional circuitry discussed in the first part of this paper, changes to the MMIC itself are needed. To address some of these issues, we initially developed a simple shielded MMIC with a ground plane over the entire chip except for bond pad regions. That "capped MMIC" is shown conceptually in Figure 5. It allows on-chip interconnects to be compatible with wire-bond-free chip interconnection. A consequence of the elimination of wire bonds is that the chip surface itself is in close proximity to the traces on the circuit board. Therefore, a means to avoid or compensate for interactions between on-chip signals and signals on the interconnect circuit are needed.

The simple capped MMIC was subsequently modified to optimize on-chip transmission lines and interconnects for best performance and processing ease. The result is a multilayer

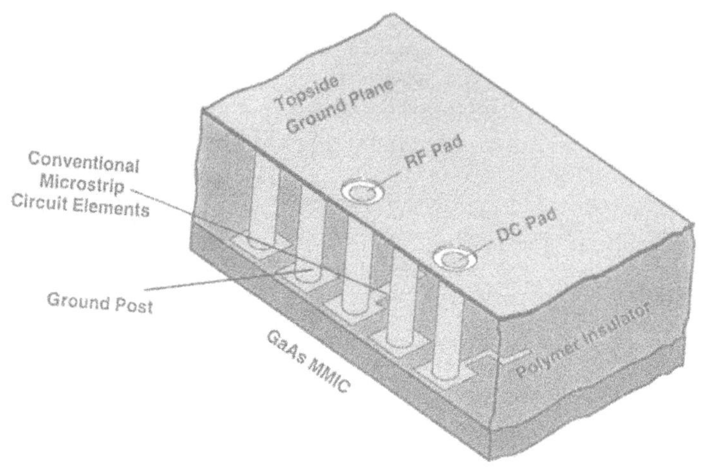

Figure 5. Capped MMIC Concept

technology using thick layers of polymer and metal. Features of this embedded transmission line (ETL) MMIC are shown in Figure 6. This structure uses two thick layers of polymer insulator and three thick metal layers making possible transmission lines both on the GaAs surface and suspended half way between the GaAs surface and the topside ground plane. Thick metal vias connect both ground and signal pads to the topside. For flip-chip mounting, the ground vias are made large enough to provide excellent heat sinking of the active devices.

The ETL MMIC improves upon simple capped MMIC in that the matching elements are designed specifically for the ETL medium. No substrate thinning or backside vias are used, and the device is suitable for flip-chip or upright mounting. Excellent RF isolation from the packaging environment and improved ruggedness with respect to chip handling are achieved. Complete RF testing can be performed in wafer form prior to packaging.

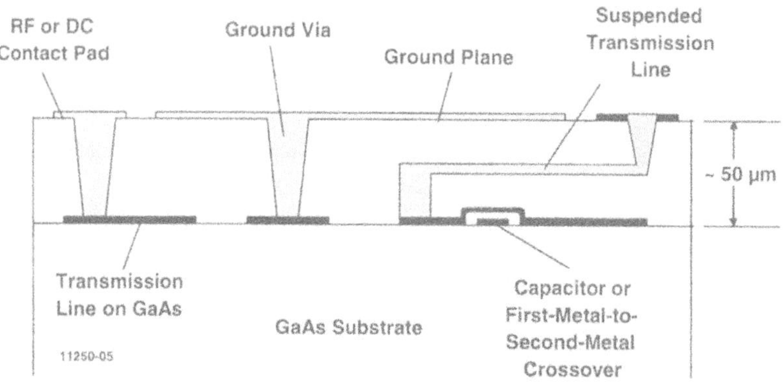

Figure 6. Features of the Embedded Transmission Line MMIC

Combined with multilayer wire-bond-free packaging, an increase in routing density is achieved.

Several polymers are under consideration for the thick dielectric. Low dielectric constant and dissipation factor, low water absorption and weight loss, and good CTE match to GaAs are desirable. Table 2 shows properties of candidate polymers that are being evaluated. We are simultaneously exploring means to fabricate the ETL MMIC on-chip interconnects. Two basic flows are envisioned, as illustrated in Figure 7. On the left via posts are plated prior to coating with polymer, a more practical process, at present, than that shown on the right, where the thick polymer is etched and then plated. In either case, some means to planarize the surface prior to forming the next layer of metal and polymer is anticipated. Figure 8 shows photographs of plated via plugs on an experimental wafer just prior to applying the thick polymer (polyimide) layers.

Table 2. Candidate Dielectric Polymers for On-Chip Interconnects

Property	**Dow BCB** (benzocyclobutene)	**DuPont PI2611** (polyimide)	**Dow PIBO** (polyamic acid solution in NMP)
CTE (ppm/C)	65	3 (horiz.)	5
Dielectric Constant	2.7	2.9 (horiz.)	~3
Dissipation Factor	0.0008	0.002	NA
Water Absorption	0.25%	0.5%	NA
Weight Loss (350C)	2%	0%	NA
Planarization	90%	<50%	NA
Adhesion	Fair	Good	NA

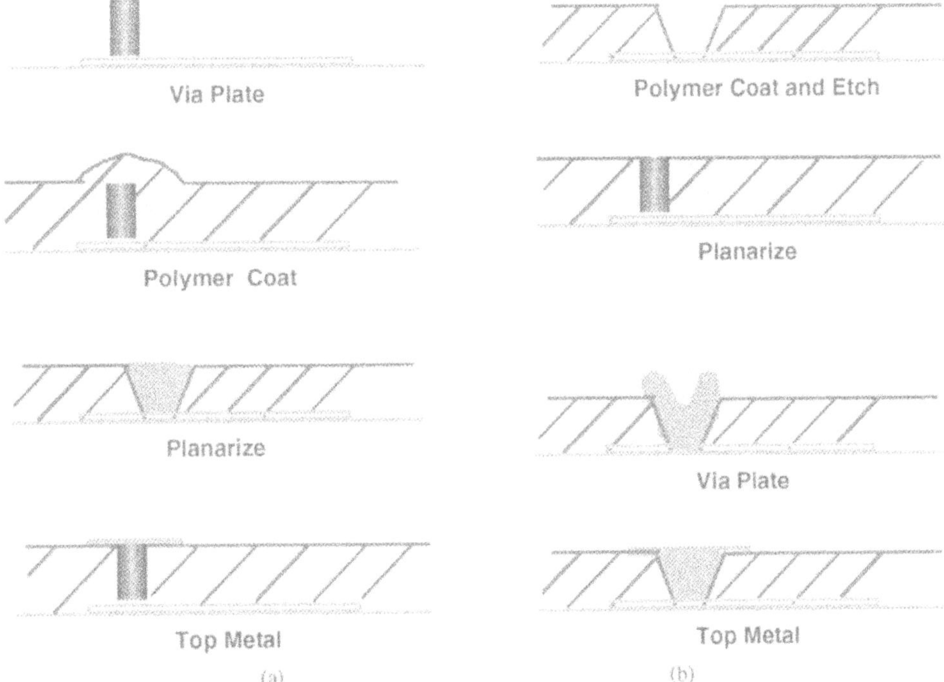

Figure 7. Alternate Interconnect Process Flows: (a) metal posts formed prior to polymer deposition, (b) polymer deposited and etched prior to metal via fill.

70-µm "Plugs" on 150-µm Pads **50-µm Thick Plated Gold Plug**

Figure 8. Plated Gold "Plugs" for Thick-Dielectric Vias (before Dielectric Coat)

Ultra-Low-Cost Module Concept

Enabled by the technologies discussed above, we envision an advanced low-cost, compact module suitable for both ultra-low-cost commercial and high-performance military applications. It incorporates a polymer/metal multilayer circuit with embedded passive components, ETL MMICs, silicon ICs, and discrete passive components attached to it with z-axis adhesives, and a molded encapsulant. Since it is manufactured in batch as a panel and contains a minimum of materials, it will be fast and inexpensive to manufacture. Features of this multi-chip assembly, shown in Figure 9, are an expansion matched, high thermal conductivity multilayer interconnect circuit, the use of z-axis material for chip attachment, the ETL MMIC, and the molded enclosure with integral heat fins. Even though a number of developmental challenges remain, this concept has the potential to revolutionize microwave circuits the way that surface mount ICs have revolutionized the low-cost computer industry.

Acknowledgments

The authors would like to acknowledge the Air Force Wright Laboratories and the Defense Advanced Projects Agency for their support of a portion of this work. We also would like to acknowledge the support of TI management and our many colleagues that have offered helpful suggestions during the course of this investigation.

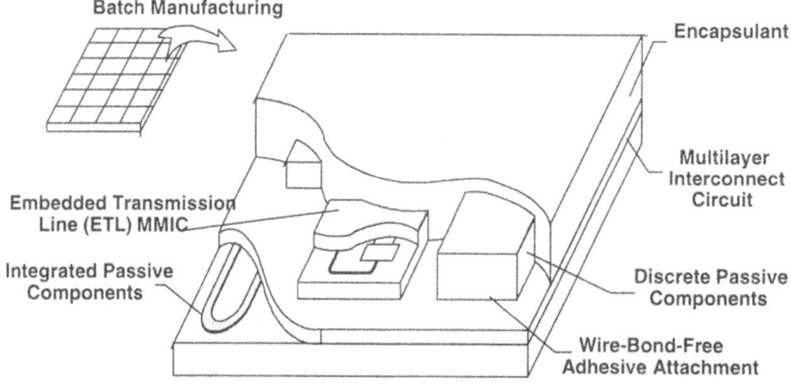

Figure 9. Concept for Batch-Manufactured Multi-Chip Assembly

Future Applications of Millimeter Technology

Robert McLaughlin[1]

Chief Scientist
WinStar Communications
Tyson's Corner, VA 22043

Abstract

As the technology that has been created for the cold war becomes more generally available the millimeter band will be more accessible. A number of applications can be developed based on the MILSTAR system, BCIS and similar systems. It is possible to develop everything from fixed, to multipoint, to mobile. Applications can be both indoor and outdoor. Areas that are being explored currently are short haul, wireless ATM, wireless cable, wireless local loop and collision avoidance radar.

Current Defense Technology

Millimeter technology has been used in the US military for over 20 years. It has been used for tactical communications under the umbrella of MILSTAR. MILSTAR uses 20, 44 and 62 GHz bands for communications. It is used in a context suggestive of low and high tier mobile and satellite communications. Other uses of the K-a and millimeter band include a Friend or Foe Identification System at 38 GHz. Work on a number of proximity radars and side-looking surveillance radars has been done. These systems operate from 30-96 GHz. Most of the solid state RF has come from the proximity radar.

The reasons for its use are the security that is provided in narrow beam and high atmospheric absorption. Antennas and Linear RF have been developed to provide these systems are near the point that they can be commercialized. This work has been done by a number of organizations including TRW, Hughes Aircraft, Lockheed-Martin, HP, Raytheon, and GEC Marconi to name a few. Work is still on going although there is a push in the DoD to use Commercial Off the Shelf (COS) Technology. Work around turning BCIS into a battlefield communications system are likely to be the most interesting in the commercial world.

Current Applications

Current applications of the K-a and Millimeter bands include:

Fixed Point to Point Service

[1] Author's Address: 7799 Leesburg Pike Suite 401S, Tyson's Corner VA, 22043. Email: rmclaugh@winstar.com.

Directions for the Next Generation of MMIC Devices and Systems
Edited by N. K. Das and H. L. Bertoni, Plenum Press, New York, 1997

Wireless Cable (LMDS)
Collision Avoidance RADAR
LEO Satellites

Currently these applications are limited by the RF power amplifiers (PAs) and Low-noise amplifiers (LNAs) commercially available. At the bottom of the K-a band solid state amplifiers can do in the range of 10 Watts Class C[2] power and about 100 Watts with a Klystron. While at the low Millimeter band solid state is limited to 1 Watt Class C solid state and with a Klystron is about 30 Watts. LNA's in the bottom of the K-a band can be found with noise figures of under 2 dB while in the millimeter band 4 dB is common. The lack of power and high receiver noise figure greatly limit applications.

In the upper K-a and millimeter at present most of the RF modules that used are based on the use of a varactor multiplier that is driven by a VCO. The VCO is used as part of a PLL that provides tuning and in some implementations modulation of the RF. The difficulty is that by using a multiply up RF source, power and the kinds of modulation techniques are limited. Generally power is not more than 18 dBm and the modulation method is QFSK.

Commercial applications as a result for the most part limited to short haul point-to-point links. These links are used as back haul for off-net locations, internal connections between a campus, and connections for an internal network. At present several companies offer services where DS1 and DS3 links are leased for these purposes. The lack of linear RF power limits links to within 6 km and the bit rates to 1 bit/hertz/second.

At 56-76 GHz collision avoidance Radars are being developed. In a short time they will become required equipment in Germany. These Radars will warn if one is too close to the car ahead that is the distance is shorter than the minimum required braking distance. Over the next several years this will become a growing area. It is possible to expand this to include analysis of road and tire conditions to allow intelligent adjustment to actual conditions. In the US the FCC has made a rule making for unlicensed vehicular radar at 46.7-46.9 GHz and 76-77 GHz while allowing unlicensed uses at 59-64 GHz for non-vehicular radar applications.

Near Term Applications

It is possible to expand to current applications to include a fixed point-to-multipoint service. This service uses a central hub that provides links to sites that are at a distance from the hub. These systems are being developed to provide a number of services is the K-a band and millimeter band. They include:

LMDS
MMDS
Wireless POTS
Satellite Links

MMDS is at 2.5 GHz and 18 GHz it is a wireless cable system. It is being deployed in the US and generally provides an analogue TV signal. LMDS downlink is at 28 GHz with an uplink at 31 GHz and is designed to allow provision of interactive wireless services. MVDS is at 44 GHz and is the European version of LMDS. Wireless POTS services are designed to provide basic POTS services in those places that do have an embedded wire infrastructure. These systems are being deployed in bands from 2-44 GHz.

It is clear, in the near term, applications in the US will be driven by the need to provide competitive access for cable and telephone systems. In the large markets of Asia, South America and the former East Block wireless communication maybe the only viable method in the near term to provide universal telecommunications services. However unless Linear RF is provided

[2] A good rule of thumb is that Class Power is about 6 dB down from Class C Or we have about 2.5 watts solid state Class A power at 28 GHz.

in the millimeter bands it is not likely these applications will occur much above the middle K-a band.

In a number of the proposed LEO and MEO satellite systems, 28 and 44 GHz are proposed for satellite to ground links. In many of these systems the switching functions are done on the ground with the gigahertz links required to provide the bandwidth needed for the up and downlinks for the carrier traffic and control of the satelite. There is some conflict between these applications and some proposed terrestrial ones. The recent FCC rule making at 28 GHz has only made these conflicts worse by not allocating enough bandwidth for either satellite or terrestrial applications while trying to make both sets of petitioners satisfied.

Medium Term Applications

As we have mentioned the deployment of millimeter technology much beyond its current uses require the development of Linear RF in the 24-50 dBm range. It is possible by the use of Klystron's and TWTs to get Linear RF in the range of 36 dBm. There are military systems that are based on GaAs that provide over 24 dBm linear RF at 44 GHz and hints of greater powers have been made. Several have mentioned the possibility of $5 MMICs[3] in the millimeter if there is a great enough demand. Delivering on this promise implies that the following applications can be developed in the next 5 years:

LMDS like services in the millimeter band
MVDS
Long Haul, high bandwidth fixed point-to-point
Medium Haul, high bandwidth fixed point-to-multipoint
Microcell Mobile
Campus Area LANs

The development of wireless interactive video services seems to be an area that will expand greatly as markets become competitive or begin to develop in Asia, South America and the former East Block. These markets will in short time consume a great amount of equipment. Since wireless has both a quick deployment time and a lower capital cost than wired solution capital cost.

With Linear RF comes both distance and dense modulation schemes. As much as 20 kilometers with 12 bits/hertz/second seems possible. This would allow services at rates in the range of OC-9. In the next five years it seems likely that we can expect services in the range of OC-3 at distances of 10 kilometers.

These distances and bit rates can be divided up to provide hubbed or multipoint services. These systems will allow the deployment of a wireless backbone for competitive local access, wireless interactive services, and wireless cable. Some applications like a backbone will require symmetric bandwidth. While the wireless interactive and cable services will require asymmetric bandwidth.

Mobile applications seem to be a natural extension of MILSTAR. It seems likely in the next 6 years a microcell system will be deployable. Given the large bandwidths possible in the millimeter bands the promisees of High Data Rate services on PCS and PACS will be delivered. Combine the bandwidth with something like a JAVA terminal and one has a very powerful service.

---

[3] TRW GaAs Products Division at a rump session of MTT-S '96 announced that by 2001 it expects runs of 2000 wafers to cost $0.75/mm^2. With a 50% yield rate for 1.0 mm^2 chip (typical PA size) this translates into $3.50 per MMIC in a plastic package for a 1 Million piece run. Such volumes seem likely only when millimeter technology becomes a consumer product. At such volumes RF Units in the realm of $50 seem possible.

Long Term Applications

In the next ten years it is possible a large number of commercial applications some of which presently seem impossible will be developed. These include but are not limited to:

Intelligent Highway
Extension of Optical Systems
Backbone
Long Haul
High Tier Mobile
Satellite Based Mobile

These services will require a greater understanding of propagation and greater density power sources. Mobile applications seem to be very attractive for some businesses as they will allow delivery of data while in transit. Millimeter based intelligent highway systems will improve trunking and the use of roads in general. The satellite application would be what the MEO and LEO satellite systems want to deliver. The idea that high bandwidth services can be delivered from any point on the earth to any other point seems to have a great utility.

As Photonic Microwave technology develops (see reference 14) it will be possible to develop systems where the radio is directly tied to a fiber link. Providing a continuous path. Systems based around this for the delivery of cable and similar services that are inherently analog are being developed in Europe at lower frequencies. Photonic Microwave has the possibility of proving radios with greatly reduced phase jitter and improved signal to noise ratios over that which is currently considered possible with electronics. These improvements will lead to new applications that have yet to be considered.

Technology Sources

At this time a number of companies are making Power Amplifiers (PA) and Low Noise Amplifiers (LNA) for the millimeter region. There needs to be some improvement before many of the applications discussed can be made real. A summary of PA's and LNA's at 38 GHz :

Company[4]	LNA Noise Figure in dB	PA 1 dB Compression Point in dBm
Alpha	4	13
DBS Microwave	4	13
Endgate	4	13
GEC Marconi	4	No Product
HP	Entering market in 1997	
Hughes	No Commercial Product.	
Lockheed-Martin	No Commercial Products	
MA/Com	Entering Market end of 1997	
Millimeter	4	17
Raytheon	Will build to suite.	
TI	No Commercial Products above 28 GHz	
TRW	4	23

There a number of companies that make multiple up-convertors and down-convertors in the millimeter band. They are used extensively in the nFSK radios that are commercially sold. To develop the potential of this spectrum use of QAM is required. This means the development of

[4]This was as of June 1996. Updated information is always welcomed. Many of the companies have expressed interested but have no dates for commercial products.

PA's in the 30 dBm realm and RF receivers whose antenna to IF output noise figure is less than 5 dB. Something that in commercial product is a few years off. The development of InP will improve this in the millimeter band in about 7 years.

A generally understated problem is the Modem. To develop wireless applications in general will require cheap single chip modems. Satellite TV and digital wireless applications have caused the development of 4 and 8 PSK modem chips that can operate at over 20 MegaBaud. For digital wireless to meet its potential chips that can deliver 1024 QAM at 25 MegaBaud or higher need to be developed. Work that has been done for the NSA and DoD in GaAs has the potential to be turned into Silicon based ASICs in the next 2 years.

Required Research

To provide these services there is a lot of research that must be done. For the most part they are all in the area of propagation. A review of the literature indicates that there are a number of holes in the studies that have been done. It is clear that millimeter is effected by rain but exactly which aspect of rain? Is the aspect drop size, rain rate, or sheeting?

These questions can be answered only with extensive study. However without that study it is not likely that all the economic potential in this spectrum will be developed. It will take a program that is national in scope about 5 years with a modeling ability to provide answers. Some studies have been done by the military but it is not likely they will provide all the required answers. As one can trade power (money) for understanding of propagation.

Microwave Photonics is developed enough that one can think in terms of all optical radio. Such a radio will have extradionary characteristics.

Self-Induced Transparency[5] is a possible means to get around the losses induced by rain in the millimeter. This would be done by having the pulse length shorter than the relaxation time of water. The modulation would be in the pulse edge. It is not clear if this can be commercialized but if the effects of rain can be mitigated this would allow a board set of applications for millimeter.

New modem designs using advances in decision theory need to be constructed. These modems should allow higher bit rates in noisy enviroments that convential modems allow. This however will also require very cheap DSPs.

Conclusion

The millimeter spectrum has the potential to deliver several commercially important wireless high-bandwidth services. With the commercialization of defense technologies it is likely the technology required for the delivery of this promises will occur. It will however require an investment in development of modems, GaAs and InP technologies and propagation research.

References

1. Theo G. Van De Roer, *Microwave Electronic Devices*, Chapman and Hall, London UK, 1994.
2. Kaveh Pahlavan and Allen Levesque, *Wireless Information Systems*, Wiley-Interscience, NYC NY, 1995.
3. Roger L. Freeman, *Radio System Design for Telecommunications (1-100 GHz)*, Wiley-Interscience, NYC NY, 1987.
4. Merrill Skolnik, *Introduction to Radar Systems*, McGraw-Hill, NYC NY, 1980.
5. Roger Lhermitte, *Attenuation and Scattering of Millimeter Wavelength Radiation by Clouds and Precipitation*, **Journal of Atmospheric and Oceanic Technology**, Volume 7, Number 3, June 1990, page 464.

---

[5]Suggested to me by Jim Frazier of the University of Texas.

6. M. Hata and S. Doi, *Propagation Tests for 23 GHz and 40 GHz,* **IEEE J. on Selected Areas in Communications**, September 1983, page 654.
7. Edmond Violette, Richard Espeland and Kenneth C. Allen, *Millimeter-Wave Propagation Characteristics and Channel Performance for Urban-Suburban Enviroments*, NTIA Report 88-239.
8. R. Espeland, E. Violette, and K. Allen, *Atmospheric Channel Performance at 10 to 100 GHz*, NTIA Report 84-149.
9. David Jones, Richard H. Espeland and Edmond J. Violette, *Vegetation Loss Measurements at 9.6, 28.8, 57.6 and 96.1 GHz Through a Conifer Orchard in Washington State*, NTIA Report 89-251.
10. J. Shapiro, *Adaptive Diversity Combining for Improved Millimeter Wave Communication Through Rain*, **IEEE J. on Selected Areas in Communications**, September 1983, page 686.
11. James Fawcette, *Milstar: Hotline in the Sky* , **High Technology**, November 1983.
12. G. Brussaard and P. A. Watson, *Atmospheric Modeling and Millimetre Wave Propagation*, Chapman and Hall, London UK, 1995.
13. Michael J. Marcus, **Recent Prgress in US Millimeter Wave Spectrum Management Policy**, *MTT-S 1996 Digest*, Volume 2, page 505.
14. Alwyn Seeds and Charles Cox, **Microwave Photonic Systems**, *Workshop at IMS '96.*

About the Author

Robert McLaughlin is the author of 40 papers and books and holds patents in switching and multimedia. He is currently responsible for new technology at WinStar Communications, Inc. a provider of services at 38 GHz in the US. He is a member of the IEEE, the ACM, the American Society for Quality Control, and the Association of Old Crows.

DISTRIBUTED POWER COMBINING AND SIGNAL PROCESSING IN A 2D QUASI-OPTICAL SYSTEM

James F. Harvey[1], Michael B. Steer[2], Huan-Sheng Hwang[2], Todd. W. Nuteson[2], Chris W. Hicks[2], James W. Mink[2]

[1]US Army Research Office
Research Triangle Park, NC 27709
[2]North Carolina State University
Raleigh, NC. 27695

INTRODUCTION

Quasi-optical power combining is a circuit integration technique that holds the promise of moderate to high powers in the high RF frequencies using a planar semiconductor integrated circuit technology, providing the traditional advantages of solid state technology: low cost, low weight, reduced size, low phase noise, high reliability, and long circuit life, to mm-wave systems. These potential advantages are extremely important for applications in satellite communications, in vehicle warning radar, in cellular radio base stations, and for primarily military applications such as missile seekers, unmanned aerial vehicles, and millimeter wave radar. Currently the requirement for moderate or high powers at mm-wave frequencies for these applications is met by commercial traveling wave tubes. A solid state technology compatible with commercial planar IC fabrication techniques could provide lower cost fabrication of circuits with the inherent advantages of small size, low weight, low phase noise, and high reliability and operating life. Unfortunately, the size of static semiconductor devices is limited by capacitive delay effects and by wavelength effects (a non-traveling wave device larger than about half a wavelength begins to introduce negative phase interference in the collected current). Since the device size is proportional to power capacity, high frequency mm-wave devices produce very small power outputs. In order to address applications requiring mm-wave power, many semiconductor devices must be combined together. However, traditional corporate combining structures are limited in the number of devices which can be combined together. In addition, losses due to radiation, substrate modes, and conductor loss per unit length due to skin effect all increase significantly at higher mm-wave frequencies. Quasi-optical techniques avoid much of these losses and problems of corporate combining circuits by doing the power combining in the extended electromagnetic field in free space or in a dielectric substrate. The size of the combining region is typically greater than several wavelengths and less than 100 wavelengths, so in the combining region the electromagnetic

Directions for the Next Generation of MMIC Devices and Systems
Edited by N. K. Das and H. L. Bertoni, Plenum Press, New York, 1997

field is controlled by devices such as lenses and mirrors rather than devices such as transmission line or single mode waveguide, resulting in the name "quasi-optics."

NEW APPROACH FOR QUASI-OPTICAL POWER COMBINING

In this paper, we describe a new approach to quasi-optical power combining using the extended EM field in a two dimensional dielectric substrate, rather than the free space combining favored by quasi-optical grid or quasi-optical array architectures. The previous paper in this review, by Prof. Rutledge, describes an amplifier grid architecture, in which the devices can be spaced much closer than half a wavelength in a planar array and the metallic connectors for the inputs and outputs form extended multiwavelength antennas. An alternative architecture uses a planar array of active antennas, each antenna circuit having an active device or IC integrated directly into the array.[1] Amplifier arrays and oscillator arrays have been demonstrated in both the grid and the antenna array architectures.[1-5] The grid architecture and the antenna array architecture both utilize two dimensional planar arrays of circuits, which are individually compatible with planar solid state fabrication technology, but which must be individually mounted and aligned in a system structure. However, the combining is done in the three dimensional free space surrounding and between planar arrays. Polarization is generally used to isolate the input from the output of these circuits, and polarizing grids are generally used to control the polarization between device arrays.

An alternative type of quasi-optical architecture combines the power in the EM field confined within a dielectric substrate. Since the substrate confines the EM field in a region of less than a wavelength in one of the dimensions, the combining is essentially in a two dimensional field distribution. (Reference 6 is an analysis which demonstrates that lenses within the substrate can guide the propagating field in Gauss-Hermite beammodes.) Two approaches have emerged for quasi-optical power combining in a two dimensional substrate. The first reported 2D combining structure used oscillator and amplifier active antenna elements at the surface of the substrate to add incrementally to the increasing EM field within the substrate.[7-8] An alternative approach couples out essentially all the power from the substrate mode, dividing it among an array of amplifier circuits and delay lines. The power is amplified in separate amplifier chains, with the phase in each controlled by separate delay lines so that the power can be coupled back into the substrate mode with the appropriately shaped wave front.[9] The first technique attempts to keep the EM field primarily within the substrate, performing as much processing functions within the substrate as possible, for example lenses and mirrors within the substrate are used to control the Gaussian lateral field distribution. The second technique takes the approach of doing most of the amplification and processing outside the substrate. There are several potential advantages for these 2D combining architectures over the more traditional 3D combining architectures described above. Since the combining is done within a 2D confined substrate and since the critical wave dimensions within the substrate are reduced by a factor of the index of refraction, 2D substrate combining can be much more compact. Since an entire 2D system can consist of a single substrate, with metallization and possibly micromachining on each side, fabrication is compatible with planar solid state photolithographic IC technology, without needing external support and spacing structure. Manufacture can be relatively simple and inexpensive. This system structure is solid and therefore rugged and resistant to mechanical shock and vibration. The absence of external supporting structure may also result in lighter system weight than for 3D systems. Since the active device elements are on or near the substrate surface, metallization on the surface can improve heat dissipation compared with many of the 3D system architectures. Finally, because the wave impedance in the substrate is less than the impedance of free space, the impedance transformation required to match individual devices to the propagating modes is less than for free space combining. In this 2D architecture, substrate modes are used for the combining and signal routing. So in contrast to conventional MMIC's, components located on the substrate surface are optimized to couple into the substrate modes.

THE 2D SUBSTRATE POWER COMBINING SYSTEM

A conceptual 2D substrate power combining system is shown in Fig. 1. A low phase noise signal source or quasi-optical oscillator array can provide the initial source of the EM field. The propagating field is shaped and controlled by optical elements such as lenses or mirrors within the substrate. Amplifier arrays increase the power in the EM field and arrays of nonlinear devices can perform various signal processing functions such as frequency tripling. Leaky wave antennas provide a very natural way to couple the EM power out of the substrate system. Over the last 3 years we have separately demonstrated most of the individual elements of this architecture at X-band frequencies. A 2D oscillator array combiner produced efficient power combining. A 2D amplifier array combiner yielded high gain. The lens system for substrate mode control was successful, and the leaky wave antenna structure proved to be an efficient way to couple out the electromagnetic energy with a frequency scannable structure. These experiments were designed to demonstrate concept viability and the circuits and devices were not fully optimized.

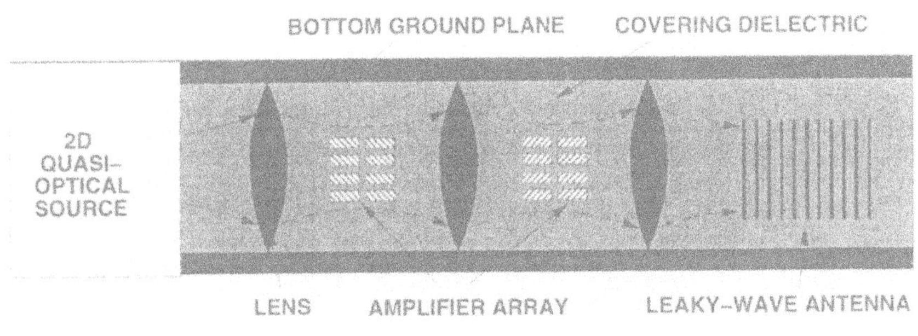

Figure 1. 2D quasi-optical power combining system.

The combining of a 2D oscillator array was successfully demonstrated for the first time. Fig. 2 shows the architecture of this experiment. An array of four active antenna elements, using GaAs MESFET oscillators directly incorporated into planar tapered slot or Vivaldi active antennas, was placed on the surface of the substrate. The Vivaldi antennas effectively isolated the active device from EM fields behind the antenna. Metal reflectors were placed at the front and back edges of the substrate, shaped to form an open resonant cavity within the substrate. The curved reflector was circular to approximate the parabolic wave front of the resonant modes, while the planar reflector is at the mode waist.

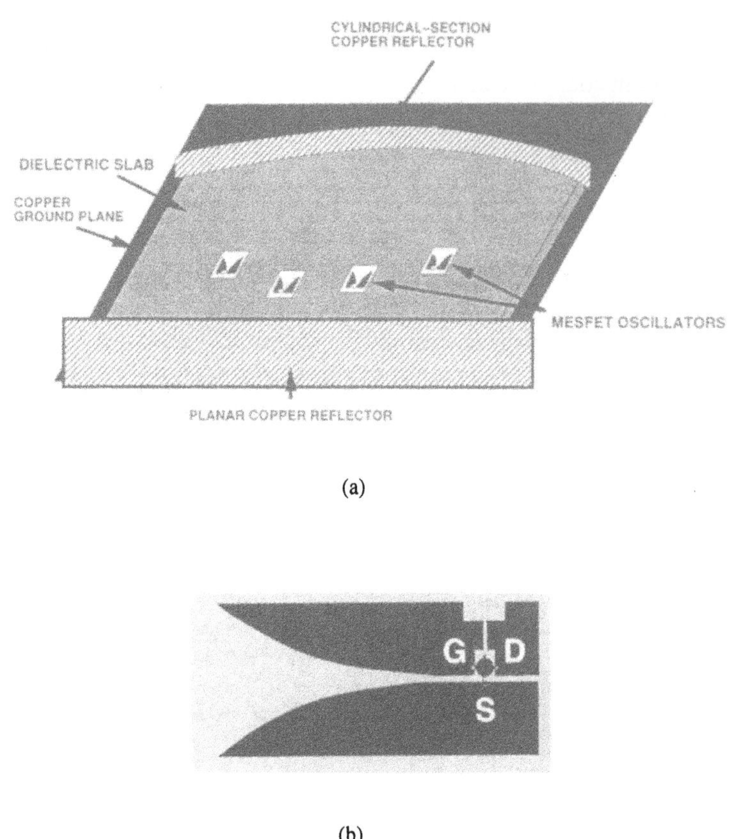

Figure 2. (a) 2D quasi-optical oscillator array, (b) oscillator unit cell.

The oscillators were turned on one at a time, each one locking to the mode within the open cavity. The oscillator spectra are shown in Fig. 3 with the power level rising 3 dB for two oscillators and an additional 3 dB for four oscillators. This result indicates highly efficient power combining for the 2D substrate architecture and may point to a major advantage of 2D quasi-optical power combining. The linewidth for four oscillators was less than 6 kHz at 30 dB below the peak. The quasi-optically combined oscillators could be injection locked using an external synthesized source with a power level 35 dB below the oscillator peak power level. The resulting injection locked signal had a reduced linewidth less than 3 kHz at 30 dB below peak. The lock-in bandwidth was 350 kHz and the locking bandwidth was 470 kHz. Increasing the injection locking signal by 3 dB increased the lock-in bandwidth to 590 kHz and the locking bandwidth to 700 kHz.[7]

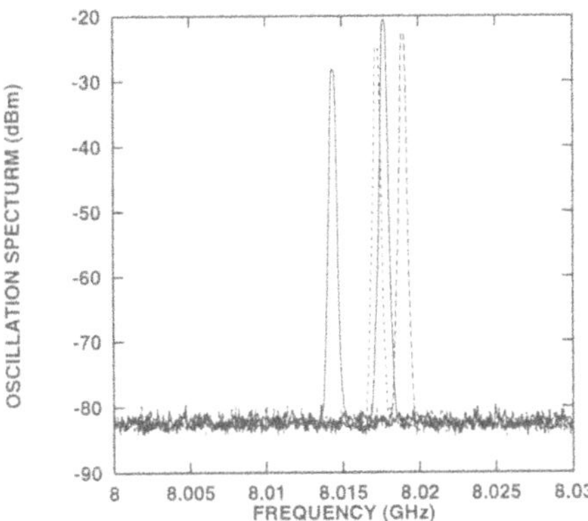

Figure 3. The oscillation spectra.

A 2D substrate quasi-optical amplifier combining array was demonstrated with four MESFET amplifier elements, each element consisting of a MESFET integrated directly into an active antenna, with input and output Vivaldi antenna sections, as shown in Fig 4. The Vivaldi antenna segments were effective in isolating traveling waves going in opposite directions within the substrate. The array produced 15 dB small signal gain and roughly 10 dB saturated gain (see Figs. 5 and 6). The MESFET amplifier circuits were biased at low current (with only milliwatts of output power) to keep the amplifiers in stable operation. The active antenna elements were located under the substrate and above the ground plane, as shown in Fig. 4, to minimize distortion of the EM fields and scattering from the top surface. The dielectric substrate material was Rexolite (ε_r=2.57) and the lenses were Macor (ε_r=5.9), with focal lengths of 28.54 cm. The substrate width was 30 cm. The substrate thickness was roughly half a wavelength, with the propagation in the TE Gaussian beam mode. The use of the TE mode reduces loss due to scattering at the surface because the field strength is very small there. The placement of a metal cover on top of the substrate decreases system losses by 3.5 dB, see Fig. 6. Other sources of system loss have been identified, and prospective countermeasures are proposed to reduce them. There is a 10 dB insertion loss for this system. An additional 4 dB in gain can be achieved by using concave lenses of air, rather than convex lenses of Macor, because the concave shape reduces scattering at the lens. An additional 2 to 4 dB can be recovered by reducing scattering at other places in the system. The amplifier array saturated at -15 dBm input power. At saturation, the gain vs input power curve for this type of quasi-optical system has a different characteristic shape than for a conventional amplifier. This is due to the unique quasi-optical structure in which each amplifier only samples a small fraction of the total electromagnetic field. As input power increases beyond amplifier saturation, the power simply bypasses all the amplifier devices and the output power increases linearly with the input power with a slope of 1.[8]

The leaky wave antenna structure shown in Fig. 2 was demonstrated as a method of coupling the radiation out of the substrate. Ten 1/8 inch wide metallic strips were placed at half wavelength intervals on the surface of the substrate. The output coupling was about 50% into the far field beam, with the remainder scattered. The 3 dB beam width was 10° and remained constant as the frequency was varied from 9 to 11 GHz. The beam direction scanned an angle of 30° during this frequency change. No attempt was made at this time to optimize the leaky wave structure.

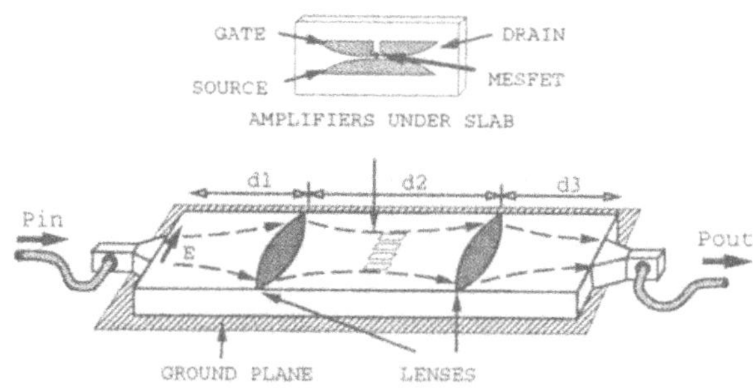

Figure 4. 2D quasi-optical amplifier array.

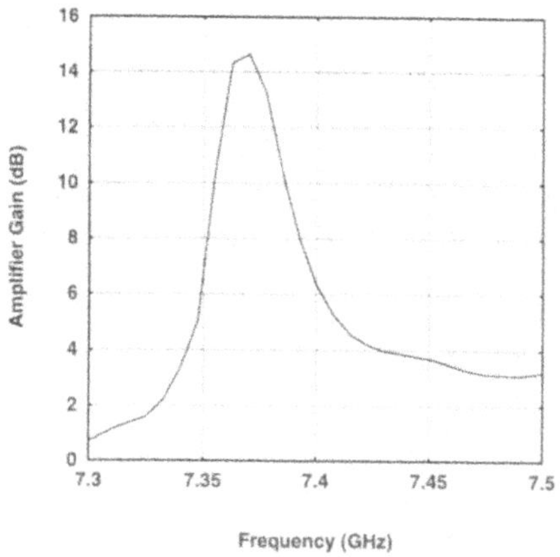

Figure 5. Small signal gain vs. frequency (GHz) for 2D quasi-optical amplifier array.

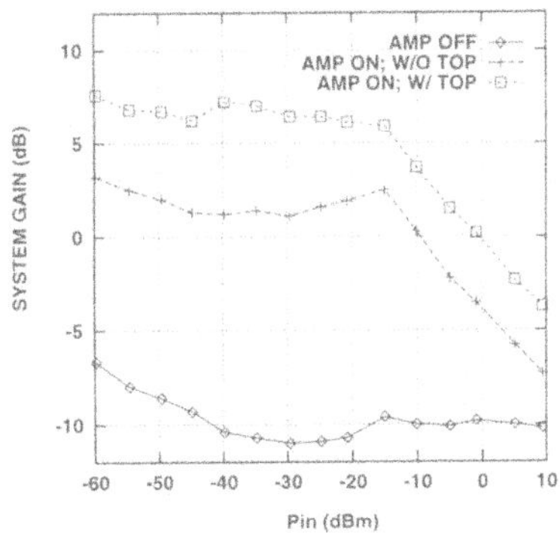

Figure 6. Gain vs. input power at 7.35 GHz for 2D quasi-optical amplifier array.

CONCLUSION

As frequency increases the leakage of millimeter wave electromagnetic energy from metallic microstrip into substrate modes or free space radiation appears to be telling us that new circuit integration techniques are needed. Quasi-optical techniques take their direction from the natural inclination of the electromagnetic energy and perform power combining and other signal processing functions in three dimensional free space or in two dimensional substrate modes. In this paper power combining of oscillators and amplifiers, substrate beam control, and output beam steering using a leaky wave structure have been discussed. Other researchers, at the California Institute of Technology, have reported the modulation of optical waveguide signals using millimeter wave energy coupled from substrate modes into an electro-optic modulator via an array of antenna elements on the substrate surface along the modulator.[10] These results suggest that a similar structure might couple detected millimeter wave energy from a waveguide-type photodetector, such as discussed in reference 11, into substrate modes. The way is opened for a circuit integration technology which keeps the millimeter wave EM energy in substrate modes to maximum extent possible, using quasi-optical techniques for power combining, signal processing, and signal routing within the substrate. The millimeter wave energy is coupled into or out of the substrate modes using leaky wave structures to interface with free space radiation or surface antenna-waveguide electro-optic structures to interface with optical fiber optical radiation. In this type of circuit integration the millimeter wave energy is in a highly distributed form. New distributed devices and distributed structures may play a significant role. Photonic crystals may be effectively used to guide and isolate propagation paths, to filter the RF signal, and to control polarization. Non-linear material may be used to perform frequency multiplication and other signal processing functions, in a manner similar to non-linear optical processing functions. Distributed devices such as waveguide electro-optic modulators and photodetectors, distributed amplifier structures, and non-linear RF waveguides may be the key functional electronic elements.

REFERENCE

1. J. Hubert, J. Schoenberg, and Z.B. Popovic, "High-Power Hybrid Quasi-Optical Ka-Band Amplifier Design," *IEEE MTT-S Int. Microwave Symp. Dig.*, 585 (1995).
2. C.-M. Liu, E.A. Sovero, W.J. Ho, J.A. Higgins, M.P. DeLisio, and D.B. Rutledge, "Monolithic Ghz 670-mW HBT Grid Amplifier," *IEEE MTT-S Int. Microwave Symp. Dig.*, 1123 (1995).
3. R.M. Weikle, M. Kim, J.B. Hacker, M.P. DeLisio, and D.B. Rutledge, "Planar MESFET Grid Oscillators Using Gate Feedback," *IEEE Trans. Microwave Theory and Techniques* **40**, 1997 (1992).
4. R.A. York and R.C. Compton, "Quasi-Optical Power Combining Using Mutually Synchronized Oscillator Arrays," *IEEE Trans. Microwave Theory and Techniques* **39**, 1000 (1991).
5. W.A. Shiroma, B.L. Shaw, and Z.B. Popovic, "A 100-Transistor Quadruple Grid Oscillator," *IEEE Microwave and Guided Wave Lett.* **4**, 350 (1994).
6. J.W. Mink and F.K. Schwering, "A Hybrid Dielectric Slab-Beam Waveguide for the Sub-Millimeter Wave Region, *IEEE Trans. on Microwave Theory and Techniques*, **41**, 1720 (1993).
7. F. Poegel, S. Irrgang, S. Zeisberg, A. Schuenemann, G.P. Monahan, H. Hwang, M.B. Steer, J.W. Mink, F.K. Schwering, A. Paolella, and J. Harvey, "Demonstration of an Oscillating Quasi-Optical Slab Power Combiner," *IEEE MTT-S Int. Microwave Symp. Dig.*, 917 (1995).
8. H. Hwang, G.P. Monahan, M.B. Steer, J.W. Steer, J. Harvey, A. Paolella, and F. K. Schwering, "A Dielectric Slab Waveguide with Four Planar Power Amplifiers," *IEEE MTT-S Int. Microwave Symp. Dig.*, 921 (1995).
9. A. R. Perkons and T. Itoh, "A Dielectric Slab Waveguide Lens Realized from Yagi-Uda Slot Array Antenna Elements and Microstrip Delay Lines ," *IEEE MTT-S Int. Microwave Symp. Dig.*, 1119 (1996).
10. W.B. Bridges, L.J. Burrows, U.V. Cummings, J.H. Schaffner, and F.T. Sheehy, "Progress in Antenna Coupled Millimeter Wave Electro-Optic Modulators," *Proceedings of the DoD Fiber Optics Conference, McLean,* Vol, **22-24** March, 289 (1995).
11. L. Y. Lin, M. C. Wu, T. Itoh, T. A. Bang, R. E. Muller, D. L. Sivco and A. Y. Cho, "Velocity Matched Distributed Photodetectors with High Saturation Power and Large Bandwidth," *IEEE Photonics Technology Letters* Vol., **8** Oct. 1376 (1996).

Multilayer Integration of Microwave and Millimeter-Wave Circuits: New Interconnect Methods and Design Considerations

N. K. Das[1], N. Herscovici[2], G. Kwan[1], and D. M. Bolle[1]

[1] Weber Research Institute
Department of Electrical Engineering
Polytechnic University, Route 110
Farmingdale, NY 11735, USA

[2] anTeg. Inc.
52 Agnes Drive
Framingham, MA 01701, USA

Abstract

In this paper we propose new designs of transmission lines and transition geometries for in-layer and layer-to-layer interconncetions in advanced multilayer MMICs. Use of conventional transmission line concepts, such as a microstrip line, a coplanar waveguide and a slotline (that are normally fabricated on a single substrate,) would fail to function properly in a general multilayer environment. The proposed designs are described, with discussion of the basic principles of operation and design considerations. Avoidance or suppression of multilayer parallel-plate modes is one of the critical issues. Theoretical and experimental results are presented in order to demonstrate the merits and proof-of-concepts of selected devices.

1 Introduction

Microwave and millimeter wave integrated circuits (MMICs,) in which active and passive components, as well as transmission-line distribution network, are fabricated on a single semi-insulating semiconductor substrate, are finding extensive applications in modern satellite, mobile and personal wireless communications. Considerable research and development efforts have been directed in the past towards the single-layered MMIC technology. However, from various technical and economic considerations, the single-layered MMIC technology may be less than optimum in many applications, and could sometimes pose serious fundamental limitations. In many MMICs, the passive components and power distribution network constitute a large fraction of the total circuitry, as compared to the active components. Because the passive components could be printed on a cheaper dielectric substrate, not necessarily on the same semiconductor substrate as the active devices, this results in an inefficient use of the expensive semiconductor substrate. Besides this economic consideration, sometimes circuit components of different functionality may require substrates of different thicknesses, dielectric constants, or other parameters for optimum performance. This is particularly true when circuits and printed antenna elements, which have potentially conflicting design reguirements, are integrated together. In order to avoid spurious radiation from the circuitry, and also to achieve a high integration density due to the resulting smaller component size, the circuit components require a thin, high-dielectric constant substrate. On the other hand, antenna elements require a relatively thicker and lower dielectric constant substrate to achieve larger bandwidth and wider scannability [1,2]. Also, considering the possible integration of high T_c superconductors, ferrite films, or other novel substrate materials, that are attractive for certain advanced circuit functions (extremely narrow band filters using high T_c superconductors, for example), it may be necessary to use a separate substrate layer for fabrication of the specific circuit component. Further, functional modularity and the design flexibility resulting from using separate layers for distinct functions, is always a desirable feature.

Separate from these general considerations of distinct functional requirements or functional modularity, another class of technological limitations of the single layer MMICs may justify further the preference for multiple substrates. Depending on the substrate dielectric constant and thickness, as well as the frequency of operation, there is an upper limit to the circuit size and integration density per unit base area. Multiple layers of integration would allow us to overcome this fundamental limitation. From topological considerations, it may be inefficient or impossible to layout certain types of circuits on the surface of a single substrate because of cross-over problems. Such topological problem of circuit layout can also be solved by routing through more substrate layers.

In summary, in contrast to a single layer MMIC, a multilayer MMIC is significantly more attractive because, a) it provides increased integration space per unit base area, and hence increased effective integration density, b) it allows significant improvement in topological flexibility of circuit routing through more substrate layers, c) it could maintain functional modularity of different layers allowing design flexibility and the optimal choice of substrates or fabrication process for separate sub-functions, and d) it permits convenient integration of other technologies, such as superconductor, optical or digital circuits, or ferrite devices on independent layers. Such a multilayer MMIC architecture can be configured as follows [3]: sub-circuits performing separate electrical functions can be integrated on different layers of a multilayer substrate configuration, electrically isolated from each other by conducting ground planes in between or by maintaining adequate physical separation. The circuits on different layers can maintain the needed electrical connectivity between each other only at selected locations, via coupling slots [4] or slotline transitions [5] across the separating ground plane, or via capacitive coupling between printed strips placed in close proximity. The above forms of "electromagnetic coupling" between layers, where coupling is established not by physical contact between the circuit components but by close proximity, is often preferred from fabricational considerations over "physical coupling" through via-holes . Fig.1 shows a possible multilayer architecture for integrated phased array applications, showing multilayer MMIC modules integrated with antennas as well as optoelectronic and digital circuits configured on independent layers employing slot-coupling between layers. Similar multilayer architectures have also been proposed for mobile communication [6].

New geometries of multilayer transmission lines, and new surface-to-surface transitions for electromagnetic coupling between layers, constitute the basic elements for the future development of multilayer MMIC components and subsystems. New fundamental problems would arise due to the multilayering, which could pose serious limitation on the type of transmission lines, transitions, or devices one uses for a multilayer MMIC. In this paper we propose new interconnects and surface-to-surface transitions that are candidates for multilayer integration. The basic principles of operations and design considerations are discussed, with selected computational or experimental results to demonstrate their operations. The candidate geometries of interconnects and surface-to-surface transitions are discussed separately in the following sections.

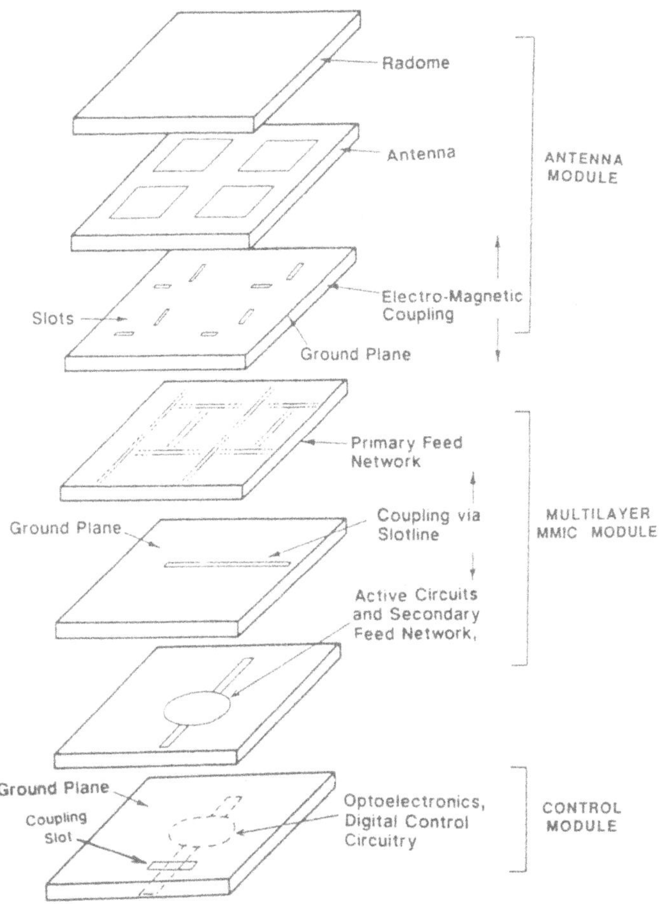

Figure 1. Multilayer integration of microwave active and passive circuits, microstrip antennas, distribution network, and control circuits, employing aperture coupling between circuit layers.

2 New Interconnects for Multilayer MMIC

Multilayer MMICs need to use metal planes between circuit layers in order to provide the necessary electrical isolation between each other. Presence of these metal planes can introduce parallel-plate modes, which can result in transversal power leakage [7,8] problems. If common types of metal lines, such as striplines, conductor-backed slotlines or coplanar waveguides, are used for signal transmission in such multilayer dielectric-metal environment, power can escape from the lines sidewise due to propagation of "leaky modes," resulting in severe power loss or cross talk between neighboring circuits. Unlike in single-layer circuits, where the standard microstrip, slotline, or coplanar waveguides do not suffer from such problems under normal operating conditions, in a multilayer MMIC environment the lines can normally suffer from the leakage problems. This necessitates use of new types of transmission media. We propose in the following three types of designs: (1) dielectric guide-coupled slotlines or coplanar waveguides, (2) transmission lines with shorting pins, and (3) parallel-plate dielectric guide.

2.1 Dielectric Guide-Coupled Slotlines or Coplanar Waveguides

Fig.2 shows the geometry of a dielectric guide-coupled slotline or coplanar waveguide. Here the central metal transmission line is surrounded by a dielectric medium of sufficiently higher dielectric constant compared to that of the outside medium. As a result, any power that may be escaping out of the central slotline or coplanar waveguide to the parallel-plate mode, is reflected back at the outer dielectric interfaces due to total internal reflection. With proper selection of the relative values of the dielectric constants, together with proper design of the central transmission line, one can completely eliminate the lekage problem. Though we specifically show in Fig.2 a slotline or coplanar waveguide geometry with conductor backing, such technique is also applicable to stripline-type geometries, which would otherwise suffer from similar power-leakage problem under practical situations [9,10]. It may be noted, that such dielectric guide-coupled geometries are more difficult to fabricate, compared to standard planar transmission lines, due to its non-planar dielectric guide structure. Though this is an issue for fabrication of hybrid-type of multilayer circuits using phtoetching process, it should not pose any serious restriction when fabricated together with semiconductor devices using a deposition/etching process.

Fig.3 shows computed results for a conductor-backed coplanr waveguide in a dielectric guide-coupled configuration. As the results show, for a range of values of the outside dielectric constant, ϵ_r', the transmission line can avoid the power-leakage problem discussed before. For the total internal reflection to occur, it is required that $\epsilon_r' < \epsilon_{eff} = \beta^2/k_0^2$, where β is the propagation constant of the transmission line, k_0 is the free-space wave number, and ϵ_{eff} is the effective dielectric constant of the transmission line.

2.2 Transmission Lines with Shorting Pins

When fabrication of the dielectric guide-coupled geometries can be difficult for hybrid-type circuits, transmission line geometries with shorting pins on both sides can also

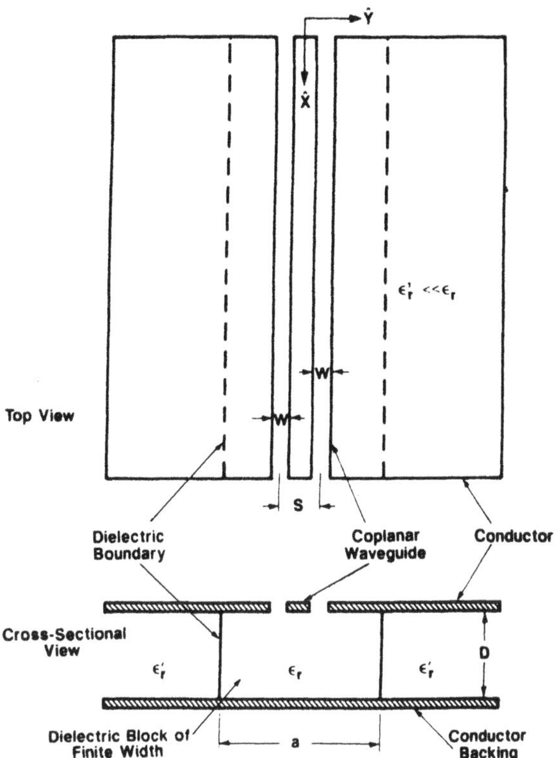

Figure 2. Geometry of a dielectric guide-coupled conductor-backed coplanar waveguide. Simialr geometry of a slotline can be realised by replacing the dual slots of the coplanar waveguide by a single slot of width W centered at $y = 0$. Similar design can also be used for the suppression of leakage for a stripline geometry.

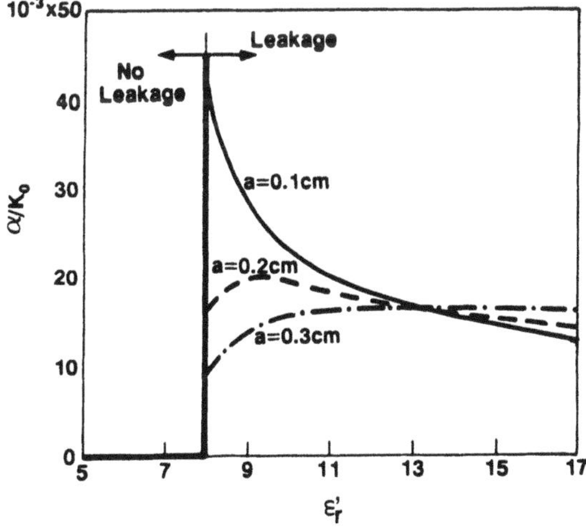

Figure 3. Attenuation constant, α, normalized to the free-space wave number, k_0, of a dielectric guide-coupled conductor-backed coplanr waveguide of Fig.2 with $W = 0.1$mm, $S = 0.2$mm, $D = 0.2$mm, frequency = 10GHz, $\epsilon_r = 13$, for different values of central dielectric width a, as a function of the external dielectric constant, ϵ_r'. The real part of the propagation constant does not change significantly for this set of parameters.

serve to reflect most of the power escaping from the central line. The effect is similar to the dielectric guide-coupled design discussed earlier. However, the shorting pins can only suppress the power leakage to an acceptable low level, but can not totally eliminate the problem. This is because, unlike the total reflection in the case of a dielectric guide-coupled geometry, a fraction of the total power always escapes through the space between shorting pins. Through proper design of placement of the shorting pins, one can achieve significant suppression of the power leakage by using only a minimal number of the shorting pins.

Fig.4 shows the basic geometry of a transmission line with shorting pins. Though here we show a slotline or coplanar waveguide geometry, similar designs are applicable to strip-type transmission lines as well. Fig.5 shows the attenuation constant of a selected geometry with shorting pins, which shows the effects of the spacing and size paramters of its shorting pins. In Fig.5, for the convenience of analysis and formulation, we have used ideal shorting strips with zero thickness and a non-zero width, δ, instead of a more realistic cylindrical shape for a shorting pin with non-zero diameter, d. (A shorting pin or stip is, in general, interchangeably referred to as a shorting post.) The effect of the shorting strips should be funtionally similar to shorting pins, where the width, δ, of shorting strips can be viewed analogous to the diameter, d, of equivalent pin geometries. As seen from the Fig.5, the loss can be reduced significantly by reducing the distance between consecutive shorting posts or by increasing W of the shorting strips. This should be expected, beceuse under the above situations less space is available between stips for power to escape through. The other governing paramter is the distance of the shorting posts from the center of the transmission line. Computations similar to Fig.5 show that the optimum distance for placement of the shorting posts from the center can be different for different types of lines. For a conductor-backed slotline, the optimum distance is about $\lambda_g/4$, where λ_g is the guide wavelength of the leakage wave in the transverse ($\hat{y}$) direction. Whereas, in order to achieve maximum suppression of leakage loss, the shorting posts in a conductor-backed coplanar waveguide or a stripline should be placed as close to the center as possible without affecting the dominant "bound fields" around the central guiding region.

2.3 Parallel-Plate Dielectric Guide

Fig.6 shows the geometry of a dielectric guide geometry placed between two parallel metal plates. The fundamental TE-to-x mode of this waveguide, referred to as a "parallel-plate dielectric-guide mode" having electric field perpendicular to the parallel plates, can propagate down to DC. This mode has not drawn attention for use in microwave integrated circuits. The mode is non-TEM, but it turns out one can use it in a way similar to a TEM or quasi-TEM line. Though voltage is strictly meaningless for such a non-TEM mode, an equivalent definition of of voltage $V = Ed$, where E is the electric field at the center of the guide, and d is the separation between the parallel plates, can be used for practcal purposes. A characteristic impedance, Z_c, can then be defined as $Z_c = V^2/P$, where P is transverse power propagating along the guide for a given V. Interestingly, the parallel-plate waveguide mode can be fed directly from a coaxial probe at the center, with the probe connecting across the parallel metal planes, without any elaborate feeding arrangement or matching circuits. This is quite similar to a coaxial feeding of a microstrip line or a stripline, for example. The coaxial feed sees approximately the above characteristic impedance at the input of the waveguide.

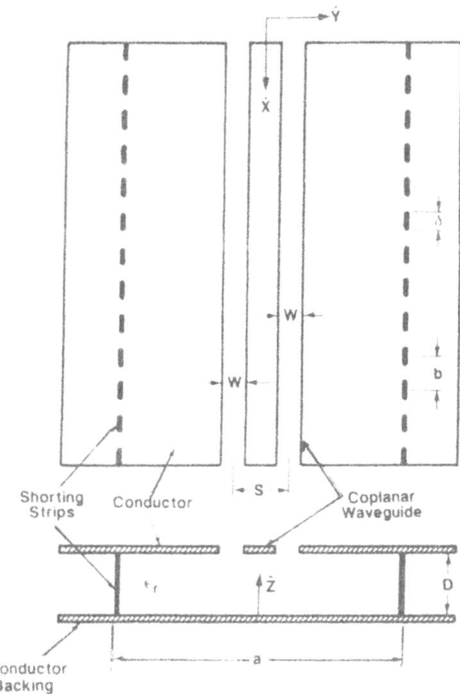

Figure 4. Geometry of conductor-backed coplanar waveguide with shorting pins used to suppress leakage to the parallel-plate mode. A conductor-backed slotline can be similarly realized by replacing the dual slots at the center by a single slot of width W centered at $y = 0$. Similar design can also be used for a leaky stripline geometry.

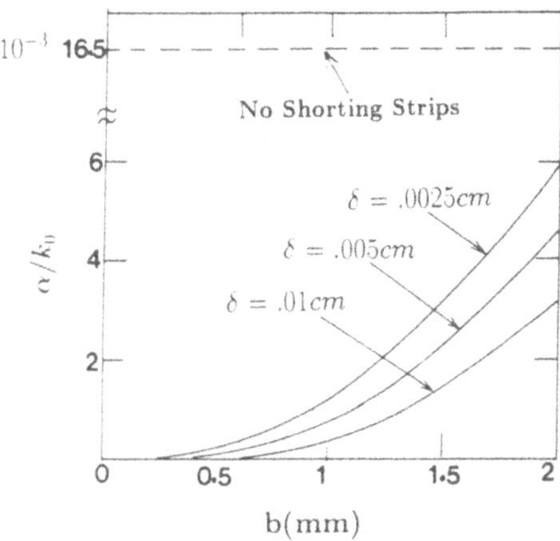

Figure 5. Leakage suppression in a conductor-backed coplanar waveguide using shorting pins, as a function of separation between the strips, b, for different values of strip-width δ. $W = 0.1$mm, $S = 0.2$mm, $\epsilon_r = 13$, $D = 0.2$mm, frequency = 10GHz, $a = 0.1$cm.

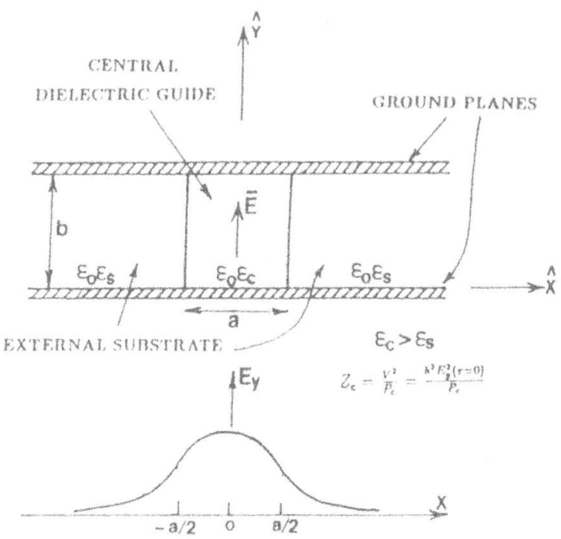

Figure 6. Geometry of a parallel-plate dielectric waveguide, proposed for multilayer application. The specific mode of operation of interest here has an electric field perpendicular to the parallel plates (the fundamental TE-to-x mode.)

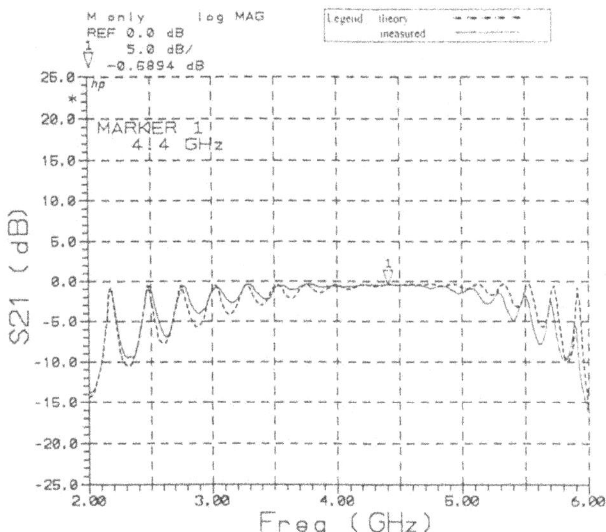

Figure 7. Calculation and measured results of insertion loss of a parallel-plate dielectric waveguide of length Ld=22.5cm, connected at the two ports using a coaxial probe feed. The feed point is terminated by a short-circuit stub of length L1=L2=10mm, which is about quarter guide wavelength at the design frequency of 3.9GHz, resulting in a open circuit at the feed points. a=5 mm, b=2.54 mm, characteristic impedance Z_c=50Ω at 3.9GHz. Cut-off frequency of the next higher order mode of the waveguide=9.58GHz. Dielectric constants ε_c=10.6, ε_s=1.0(air) . Radiation loss at the coaxial probes has been accounted for in the calculation. Metal plates and dielectric substrates are assumed to be lossless.

Besides the above consideration, the parallel-plate dielectric guide is compatible to multilayer integration where other circuits can be integrated above and below its two groundplanes. More importantly, the waveguide is non leaky, and can propagate signals without excitation of the parallel plate mode in the surrounding structure. Also, as discussed in the next section, excitation of the parallel-plate mode at a transition discontinuity of a parallel-plate dielectric waveguide can be trapped inside the waveguide without escaping to the outside parallel-plate structure. The above properties make the geometry particularly suitable for multilayer applications.

Fig.7 shows the insertion loss of a parallel-plate dielectric guide connected at two ends by coaxial probes, as compared to theoretical circuit simulation using the above definition of characteristic impedance (designed close to the input port impedance = 50Ω.) The coaxial probe has been rigorously modeled in the simulation as an equivalent resistance and inductance in series with the port. The series resistance accounts for any radiation loss from the probe. From Fig.7 it is seen that the measured insertion loss matches closely with simulation in the entire frequency band, with some deviation towards the higher frequency end. Clearly, the waveguide can be seen to perform well at the center of the band, with an insertion loss of about -0.7dB, without any significant power loss to parallel-plate mode.

3 New Surface-to-Surface Transitions for Multilayer MMIC

Electrical connection between circuit layers in a multilayer MMIC can be, in principle, established in a few different ways: (1) using feed-through connections, (2) capacitive coupling between two layers if there is no separating metal planes between them, or (3) aperture coupling thorugh small slots on a separating metal plane. However, problem of excitaiton of parallel-plate at a discontinuity arises when multiple metal planes exist. A vertical feed-through connection across, or a coupling aperture printed on, one of the metal planes of a parallel-plate geometry would excite the parallel-plate mode that can often be excessive and serious. We propose in the following two possible geometries that can overcome the potential problem.

3.1 Multilayer Surface-to-Surface Transision Using a Dielectric Plug Configuration

Fig.8 shows a signal transition from the bottom microstrip layer to the top microstrip layer across a stripline circuit layer sandwiched in between. The geometry uses apertures for coupling across metal planes, which, as mentioned earlier, would not normally work due to strong coupling to the parallel-plate mode. However, we use here a "dielectric plug" in order to overcome the problem. As shown in Fig.8, a dielectric plug is placed around the coupling slot, having its dielectric constant sufficiently higher than that of the surrounding media. As a result, the field excited by the coupling slot is mostly confined inside the plug due to internal reflection at its outer dielectric interface. It may be noted, that the stripline circuit that connects the two slots consists of two types of lines. The sections of the stripline in the dielectric-plug region see two different dielectric media above and below its strip, whereas the section outside the dielectric-plug region see a uniform dielectric medium around it. As reported in [9,8,10], in both

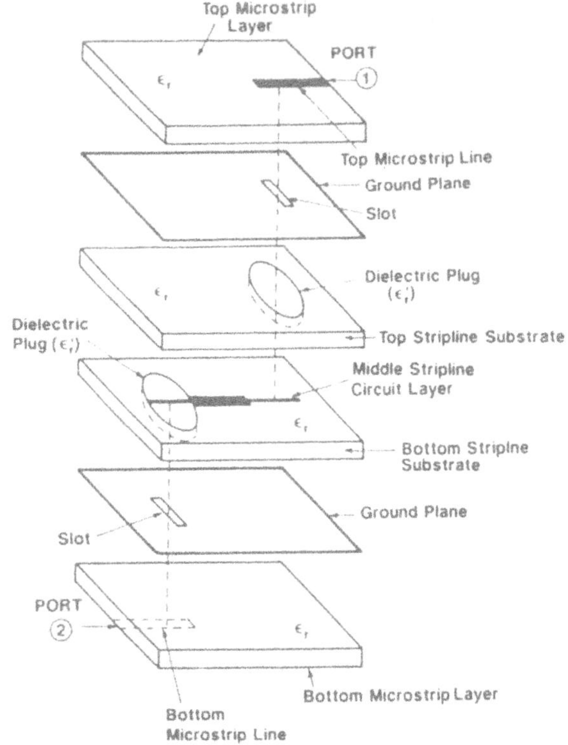

Figure 8. A microstrip-stripline-microstrip transition for signal transfer from the bottom microstrip to top microstrip layer via the central stripline layer. Dielectric plugs are used in order to avoid the problem of the power leakage by the coupling slots to the parallel-plate mode.

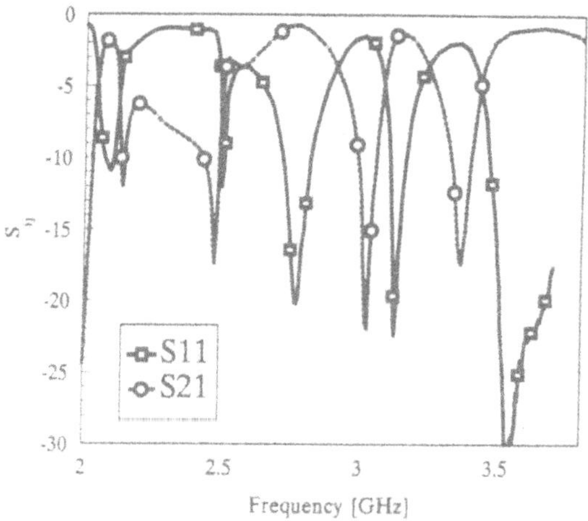

Figure 9. Experimental results for the multilayer transition of Fig.8, with $\epsilon_r = 2.1$, and thickness = 31mil for all substrates, slot dimensions 2.4x0.1cm, all stub lengths (open) = 1.5cm, diameter of dielectric plugs = 3cm, characteristic impedance = 50Ω for all lines.

these cases the stripline is found to be non leaky. Consequently, both the slot discontinuity, as well as the connecting transmission lines of Fig.8, should perform as desired without the problem of excessive excitation of the parallel-plate mode.

Fig.9 shows the experimental insertion loss of one such layer-to-layer transition, which demonstrates proper functioning of the device. Most of the non-zero return loss (about 1.0dB) at around the operating frequency of about 2.7GHz can be attributed to material loss of the entire circuit.

3.2 Aperture Transition Between a Parallel-Plate Dielectric Guide and a Microstrip Antenna

Fig.10 shows an aperture-transition geometry for power transfer between a parallel-plate circuit layer and a microstrip antenna on top across a common metal plane. The problem of excessive excitation of the parallel-plate mode by the coupling aperture is significantly minimized here by the use of a parallel-plate dielectric waveguide structure. First, the parallel-plate dielectric waveguide itself is non-leaky, as discussed earlier. Next, the coupling aperture is designed with its length sufficiently smaller than the width of the parallel-plate dielectric waveguide. As a result, most of the power excited from the coupling aperture to the parallel-plate structure is totally reflected from the outer dielectric interface of the waveguide. This reflected power is partially coupled back to the aperture, and the rest is guided along the parallel-plate dielectric waveguide. In other words, the power that would have otherwise escaped to the unwanted parallel-plate mode is now converted to useful power in the feeding waveguide. This power can be accounted for in the design of the feed structure in order to achieve proper impedance matching condition.

Fig.11 shows experimental results for insertion loss of one such transition, as compared to an approximate theoretical model that accounts for the characteristics of the feeding waveguide, the coupling mechanism and the microstrip antenna characteristics. The details of the theoretical analysis of the geometry is beyond scope of the present paper, and would be presented elsewhere. The general agreement between the theory and experiment in Fig.11 suggests proper functioning of the feeding waveguide as well as the proposed coupling arrangement.

4 Conclusion

The transmission lines and surface-to-surface transitions discussed in this paper constitute a class of building-block devices for multilayer implementation. Several variations of the particular concepts we have discussed in the paper can be possible. With development of suitable design techniques and analytical/computational tools, and reliable fabrication process for the proposed devices, increasingly complex multilayer, multi-functional integrated circuits can be realised for advanced applications.

Ackowledgement

The work was funded by the US Army Research Office, Research Triangle Park, NC, under grant DAAH04-95-1-0254.

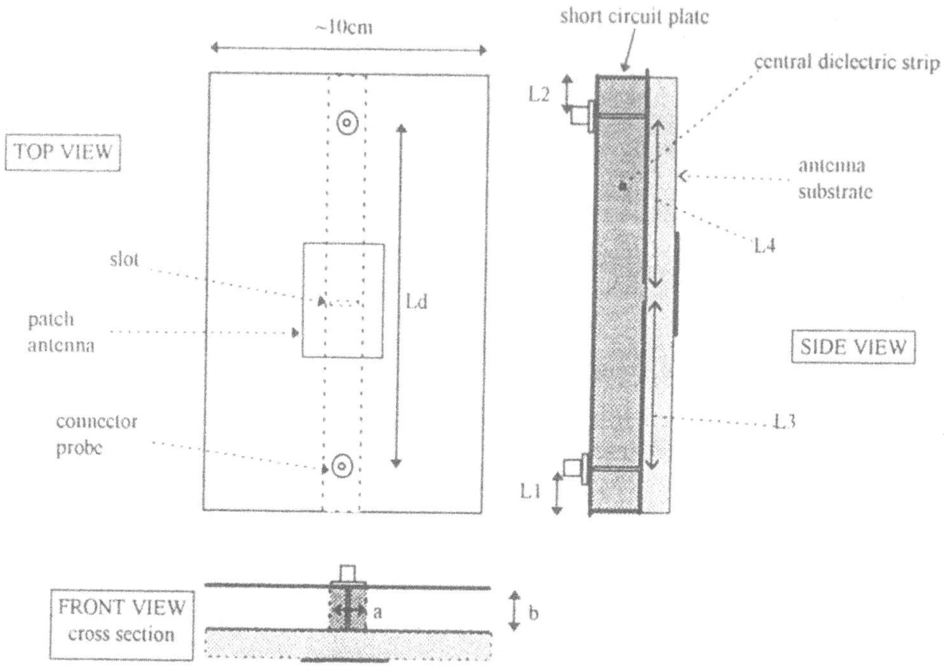

Figure 10. Slot coupling between a parallel-plate dielectric waveguide to a microstrip antenna across the common ground plane. The feeding waveguide is connected at the two ends by coaxial feed probes for two-port measurements. One port may be terminated using a suitable stub for impedance tuning.

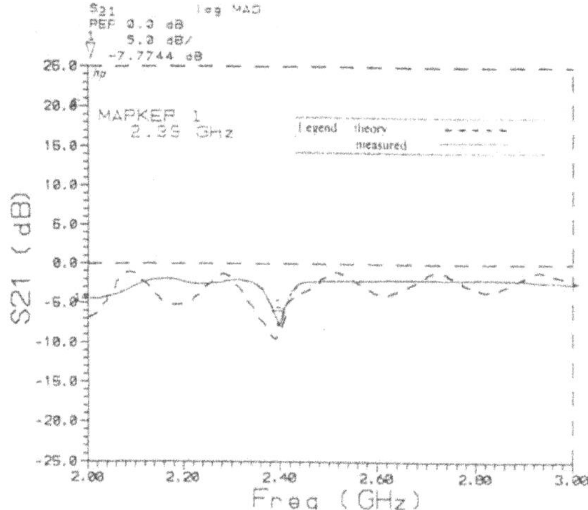

Figure 11. Measurement of the insertion loss of the transition geometry of Fig. 10, as compared with computed results. a=14.4mm, b=5.08mm, Ld=20.6cm, short circuit stubs L1=L2=10mm, which are about quarter guide wavelengths at the design frequency of about 2.4GHz, characteristic impedance Z_c=50Ω at 2.4GHz. Cutoff frequency of the next higher order mode =3.3GHz. Patch antenna dimensions 40(resonant)x37.5mm, antenna substrate ε_r=2.33, thickness 1.53mm, coupling slot dimensions 12x1mm. Dielectric constants ε_c=10.6, ε_s=1.0(air). Radiation loss at the coaxial probes has been accounted for in the calculation. Metal and dielectric losses are also calculated based on a conductivity about 1.58e7 S/m (brass) and a loss tangent of 0.0028.

References

[1] D. M. Pozar and D. H. Schaubert. Comparison of architectures for Monolithic Phased Array Antennas. *Microwave Journal*, 93–103, March 1986.

[2] D. M. Pozar and D. H. Schaubert. Scan Blindness in Infinite Phased Arrays of Printed Dipoles. *IEEE Transactions on Antennas and Propagation*, AP-32(6):602–610, June 1984.

[3] N. K. Das and D. M. Pozar. Multiport Scattering Analysis of Multilayered Printed Antennas Fed by Multiple Feed Ports, Part I: Theory; Part II: Applications. *IEEE Transactions on Antennas and Propagation*, AP-40(5):469–491, May 1992.

[4] D. M. Pozar. A Reciprocity Method of Analysis of Printed Slots and Slot-Coupled Microstrip Antennas. *IEEE Transactions on Antennas and Propagation*, AP-34(12):1439–1446, December 1986.

[5] J. B. Knorr. Slot-line Transitions. *IEEE Transactions on Microwave Theory and Techniques*, MTT-22(5):548–554, May 1974.

[6] Personal Communications, and Motor Vehicle and Highway Automation Spark New Microwave Applications. August 1991.

[7] H. Shigesawa, M. Tsuji, and A. A. Oliner. Conductor Backed Slotline and Coplanar Waveguide: Dangers and Fullwave Analyses. *IEEE Microwave Theory and Techniques Symposium Digest*, 199–202, 1988.

[8] N. K. Das and D. M. Pozar. Full-Wave Spectral-Domain Computation of Material, Radiation and Guided Wave Losses in Infinite Multilayered Printed Transmission Lines. *IEEE Transactions on Microwave Theory and Techniques*, MTT-39(1):54–63, January 1991.

[9] J. T. Williams N. Nghiem and D. R. Jackson. Proper and Improper Modal Solutions for Inhomogeneous Stripline. *IEEE Microwave Theory and Techniques Society Simposium Digest*, 567–570, 1991.

[10] L. Carin and N. K. Das. Leaky Waves in Broadside-Coupled Microstrips. *IEEE Transactions on Microwave Theory and Techniques*, MTT-40(1):58–66, January 1992.

THE HYBRID INTEGRATION TECHNIQUE OF PLANAR AND NRD-GUIDE CIRCUITS FOR MILLIMETER-WAVE APPLICATIONS

Ke WU and Liang HAN
Poly-Grames Research Center
Dept. de génie électrique et de génie informatique, École Polytechnique
C. P. 6079, Succ. Centre-Ville, Montréal, Canada H3C 3A7

ABSTRACT

We have recently proposed the hybrid integration technique of planar and NRD-guide circuits suitable for microwave and in particular millimeter-wave applications. This new integration technique makes it possible to exploit complementary distinctive advantages of planar structures and NRD-guide. Theoretical and experimental studies suggest that this three-dimensional hybrid scheme promise to be an alternative in the design of low-cost millimeter-wave circuits and systems. This paper reviews briefly our research progress on this new technique with emphasis on the state-of-the-art of the aperture-based transitions/baluns developed to date which interconnect the two subsets of circuits. Potential problems are discussed and future directions are indicated on this particular subject.

INTRODUCTION

Any successful deployment of a wireless technology at microwave and millimeter-wave frequencies intended for widespread and extensive commercial applications depends heavily on the availability of a technology having properties such as low-cost, compact-size, low-power consumption and mechanic rigidity. It is also required that the applied technology presents low-loss signal transmission which is essential for realizing high-Q circuits.

Recent progress made in the R&D of RFICs indicates that multilayer planar technology provides a high-level module integration achieving some of these stringent requirements such as low-cost and compactness[1]. Nevertheless, such a technology has a number of limitations in the design of high-performance millimeter-wave circuits and systems such as the vulnerable high loss (low-Q) of signal transmission. This is especially painful for the filter and multiplexer/diplex designs. To a great extent, it is difficult to achieve simultaneously overall required circuit performance under a single technology framework. This argument

Directions for the Next Generation of MMIC Devices and Systems
Edited by N. K. Das and H. L. Bertoni, Plenum Press, New York, 1997

suggests that an appropriate hybrid scheme involving two or more technologies provide a possibility of accomplishing all desired features by combining their advantages while each individual inherent shortcomings are eliminated. In view of its advantages and disadvantages, the non-planar technology has been known by its complentarity with respect to its planar counterpart. Obviously, the hybrid scheme based on combined planar and non-planar technologies is more appealing.

A module integration of microstrip line with the metallic waveguide has been reported[2] which was essentially related to the design of wideband transition between the metallic waveguide and microstrip line. Judging from its geometry and its compatibility with the planar circuits, the metallic waveguide is bulky and cannot provide a high-level integration. It is unfortunate that very limited information is available to date for building up a hybrid scheme that is operational at millimeter-wave frequencies. On the other hand, very little attention has been directed to the potential integration of a planar structure with dielectric waveguide even though a large class of dielectric waveguides have already been proposed for millimeter-wave and submillimeter-wave applications. Results have been reported only for planar patch antennas fed by dielectric image line[3]. It has been known that the fundamental limitation of using a dielectric waveguide is its severe radiation loss once circuit bends and discontinuities are encountered, which jeopardizes useful applications of the dielectric waveguide. This perception holds until the invention of a nonradiative dielectric (NRD) waveguide[4].

This paper reviews briefly our research progress on this new technique with emphasis on the state-of-the-art of the aperture-based transitions/baluns developed to date. This is because the integration between the two dissimilar structures is achieved by an aperture coupling or a balun geometry which presents the key to successful applications of the new technique. Particular applications can then be made in a straightforward manner. Potential problems are also discussed and future directions are indicated on this particular subject.

INTEGRATION OF NRD-GUIDE AND PLANAR STRUCTURE

The hybrid integration of NRD-guide and planar circuit is made to eliminate the underlying disadvantages of the both building blocks at millimeter-wave frequencies while their technical benefits and design freedom can be maintained. In contrast to the previously proposed integration of planar circuits with the NRD-guide [4, 5, 6], we have proposed a new scheme that removes effectively the space constraint of a planar circuit inserted into NRD-guide whose lateral extent is limited by a half of free-space wavelength. The planar structure may be in the form of microstrip line or coplanar waveguide or even slot line. In Fig. 1, an integrated transition of microstrip line to NRD-guide is made through a magnetic aperture coupling. The microstrip line is placed at the perpendicular direction to the dielectric strip of the NRD-guide. The microstrip line can be relocated at either side of the parallel metallic plates of the NRD-guide. In this way, the microstrip line shares the common ground plane with the NRD-guide which is actually one of the parallel metallic plates. The coupling aperture is made on the ground plane (the parallel plate).

Obviously, a number of microstrip lines may be attached on the both sides of the NRD-guide simultaneously. This consideration gives rise to some interesting feature of the proposed hybrid technology such as space saving and interference reduction. Actually, a self-packaged circuit design can be easily achieved. It is obvious that the passive components made of the NRD-guide will present unmatched performance such as high-Q, low-loss transmission, radiationless, and potential cost-effective. The advantages of the planar structures are exploited for the design and realization of two- or three-terminal based active devices such as easy integration and use of M(H)MICs. This new technology presents a high-level integration involving dielectric waveguide and planar circuits which may be in the form of MICs, MHMICs and MMICs.

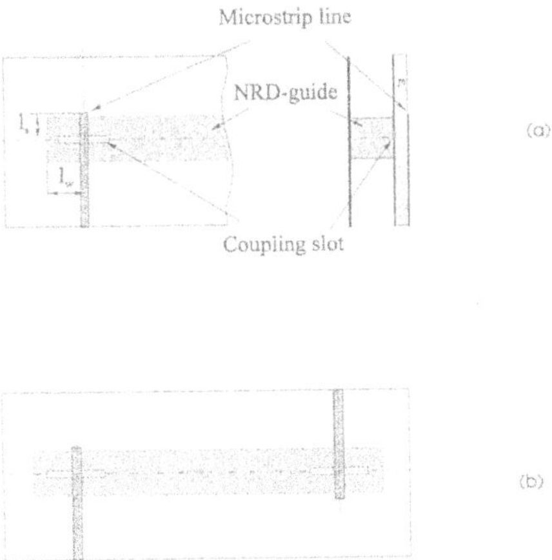

Fig. 1. (a) Geometry and (b) testing setup of the transition from microstrip line to NRD guide[7] (reprinted with permission).

APERTURE-BASED TRANSITIONS/BALUNS

The integrated transition between the NRD-guide and planar structure is the key to a successful application of this new hybrid technology. The preliminary design issue and electrical performance have been presented[7] for the transitions between the NRD-guide and microstrip line with modeling and experimental results. Obviously, the proposed hybrid technology is potentially low-cost since the basic design of the NRD-guide based components is related to a series of mechanic fabrications and assembling of integrated and discrete devices and components.

Note that the following preliminary experimental examples do not present an optimized electrical performance under their geometrical conditions.

A. NRD-guide-to-Microstrip Transition

The design of a transition of microstrip line to NRD-guide is actually focused on the impedance matching between two dissimilar structures which are coupled to each other through a rectangular aperture having a magnetic coupling. To verify electrical performance of the proposed transition, an experimental prototype is made which uses two identical transitions of microstrip line-to-NRD-guide which are interconnected through a NRD-guide terminated with two open ends having a length of 82 mm (see Fig. 1). The distance between two microstrip feed lines is 76.42 mm which are fabricated on a 60×98 mm^2 substrate (RT/Duroid 5880, $\varepsilon_r = 2.3$) with a thickness of 20 mil. The line impedance of 50 Ω is designed with a strip width of 1.53 mm. The NRD-guide is made of a rectangular dielectric strip (Rogers TMM®-3, $\varepsilon_r = 3.27$) and designed to operate around 20 GHz with a = 6.1 mm and b = 6.5 mm. The coupling aperture on the common ground plane is a narrow rectangular slot with a 7.5×0.5 mm^2.

Experimental and theoretical results are shown in Fig. 2 for the designed microstrip line-NRD-microstrip line as shown in Fig. 1 which operates at frequencies from 17.5 GHz to 21.2 GHz with 15% effective bandwidth. This coincides with about the same bandwidth as the operational monomode bandwidth of the NRD-guide based on our design. The losses of two microstrip feeding lines that are not calibrated out are involved in the measurement results, and the transitions are not optimized yet. The best insertion loss for the complete circuit is observed with less than 1.5 dB over 10% bandwidth and less than 3 dB over 30% at the center frequency of 20 GHz.

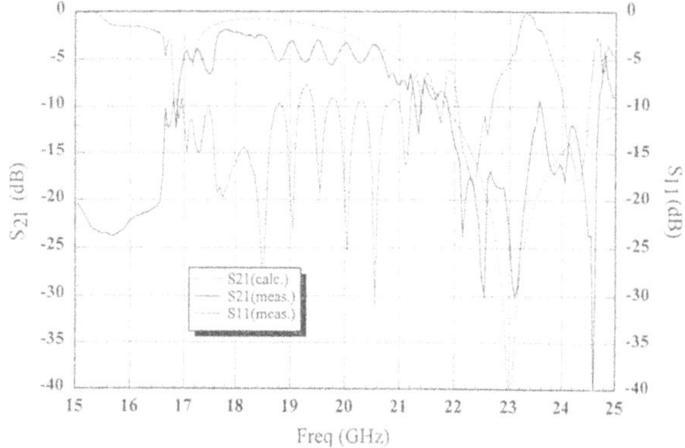

Fig. 2. Measured and predicted performance of two back-to-back transitions between microstrip lines, separated by 82 mm of NRD-guide[7] (reprinted with permission).

It can be observed that the ripple appearing in the band can be attributed to the internal impedance matching problems of the transitions and a small standing wave is formed along the NRD-guide between the two transitions. These preliminary results are encouraging, considering the fact that the structure is not optimized yet and the losses of two connection lines are not removed from the measurement. A simple and approximate model was developed[7] to predict electrical performance of the microstrip line to NRD-guide transition. The accuracy of that model can be further confirmed[8] by our TLM modeling results as shown in Fig. 3.

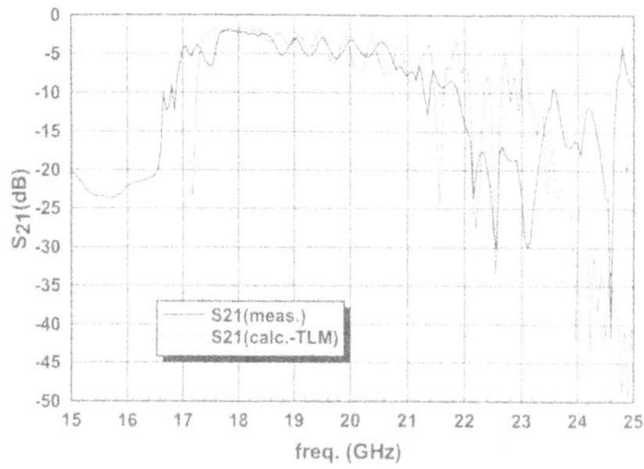

Fig. 3. Comparison of experimental results with predictions made by a TLM modeling tool[8].

B. NRD-guide-to-CPW Transition

Similar to the design consideration of the transition of a microstrip line-to-NRD-guide, a backside-grounded CPW-to-NRD-guide transition was designed, fabricated and characterized in the same frequency band. The geometry of the fabricated experimental prototype is shown in Fig. 4. Unlike the above proposed transition of microstrip line-to-NRD-guide, the present transition of the proposed structure is realized, however, through an electrical coupling via the small aperture at the ground plane instead of the magnetic coupling.

A compensated grounded CPW with Z_0=50 Ω connected to a bilateral slot line balun is designed with the slot line opened at both sides. A series of via holes are used along both sides of the grounded CPW input lines and the open-ended stub instead of the conventional air bridges. The successive procedure is to determine the optimum length of the two open-ended stubs of the coplanar waveguide, which surely effect the insertion loss and the bandwidth of the transition. In this design example, the length of the open stubs is designed to be short around $\lambda_g/8$. It should be noted that the position of the open end of the dielectric strip affects also the frequency bandwidth of the transition. The proper lengths of open end of the slot line located at each side of the grounded CPW's center conductor are determined empirically to form a resonator operating at the designated frequency range. The grounded CPW was then fed with an incident wave from a network analyzer through an SMA connector.

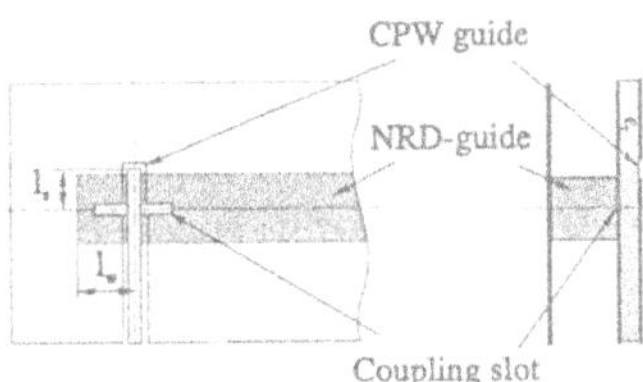

Fig. 4. Transition from coplanar waveguide (CPW) to NRD-guide.

The experimental set-up of a CPW-NRD-guide-CPW structures involving two transitions is shown in Fig.5 together with S-parameter measurement results for the insertion loss. The effective bandwidth of the designed transition extends from 18 GHz to 22 GHz which covers almost all the useful working bandwidth of the NRD-guide.

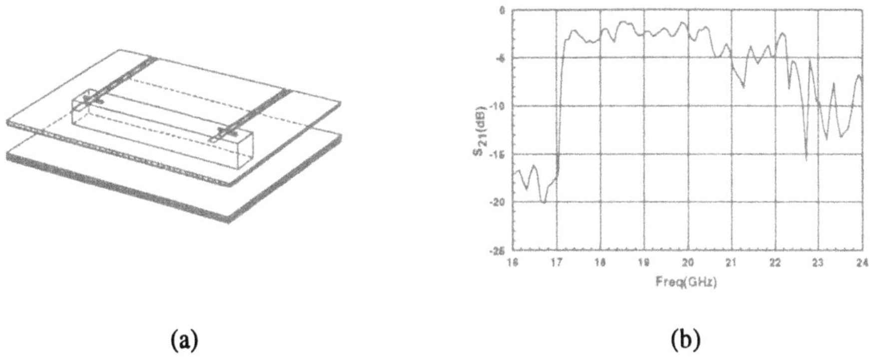

Fig. 5. (a) Experimental setup of a CPW-NRD-guide-CPW transmission, and (b) measured electrical performance of the transmission structure with two transitions.

POTENTIAL PROBLEM AND FUTURE DIRECTION

The potential problems of a design may be related to possible radiation loss and mode conversion problems since the transition usually presents an unbalanced geometry from the view point of both NRD-guide and planar line. In particular, the unbalanced (or asymmetrical) topology may be a critical factor for designing NRD-guide. In our opinion, the aperture-based coupling will not lead to any visible leakage of power or mode conversion since it will not disturb the electric field profile of the fundamental nonradiative mode (LSM mode). On the other hand, the open- or short-ends of the planar line may potentially generate radiation loss and mode conversion (surface wave, for example). This problem can be solved by an appropriate design and selection of topology. Other problems are more or less related to thermal and mechanical rigidity of planar circuit sheets.

Design and optimization of the interconnecting transitions between the NRD-guide and planar structures can be made by the use of field theory-based tools. Since the topologies of the transitions are related to both NRD-guide and planar circuit, the modeling and design parameters are much more involved and should be adequately characterized. A systematic analysis and optimized design should be driven towards the realization of flat wideband response with low insertion loss. Usually, the NRD-guide presents a high impedance which may cause the serious problem of an impedance matching with the planar circuit if a low dielectric material is used for the NRD-guide. Fig. 6(a) shows the characteristic impedance of the fundamental LSM mode of NRD-guide which is defined by a modal voltage over the transmitted power. The results indicate clearly that higher dielectric permittivity may lead to a better matching condition and therefore lower insertion loss and flatter frequency response. This is confirmed in Fig. 6(b) that predicts frequency response of NRD-guide made of different dielectric permittivity ranging from 2.56 to 6.00 with the developed approximate model presented[7] for a normalized NRD-guide open-end length of l_w/λ_g = 0.598.

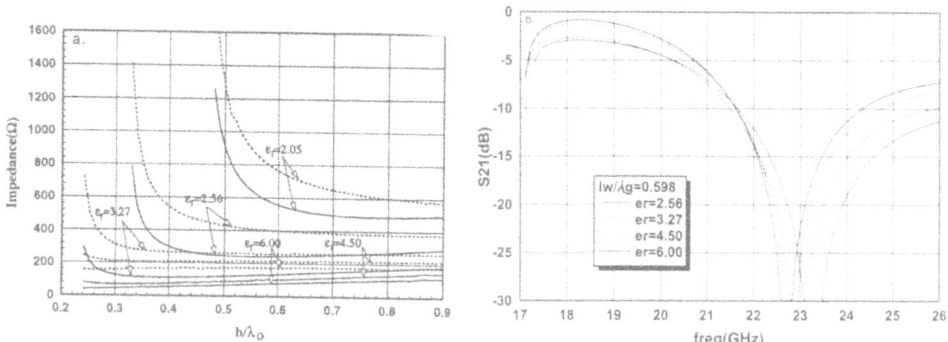

Fig. 6. (a) Characteristic impedance of NRD-guide made of a dielectric material having different relative permittivity, and (b) frequency response of the microstrip-to-NRD-guide transition that uses different NRD-guide made of different dielectric materials.

To improve the effective bandwidth performance and coupling efficiency (reduction of insertion loss), alternative aperture geometries may be used as shown in Fig. 7. Different topologies were studied for multilayer based aperture coupled antenna and circuit applications[9,10]. Nevertheless, their application to the hybrid integration of NRD-guide/planar circuit should be examined with the consideration of various parametric effects.

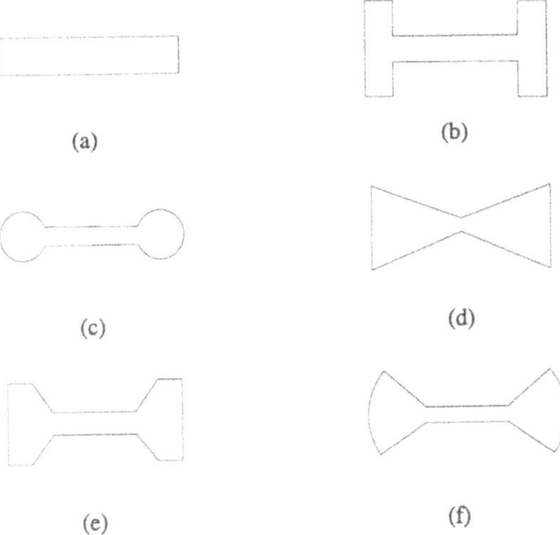

Fig. 7. A number of potential aperture geometries that can be used to improve the bandwidth and frequency response performance of the NRD-guide/planar line transitions.

Fig. 8 shows a systematic view of transitions/baluns for connecting NRD-guide with planar line: (a) microstrip line to NRD-guide, (b) CPW to NRD-guide (the fundamental mode operation), (c) slot line to NRD-guide with an inverted (insulated) NRD-guide configuration, and (c) CPW to NRD-guide (the first higher-order mode operation) with an inverted (insulated) NRD-guide configuration. Clearly, alternative NRD-guide may be exploited for the hybrid integration of these two set of dissimilar structures. The insulated NRD-guide was in particular proposed for the use of high dielectric permittivity materials. Further study on this type of NRD-guide transitions to planar circuit shoudl be considered since the slot line has been so far ignored in our results discussion and theoretical studies. This will extend the proposed hybrid integration into a wide range of applications.

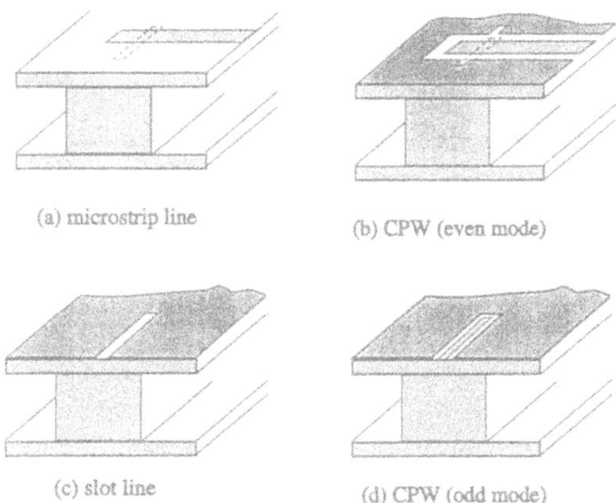

Fig. 8. Global view of transitions/baluns that are used to connect NRD-guide with a class of planar circuits including slot line and CPW's first high-order mode.

Alternative transitions can also be designed and made for an NRD-guide to connect a shielded strip line and/or another NRD-guide as shown in Fig. 9. In this case, a number of applications can be expected such as filters and couplers. In particular, the vertical NRD-guide-to-NRD-guide coupling may provide an alternative way to achieve tight coupling that is known to be difficult in the horizontal coupling scheme.

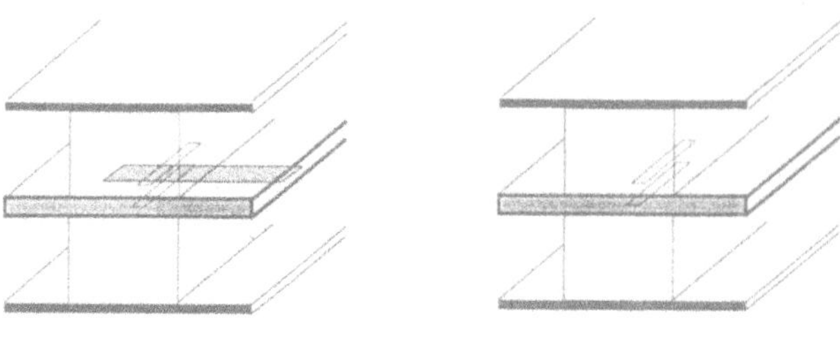

(a) NRD-stripline-NRD coupling (b) NRD-to-NRD aperture coupling

Fig. 9. Transitions for (a) NRD-guide-strip line-NRD-guide interconnects, and (b) NRD-guide-to-NRD-guide coupling.

CONCLUSION

This paper reviews the hybrid integration technique of NRD-guide and planar circuit that has been proposed for microwave and especially millimeter-wave applications with emphasis on the transitions/baluns between two sets of dissimilar structures. It has been demonstrated that the proposed hybrid scheme has a number of attractive features for designing passive components and active devices which can be summarized as low-cost, compactness, flexibility of design and reduction of interference. Potential problems and possible solutions have also been presented in the paper. Future research direction on this particular subject is also indicated. It is believed that this new technique offers a potentially cost-effective and performance promising solution for widespread applications.

REFERENCES

1. Holger H. Meinel, "Commercial applications of millimeterwaves history, present status, and future trends", *IEEE Trans. Microwave Theory Tech.*, vol. 43, pp. 1639-1653, July, 1995.
2. Wilfried Grabherr, Bernhard Huder, and Wolfgang Menzel, "Microstrip to waveguide Transition compatible with mm-wave integrated circuits", *IEEE Trans. Microwave Theory Tech.*, vol. 42, pp. 1842-1843, July, 1995.
3. S. Kanamaluru, M. Li, and K. Chang, "Analysis and design of aperture-coupled microstrip patch antennas and arrays fed by dielectric image line," *IEEE Trans. Antennas Propagat.*, vol. AP-44, pp. 964-974, July, 1996.
4. T. Yoneyama, "Nonradiative dielectric waveguide", *Inferred and Millimeter wave*, Academic Press, Inc. Vol. 11, Ch.2, pp. 61-98, 1984.
5. T. Yoneyama, "Recent development in NRD-guide technology," *Ann. Télécommun.*, 47, no.11-12, pp.508-514, 1992.
6. F. Furoki and T. Yoneyama, "Nonradiative dielectric waveguide circuit components using beam-lead diodes," *Electronic and Communications in Japan, Part 2*, vol. 73, pp.35-40, No.9, 1990.
7. L. Han, K. Wu and R.G. Bosisio, "An integrated transition of microwave to nonradiative dielectric waveguide for microwave and millimeter-wave circuits," *IEEE Trans. Microwave Theory Tech.*, vol. 44, pp. 1091-1096, July, 1996.
8. L. Han, N. Simons, K. Wu, A. Ittipiboon, and M. Cuhaci, "Experimental and numerical investigation of a microstrip to NRD transition," *Conf. Proc. of ANTEM'96*, pp. 15-18, Aug. 6-9, Montreal.
9. V. Rathi, G. Kumar, and K. P. Ray, "Improved coupling for aperture coupled microstrip antenna," *IEEE Trans. Antennas Propagat.*, vol. AP-44, pp. 1196-1198, Aug., 1996.
10. C. Chen, M-J. Tsai, N. G. Alexopoulos, "Optimization of aperture transitions for multiport microstrip circuits," *1996 IEEE MTT-S Inter. Microwave Symp. Digest*, pp. 711-714, June 17-21, San Francisco.

SOLDER FREE INTERCONNECTS FOR HIGH DENSITY PHASED ARRAY INTEGRATION

D. Strack, H. Fudem, F. Kuss, and W. Marsh

Northrop Grumman Corporation
1212 Winterson Road
Linthicum, MD 21090

ABSTRACT

A method of achieving high density interconnections is demonstrated through the use of solder free interconnects, between multilayer substrates in which a common technology can be used for both microwave and DC interconnects of large electronic systems.

INTRODUCTION

The ever continuing push towards lower cost, weight and size, electronic system requirements for higher density interconnects and integration become a driving factor in many designs. The goal of this effort was to design, develop and test a microwave interconnect method that would simultaneously meet a number of these parameters, low cost, low loss microwave transmission, high density, and be producible. The design was aimed at insertion in active aperture arrays for military radar [1], [2], but the approach is applicable to any system needing to interconnect large numbers of both RF/microwave and DC/logic signals (mixed signals).

The need for low cost and high density precluded the use of separate coax connectors for microwave and pins or ball grid arrays for DC/logic. A uniform technology that could accommodate both microwave and DC/logic interconnects was needed. The most straight forward interconnect technique is to use a ball grid array interconnect between the two substrates. This approach would be impractical over large areas (as in an active array antenna) for two reasons. First, the vertical tolerance stackup would be too large to support a ball grid array. Secondly, the ability to repair or replace components in the array would be extremely difficult. For these reasons, we pursued a solder free interconnect (SFI) approach, applying the technology already demonstrated on the Commanche program in which thousands of DC and logic interconnects are made using SFIs. The next step was to demonstrate microwave interconnects, as shown in Figure 1, using the same grid as the DC/logic SFI, thus allowing the same SFI to interconnect mixed signal multichip modules (T/R Modules). To achieve small size (i.e. high density) and low cost, the microwave connector was designed to the minimum standard SFI pitch of 50 mils. The microwave interconnect shown in Figure 1 connects two ceramic substrates in a coaxial manner.

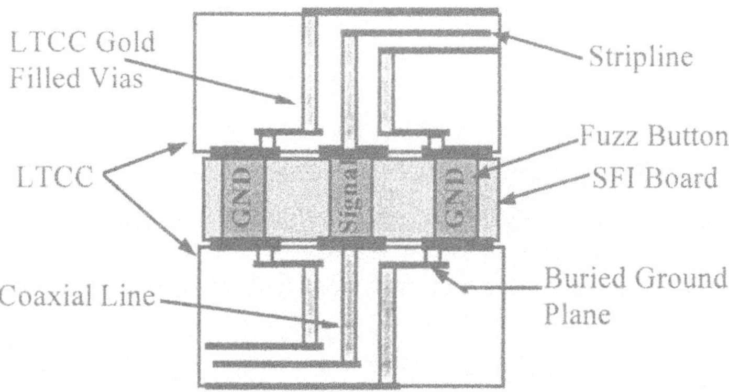

Figure 1 Cross Section of Microwave SFI Coaxial Interconnect.

Figure 2 shows a picture of an SFI board in which both microwave and DC/logic can be interconnected, using the same 50 mil grid spacing for both. The SFI shown in Figure 2 is used in an existing program for power and logic interconnects, but could also be used for microwave interconnects based on the design presented in this paper. The SFI structure consists of only two parts, the metal contacts (fuzz buttons) manufactured by Cinch and the dielectric support. Figure 3 shows a cross section of the two components (metal contacts and dielectric board) assembled into an SFI. The metal contact can be viewed as a 'brillo pad' made out of molybdenum, then nickel and gold plated. Using a refractory metal (molybdenum) as the base allows the metal contacts to be temperature cycled without losing their elasticity. The cost per contact is only cents (in quantity). The fabrication of the dielectric board is simply a two sided drilling operation which creates an 'hour glass' in the dielectric board to hold the fuzz button.

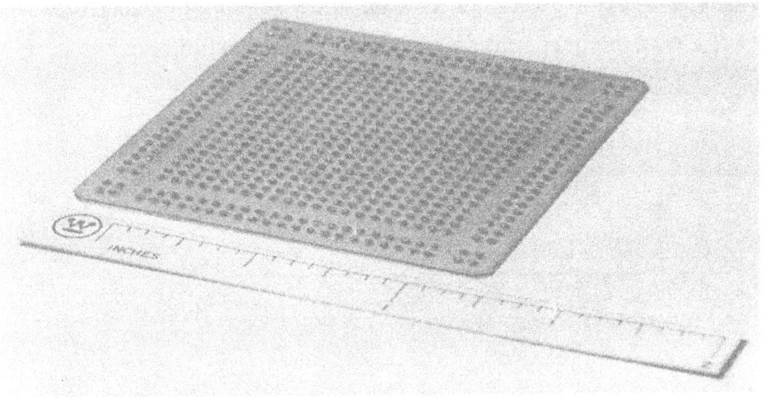

Figure 2 Grid Array of SFIs on 50 mil Pitch for Use with Mixed Signal Systems.

Figure 3 Cross Section of SFI, Including Metal Contacts and Dielectric Board.

DESIGN APPROACH

The SFI development at Northrop Grumman, demonstrated the SFI approach as a microwave interconnect between two ceramic multichip modules (MCMs), using a standard low cost SFI board previously used for DC and logic interconnects. Figure 4 shows a cross section of a planar array using SFIs to connect the DC/logic substrate with the microwave substrate. Test results will be shown for an interconnect between an SFI and a low temperature cofired ceramic (LTCC) substrate. The interconnect between high temperature cofired ceramic (HTCC) substrates is still under development, although, the approach described here is applicable to SFI interconnects between all types of multilayer substrates. The LTCC/SFI interconnect represents the microwave I/O that could be used for T/R modules. In addition to the microwave connection required of the SFI, one of the primary functions the SFI must perform is to allow for misalignment between the two ceramic MCMs, with minimal degradation in microwave performance.

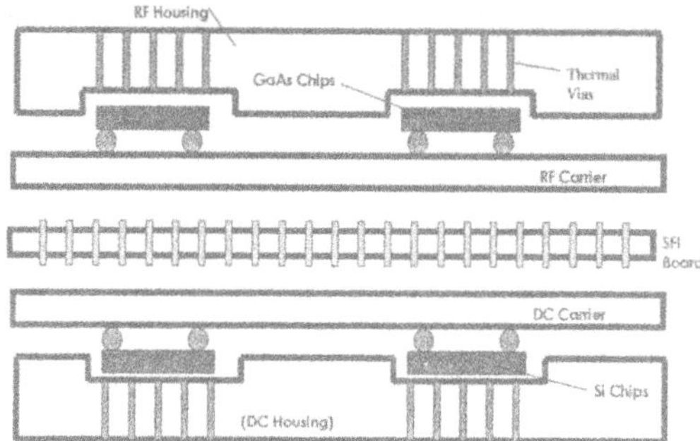

Figure 4 SFI Interconnect for T/R Modules in a Planar Array.

The first step undertaken was to model the SFI/LTCC transition based on the minimum grid array spacing presently available for the SFI which was a 50 mil pitch. The metalization on the top surface of the LTCC consisted of 30 mil diameter contact pads for the SFI, on a 50 mil pitch. The large pads on the LTCC served two functions. First, the SFI itself is a series inductive element of approximately .6 nH per fuzz button, in addition, at the 50 mil grid spacing used, the coaxial nature of the SFI presented an impedance of > 50 Ω (i.e. inductive). The large metal pads create a shunt capacitive effect helping tune out the SFI inductance, thus maintaining good return loss over the microwave bandwidth. Figure 5 shows simulated return loss of the LTCC/SFI interface. As can be seen from Figure 5, the bandwidth of the modeled LTCC/SFI transition clearly supports microwave operation which makes the structure suitable for most airborne active aperture applications.

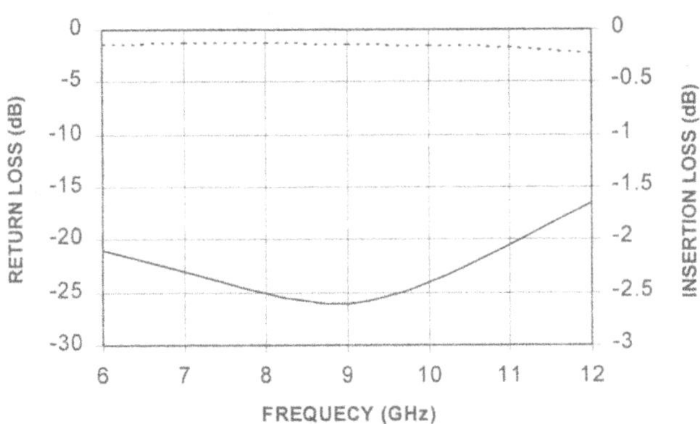

Figure 5 Simulated Insertion Loss and Return Loss of LTCC/SFI Transition.

MEASURED RESULTS

Measurements involved tests of a number of different structures and configurations. It includes two types of dielectric boards, one 35 mils thick with an $\varepsilon r = 3.0$, the other 20 mils thick with an $\varepsilon r = 6.0$, the LTCC structure, included a vertical 'micro' coax transition. Contact repeatability tests were performed and measurements taken to determine electrical degradation due to misalignment between the two substrates. All testing was performed using SMA bulkhead connectors (P/N 2052-1215-00) modified to allow for a flush connection to the SFI, except for a small portion of the center pin used to 'key' the center contact of the SFI.

In order to validate the test structure, the connectors were first joined with a single SFI to verify that a low return loss was possible with the test structure. This test structure provided a return loss of better that -20 dB through 12 GHz. Figure 6 shows the configuration of both the validation/calibration structure and the test structure itself. With the test structure validated, the SFI/LTCC transitions could be tested with a high degree of confidence. The preliminary 'calibration' measurement of the two connectors and single SFI was made and stored in the network analyzer as a through calibration, so the SFI/LTCC measurement could subsequently be de-embedded. The calibration measurement provided a return loss of better than -20 dB through 12 GHz for both SFI structures. After the 'through' calibration was performed, an additional SFI and LTCC piece was inserted in the test structure, also shown in Figure 6. Network analyzer measurements of the SFI/LTCC transition using both the 35 mil thick $\varepsilon r = 3.0$ SFI board and the 20 mil thick $\varepsilon r = 6.0$ SFI board are shown in Figure 7. The transitions provide a well matched (less than -20 dB return loss) interconnect through 11 GHz with an extremely low insertion loss of nominally .05 dB through 11 GHz.

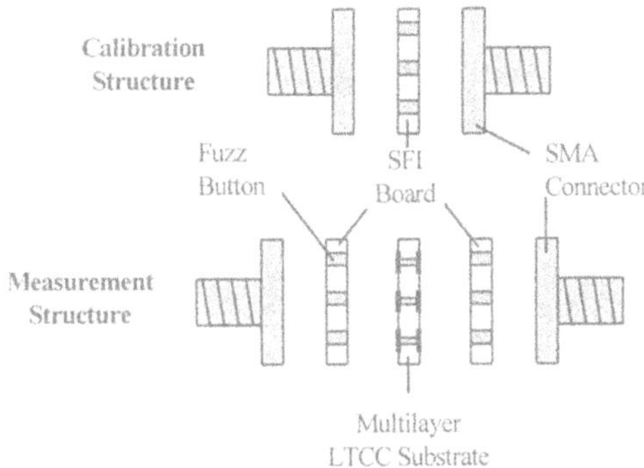

Figure 6 Calibration 'Thru' and Test Structures

In order for the microwave SFI approach to be viable in production, two additional attributes needed to be verified, first the repeatability of the microwave connection, and secondly, the ability of the microwave connection to withstand variations in alignment without degradation in microwave performance.

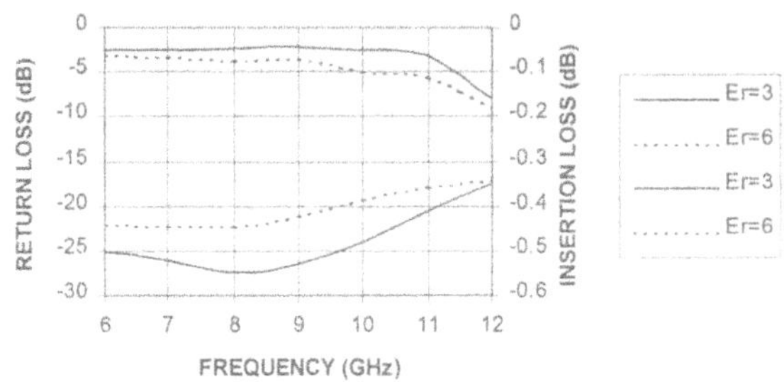

Figure 7 Measured Results of LTCC/SFI Transition.

To test the repeatability of the microwave connection, the test structure was disassembled and reassembled five times, Figure 8 shows all five plots. The data in Figure 8 is for the $\varepsilon r = 3.0$ SFI board. A similar test was performed using the 20 mil thick $\varepsilon r = 6.0$ SFI material with similar results.

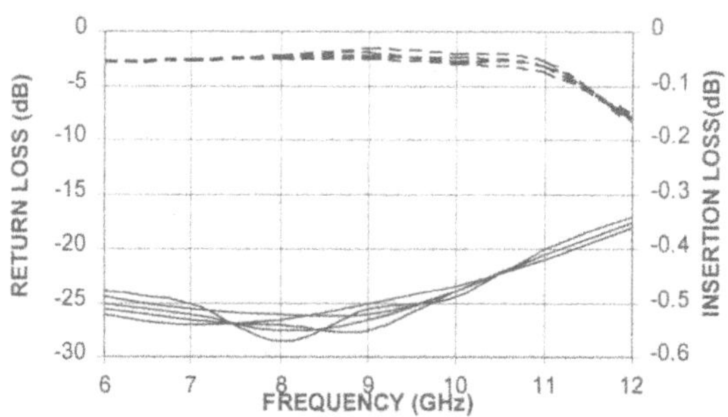

Figure 8 Measured Repeatability of LTCC/SFI Transition after five 'Makes and Breaks'.

To demonstrate the ability of the SFI/LTCC to withstand misalignment, LTCC pieces were cut with offsets, from 0 to 20 mils in 5 mil increments, from center. Each of these LTCC structures were tested in the fashion stated above and plotted. The series of plots is shown in Figure 9. Even at 20 mils offset from center the return loss degraded by less than 3 dB at 12 GHz.

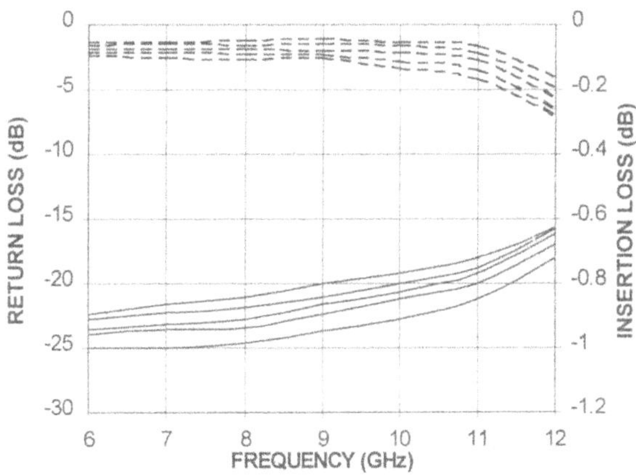

Figure 9 Measured Results of LTCC/SFI Transition, Offset in 5 mil Increments.

CONCLUSION

The ability to used a singular type of I/O for a mixed signal (microwave/RF and DC/logic) interconnect has been demonstrated. The microwave performance of the SFI to LTCC interconnect has been demonstrated, most importantly, it has been demonstrated that the interconnect can withstand large (20 mils) misalignments without significant degradation in microwave performance. The use of this type of solder free interconnect should help reduce cost, weight and increase the density of future phased array radar systems.

ACKNOWLEDGMENTS

The Authors would like to acknowledge and thank ARPA and the HDMP Program, in particular, for the support given in this effort.

REFERENCES

[1] J. Costello, et al., “The Westinghouse High Density Microwave Packaging Program”, MTTS-Digest, May 1995, pp. 177-180.

[2] J. Wooldridge, “High Density Microwave Packaging for T/R Modules”, MTTS-Digest, May 1995, pp. 181-184.

[3] D. Strack, H. Fudem, F. Kuss, W. Marsh, J. Costello et al., “Solder Free Interconnects for Mixed Signal (DC/Microwave) Systems”, MTTS-Digest, June 1996, pp. 231-234.

SILICON BASED ON-WAFER AND DISCRETE PACKAGING

Rashaunda M. Henderson and Linda P.B. Katehi

The Radiation Laboratory
EECS Department
The University of Michigan
Ann Arbor, MI 48109

INTRODUCTION

Advanced packaging concepts for three dimension (3-D) microwave monolithic integrated circuits (MMIC's) are being investigated at The University of Michigan. As high frequency circuit design complexity increases so must the design and development of the housing which supports these components. In the past more emphasis was placed on improving integrated circuit (IC) processing technology while research in the area of electronic packaging was minimal. Now that IC processing technology has reached a mature level and it is possible to fabricate excellent components which function well at high frequencies, it has been determined that circuit/system performance is being limited due to the lack of performance on the part of the electronic package. High frequency circuits require electronic packages which provide excellent electrical, mechanical, and environmental support while remaining non-invasive.

This paper reports on efforts at The University of Michigan which offer a method to simulate the performance of high frequency packages using the finite element method (FEM), as well as experimental development of high frequency packaging options using thin film technology with silicon as the base substrate. These advances lend themselves to the objective of developing microwave and millimeter wave packaged components which are low in cost, weight, and volume. Applications for this research lie in the area of communication systems, defense and avionics, industrial and consumer equipment.

SILICON AS A PACKAGING MATERIAL

There is no doubt about the excellence and performance of silicon in developing electronic circuits and one can note that clearly by observing the computer industry and how VLSI has soared. Petersen reports on the validity of silicon as a rugged mechanical material. He sights many applications where silicon is used to develop micromechanical devices[1]. If one considers designing a silicon package for a silicon electronic component, there is no need to concern themselves with the thermal mismatch issue often encountered in packaging. Micromachining techniques have shown that miniaturized, high-precision silicon devices are capable of being developed.

SELF-PACKAGED COMPONENTS

It has been proven that it is advantageous to use planar structures in broadband high frequency circuit design[2]. Micromachining techniques have been used to compensate for the electrical performance limitations due to the planar character of the circuits such as substrate mode effects. Coupling and radiation have been successfully reduced by way of monolithic packages developed using micromachining and standard IC processing techniques. Self-packaged circuits are miniaturized with cavities that reduce size, volume and cost.

Micromachining is a mature technology for sensors and much effort has been shown to indicated its usefulness in high frequency circuit design. A two wafer system has been developed where an upper and lower cavity follow the path of a circuit to provide isolation and minimize cross coupling[2]. The lower wafer reduces surface waves within the substrate and the upper shield reduces free space radiation. By conforming to the shape of the structure its protecting, additional layout space is available which allows for increased circuit density.

The applications for these miniaturized structures are personal communication systems and complex wireless systems. On wafer self-packaged components have been designed and developed which offer enhanced electrical performance compared to their conventional "open" counterparts[2]. Figure 1 shows the cross section of conventional and packaged microstrip structures.

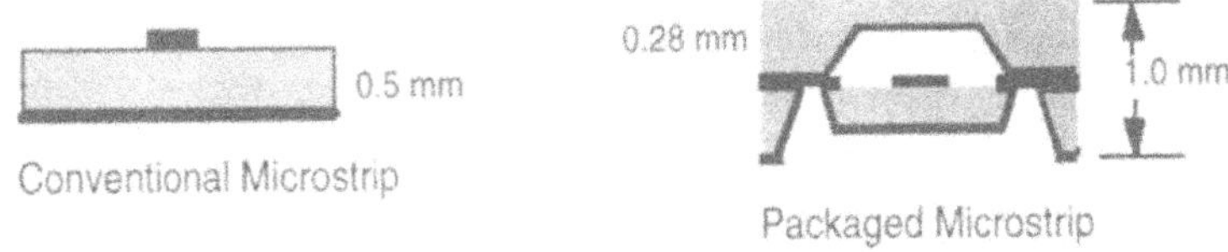

Figure 1. Cross section of microstrip structures.

Cross talk between neighboring interconnect lines can be reduced by using the self-packaging technology as well. Figure 2 shows two back-to-back right angle bends which are separated by 3 mm and their corresponding top and bottom wafer views.

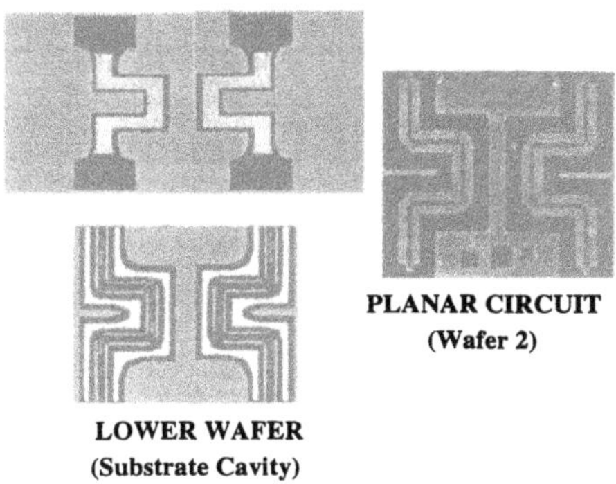

Figure 2. Back-to-back right angle bends.

Figure 3 shows the cross coupling between the two bends. By shielding the two lines the surface waves which couple them have been successfully eliminated and the cross coupling is reduced.

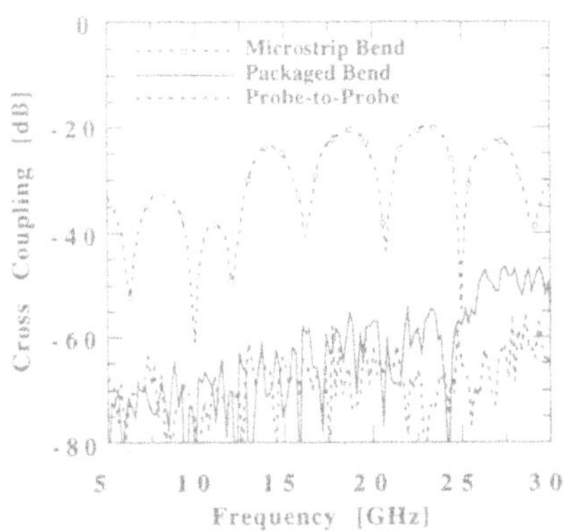

Figure 3. Measured cross coupling between open and packaged bends.

By providing upper and/or lower shielding cavities, it has been shown that these self-packaged structures enhance power transmission in feedlines for microstrip patch antennas. Figure 4 shows the return loss of a packaged and open microstrip fed patch antenna. One can see the improvement in bandwidth which is attributed to the packaged feedline maintaining the input impedance over the frequency range as compared to the open structure.

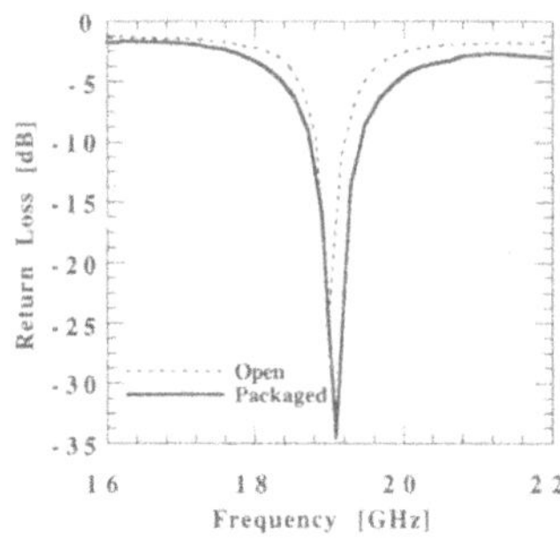

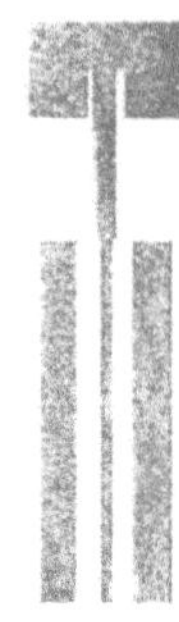

Figure 4. Return Loss for open and packaged microstrip fed patch antennas.

NUMERICAL MODELING OF MMIC PACKAGES

In the design process of electronic devices there exists a circuit design and analysis phase. It would be ideal if an engineer were able to use commercially available software which not only simulated the performance of his circuit but also considered the effects of the components which will be used to complete the packaging. A simulator of this kind would greatly reduce the cost associated with preliminary design and prototype iterations. Currently available software is not capable of giving the engineer information necessary in high frequency design.

An effort is underway where the FEM is being used to model a high frequency MMIC package for a phase shifter[3]. Using this code, the effects of package features (via holes, bonding wires, CD bias lines, symmetry and location of IC) are taken into account and able to provide much needed information with respect to these issues. The K/Ka-band hermetic package being modeled in Figure 5 has been fabricated by Hughes Aircraft Company for NASA Lewis Research Center.

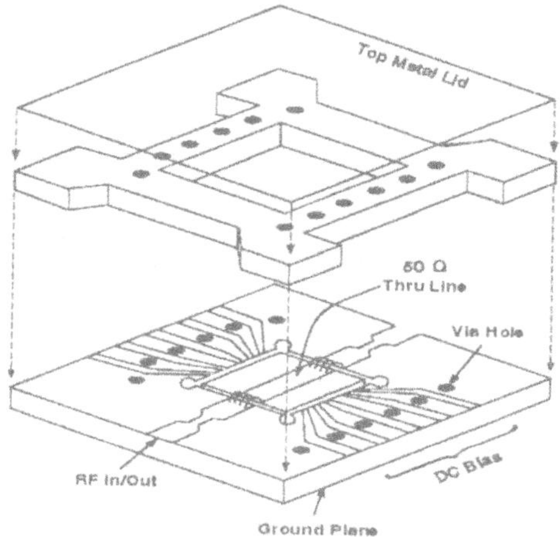

Figure 5. K/Ka-band hermetic package.

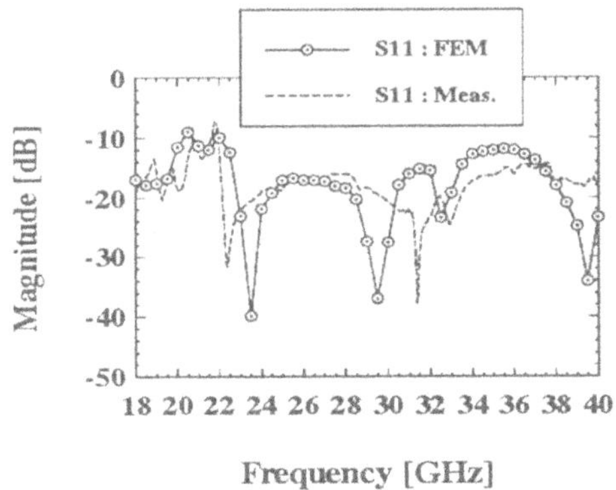

Figure 6. Return Loss for hermetic package.

Using HTCC processing techniques the (7.112 x 7.112 x 1.27) mm package is fabricated from alumina (95% pure, $\varepsilon_r = 9.5$). The experimental data compares well with the numerical results as seen in Figures 6 and 7 where a through line was used to characterize the package. This code also predicts how the package peripherals affected the response of the package[3].

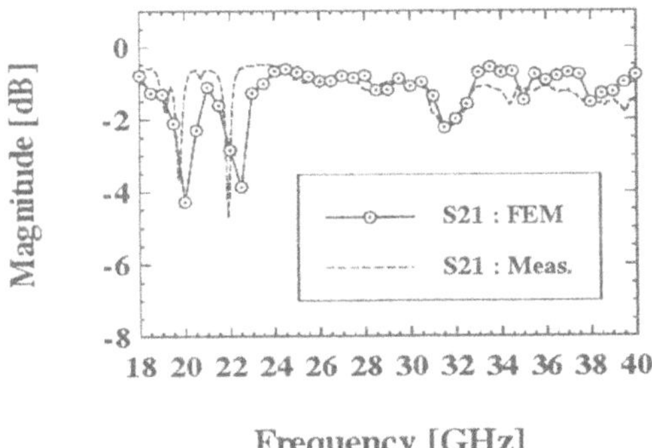

Figure 7. Insertion Loss for hermetic package.

SILICON BASED MMIC PACKAGE

Current research efforts are being conducted in the area of developing single chip carriers such as the one being modeled using the FEM code. We propose to use thin film technology to develop silicon based packages. There are many advantages to using an established silicon technology to develop high frequency high density electronic packages compared to other materials and assembly techniques. The smallest linewidth and via diameter realization is achievable and high frequency requirements can be easily met when employing this technique[4]. When compared to laminate and ceramic processing technologies thin film packaging offers increased circuit density and the capability to accommodate multilayer tile arrangements which are important in space applications where low volume, high performance, high frequency packaged structures are required[4]. Research shows that space communications can benefit from this packaging scheme to achieve reduced volume and weight[5]. Figure 8 shows the top views of the HTCC alumina and the thin film silicon phase shifter packages.

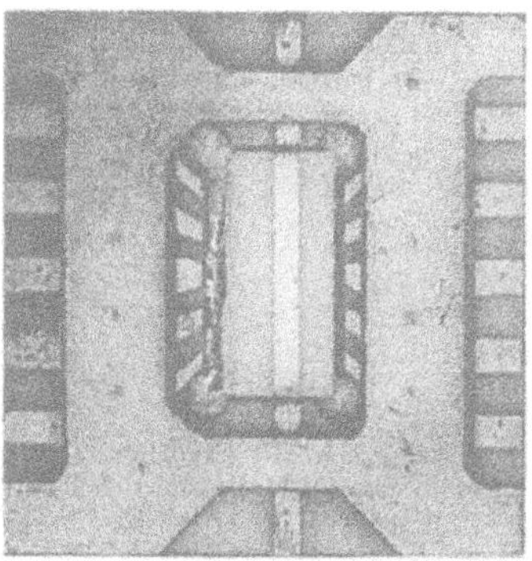

Figure 8. Photograph of HTCC and Thin film hermetic packages.

The processing techniques used to fabricate this silicon structure were standard IC steps such as metallization by evaporation and plating, photolithography and wet chemical etching to generate the 3-D structure. Four wafers comprise the package itself. They consist of metallized silicon as a holder for the package, two processed wafers which realize the transmission lines (open microstrip, stripline, and shielded microstrip) and one final wafer for the top cover to offer hermetic shielding. Figure 9 shows a plan view of the layers used in developing the package in addition to the through line being used for electrical characterization.

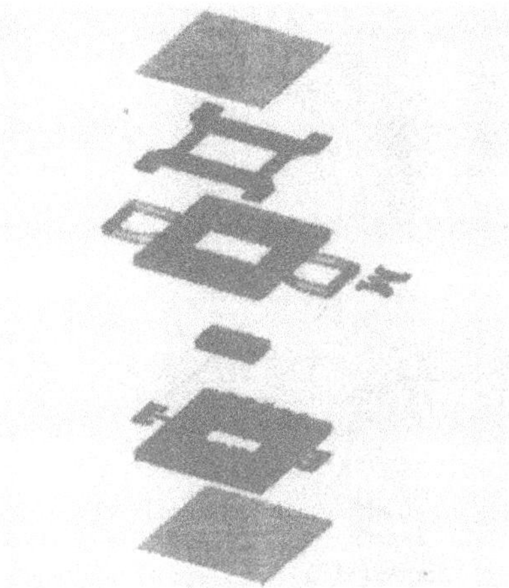

Figure 9. Thin film (silicon based) MMIC package layout.

FUTURE WORK/CONCLUSIONS

Research is underway at The University of Michigan in an effort to create the building blocks required for the development and design of high frequency circuits and packages which offer excellent electrical performance while achieving the cost and weight requirements proposed by industry. The theoretical characterization aids in reducing research dollars by providing the design engineer an alternative to expensive iterations on prototype designs. Figure 10 shows an example of the proposed future on-wafer packaging structures which has a hundred-fold reduction in cost.

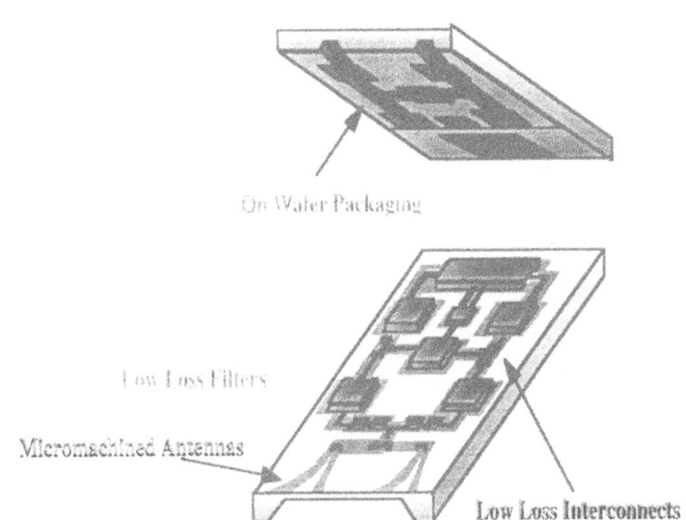

Figure 10. Future on-wafer packaging.

ACKNOWLEDGMENTS

This work has been supported by contracts from the Army Research Office and the Office of Naval Research.

REFERENCES

1. K. E.Petersen, "Silicon as a Mechanical Material," *Proceedings of the IEEE*, Vol. 70, No. 5 May 1982.
2. R. F. Drayton, et al., "Advanced Monolithic Packaging Concepts for High Performance Circuits and Antennas," in *1996 IEEE MTT-S Int. Microwave Symp. Dig.*, pp. 1615-1618.
3. J. G. Yook, L. P. B. Katehi, et. al., "Experimental and Theoretical Study of Parasitic Leakage/Resonance in a K/Ka-Band MMIC Package" Submitted to *IEEE 1996 MTT-S Special Issue.*
4. J. E. Sergent and C. A. Harper, *Hybrid Microelectronics Handbook*, 2d ed.,Mc-Graw-Hill, Inc., New York, 1995.
5. M. I. Herman, K. A. Lee, et. al., "Novel Techniques for Millimeter-Wave Packages," *IEEE Trans. on Microwave Theory and Techniques*, Vol. 43, No. 7 July 1995.

PATCH-MMIC-FERRITE INTEGRATION IN NOVEL PHASED ARRAY TECHNOLOGY

Ernst F. Zaitsev,[1] Yu. P. Yavon,[1] Anton B. Guskov,[1] and George A. Yufit[2]

[1] St. Petersburg State Technical University
St. Petersburg 195251, Russia

[2] Beltran, Inc.
Brooklyn, NY 11210

The proposed new class of electrically controlled antennas is an advancement on the integrated phased arrays (IPA) with ferrite control published by Zaitsev et al.[1-3]

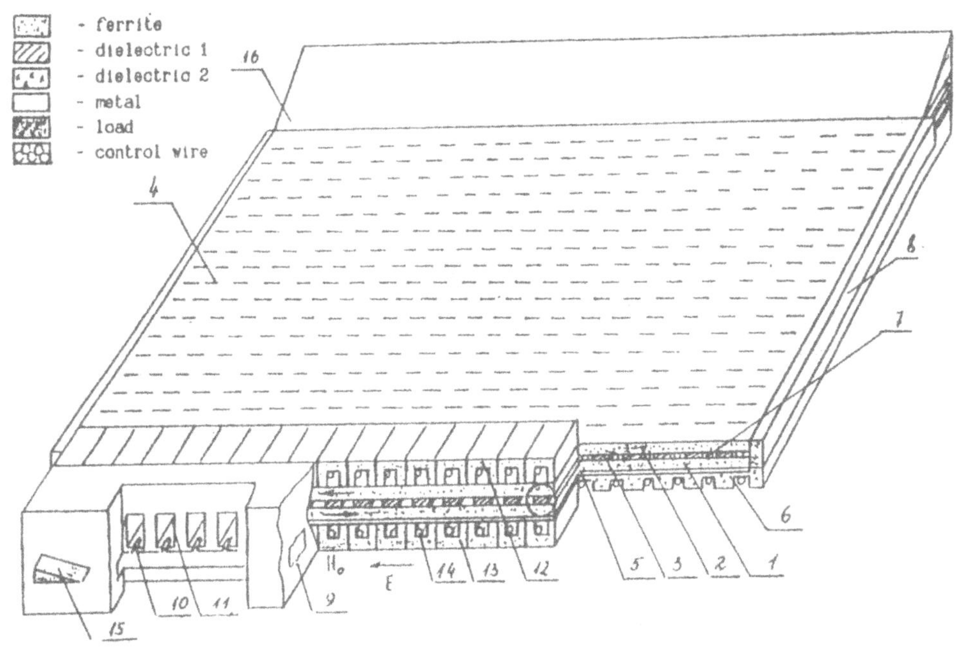

Figure 1. Design of planar passive IPA.

Directions for the Next Generation of MMIC Devices and Systems
Edited by N. K. Das and H. L. Bertoni, Plenum Press, New York, 1997

The design of planar passive IPA is shown in Figure 1. The main element of the antenna is a waveguiding ferrite-dielectric-ferrite (FDF) structure. The bottom surface of this FDF structure is metallized; on its top surface radiating dipoles are disposed. Wires for control winding are placed between the ferrite layers. The variation of ferrite magnetization by the control current causes alternation of phase velocity of the waves in the FDF structure and scanning of antenna beam.

Significant results are achieved in both theoretical analysis of such antennas and the creation of experimental samples. One of the samples (Figure 2) contains 22 x 25 = 550 radiating dipoles; these antenna parameters are shown in Figure 3.

Figure 2. 8-mm 550 element plane phased array.

Wavelength range	8 mm
Number of radiating elements	550
Gain	25 dB
Beamwidth	4*5 degrees
Sidelobes	-13 DB (H-plan), -20 dB (E-plan) It is possible to decrease beamwidth to 1-2 degrees and SL to -20 dB.
Scanning sector	20*40 degrees or 40*40 degrees
Losses	3-4 dB
VSWR input	less than 1.5
Radiating power	pulse 10 kW, mean 20 W
Energy of one beam switching	4-5 mJ
Beam switching time	4 microsec.
Control circuit consuming power at fixed beam position	8-12 W

Figure 3. Design Parameters of 8-mm 22x50=550 dipole antenna

The most important advantages of this antenna are:

a) possibility of operation in mm-wave band;

b) simple and economical beam control (single current is needed to control beam position of the entire antenna; 2-D scanning demands two control currents);

c) planar integrated design, in which the FDF structure simultaneously performs functions of power distribution, phase shifting and beam control; the structure thickness of the antenna is about λ;

d) as a result of the previous item, low cost of such antennas, which opens the perspectives of their wide application.

But there are some particularities and disadvantages:

a) increased angle vs. frequency dependence:

$$\Delta\theta = \left(\frac{c}{v_g \cos\theta_0} - \mathrm{tg}\theta_0 \right) \frac{\Delta f}{f}$$

where θ_0 – angle position of pattern diagram maximum; c – velocity of light; v_g – group velocity of wave in FDF structure. Usually c/v_g - 4...5; accordingly 1% frequency variation causes $\Delta\theta$ = 2...3 deg. This is several times greater than for conventional phase array (in this case the first term in parenthesis is absent).

b) The antenna efficiency drops if the radiating aperture increases because of dissipation losses in the FDF structure. Therefore, a limiting value of gain exists; the greater value $\alpha\lambda$, the less limiting the gain value, where λ – wavelength in vacuum, α – decrement.

c) The greater the operating wavelength, the greater the dimensions and mass of ferrite slabs; accordingly, power of control increases. Due to this, it is not expedient to use the antenna at frequencies lower than 10-12 GHz.

In this paper a new class if IPA is proposed – active IPA, in which the above disadvantages are eliminated. At the same time all the advantages of passive IPA are retained. As a result, the region of practical application of the antenna expands radically in many directions: decreasing beam width, enlarging of band width, motion towards lower frequencies, etc..

The basis of the antenna is a combination of FDF structure and microstrip array having patch radiators with microwave amplifiers.

Two versions of active IPA are proposed.

For the first version, let us consider the example of 1-D scanning antenna. In Figure 4 the schematic diagram of the antenna is presented, and in Figure 5, the possible practical realization. The antenna consists of an electrically controlled FDF power dividing unit with slot coupling elements, MMIC amplifiers and radiating patches. Coupling slots are etches in the metallization of the top ferrite plate; the opposite side of the FDF structure is covered with a metal screen.

Amplifiers compensate losses in the FDF structure. This makes it possible to decrease beam width by raising the dimensions of the radiating structure, without taking efficiency drop into account, since this value is not of great importance at this time. Clearly the values of the coupling coefficients of microstrip line and FDF waveguide will vary along the waveguide to realize distribution that is uniform or decreases to the edges to obtain low side lobe levels. The gain factor G of the active antenna depends not only on the directivity of radiation, but also on gain of active elements. The ratio S/N of the receiving antenna is in direct proportion to the ratio G/T, where $T = T_A + T_R$, T_A and T_R being respectively noise temperature of the antenna and the receiver. In the case of a transmitting antenna the density of power flux is in straight proportion to $P_{rad}D$, where P_{rad} – total radiated power, D – directivity ratio.

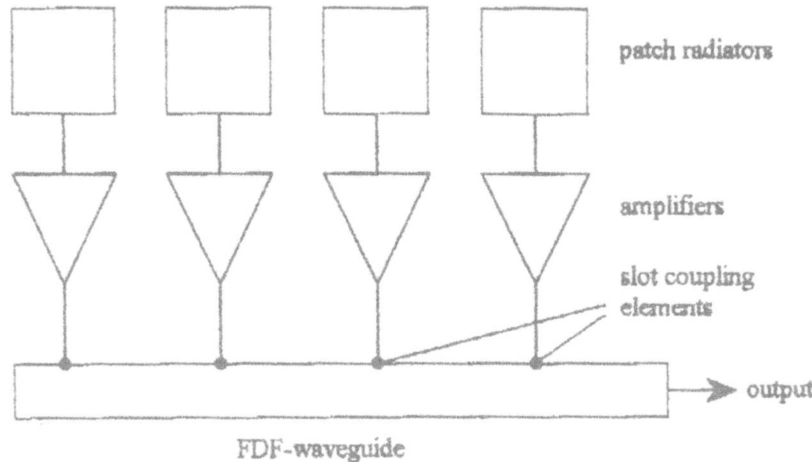

Figure 4. Scheme of an active linear IPA.

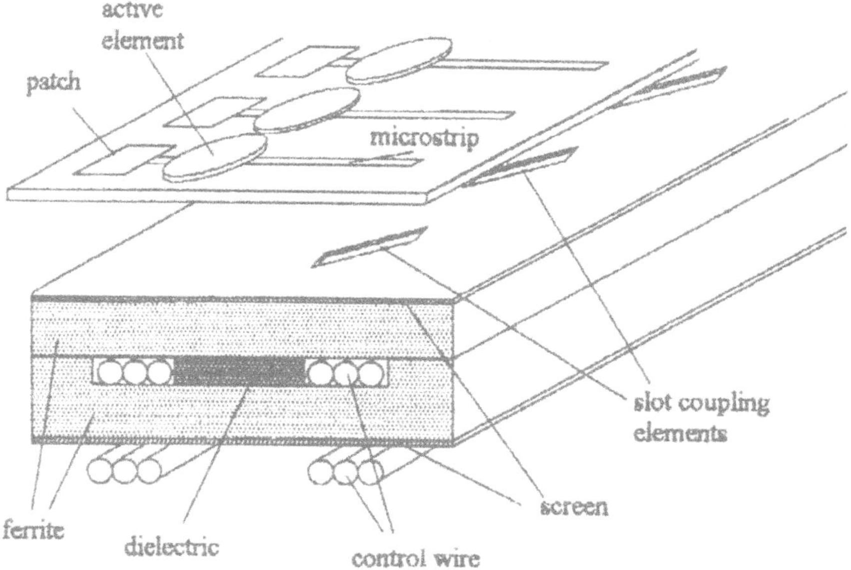

Figure 5. Active linear IPA design.

An analysis of active IPA is very simple when the following conditions are valid: a) all amplifiers are equal; b) all radiators are equal and are in equal rounding ('large array" concept); c) losses in radiators are negligible; d) feedback from amplifier output to input is absent. Then, considering the transmitting antenna, we obtain a circuit equivalent to the one shown in Figure 6. Here A_1 is antenna with no losses having the same pattern diagram as the entire array. Its input impedance Z coincides with the resistance of a single radiator taking into consideration the influence of other radiators; its value depends on beam direction. A_2 is the same amplifier as in each channel, having power gain factor K (K is determined in such an amplifier regime when the internal resistance of source and load are equal to the wave resistance of the microstrip line Z_0). A_3 is a two-port circuit with transfer factor η, which is equal to the ratio of the sum of the available powers in all the microstrip inputs of the FDF power divider and available power in the common input of the FDF waveguide.

It may be shown, using the generalized reciprocity principle,[1] that this equivalent circuit is also appropriate for the receiving antenna. The meaning of all the elements stays the same, but the direction of transfer of A_2 changes to the opposite.

As follows from the circuit in Figure 6, the gain factor of the antenna in both cases (receiving and transmitting) is equal to

$$G = KD\eta\left(1-\left|\Gamma^2\right|\right), \quad \Gamma = \frac{Z-Z_0}{Z+Z_0}$$

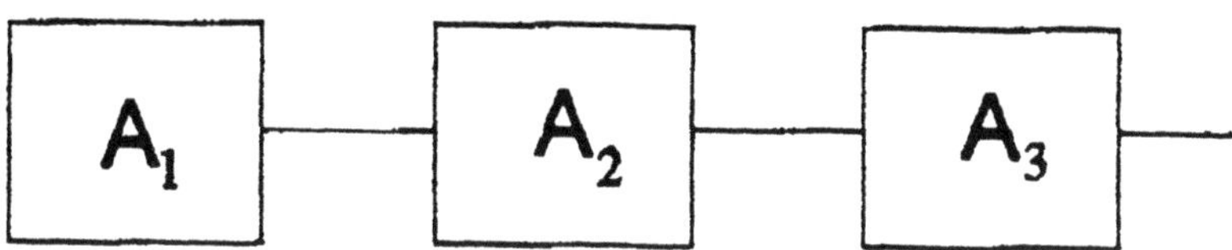

Figure 6. Equivalent scheme of the IPA

The multiplier in parenthesis depends on the scanning angle and determines the losses of mismatching of radiating elements due to their mutual coupling. The multiplier η is caused by dissipation losses in the FDF waveguide.

Let the power in the input of the transmitting antenna be P, then,

$$P_{rad} = PK\eta\left(1-\left|\Gamma^2\right|\right), \quad \text{hence } P_{rad}D = PG.$$

For the receiving antenna, the noise temperature of the entire system, recalculated to its input, is

$$T = \left[T_E\left(1-\left|\Gamma^2\right|\right)+T_K\right]K\eta + T_W(1-\eta)+T_R$$

where T_E is equivalent to noise temperature of external sources of antenna A_1 in Figure 6, T_K – noise temperature of amplifier A_2 (generally speaking it depends on Z and changes while scanning), T_W – physical temperature of FDF waveguide (usually close to T_0 = 290K).

Hence we obtain (marking $L = \eta^{-1}$)

$$\frac{G}{T} = \frac{D\left(1-\left|\Gamma\right|^2\right)}{T_E\left(1-\left|\Gamma\right|^2\right)+T_K+\dfrac{T_W(L-1)+T_R L}{K}} = \frac{D\left(1-\left|\Gamma\right|^2\right)}{T'}$$

The denominator of this expression, T' can be treated as the noise temperature of the antenna recalculated to amplifier inputs.

For the passive version, we should assume $T_K = 0$, $K = 1$.

Note that the parameters D and T' are more convenient than G and T, because D depends only on antenna geometry and amplitude-phase distribution on its aperture and does not depend on amplifier parameters. On the other hand, T' depends on amplifiers, receiver, and external noise, but hardly depends on the other antenna parameters (except the L value).

Let us consider a numerical example. Assume

$$T_E\left(1-|\Gamma|^2\right) = 40\text{K}, \quad T_K = T_R = 120\text{K}, \quad T_W = 290\text{K}, \quad L = 2.5(4\text{dB})K = 100(20\text{dB})$$

Then

$$T = 40 + 120 + \frac{290 \cdot 1.5 + 120 \cdot 2.5}{100} = 167\text{K}$$

In the same conditions passive IPA would have

$$T' = 40 + 290 \cdot 1.5 + 120 \cdot 2.5 = 775\text{K}$$

i.e., we have a 4.6 times advantage of sensitivity.

The main benefit here is that if increasing the antenna dimensions (to raise D), the T' value stays close to T_K of the amplifier in spite of the increment of losses L.

In active IPA it is easy to realize a compensation of angle vs frequency dependence. To achieve this the lengths of the microstrip lines between amplifiers and the FDF waveguide have to be done differently (see Figure 7). Namely, it is necessary that the group delay time of the signal from each radiator to the common output in the middle of the scanning sector be equal for all channels. In this case, angle vs frequency dependence is absent for the normal beam position; for other beam position, it would be of the same order as for conventional phased array.

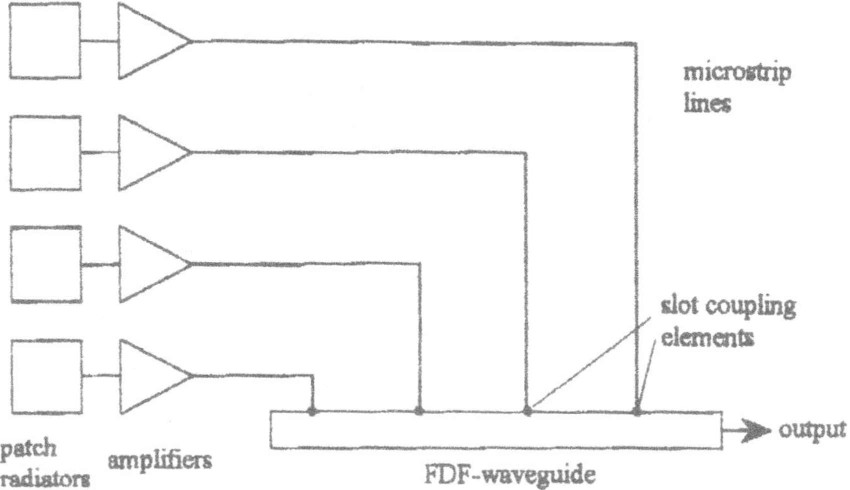

Figure 7. Active IPA with compensation of angle-frequency dependence

The second version of active IPA proposed in the present work uses frequency conversion. This version is suitable for realization in UHF and in microwave bands. Figure 8 shows the schematic diagram of the receiving antenna of such a version with 2D-scanning.

There are converters in each channel at the outputs of amplifiers to higher intermediate frequency $f_i = f + f_k$, where f – frequency of input signal (for example 1-2 GHz), f_k – heterodyne frequency (about 35-40 GHz). The heterodyne signal comes to the converters through the FDF waveguide A and the microstrip dividing units. The same microstrip circuits and other FDF waveguide B are used for adding signals of intermediate frequency.

The initial frequency f is restored in output after additional amplification. As is seen from Figure 8, waveguide A provides phase shifting of signals among horizontal rows of radiators and waveguide B – among vertical rows. As a result, an extremely simple control is achieved, only two control currents for 2D scanning. The antenna contains all together two FDF waveguides with very small dimensions, because they operate in mm-wave bands. Correspondingly, the power consumption for control is very small. At the same time, the realization of rather great phase shifts does not demand great variation of phase velocity because of the large distance between coupling elements, 6.

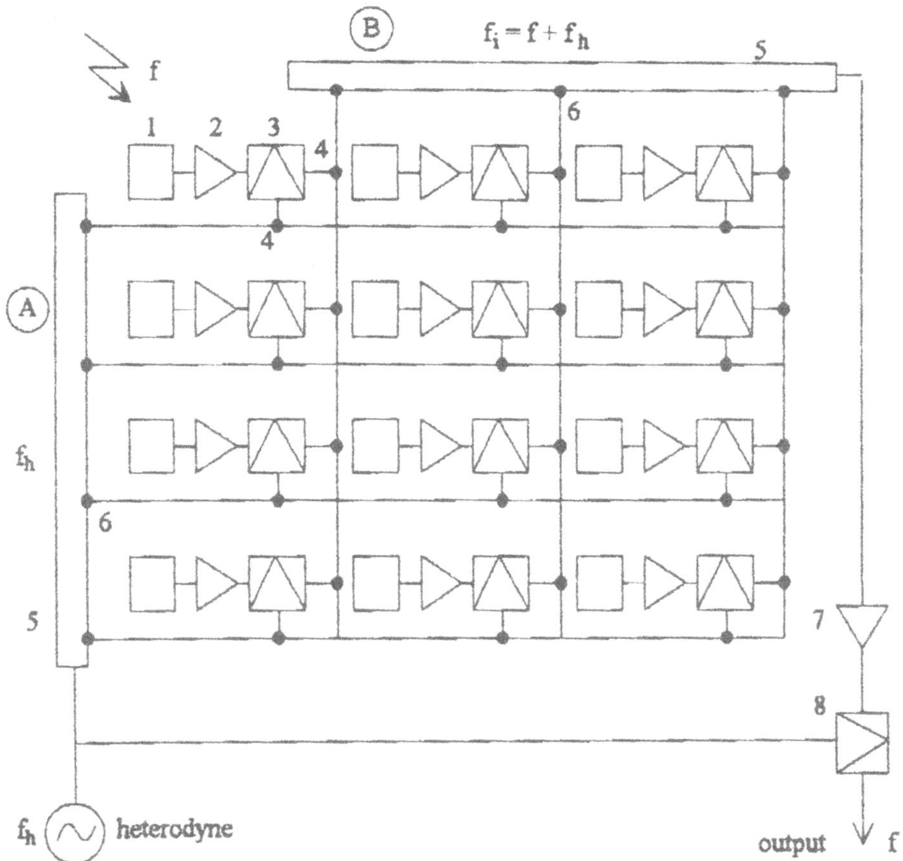

1. patch radiator
2. microwave amplifier
3. up-converter
4. microstrip coupling element
5. FDF waveguide
6. slot coupling element
7. MM-wave amplifier
8. down-converter

Figure 8. 2D-scanning active IPA with frequency conversion

Thus, the utilization of frequency conversion makes it possible to create active phased arrays in any frequency range with extremely simple and power saving control.

The class of scanning antennas we have presented is particularly well adapted to modern technology: PCB (Printed Circuit Block), MMIC (Microwave/MM-Wave Monolithic Integrated Circuit), SPS (Substrate Plasma Spraying), that will make it possible to provide low cost serial production.

REFERENCES

1. E.F. Zaitsev, Yu. P. Yavon, Yu. A. Komarov, A.B. Guskov, and A. Yu. Kanivets, MM-wave integrated phased arrays with ferrite control, *IEEE Transactions on Antennas and Propagation* 42:3:1362-1368 (1994).
2. E.F. Zaitsev, Yu. P. Yavon, Yu. A. Komarov, A.B. Guskov, and A. Yu. Kanivets, Millimetre wave integrated phased arrays with one current and two current control, in: *Proc. of 1994 Microwaves Conf.* London, UK (Oct. 25-27, 1994).
3. . E.F. Zaitsev, Yu. P. Yavon, and Yu. A. Komarov, Electrodynamic analysis of electrically controlled FDF waveguide, in: *Proc. 15th Intl. Symp. of Electromagnetic Theory*, St. Petersburg, Russia, (May 23-26, 1995).

NOVEL ANTENNAS AND DEVICE TECHNOLOGY

BROADBANDING GUIDE LINES OF STRIP-ELEMENT MICROSTRIP PHASED ARRAYS

A. Hessel

Weber Research Institute
Department of Electrical Engineering
Polytechnic University
Route 110, Farmingdale, NY 11735

ABSTRACT

The probe-fed microstrip phased array strip-element dates back to the 1970's. Although, as a result of the work by Byron and Laughlin, the element is known to possess broadband properties, little or no specific information has been forthcoming, about the strip-element phased arrays scan-frequency characteristics, capabilities and radiating aperture design.

This paper will survey the Polytechnic effort to address these issues. Initially, characteristics of the harmful scan resonances of these arrays will be considered and means to avoid them, an essential component of the broadbanding procedure. While the E-plane scan response confirms the previous findings, the H-plane scan exhibits a novel type of array blindness which is due to excitation of a strip-line mode of the mutually coupled strip-element array. It will then be shown that the broadband property of the probe-fed phased array strip-element is due to an inherent double-tuning mechanism that is revealed only after the scan resonance effects have been eliminated from the desired scan-frequency coverage. A systematic, computer-aided broadbanding design procedure, not detailed here, has been developed. Examples will be presented of scan-frequency performance of two phased array radiating apertures designed for 40° conical scan coverage over a ±15% bandwidth, and for 50° conical scan coverage over ±10% frequency band, with SWR < 2.

I. INTRODUCTION

The use of the phased array strip element was suggested by Cheston [1]. Byron [2], using simulators and small test arrays, realized bandwidths on the order of ±6% with SWR≤3 over a 45° E- and H-plane scan coverage. When employing a "single probe" feed per element, as opposed to a balanced feed (probe pair) arrangement, he found it necessary to resort to grounding screws in order to eliminate undesirable H-plane scan resonances.

Best results were obtained by Laughlin [3]. Using an approximate analysis for the H-plane scan, along with simulators and small test arrays he achieved with a balanced probe-feed, using a WAIM sheet an octave bandwidth (0.5 GHz to 1 GHz) with SWR≤3 over a 60° E and H-plane coverage. E-plane scan analysis was carried out at the Polytechnic [4] and scan-bandwidth trade-offs were found: the larger the bandwidth the smaller the scan range and vice versa. Until recently no comprehensive investigation has been reported of the scan-frequency characteristics and limitations of the probe-fed strip element phased arrays. In 1995 this void was filled by [5].

Directions for the Next Generation of MMIC Devices and Systems
Edited by N. K. Das and H. L. Bertoni, Plenum Press, New York, 1997

A subsequent paper submitted for publication [6] deals with a systematic computer-aided radiating aperture design of such arrays. We shall describe the broadbanding guidelines based on [5], and only touch upon some material from [6].

II. EXAMPLES OF STRIP-ELEMENT ARRAYS, PHASED ARRAY MODEL

Figure 1 shows an example of a strip-element array. It is taken from a paper by Byron [2]. The radiating elements are thin, conducting strips on a grounded dielectric substrate. Each strip is excited by a linear array of coaxially-fed probe pairs in antiphase (balanced feed). In other test arrays (Fig. 2) Byron used, instead of probe pairs, "single" feed probes (unbalanced feed). In this case he found it necessary to resort to grounding screws to eliminate undesirable H-plane resonances. The strip-element phased array model that was used for analysis in [5] is shown in Fig. 3. It consists of periodically spaced (spacing d) conducting strips of infinitesimal thickness on an infinite, grounded, lossless dielectric substrate of thickness h and relative permittivity ϵ_r. Each strip is excited by a periodic linear array of offset probes with spacing b. The overall probe grid is rectangular. The probes are modelled by ribbon currents with no variation in z (thin substrate approximation). These currents are of equal amplitude and progressive phase. By varying the phase across the strips (E-plane) and along the strips (H-plane) the beam can be pointed in any direction within the hemisphere, Fig. 4 defines the scan angles with respect to the array geometry. The analysis, presented in [5] will not be repeated here. Instead we shall concentrate on the relevant issues.

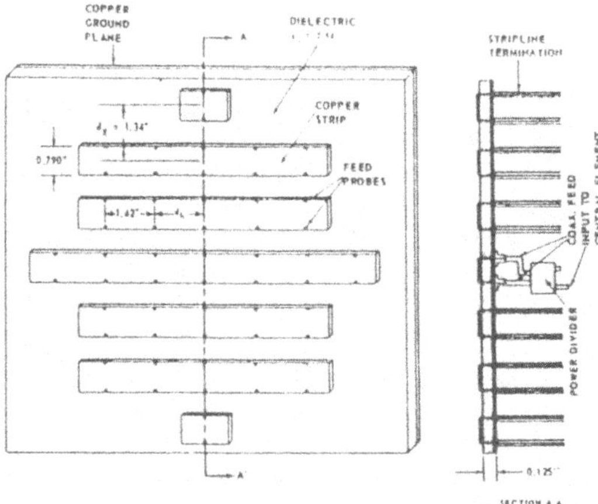

Fig. 1: 29 element array of strip radiators with balanced feed. From E.V. Byron, "A New Flush Mounted Antenna for Phased Array Applications," in "Phased Array Antennas," Oliner and Knittel, Eds., Artech House, 1972.

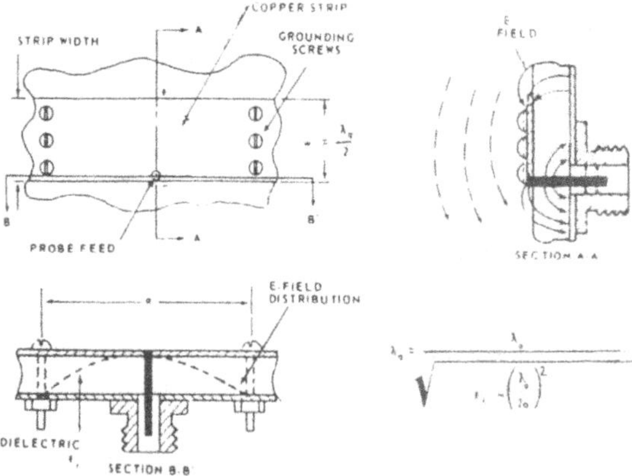

Fig. 2: Single probe strip array radiator. From E.V. Byron, "A New Flush Mounted Antenna for Phased Array Applications," in "Phased Array Antennas," Oliner and Knittel, Eds., Artech House, 1972.

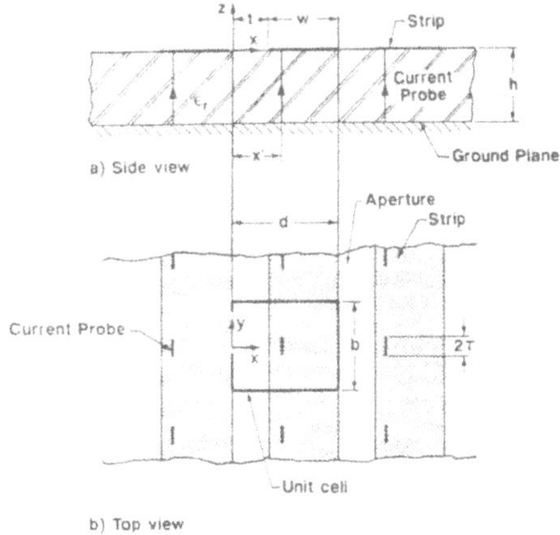

Fig. 3: Strip-element microstrip array model.

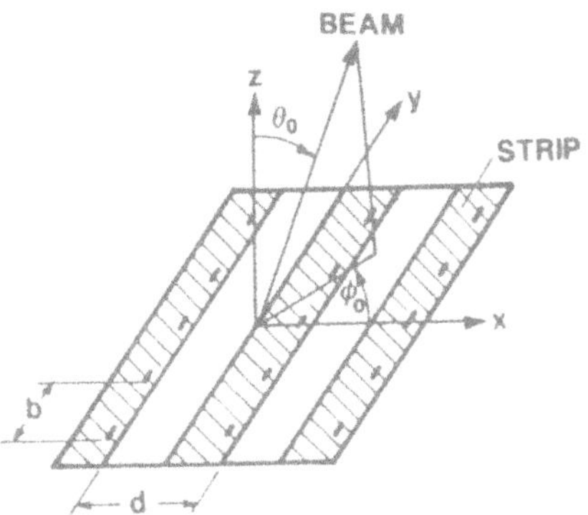

Fig. 4: Strip-array geometry and scan angles.

III. SCAN RESONANCES AND GUIDED WAVES

1. Scan Resonances and Phase Synchronism

In order to broadband a phased array it is first necessary to consider the scan range limitations. In addition to paying attention to the grating lobe's incipience, it is essential to exercise control over any large mismatch producing scan resonances falling within the desired scan-frequency coverage. The origin of such resonances should be identified in order that appropriate countermeasures can be taken. Scan resonances can be traced to a phase synchronism of one of the Floquet modes, i.e., one of the harmonics of the impressed steering phase, with a guided wave supported by the passive array structure. A scan resonance is accompanied by a sharp amplitude increase of the synchronous Floquet mode. This, in turn, distorts the desired radiating element aperture distribution giving rise to a large mismatch. A one dimensional scan phase synchronism condition is

$$k_0 \sin\theta_0 \pm \frac{2\pi}{p} = \beta_{GW}$$
$$or \qquad (1)$$
$$\sin\theta_0 \pm \frac{\lambda_0}{p} = \pm\frac{\beta_{GW}}{k_0}$$

where θ_0 is a scan angle off broadside, d the element spacing and β_{GW} the phase constant of a guided wave propagating in the scan plane on the array structure. For a microstrip element array the relevant guided waves are slow so that $\frac{\beta_{GW}}{k_0} > 1$.

Figure 5 is a graphical illustration of the phase synchronism in sin θ space. When the main beam is at broadside, the ± 1 grating lobes are at ± λ/d. As the beam is scanned to an angle θ_0, the main beam moves a distance sin θ_0 to the right and all grating lobes follow in unison by the same distance. If the array supports a slow guided wave with a propagation constant β_{GW}, then at a scan angle θ_0^{SR} the n=-1 grating lobe intercepts $\frac{\beta_{GW}}{k_0}$ and a scan resonance occurs.

2. Guided Waves in Strip-Element Arrays

In order to facilitate the physical interpretation of the computed results let us intuitively consider the guided waves supported by microstrip strip-element arrays, and relevant to the E and H-plane scan resonances.

Fig. 6 considers the guided waves associated with E-plane scan resonances. Since, for E-plane scan the E-field is across the strips the strip loading is light and the dominant guided wave is expected to be a TM_0 surface wave supported by the unloaded, grounded dielectric substrate slab, and perturbed by periodicity. Therefore

$$\beta^E_{GW} \sim \beta_{SW} \tag{2}$$

Actually this guided wave is a leaky wave. Fig. 7 considers the expected guided waves propagating along the strips and relevant to the H-plane scan resonances. In this case the current flow is along the strips, the strip loading is heavy, no matter how narrow the strips, and the field is localized to the vicinity of the strips. Thus, we expect the guided wave relevant to the principal H-plane scan resonance to be essentially the dominant mode of an isolated strip-line perturbed by mutual coupling. Thus,

$$\beta^H_{GW} \sim k_0\sqrt{\epsilon_r}\ . \tag{3}$$

3. Scan Resonances. Computed Results

Figure 8 shows an example of guided wave loci in the normalized (k_x, k_y) plane for two strip width w=0.1d and 0.5d, with other parameters listed in the legend. The wavenumber k_x corresponds to variation across the strips, k_y along the strips. The figure also includes two circles, the smaller with radius β_{SW}/k_0 corresponds to the TM_0 surface wave supported by the unloaded, grounded substrate slab. The larger circle has the radius $\sqrt{\epsilon_r}$. It is seen that the guided wave loci relevant to the E-plane scan resonances are nearly concentric circular arcs close to the surface wave circle. For the narrower strips the arcs are practically coincident with the SW circle. For the wider strips the guided wave is somewhat slower due to additional strip loading. These guided waves are leaky waves and the arcs correspond to the real part of their propagation constant.

Figure 8 also includes the guided wave loci relevant to the H-plane scan resonances. These are essentially straight lines indicating a weak dependence of the E-plane phasing of the propagation constant along the strips. This confirms our physical intuition of low mutual coupling between the strips and of the localization of periodic strip line mode fields. Furthermore, the location of the guided wave loci in proximity of the $\sqrt{\epsilon_r}$ circle confirms our expectation that the propagation constant of the dominant periodic strip-line mode is

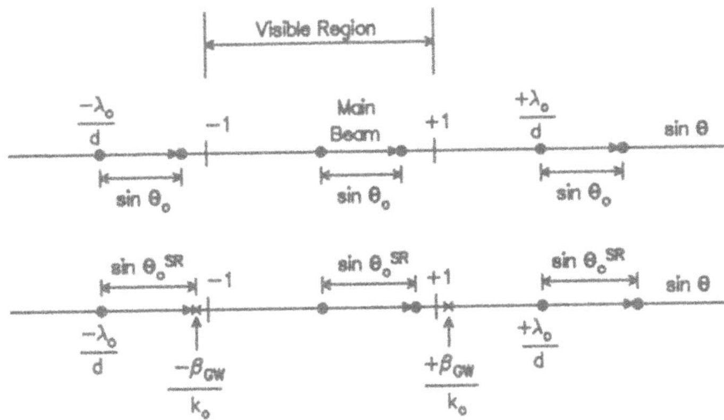

Fig. 5: Graphical interpretation of scan-resonances.

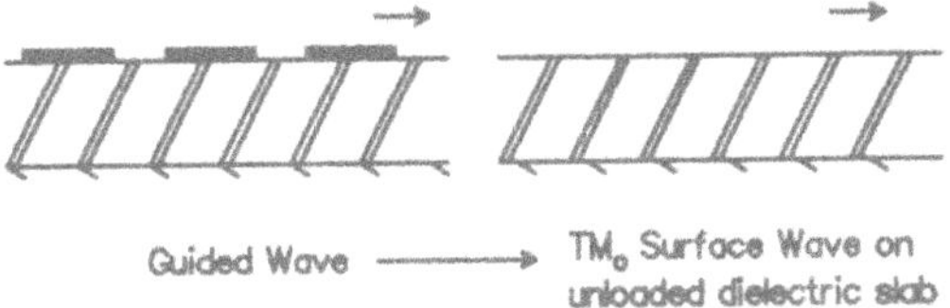

Fig. 6: Dominant guided waves relevant to E-plane scan-resonances.

Fig. 7: Dominant guided waves relevant to H-plane scan-resonances.

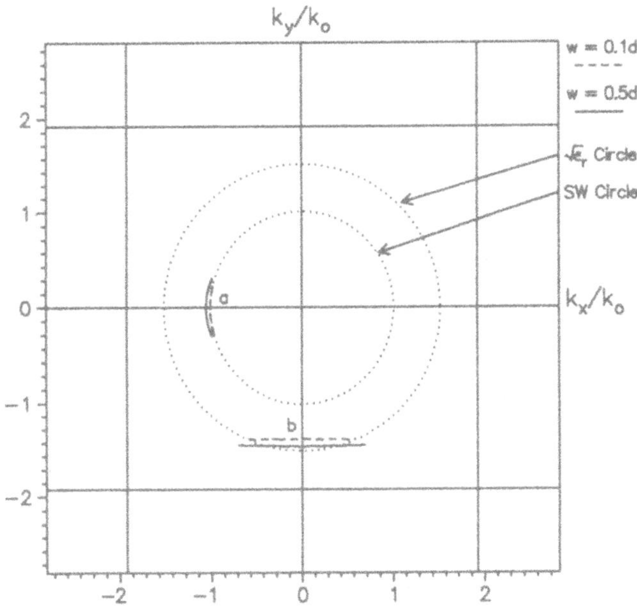

Fig. 8: Loci of guided wave wave-numbers in the normalized (k_x, k_y) plane. $d=b=0.52\lambda_0$, $h=0.04\lambda_0$, $\epsilon_r=2.3$, for $w=0.1d$ and $0.5d$.

'a' dominant leaky wave

'b' dominant periodic strip-line mode

$\frac{\beta_{GW}^{H}}{k_0} \sim \sqrt{\epsilon_r}$. The periodic strip-line mode and its relevance to the phased array scan resonances does not seem to have been previously recognized. These modes may also be relevant to patch element phased arrays with a large H-plane patch dimension.

We now realize that the E-plane scan resonances are expected to occur approximately whenever the (-1,0) grating lobe intersects the surface wave circle, while their H-plane scan counterparts should occur approximately when the (0,-1) grating lobe crosses the $\sqrt{\epsilon_r}$ circle. This is shown in Figs. 9 and 10.

These features can be utilized to eliminate the scan resonances from a given scan-frequency coverage by appropriately reducing the element spacing at the highest operating frequency in order to move the grating lobes away from the unit circle. This will be discussed later on. We further observe that the E-plane element spacing can be made significantly larger than the H-plane spacing, because the surface wave circle radius is smaller than that of the $\sqrt{\epsilon_r}$ circle. This feature becomes significant in broadbanding design procedure.

We shall now illustrate the effects of scan resonances. Figure 11 shows the E-plane scan variation of the unmatched active resistance R_a and reactance X_a for a number of substrate thickness values h/λ in a square lattice array with $d/\lambda=0.52$ spacing, $\epsilon_r=2.3$ and strip width $w=d/2$. For each value of h/λ, R_a exhibits a peak-dip combination which tends to broadside with increasing amplitude as h/λ increases. These scan resonances occur closer to broadside than the grating lobe incipience at 67°. The active reactance X_a also exhibits a resonant behavior. The reason that the X_a locus does not pass through zero at resonance is due to the higher Floquet mode contributions which are nonresonant. The dark arrows correspond to the (-1, 0) grating lobe crossing the SW circle. The light arrows indicate the grating lobe crossing the leaky wave arc. Fig. 12 exhibits the E-plane scan variation of the magnitude of the active reflection coefficient Γ_a^m for a broadside matched array with geometry of Fig. 11. One observes that the E-plane scan resonances do not produce a (near) total reflection, as in the case of microstrip patch phased array element, but the mismatch is severe. The matched array Γ_a^m maxima occur close to the (-1, 0) grating lobe interception of the SW circle.

A drastically different response is found in the H-plane scan (Fig. 13). For the array geometry of Fig. 11 one finds that the active resistance R_a is relatively flat. But the active reactance exhibits for each value of h/λ an extremely sharp peak in the scan range from 25° to 28°, which compares well with the expected 23.9° according to (1) and (3).

Fig. 14 shows the resonant peaks on expanded scale. In each case X_a changes sign across the resonance and goes to infinity at resonance, giving rise to total reflection. Since the associated system determinant goes through zero at each peak and changes sign across the peak, one deduces that the resonance corresponds to a lossless guided wave which is termed periodic strip line mode. In contrast to the E-plane scan resonances, the angular location of the H-plane scan resonances is weakly dependent on the substrate thickness.

Figs. 15 and 16 show that scan resonances are suppressed by reducing the element spacings e.g. in such a way that the (0,-1) grating lobe intercepts the $\sqrt{\epsilon_r}$ circle at $\theta_0=90°$. Here again, for simplicity, a square grid array is used with $d=0.395\lambda$, corresponding to $\epsilon_r=2.3$ and with the strip width $w=d/2$. We observe that the E-plane scan resonance has been suppressed with both R_a and X_a are relatively flat (Fig. 15). This is not unexpected because for the chosen spacing the (-1,0) grating lobe does not reach the SW circle. In the H-plane scan both R_a and X_a are well behaved (Fig. 16), except near endfire where the (0,-1) grating lobe just reaches the $\sqrt{\epsilon_r}$ circle at $\theta_0=90°$. Most importantly it is seen that the reduction of H-plane spacing eliminates the sharp peaks of X_a. We conclude that by appropriately reducing the element spacing at the highest operating frequency, the effects of scan resonances can be suppressed from the required scan-frequency coverage.

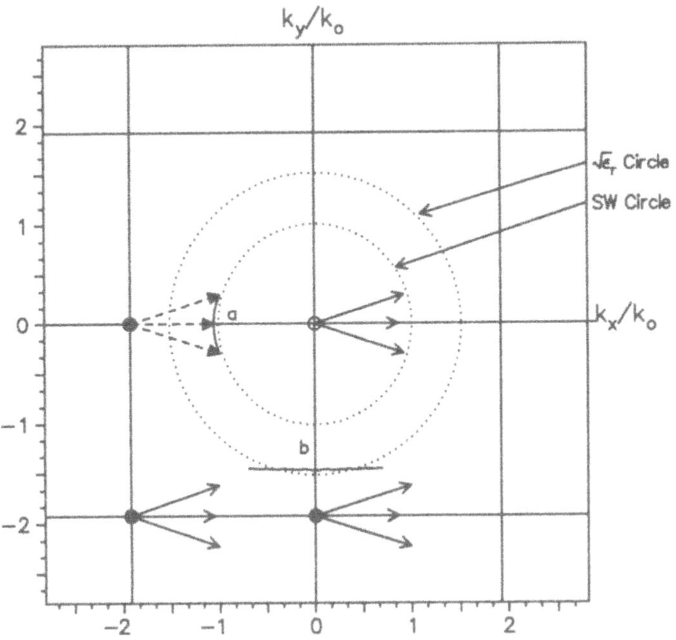

Fig. 9: Loci of scan-resonances in and off E-plane. $d=b=0.52\lambda_0$, $h=0.04$, $e_r=2,3$, $w=0.5d$.

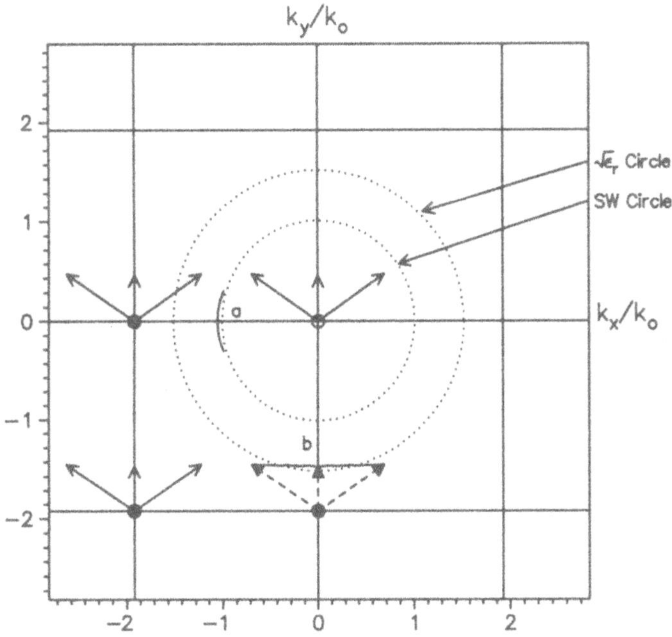

Fig. 10: Loci of scan-resonances in and off H-plane. $d=b=0.52\lambda_0$, $h=0.04\lambda_0$, $e_r=2.3$, $w=0.5d$.

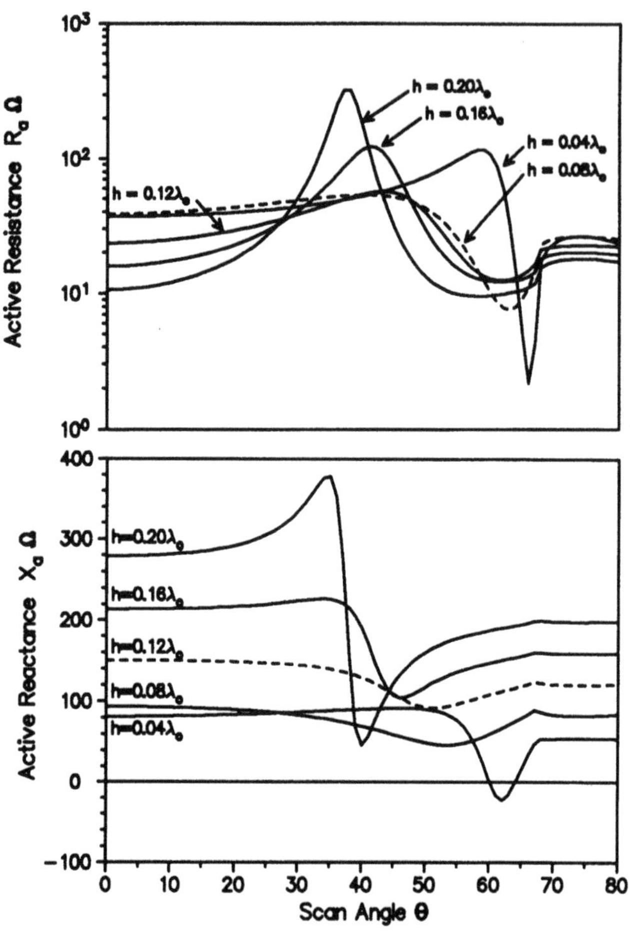

Fig. 11: E-Plane scan variation of the unmatched active resistance (R_a) and active reactance (X_a). Independent parameter h. $d=0.52\lambda_0$, $b=0.52\lambda_o$, $w=d/2$, $d_p=w/8$, $e_r=2.3$.

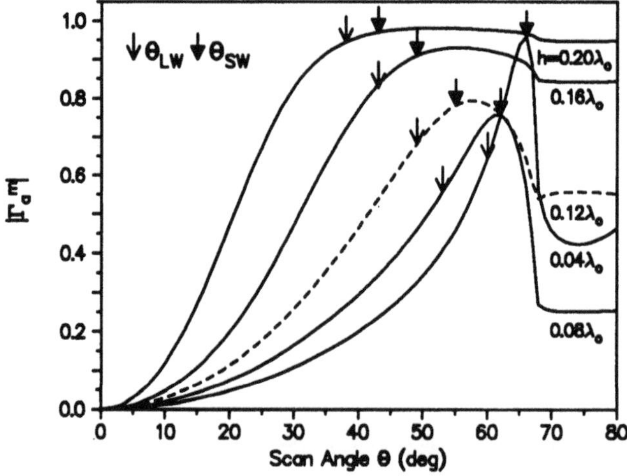

Fig. 12: E-plane scan variation of magnitude of broadside matched active reflection coefficient. $d=0.52\lambda_0$, $b=0.52\lambda_0$, $w=d/2$, $e_r=2.3$.

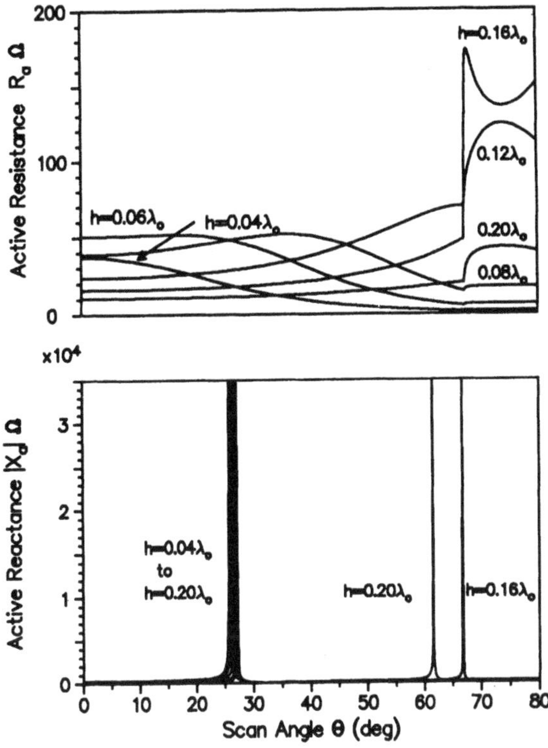

Fig. 13: H-plane scan variation of the unmatched active resistance (R_c) and active reactance (X_a). Independent parameter h. $d=0.52\lambda_0$, $b=0.52\lambda_0$, $w=d/2$, $\epsilon_r=2.3$.

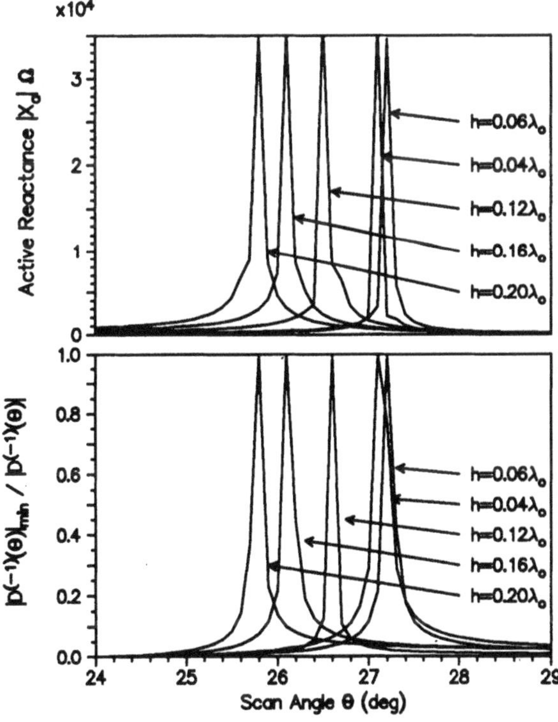

Fig. 14: Expanded view of H-plane scan variation of unmatched active reactance (top) and the corresponding inverse of the normalized system determinant (bottom). Independent parameter h. $d=0.52\lambda_0$, $b=0.52\lambda_0$, $w=d/2$, $\epsilon_r=2.3$.

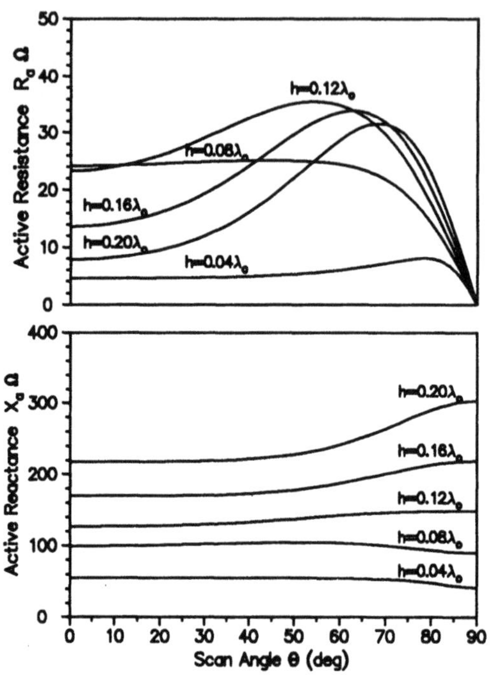

Fig. 15: Smaller probe spacing, E-plane scan variation of the unmatched active resistance (R_a) and reactance (X_a). Independent parameter h. $d=0.395\lambda_0$, $b=0.395\lambda_0$, $w=d/2$, $e_r=2.3$.

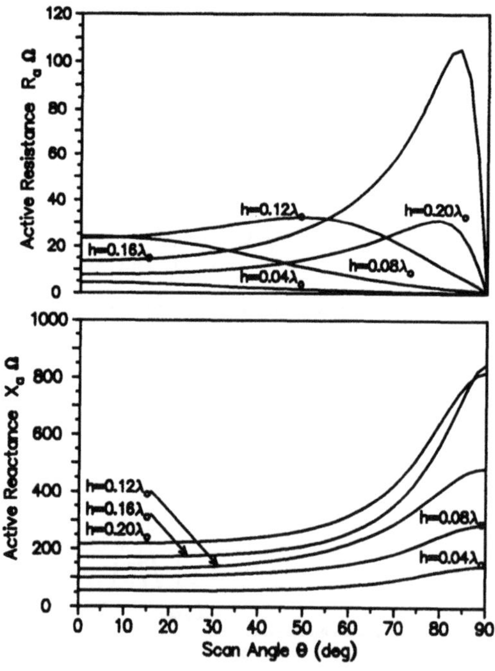

Fig. 16: Smaller probe spacing, H-plane scan variation of the unmatched active resistance (R_a) and reactance (X_a). Independent parameter h. $d=0.395\lambda_0$, $b=0.395\lambda_0$, $w=d/2$, $e_r=2.3$.

IV. ELEMENT BROADBANDING

At this point we will consider element broadbanding proper. Since at the outset the substrate thickness and consequently β_{SW} is unknown, the initial E- and H-plane spacing is selected independently of h to allow the (0,-1) grating lobe to intercept the $\sqrt{\epsilon_r}$ circle at $\theta_0=90°$, at the highest operating frequency, and thereby eliminate the scan resonances from essentially the entire hemisphere via

$$\frac{d}{\lambda^+}=\frac{b}{\lambda^+}=\frac{1}{1+\sqrt{\epsilon_r}} \; . \tag{4}$$

With the scan resonances excluded, the dynamic range of the unmatched active reflection coefficient Γ_a^u should not be excessive. In order to further reduce the frequency variation of Γ_a^u, the broadbanding approach concentrates now on the equalization of $\Gamma_a^u(f)$ at the high (f^+) and the low (f^-) frequency ends of the desired band. Basic to the success of the broadbanding procedure is the property of the probe-fed phased array strip element that at a given scan direction and for element spacing for which the effects of scan resonances have been suppressed, a substrate thickness-stripwidth combination (h,w) can be found for which $\Gamma_a^u(f^+)=\Gamma_a^u(f^-)$, provided that the bandwidth is not excessive. To illustrate this point Fig. 17 exhibits the effect of h and w on the complex Γ_a^u over $\pm 10°$ frequency range. Again, for simplicity, a square grid is chosen for $\epsilon_r=2.3$ according to (4), whereby $d/\lambda_0{}^c=0.36$. The figure shows loci of complex Γ_a^u (with respect to 50Ω line, at broadside scan as a function of frequency $0.9f_c \le f \le 1.1f_c$ for strip width $w=0.224\lambda_0{}^c$ and three values of substrate thickness. The frequency increases clockwise. One observes that double-tuned loops are formed. For $h/\lambda_0{}^c=0.075$ the loop is still open, while for $h/\lambda_0{}^c=0.085$ it closes prematurely. Only for $h/\lambda_0{}^c=0.080$ the loop closes so that $\Gamma_a^u(1.1f^+)=\Gamma_a^u(0.9f^-)$, with an associated reduction of the dynamic range of $\Gamma_a^u(f)$. Similar behavior is found where h is kept fixed and w varies. One further observes that double-tuned loops are formed at $\theta_0=40°$ in either the H- or the E-plane, but that the (h,w) combination required for loop closure is scan dependent. The design procedure that is the topic of [6] minimizes the spread of the three (h,w) combinations required for the loop closure, i.e. $\Gamma_a^u(f^+)=\Gamma_a^u(f^-)$, at each of the extreme scan directions. This is accomplished by increasing the E-plane spacing which originally was chosen too small via relation (4). After the spread of the three (h,w) combinations has been reduced one proceeds to choose a <u>single</u>, compromise, (h,w) combination which <u>equalizes</u> the magnitude

$$|\,\Gamma_a(f^+)-\Gamma_a(f^-)\,|$$

along the three extreme scan directions. Fig. 18 (top) shows the scan-frequency performance of two strip-element phased arrays designed by the described procedure for conical scan coverage of 40° over a ±15% frequency band (top) and 50° over ±10% band (bottom). The H-plane spacing was selected to exclude the periodic strip-line mode scan resonance from the entire hemispherical coverage. The E-plane spacing is larger and was chosen to minimize the spread of the three (h,w) combinations required for loop closure i.e. $\Gamma_a(f^+)=\Gamma_a(f^-)$ at broadside, $\theta_{0,\max}^E$ *and* $\theta_{0,\max}^H$. The final (h,w) combination was chosen to equalize the error, i.e.,

$$|\,\Gamma_a(f^+)-\Gamma_a(f^-)\,|$$

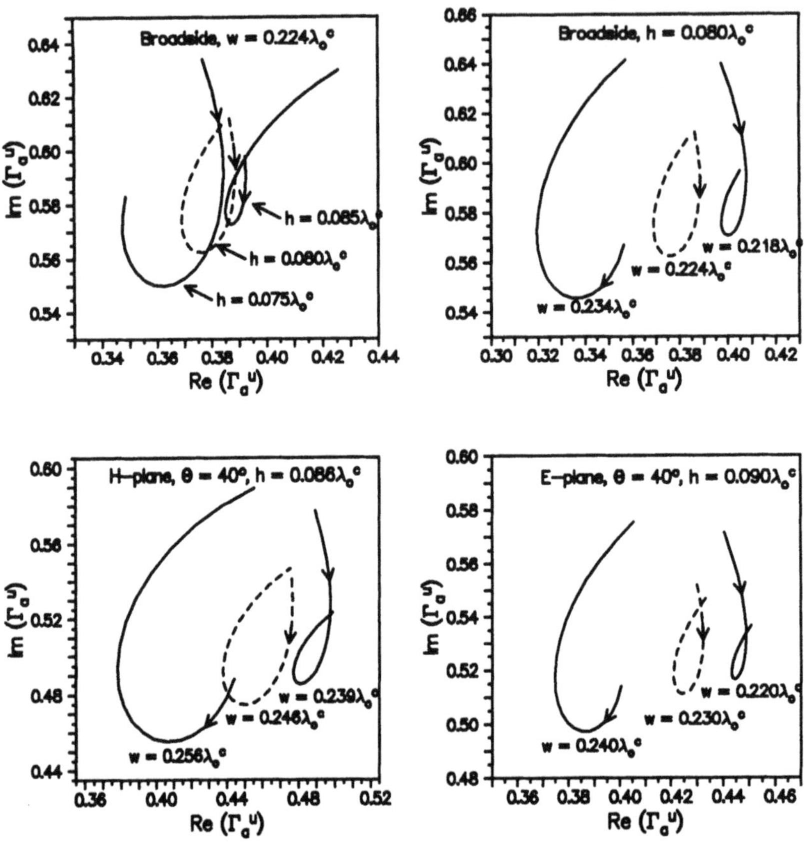

Fig. 17: Loci of complex $\Gamma_a{}^u$ with variable frequency $0.9f^c \leq f \leq 1.1f^c$. Broadside scan: w fixed and parameter h (top left), h fixed and parameter w (top right). Fixed h and parameter w: θ=40° in H-plane (bottom left) and E-plane (bottom right). $d=0.36\lambda_0{}^c$, $b=0.36\lambda_0{}^c$, $\epsilon_r=2.3$.

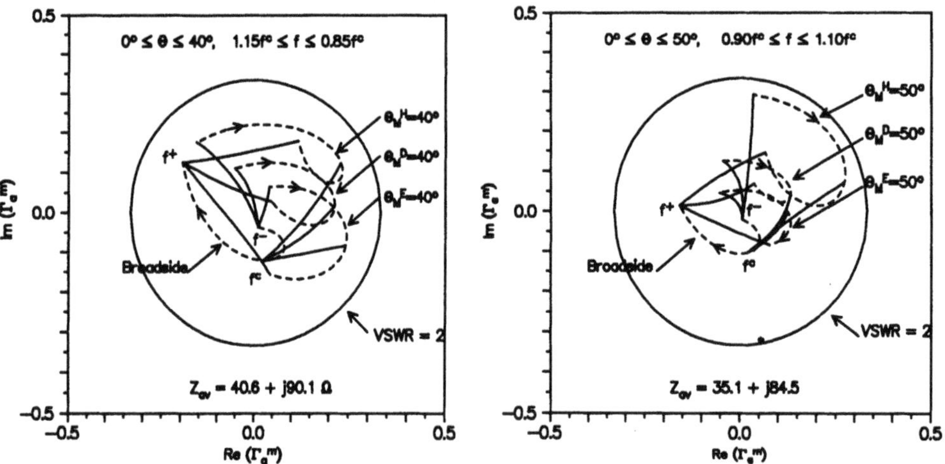

Fig. 18: Scan-frequency performance in complex Γ-plane. Loci of $\Gamma_a{}^m$ in E-, D-, and H-scan plane with variables scan angle θ (full lines) and frequency f (dashed lines).

$d=0.44\lambda_0{}^c$, $b=0.35\lambda_0{}^c$, $h=0.090\lambda_0{}^c$, $w=0.249\lambda_0{}^c$, $\epsilon_r=2.3$ (top).
$d=0.36\lambda_0{}^c$, $b=0.28\lambda_0{}^c$, $h=0.077\lambda_0{}^c$, $w=0.156\lambda_0{}^c$, $\epsilon_r=5.0$ (bottom).

for the above three scan directions. We see here (Fig. 18 top) loci in the complex Γ-plane loci of matched active reflection coefficient (with match to the average Z_a^u over the scan volume and frequency band). We observe three "spiders" for E, D and H-plane scan at center frequency f_c and at $0.85f_c$ and $1.15f_c$. The dashed loops joining the corresponding points of the "spiders" yield the frequency variation at broadside, $\theta^E = 40°$ *and* $\theta^H = 40°$. The loci lie within the SWR=2 circle. The frequency loci have a double tuned loop character, but the loops do not close, a consequence of the compromise choice of the final (h,w) combination. Fig. 18 (bottom) presents the scan-frequency performance of an array with $\epsilon_r=5.0$, $d=0.38\lambda_0^c$, $b=0.27\lambda_0^c$, $h=0.080\lambda_0^c$, $w=0.153\lambda_0^c$ designed for 50° conical scan coverage over ±10% frequency band.

V. SUMMARY

It was shown that:

1. Scan resonances occur when a guided wave is excited by a synchronous higher order Floquet mode (grating lobe).
2. The relevant guided waves are:
   a) For the E-plane scan they are of the leaky wave type (perturbed TM_0 surface wave on the unloaded, grounded dielectric substrate slab).
   b) For the H-plane scan they are (novel) periodic strip line modes (perturbed by mutual coupling dominant mode of an isolated strip line).
3. Initial choice of substrate-thickness-independent element spacing to suppress the E- and H-plane scan resonances was presented.
4. Suppression of scan resonances allows for double-tuning by substrate thickness-strip width interplay (without recourse to mode suppressor pins).
5. The (h.w) combinations that yield double-tuning for a given bandwidth are scan dependent.
6. The spread of three (h,w) combinations for three extreme scan directions is reduced by increasing the E-plane spacing.
7. The final, single (h,w) combination is chosen to equalize

$$|\Gamma_a^u(f^+,\theta,\phi) - \Gamma_a^u(f^-,\theta,\phi)|$$

   along the three extreme beam pointing directions.
8. Scan-frequency performance of two probe-fed strip-element microstrip phased arrays was presented. The first was designed for 40° conical scan coverage over ±15% bandwidth, the second for conical scan volume of 50° over ±10% band. The SWR is less than 2:1.
9. The phased array strip element is linearly polarized, changing from vertical in the E-plane to horizontal in the H-plane scan.

Final Remarks

The broadbanding procedure is not a brute force algorithm. One first determines the physical mechanism of scan resonances and means to avoid them. After scan resonances have been removed from a given scan-frequency coverage, the double-tuned element property is disclosed and utilized for element broadbanding.

ACKNOWLEDGEMENT

The author gratefully acknowledges the IEEE permission for reprinting a number of figures from reference [5].

REFERENCES

[1] T.C. Cheston, "Collings Radiator," Johns Hopkins University, Applied Physics Laboratory, Internal Memo MRT-0-512, Sept. 1968.

[2] E.V. Byron, "A New Flush Mounted Antenna for Phased Array Application," in Phased Array Antennas, Oliner and Knittel, Eds., Dedham, MA, Artech House, 1972.

[3] G.J. Laughlin, "Wideband Solid State Phased Array Antenna Development a UHF," Johns Hopkins Univ. Applied Physics Laboratory, Technical Memorandum TG 1278, AD-A015090, July 1975.

[4] C.C. Liu, J. Shmoys, and A. Hessel, "E-Plane Performance Trade-Offs in Two-Dimensional Microstrip-Patch Element Phased Arrays," IEEE Trans. Antennas and Propagation, vol. AP-30, pp. 1201-1206, Nov. 1982.

[5] C.P. Nehra, A. Hessel, J. Shmoys, and H.J. Stalzer Jr., "Probe-Fed Strip-Element Microstrip Phased Arrays: E- and H-Plane Scan Resonances and Broadbanding Guidelines," IEEE Trans. Antennas and Propagation, Vol. AP-43, pp.1270-1280, Nov. 1995.

[6] C.P. Nehra and A. Hessel,"Computer-Aided Design of Broadband Microstrip-Element Phased Arrays," submitted for publication to IEEE Trans. AP.

U-SLOT PATCH WIDEBAND MICROSTRIP ANTENNA

K. F. Lee,[1] K. M. Luk,[2] T. Huynh[3], K. F. Tong[2], and R. Q. Lee[4]

[1]University of Missouri-Columbia
[2]City University of Hong Kong, Hong Kong
[3]Decibel Products, Dallas, Texas
[4]NASA Lewis Research Center, Cleveland, Ohio

INTRODUCTION

Bandwidth broadening techniques for microstrip antennas have been studied for a number of years. A popular method is the use of parasitic elements, either in stacked geometry[1] or in coplanar geometry[2]. The stacked geometry involves more than one patch and more than one dielectric layer, with the result that the antenna thickness is increased. The coplanar geometry has one dielectric layer but requires parasitic patches adjacent to the fed patch, thereby increasing the lateral size of the antenna.

Recently, the authors introduced a wideband patch antenna which has only one patch and one dielectric layer, in the form of a coaxially-fed rectangular patch with a U-shaped slot. Some of the results are presented in this paper.

EXPERIMENTAL RESULTS

The general geometry of a coaxially-fed rectangular patch with a U-shaped slot is shown in Figure 1. Experimental results were obtained by Huynh and Lee[3] at the center frequency of about 900 Mhz and by Lee et al.[4] at the center frequency of about 4.5 Ghz. In both cases, the dielectric material was foam with a relative permittivity close to unity. Depending on the design, the structure can either perform as wideband antenna or as a dual-band antenna. Representative dimensions of the antennas at these frequencies are given in Table 1.

Table 1 Representative dimensions (in mm) of U-Slot Patches at 900 Mhz and at 4.5 Ghz

Antenna	W	L	F	W_s	a	b	c	h
1 (wideband)	219.7	124.5	62.2	68.6	2.29	19.4	10.2	26.9
2 (wideband)	35.5	26.0	15.0	12.0	2.2	4.3	2.1	5.0
3 (dual-band)	40.0	26.0	15.0	14.0	5.2	4.3	2.1	5.0

Directions for the Next Generation of MMIC Devices and Systems
Edited by N. K. Das and H. L. Bertoni, Plenum Press, New York, 1997

Antenna 1

This antenna performs as a wideband antenna with center frequency around 900 Mhz. The measured impedance loci, VSWR versus frequency and copolarization patterns in the E and H planes were reported in Huynh and Lee[3] and are not repeated here. Suffice it to point out that, for h=26.9 mm ($\cong 0.08\lambda$), the impedance bandwidth (VSWR=2) was over 40%, with good pattern characteristics.

Antenna 2

This antenna performs as a wideband antenna with center frequency around 4.5 Ghz. The measured VSWR versus frequency is given in Figure 2. The impedance bandwidth is about 32.4%. The polarization of this antenna is linear, with the E-plane parallel to the vertical slots and the H plane parallel to the horizontal slot. The co-polarization patterns are stable: the half-power beamwidths in the E plane are 63.9° at 3.72 Ghz and 58.0° at 5.16 Ghz. In the H plane, they are 66.3° at 3.72 Ghz and 64.9° at 5.16 Ghz. The measured broadside gain versus frequency is given in Figure 3. The maximum gain is 8.5 dB occurring at 3.90 Ghz. The 3 dB bandwidth for gain is from 3.67 Ghz to 5.15 Ghz or 33.6%. In the E-plane, the cross polarization levels are below 23 dB across the VSWR=2 band. However, in the H plane, at 4.28 Ghz, the highest level is about -18 dB occurring at 45° from broadside. It increases to -12 dB at 3.72 Ghz and at 4.82 Ghz.

Antenna 3

In addition to wide bandwidth operation, the U-slot patch can also function as a dual-band antenna. It appears that, starting with a wideband design, if one of the parameters such as patch width or feed position is varied, the broadband characteristic is changed into a dual-band characteristic. An example is antenna 3 of Table 1; the measured VSWR versus frequency curve is shown in Figure 4. The separation of the center frequencies of the bands is 1.16 Ghz, which is about 30% when measured with respect to the lower center frequency. Note that the impedance bandwidths of the lower and upper bands are about 9% and 12% respectively.

SIMULATION RESULTS

We have recently obtained some simulation results of the U-slot patch antenna using two approaches. In one, the ENSEMBLE software developed by Boulder Microwave Technologies Inc. was used. This software is based on moment-method analysis. It assumes a uniform current on the coaxial probe. The VSWR versus frequency curve obtained for antenna 1 is shown in Figure 5. The impedance bandwidth is about 34%. While this is less than that measured by Huynh and Lee,[3] the simulation result confirms the wideband behaviour of the U-slot patch. The discrepancy between simulation and measured results may be due to the uniform current assumption and the idealizations in the theoretical model such as infinite ground plane.

In another approach, we developed a FDTD code for the U-slot patch antenna. Using our FDTD code for antenna 2, we obtain the VSWR versus frequency curve shown in Figure 6. This is in reasonable agreement with the measured curve of Figure 2, again confirming the wideband characteristic of the U-slot patch.

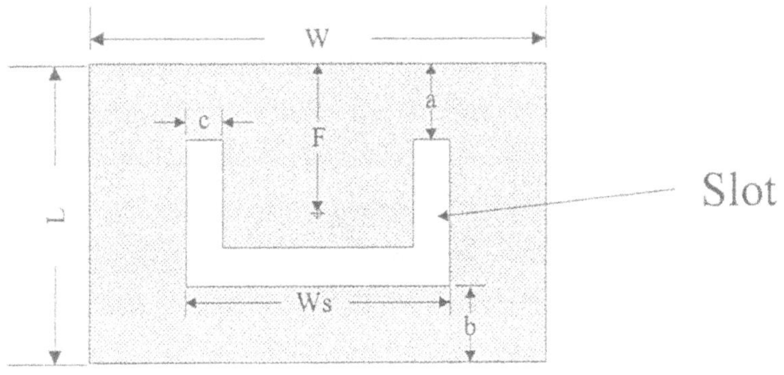

Top View

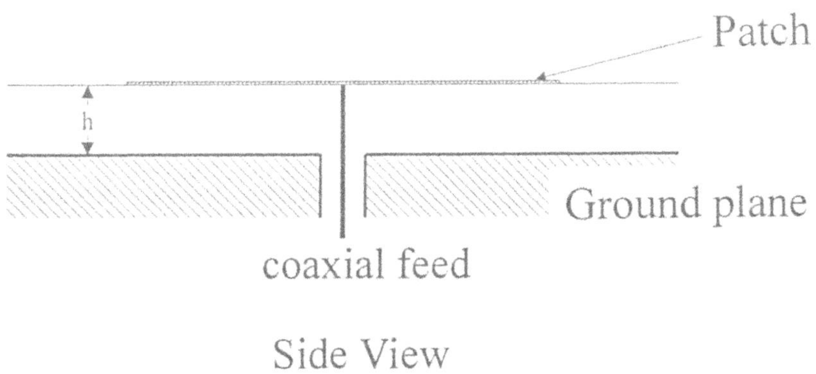

Side View

Figure 1 Geometry of the coaxially-fed rectangular patch antenna with a U-shaped slot

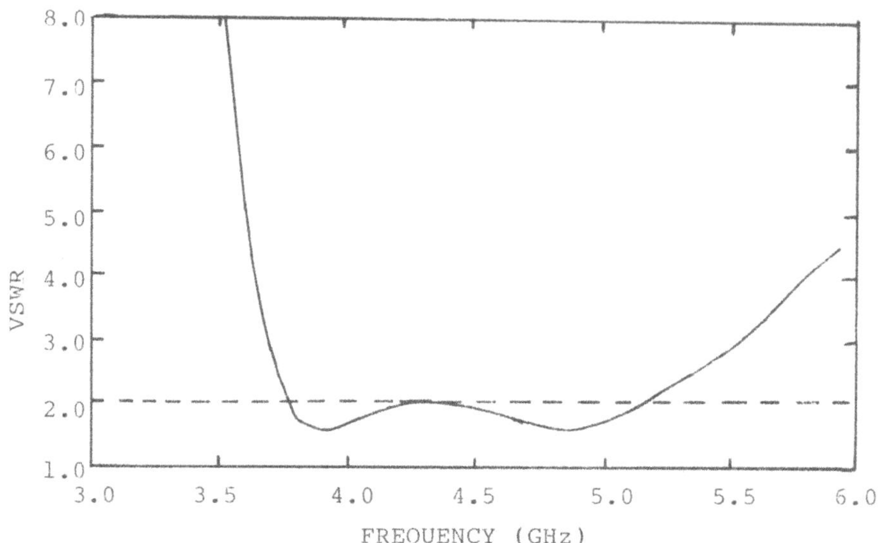

Figure 2 Measured VSWR versus frequency for antenna 2

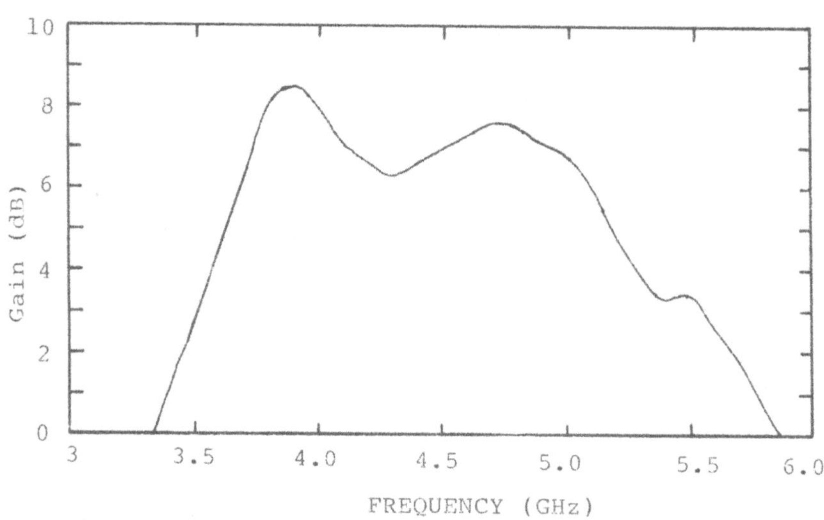

Figure 3 Measured broadside gain versus frequency for antenna 2

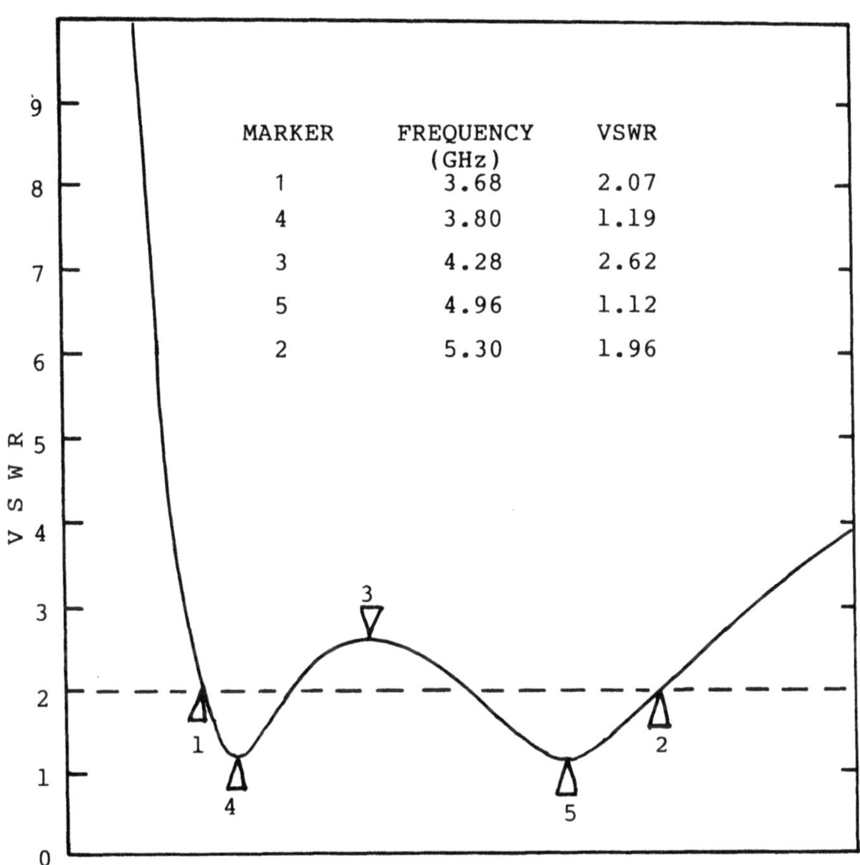

Figure 4 Measured VSWR versus frequency for antenna 3

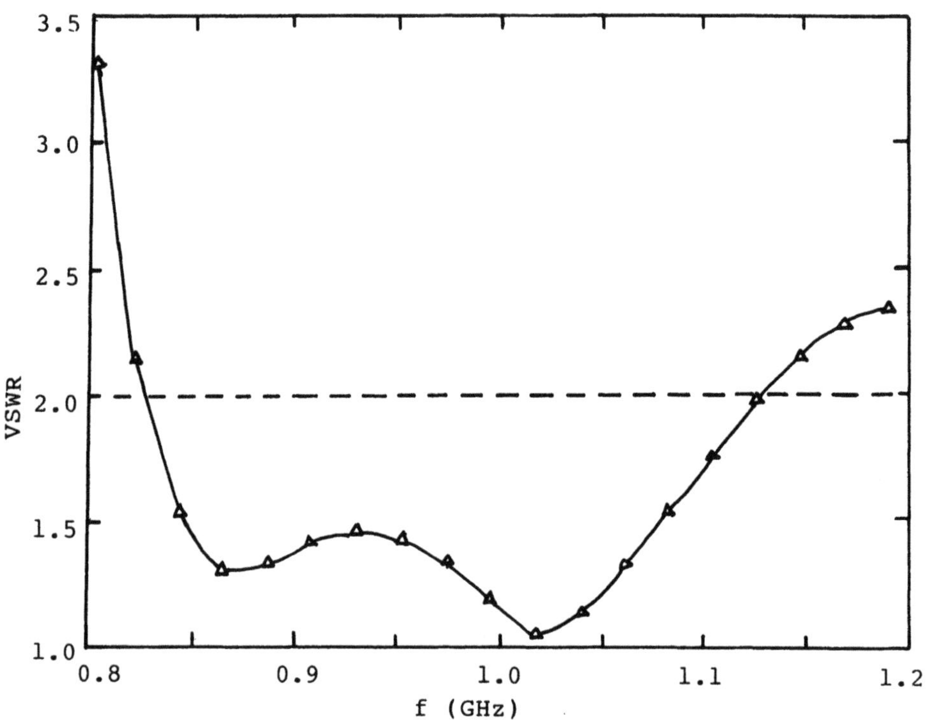

Figure 5 Simulation result of VSWR versus frequency using ENSEMBLE for antenna 1

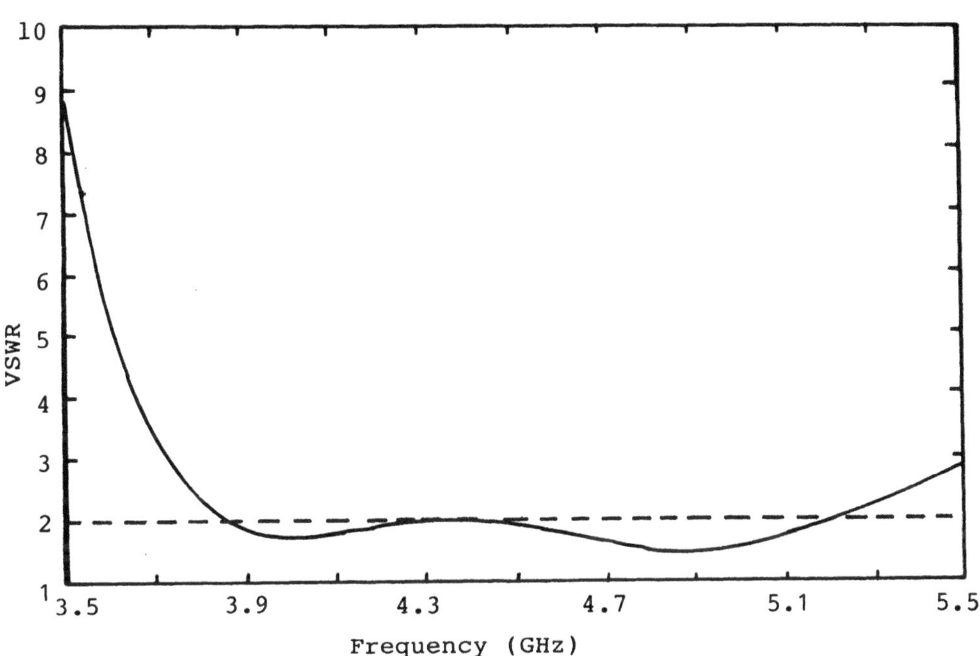

Figure 6 Simulation result of VSWR versus frequency for antenna 2 using our FDTD code

OTHER U-SLOT PATCHES STUDIED

Other U-slot patch configurations we have studied include aperture coupled U-slot rectangular patches, circular patch with a U-slot, and a circular array of such patches. The results will be reported elsewhere.

REFERENCES

1. R. Q. Lee, K. F. Lee, and J. Bobinchack, Characteristics of a two-layer electromagnetically coupled rectangular patch antenna, Electron. Lett., 23:1070 (1987).

2. W. Chen, K. F. Lee, and R. Q. Lee, Spectral-domain moment-method analysis of coplanar microstrip parasitic subarrays, Microw. Opt. Technol. Lett., 6:157 (1993).

3. T. Huynh and K. F. Lee, Single-layer single-patch wideband microstrip antenna, Electron. Lett., 31:1310 (1995).

4. K. F. Lee, K. M. Luk, K. F. Tong, and T. Huynh, Experimental study of the rectangular patch with a U-shaped slot, IEEE AP-S International Symposium Digest, 1995:10.

SURFACE WAVE MODE REDUCTION FOR RECTANGULAR MICROSTRIP ANTENNAS ON HIGH-INDEX MATERIALS

John Papapolymerou,[1] Rhonda F. Drayton,[2] and Linda P.B. Katehi[1]

[1]The Radiation Laboratory
EECS Department
The University of Michigan
Ann Arbor, MI 48109
[2]The University of Illinois at Chicago
EECS Department
Chicago, IL 60607

INTRODUCTION

Microstrip and slot antennas are used in a broad range of applications from communication systems (radars, telemetry and navigation) to biomedical systems, primarily due to their planar characteristics and low manufacturing cost[1]. With the recent development of microwave and millimeter-wave integrated circuits and the trend to incorporate all microwave devices on a single chip for low-cost and high density, there is a need to fabricate patch antennas on high-index materials such as Silicon, GaAs or InP. Because of their high index, these materials exhibit a pronounced surface wave excitation, thus leading to lower antenna efficiency, reduced bandwidth, degradation of the radiation pattern and undesired coupling between the various elements in an array design. Only a few approaches have been put forth to resolve the excitation of substrate modes in microstrip antennas, including using a substrate-superstrate configuration[2], adjusting the radius of a circular patch to a critical value[3], suspending the patch over an air cavity through the use of a membrane[4] or over closely spaced periodic holes in the substrate[5].

We have developed two different techniques in order to address the previously mentioned problems, based on the implementation of micromachining technology. In the first technique, material is removed laterally in a region under and around the patch antenna in an effort to locally reduce the dielectric constant. Experimental results have shown a significant increase in antenna efficiency, when compared with a regular rectangular patch. In the second technique, the resonant length of the patch is adjusted in order to minimize the dominant TM_0 surface mode. Because the dimensions required to suppress the surface wave are larger than the dimensions for the desired frequency of operation, dielectric material is removed in an area only under the patch. Simulation results (FDTD), have shown a significant decrease in the excited surface waves between the micromachined antenna and the regular one.

REDUCTION OF THE EFFECTIVE DIELECTRIC CONSTANT (TECHNIQUE I)

Calculation of Reduced Dielectric Constant

In this approach, material is removed underneath the antenna by using selective etching techniques and the excitation of surface waves is suppressed by creating a micromachined cavity as shown in Figure 1.

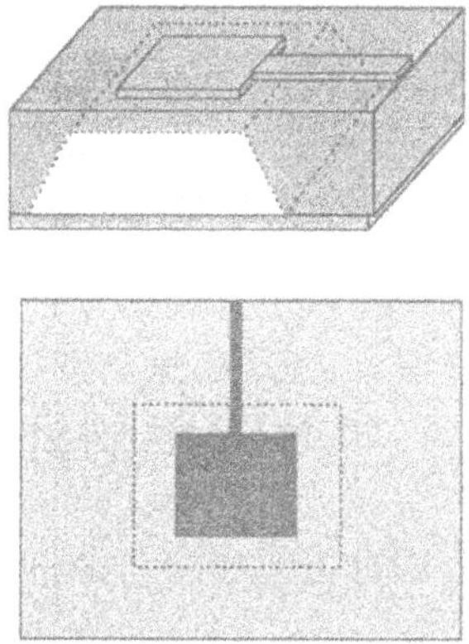

Figure 1. Geometry of the micromachined antenna with mixed air-substrate region that has been laterally etched away.

In order to predict the effective dielectric constant of the mixed air-dielectric (in our case Silicon or Duroid with ε_r=10.8) region for varying thickness ratios underneath the patch antenna, a quasi-static model based on series capacitors is used. A plot of the theoretical and measured effective dielectric constant versus the air gap thickness can be found in Figure 2, where an effective dielectric constant of approximately 2.2 is achieved for a mixed air-silicon ratio of 3:1 from the capacitor model that includes the open-end effect extension length ΔL.

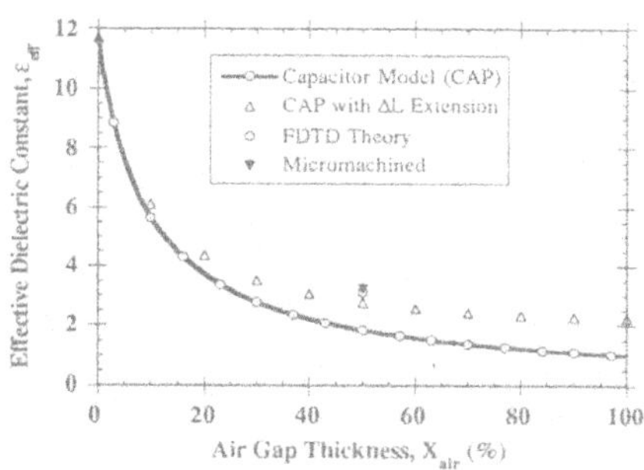

Figure 2 . Effective dielectric constant versus air-gap thickness (percent) for the silicon micromachined patch based on the capacitor model.

Return Loss, Pattern and Efficiency Measurements of Model Antennas

A micromachined patch antenna with an air-dielectric ratio of 3:1 was fabricated on a 635 μm thick Duroid substrate with ε_r=10.8, for checking the validity of our theoretical approach. The selection of this substrate is due to the fact that its dielectric constant is the closest possible to the one of Silicon (11.7). The dimensions of the antenna were 7.624x6.676 mm and those of the cavity 15.19x14.478 mm. The distance between the radiating edges of the patch and the edges of the machined cavity was 3.783 mm. Due to the specifications of the efficiency measurement system set-up the resonant frequency of the patch was between 12.5 and 13.5 GHz (f_{res}=13.165 GHz).

In order to compare the performance of the micromachined antenna with that of a regular antenna a conventional patch with dimensions of 3.75 x4.42 mm was fabricated on the same substrate (f_{res}=12.84 GHz). In addition, a regular antenna with dimensions of 7.57x7.34 mm was fabricated on a Duroid substrate with a thickness of 500 μm and ε_r=2.2. All three antennas were fed by a 50 Ω microstrip line for the best possible match and from the return loss that can be seen in Figure 3, we observe an increase in the -10 dB bandwidth of the machined antenna from 1.40% to 2.3% when compared with the bandwidth of the regular antenna on high index. In reality, the 2.3% bandwidth of the etched antenna is the same as the bandwidth of the low-index (2.2) regular antenna. Since bandwidth is inversely proportional to the quality factor, Q, defined as the ratio of total stored energy in the antenna to the energy dissipated or radiated from the antenna, the increase in bandwidth provides the first indicator of a potential increase in radiation efficiency.

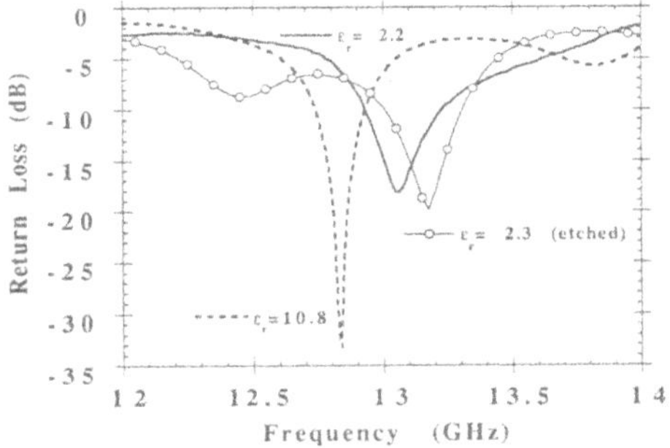

Figure 3. Return loss for the micromachined high-index patch, the regular high-index patch and the regular low-index one.

Radiation patterns were also taken for all three antennas and they can be seen in Figure 4, where we observe that the regular high index E-plane pattern has a large peak at about -50 degrees, indicating large power leakage in the form of surface waves. In contrast, the micromachined patch has a much smoother E-plane and is very similar to the one of the low index antenna. As expected, the H-plane patterns are similar in all cases.

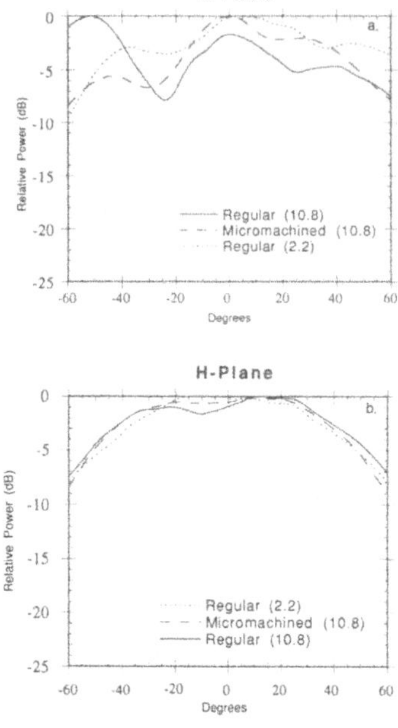

Figure 4. Radiation patterns for the regular high-index antenna, the micromachined antenna and the low-index antenna: a) E-plane and b) H-plane.

In order to obtain conclusive evidence about the increase in radiation efficiency of the micromachined antenna, efficiency measurements were performed for all three patches based on a radiometric method[6,7] and a hemispherical hot/cold load (black-body absorber Ecosorb). The three antennas were placed on a test fixture with an RF connector at the input of the feeding line. The efficiency measurement was referenced at the RF connector and ,thus, included besides the loss due to surface waves the feed-line loss (dielectric and conductor), the connector loss and the mismatch to the 50 Ω line loss. The results can be seen in Figure 5, where we observe that the efficiency of the micromachined patch at the resonant frequency is 73±3 % whereas the efficiency of the regular high-index patch is 56±3 %. The low-index regular antenna has an efficiency of 76±3 %.

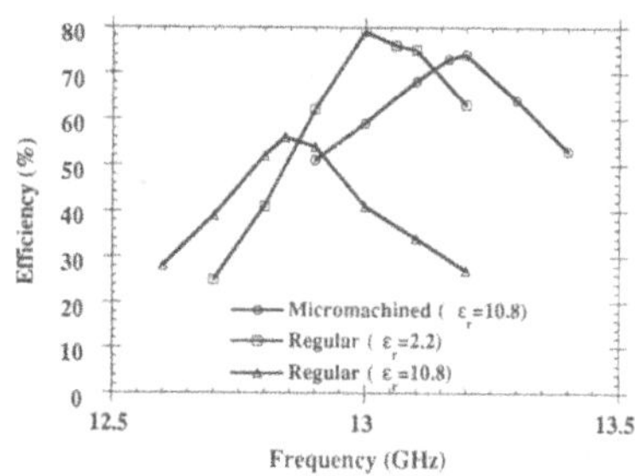

Figure 5. Measured efficiency for the micromachined and regular antennas.

From the above results, it is obvious that the micromachined antenna performs like a regular low-index antenna and gives a 30% increase in the measured efficiency when compared with the conventional high-index patch. We should also note here that since the measured efficiencies include losses irrelevant to the antenna (connector, feed-line and mismatch) a de-embedding process needs to be performed in order to get the efficiency of the antenna itself. This can be achieved by modelling the feed-line with a commercial software and calculating its losses, as well as by measuring the mismatch/return loss and using an empirical value for the RF connector loss. After de-embedding, the calculated efficiencies are 66% for the high-index regular patch, 85% for the micromachined high-index patch and 82% for the low-index regular patch. These results indicate that the micromachined antenna is successful in suppressing the losses due to surface waves.

ELIMINATION OF TMo SURFACE WAVE (TECHNIQUE II)

Thoretical Analysis

In this approach the resonant length of the rectangular patch is adjusted in order to minimize the dominant TM_0 surface mode. According to the equivalence principle and the cavity model, in terms of radiation a rectangular patch can be modeled as a rectangular loop of magnetic current. From the cavity model, the electric field of the dominant TM_{10} mode for the geometry shown in Figure 6 is given by:

$$E_z = A\cos\left(\pi\frac{y}{b}\right) \tag{1}$$

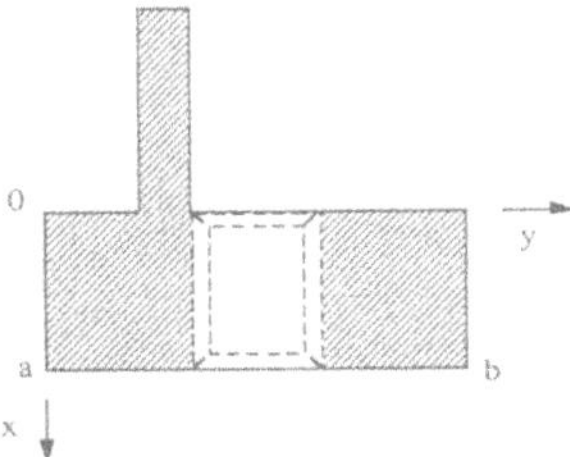

Figure 6. Micromachined antenna for the elimination of TMo mode.

where a,b are the width and the resonant length of the patch, respectively. The radiating edges are considered to be at y=0,b since only the dominant mode of the cavity is excited. The equivalent magnetic current at these two edges is:

$$\vec{M} = 2\vec{E}\times\hat{n} = 2A\hat{x} \tag{2}$$

As it is known, a single Hertzian magnetic dipole, oriented in the x direction at a height z' above the ground will give rise to a TM_0 surface wave field given by:

$$\Psi = K(z, z')H_1^{(2)}(\beta_{TMo}\rho)\sin\phi \tag{3}$$

where β_{TMo} is the propagation constant of the TM_0 mode and K(z,z') is an amplitude factor that depends on the height of the source and the observation point. By integrating over the two radiating edges of the rectangular patch for the magnetic current distribution, the total surface wave field radiated by the magnetic currents takes the form:

$$\Psi = -4A\frac{B(z)}{\beta_{TMo}}H_1^{(2)}(\beta_{TMo}\rho)\tan\phi\cos\left(\beta_{TMo}\frac{b}{2}\sin\phi\right)\sin\left(\beta_{TMo}\frac{a}{2}\cos\phi\right) \tag{4}$$

where $B(z) = \int_0^h K(z, z')dz'$.

In order to derive the above expression the far field approximation for the phase of the radiated wave was used. The pattern of equation (4) is shown in Figure 7 (dashed line). By setting $b=\pi/\beta_{TMo}$ nulls are placed at the location of the peaks of the lobes at $\phi=\pi/2, 3\pi/2$ so that four minor lobes replace the two major lobes (Figure 7, solid line). As a result, the surface wave pattern is reduced.

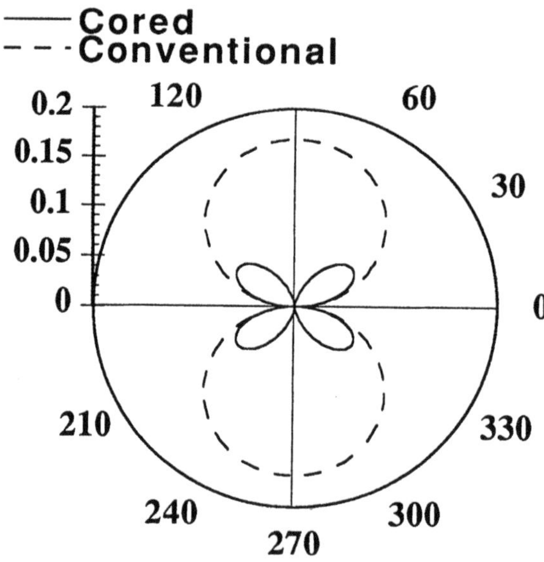

Figure 7. Surface wave field pattern for the regular antenna and the micromachined antenna.

Choosing b according to the above formula, results in a greater resonant length than the one chosen for the design frequency. In order to overcome this decrease in operating frequency, the material in a rectangular region under the patch is removed, thus creating a lower effective dielectric constant which will permit the desired increase in operating frequency. The length of this "cored" region is found from the transcedental equation, which is derived from the field expressions of the dominant mode cavity, after applying the boundary conditions at the various interfaces.

Results

In order to test the accuracy of the previous analysis a rectangular patch with a resonant frequency of 22.5 GHz was designed on a Duroid substrate with ε_r=10.8 and a thickness of 635 μm. Since for the given substrate , higher order surface modes exist above 40 GHz elimination only of the dominant TM_0 mode is sufficient. The micromachined patch was compared with a regular patch on the same substrate and with the same resonant frequency through FDTD simulations. The results can be seen in Figure 8, where measurement of the electric field inside the dielectric region shows a suppressed field for the micromachined antenna by at least 10 dB, when compared with that of the conventional patch.

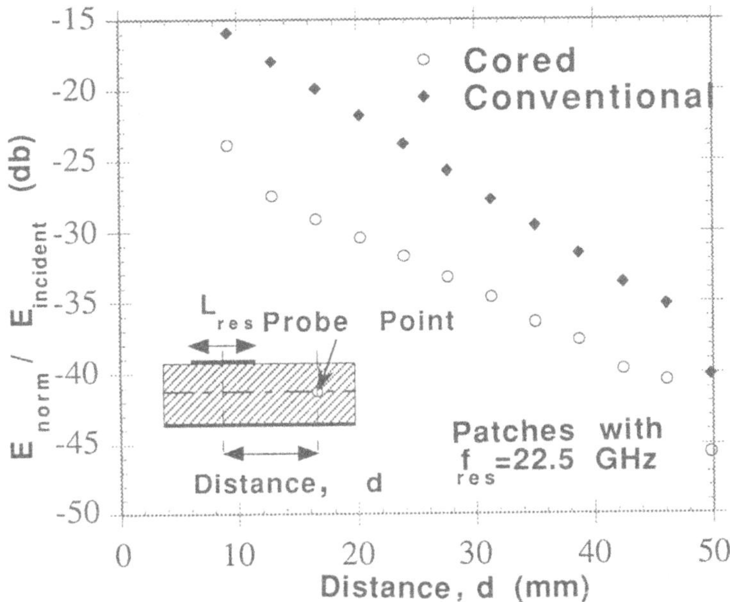

Figure 8. Normalized electric field inside the dielectric versus distance from the center of the antenna for the micromachined patch and the regular patch.

CONCLUSIONS

Two different approaches for the reduction of surface waves in rectangular microstrip antennas were presented in this paper. In the first technique, material was removed laterally underneath and around the patch and an increase of 64% and 30% was observed in the bandwidth and radiation efficiency, respectively. In the second technique, the resonant length of the rectangular patch was adjusted to a critical value in order to suppress the dominant TM_0 surface mode. Simulation results with FDTD showed a substantial decrease in the electric field measured inside the dielectric. Since both approaches require the use of micromachining techniques and result in the enhancement of antenna performance, integration of these high-efficiency rectangular patches on MMIC's operating at high frequencies will be feasible, successful and easy.

ACKNOWLEDGMENTS

The authors would like to thank the Office of Naval research for the financial support of this project under contract No. N00014-92-J-1070.

REFERENCES

1. D.M. Pozar, "Microstrip antennas," *IEEE Proceedings*, Vol. 80 79:91 (1992).
2. N.G. Alexopoulos and D.R. Jackson, "Fundamental superstrate (cover) effects on printed circuit antennas," *IEEE Trans. on Antenna and Propagation*, Vol. AP-32, No. 8 807:816 (1984).
3. D. R. Jackson, J. T. Williams, et. al, "Microstrip patch designs that do not excite surface waves," *IEEE Trans. on Antennas and Propagation*, Vol. 41, No. 8 1026:1037 (1993).
4. M. J. Vaugh, K. Hur, and R. C. Compton, "Improvement of microstrip patch antenna radiation patterns," *IEEE Trans. on Antennas and Propagation*, Vol. 42, No. 6 882:885 (1994).
5. G.P. Gauthier, A. Courtay and G.M. Rebeiz, "Microstrip antennas on synthesized low dielectric-constant substrates," Submitted to *IEEE Trans. on Antennas and Propagation*, (1996).
6. J. Ashkenazy, E. Levine and D. Treves, "Radiometric measurement of antenna efficiency," *Electron. Lett.*, Vol. 21, No. 3 (1985).
7. D.M. Pozar and B. Kaufman, "Comparison of three methods for the measurement of printed antenna efficiency," *IEEE Trans. on Antennas and Propagation*, Vol. 36, No.1 136:139 (1988).

Integrated Strip Gratings on Top of Microstrip Antennas and Arrays for Low and Ultra-Low Cross-Polar Radiation

A. Mohanty and N. K. Das

Weber Research Institute
Department of Electrical Engineering
Polytechnic University, Route 110
Farmingdale, NY 11735

1 Introduction

Antennas with very low levels of cross-polarization are useful for many applications in communication, radar and remote-sensing. Printed antennas and arrays, although attractive due to their light weight, conformability and low cost [1,2], may not meet the low cross-polarization requirements. The cross-polar radiation may be introduced due to unwanted radiation from feed networks and active circuitry. Orthogonal currents, either inherently excited on the antenna elements, or additionally induced due to mutual coupling with other elements in an array environment, can contribute to cross-polar radiation. The dielectric substrate of the antenna can also introduce cross-polar radiation in off-principal directions. It may be particularly difficult to maintain a uniformly low level of cross-polarization over a broad scan angle or a broad bandwidth.

Strip gratings can be effectively used on top of an antenna for suppression of the cross-polar radiation. The strip grating may be separately fabricated and properly spaced in front of the antenna, or it may be integrated directly on top of the antenna with a thin layer of dielectric substrate in between. In either case, the strip grating can have significant effects on radiation and/or impedance characteristics of of the antenna. When the grating is integrated directly on top, due to close proximity of the grating layer the reactive near fields of the antenna will strongly couple with the grating strips. This may significantly affect the impedance level, bandwidth, resonant frequency, as well as the radiation and scan characteristics. In addition, when the grating is either spaced above or placed directly on top, certain resonant conditions established between the grating strips and the antenna substrate may result in excessive power loss from the antenna element due to excitation of energy along the grating. This may result in poor radiation efficiency, or affect the cross-polarization or radiation pattern of an isolated element, or can introduce scan blindness in a phased array. In order to successfully use the integrated grating designs, it is essential to accurately model a printed antenna geometry rigorously including all effects of the strip grating on top. In addition, some fundamental phenomena of the strip grating configuration must be well understood, and suitable techniques should be devised in order to overcome the resulting undesired effects.

In this paper, we present a rigorous analysis of an infinite array of printed dipoles covered on top by a layer of printed strip grating. The new analysis is based on a spectral domain moment method formulation. A separate analysis of the strip grating is first performed. The results of this analysis are then used in a spectral-domain moment method procedure in order to rigorously account for the effects of the substrate and the strip grating layers. Basic design data are generated from the analysis, and the various loading characteristics due to the grating layer are studied. Interesting fundamental effects of the grating geometry on the scan and radiation performance of the antenna are discussed, providing basic guidelines to overcome their possible undesired effects.

2 Theory

Fig.1 shows the geometry of an infinite array of printed dipoles covered by a periodic strip grating. For simplicity of modeling, each dipole element is center-fed by an idealized delta-gap current source with proper phasing for scanning. In addition, we assume that the antenna periodicity, A, along the $\hat{x}$ direction is an integral multiple of the grating periodicity, D. Under this condition the total antenna structure can be assumed to be physically periodic, based on which a Floquet mode analysis is possible [3].

The analysis of the array, without the grating can be done using a periodic moment method procedure [3,4]. The unknown currents in one unit cell are expanded using known basis functions with unknown coefficients. (In a phased array, due to periodicity, the current distributions on the other unit cells are the same except for a phase factor.) Then the unknown system is solved by a moment method testing procedure. When the grating is included, this approach can be extended, in principle, by treating the currents on the grating strips as additional unknowns of the moment method system. However, since the number of strips per unit cell in a practical grating can be quite large, the number of unknowns required can be orders of magnitude larger than that without the grating. This brute-force numerical approach would have poor convergence It would not yield much information on the effect of wave-guidance by the grating.

We present here a novel two-step approach for a rigorous analysis of the geometry of Fig.1. First, we replace the antenna layer by a spectral current sheet of the form, $\vec{\mathbf{J}}_s = \hat{\mathbf{x}}_s e^{-j(xk_{sx}+yk_{sy})}$, arbitrarily directed along $\hat{\mathbf{x}}_s$. Under this current-sheet excitation the induced currents on the grating strips and the resulting total fields are modeled by a one-dimensional periodic analysis. In the second part of the analysis, any given basis current distribution on the dipole array is expressed as a superposition of a discrete number of spectral sheets via a two dimensional Floquet mode modeling [3]. The field produced due to each of the spectral currents is obtained from the first part of the analysis, the superposition of which will provide the total fields due to the known basis current. Then, the unknown current distribution on the antenna is expanded by a set of known basis functions with unknown coefficients, and the unknown coefficients are solved by a moment method procedure. The known fields of the basis currents already obtained above are employed here for the necessary boundary condition testing.

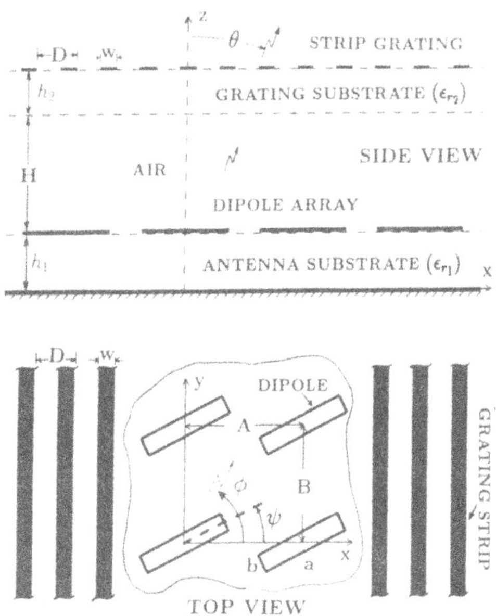

Fig.1 General geometry of an infinite array of printed dipoles, with a layer of printed strip grating on the top.

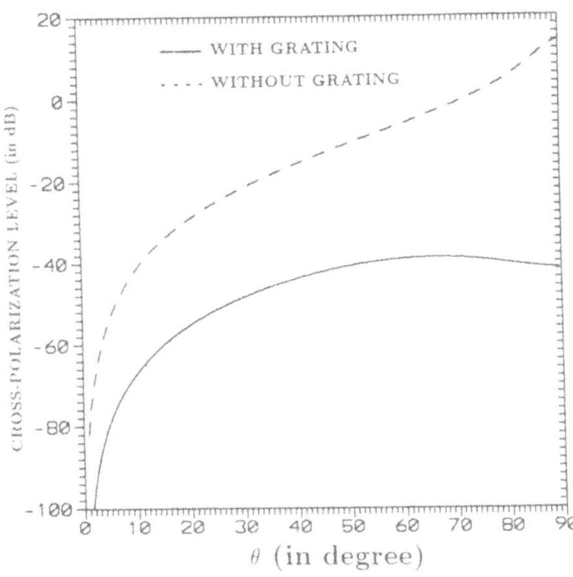

Fig.2 Effect of the grating on the cross-polarization level in the diagonal plane. Frequency 3 GHz; array unit cell: $A = 35$ mm, $B = 35$ mm; dipole parameters: $a = 31$ mm, $b = 1$ mm, $\psi = 0°$; substrate parameters: $h_1 = 1.6$ mm, $H = 0$, $h_2 = 1.6$ mm, $\epsilon_{r_1} = 2.2$, $\epsilon_{r_2} = 2.2$; grating parameters: $D = 2.5$ mm, $w = 1.0$ mm; scan plane: $\phi = 45°$.

2.1 Current Sheet Excitation of the Grating

Let the dipole array in Fig.1 be replaced by a spectral current sheet,

$$\vec{\mathbf{J}}_s = \hat{\mathbf{x}}_s e^{-j(xk_{sx}+yk_{sy})}. \tag{1}$$

The unit vector

$$\hat{\mathbf{x}}_s = (\hat{\mathbf{x}}\cos\phi_s + \hat{\mathbf{y}}\sin\phi_s), \tag{2}$$

which makes an angle ϕ_s with $\hat{\mathbf{x}}$, specifies the direction of the current on the spectral current sheet.

Using Floquet's theorem the overall grating current distribution, $\vec{\mathbf{J}}_g(x,y)$, under spectral current sheet excitation can be expressed in terms of the current distribution, $\vec{\mathbf{J}}(x,y)$ on one strip at $y=0$:

$$\vec{\mathbf{J}}_g(x,y) = \sum_{m=-\infty}^{\infty} \vec{\mathbf{J}}(x-mD,y)e^{-jmk_{sx}D} = e^{-jk_{sy}y}\sum_{m=-\infty}^{\infty} \vec{\mathbf{J}}(x-mD,0)e^{-jmk_{sx}D}. \tag{3}$$

The unknown current $\vec{\mathbf{J}}(x,0)$ is expanded as a sum of basis functions with unknown coefficients.

$$\vec{\mathbf{J}}(x,0) = \sum_{n=1}^{N_g} a_n \vec{\mathbf{f}}_n(x) \tag{4}$$

The current on the grating strip is dominantly y-directed. The y-component can have singularities at the strip edges. The x-component is usually small and vanishes at the edges. A suitable choice of entire domain modes for the basis functions is:

$$\vec{\mathbf{f}}_n(x) = \begin{cases} \hat{\mathbf{y}} f_{yp}(x) & p = \frac{n-1}{2}, n = 1,3,5,\ldots; \\ \hat{\mathbf{x}} f_{xp}(x) & p = \frac{n}{2}, \quad n = 2,4,6,\ldots; \end{cases} \tag{5}$$

where, $f_{yp}(x)$ and $f_{xp}(x)$ are zero for $|x| > w/2$, and are defined for $|x| < w/2$ as:

$$f_{yp}(x) = \frac{1}{\pi\sqrt{(\frac{w}{2})^2 - x^2}} \begin{cases} \cos(p\pi x/w) & p \text{ even}; \\ \sin(p\pi x/w) & p \text{ odd}; \end{cases} \tag{6}$$

$$f_{xp}(x) = \frac{1}{w} \begin{cases} \sin(p\pi x/w) & p \text{ even}; \\ \cos(p\pi x/w) & p \text{ odd}. \end{cases} \tag{7}$$

It is assumed that the origin of the $\hat{x}$-axis coincides with the center of the 0-th strip. Using (4) in (3) and then applying the Poisson summation formula $\vec{\mathbf{J}}_g(x,y)$ can be expressed as

$$\vec{\mathbf{J}}_g(x,y) = \frac{1}{D}\sum_{n=1}^{N_g} a_n \sum_{\mu=-\infty}^{\infty} \left[\tilde{\vec{\mathbf{f}}}_n(k_{sx\mu})e^{-j(k_{sx\mu}x + k_{sy}y)}\right]_{k_{sx\mu}=k_{sx}+\mu\frac{2\pi}{D}} \tag{8}$$

Now the grating current is expressed as a sum of spectral current sheets. The electric field, $\vec{\mathbf{E}}_g(x,y)$, produced at the grating layer due to the grating current in (8) can be obtained using spectral domain dyadic Green's functions [5,6]. The incident field on the grating due to the current sheet can also be evaluated this way. Thus the total E field can be expressed in terms of the unknown coefficients a_n's. The boundary condition all over the grating is automatically satisfied by the vanishing of the tangential component of the E field on one strip at $y=0$. When this is enforced using a Galerkin testing procedure[7] the following system of equations is obtained.

The transform of the total E field on the antenna layer in the presence of the grating can be expressed in terms of the

$$\sum_{n=1}^{N_g} a_n Q_{mn} = V_m; \quad m = 1, 2, \ldots, N_g, \tag{9}$$

$$V_m = -\tilde{\vec{\mathbf{f}}}^*_m(k_{sx}) \cdot \tilde{\underline{\underline{\mathbf{G}}}}^{Eg}_{Ja}(k_{sx}, k_{sy}) \cdot \hat{\mathbf{x}}_s, \tag{10}$$

$$Q_{mn} = \frac{1}{D} \sum_{\mu=-\infty}^{\infty} \tilde{\vec{\mathbf{f}}}^*_m(k_{sx} + \mu\frac{2\pi}{D}) \cdot \tilde{\underline{\underline{\mathbf{G}}}}^{Eg}_{Jg}(k_{sx} + \mu\frac{2\pi}{D}, k_{sy}) \cdot \tilde{\vec{\mathbf{f}}}_n(k_{sx} + \mu\frac{2\pi}{D}). \tag{11}$$

Notationally, $\tilde{\underline{\underline{\mathbf{G}}}}^{Eg}_{Ja}$ is the Green's function relating the electric field on the grating layer to the current on the antenna layer, whereas $\tilde{\underline{\underline{\mathbf{G}}}}^{Eg}_{Jg}$ is the Green's function relating the electric field on the grating layer to the current on the grating layer. The complete set of linear equations (9) can now be solved for the unknown basis coefficients, a_n's. It should be noted that in (11) Q_{mn} remains unaltered when k_{sx} is shifted by an integral multiple of $2\pi/D$. This periodicity can be used to speed up the evaluation of various spectral sums and integrals. Once the unknown current amplitude values, a_n's, are found, the field anywhere in the multilayered structure can be obtained using suitable Green's functions. Clearly, the total field produced by the current sheet $\vec{\mathbf{J}}_s$, in the presence of the grating, includes an infinite number of spectral values at $k_x = k_{sx} + \mu\frac{2\pi}{D}$, $k_y = k_{sy}$; $\mu = 0, \pm 1, \pm 2, \ldots$. The transform of the the total electric field, $\tilde{\vec{\mathbf{E}}}_{as}$, on the antenna layer, can be found by adding the transforms of the incident field due to the current sheet, and the scattered field due to the grating. Eventually it can be expressed in the form

$$\begin{aligned}&\tilde{\vec{\mathbf{E}}}_{as}(k_x, k_y; k_{sx}, k_{sy}; \phi_s)\\ &= (2\pi)^2 \delta(k_y - k_{sy}) \sum_{\mu=-\infty}^{\infty} \delta\left(k_x - k_{sx} - \mu\frac{2\pi}{D}\right) \tilde{\vec{\mathbf{e}}}_\mu(k_{sx}, k_{sy}; \phi_s),\end{aligned} \tag{12}$$

where, $\tilde{\vec{\mathbf{e}}}_\mu(k_{sx}, k_{sy}; \phi_s)$ is the amplitude of the μ-th Floquet mode E field.

2.2 Impedance of Array Covered by a Grating

In Fig.1, the (r, s)-th dipole of the infinite array is centered at $x = rA$, $y = sB$ for any integers r and s. For scanning of the array in the direction (θ, ϕ), the overall dipole array current distribution, $\vec{\mathbf{J}}_a(x, y)$ dipole array can be expressed in terms of the current distribution on the $(0,0)$-th dipole, $J(x, y)$ as:

$$\vec{\mathbf{J}}_a(x, y) = \sum_{s=-\infty}^{\infty} \sum_{r=-\infty}^{\infty} \vec{\mathbf{J}}(x - rA, y - sB) e^{-j(rAk_{xo} + sBk_{yo})}, \tag{13}$$

where $k_{xo} = k_o \sin\theta \cos\phi$ and $k_{yo} = k_o \sin\theta \sin\phi$. The unknown current $\vec{\mathbf{J}}(x, y)$, is expanded by a set of basis functions with unknown coefficients, b_q's:

$$\vec{\mathbf{J}}(x, y) = \sum_{q=1}^{N_d} b_q \vec{\mathbf{J}}_q(x, y) = \sum_{q=1}^{N_d} b_q \hat{\mathbf{x}}' C_q(x, y), \tag{14}$$

$$C_q(x, y) = \begin{cases} \frac{1}{b}\cos(q\pi x'/a) & |x'/a| < 1/2, \text{ and } |y'/b| < 1/2, \\ 0 & \text{otherwise;} \end{cases} \tag{15}$$

$$x' = x\cos\psi + y\sin\psi, \qquad y' = -x\sin\psi + y\cos\psi. \tag{16}$$

The above current expansion assumes that currents flow only along the longitudinal direction, $\hat{x}'$, of the dipole. This is known to be a good approximation for narrow dipoles [3]. The dipoles are assumed to be gap fed at the center. Using the Poisson summation formula, it is possible to express the overall array current as a sum of spectral current sheets. The field due to each current sheet in the presence of a grating can be found using the grating analysis discussed earlier. When the boundary condition on the dipole is enforced using Galerkin testing the following system of equations results.

$$\sum_{q=1}^{N_d} \zeta_{pq} b_q = v_p; \quad p = 1, \ldots, N_d, \tag{17}$$

$$v_p = \int\int_{\text{Dipole}} -\vec{\mathbf{E}}_i \cdot \vec{\mathbf{J}}_p \, dS, \tag{18}$$

$$\zeta_{pq} = \frac{1}{AB} \sum_{s=-\infty}^{\infty} \sum_{r=-\infty}^{\infty} [\tilde{C}_q(k_{xr}, k_{ys}) \sum_{\mu=-\infty}^{\infty} \tilde{C}_p^*(k_{xr} + \frac{2\pi\mu}{D}, k_{ys})\hat{\mathbf{x}}' \cdot \tilde{\mathbf{e}}_\mu(k_{xr}, k_{ys}; \psi)]. \tag{19}$$

where,

$$k_{xr} = k_{xo} + \frac{2\pi r}{A}, \qquad k_{ys} = k_{yo} + \frac{2\pi s}{B}. \tag{20}$$

The complete set of linear equations (17) can be solved for the unknown coefficients b_q's, using ζ_{pq} and v_p expressions of (19,18). Once the coefficients b_q's are known, the dipole impedance is found as

$$Z_{\text{in}} = -1/\left(\sum_{q=1}^{N_d} b_q v_q\right). \tag{21}$$

3 Results and Discussion

3.1 The Cross-Polarization Level

Fig.2 shows the cross-polarization characteristics of a dipole array integrated with a strip grating, as a function of the scan angle, θ, in the D-plane ($\phi = 45°$.) The grating is integrated close to the antenna ($H = 0$.) The corresponding result without the grating layer is also presented together for comparison. Clearly, significant improvement in the cross-polarization level has been achieved in Fig.2 due to the use of the grating. The cross-polarization level with the grating is -40dB or better over the entire scan plane, which can be improved further by proper design of the grating.

Interestingly, the situation can be quite different in Fig.3, where the grating is placed sufficiently above the antenna ($H = 55$ mm), all other dimensions remaining the same as before. The sharp increase of the cross-polarization level in Fig.3 at $\theta = 31.2°$ is due to a resonant condition established between the grating layer and the antenna ground plane. The cross-polar component (with respect to the grating) of the radiated antenna field is almost totally reflected from the grating layer, and then experiences multiple reflections between the grating and the antenna ground plane. This results in producing strong cross-polar fields between the two planes ($0 < z < (H + h_1 + h_2)$) for the resonant scan angle, θ_c, given approximately for thin substrates by:

$$\theta_c \simeq \arccos(\frac{n\lambda_0}{2(H + h_1 + h_2)}), \tag{22}$$

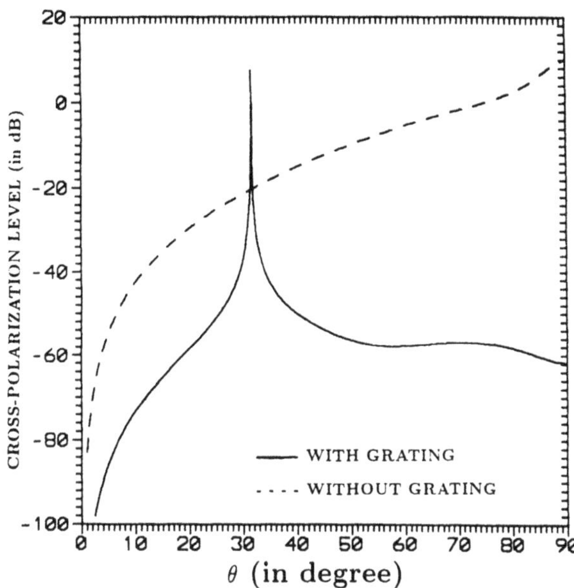

Fig.3 Spikes in the cross-polarization level due to grating resonance, when the grating is spaced above the antenna. Frequency 3 GHz; array unit cell: $A = 35$ mm, $B = 35$ mm; dipole parameters: $a = 31$ mm, $b = 1$ mm, $\psi = 0°$; substrate parameters: $h_1 = 1.6$ mm, $H = 55$ mm, $h_2 = 1.6$ mm, $\epsilon_{r_1} = 2.2$, $\epsilon_{r_2} = 2.2$; grating parameters: $D = 2.5$ mm, $w = 1.0$ mm;

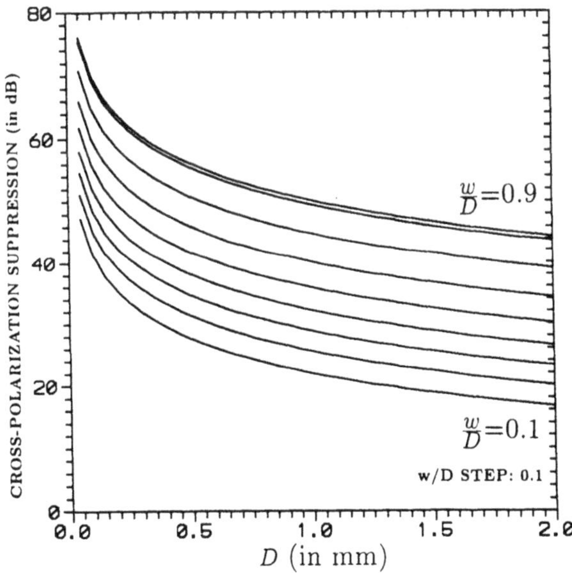

Fig.4 The level of cross-polarization suppression in the broadside scanning for different grating parameters. The dipoles are intentionally misoriented 5° in order to introduce cross-polarization in the broadside direction. Frequency 3 GHz; array unit cell: $A = 35$ mm, $B = 35$ mm; dipole parameters: $a = 30.55$ mm, $b = 1$ mm, $\psi = 5°$; substrate parameters: $h_1 = 1.6$ mm, $H = 0$, $h_2 = 1.6$ mm, $\epsilon_{r_1} = 2.2$, $\epsilon_{r_2} = 2.2$.

where n is an integer number for the order of the resonance. A small fraction of this strong cross-polar cavity field always escapes through the non-zero gap, $(D - w)$, between the grating strips to the external $(z > H + h_1 + h_2)$ radiation. As a result, a strong spike is seen in the cross-polar pattern of Fig.3. The angle of the single spike $(n = 1)$ in Fig.3 agrees well with the theoretical calculation of 30.8 degrees using (22.) It should be clear from (22), that this resonance effect can be avoided by designing the separation, H, less than $\lambda_0/2$. Further, as predicted from (22), due to the higher order resonance conditions more than one spike will occur in the cross-polar pattern when H is larger than about λ_0.

Fig.4 shows the design data relating the printed grating parameters to the level of cross-polarization suppression that can be achieved. These results have been computed for the case when the grating is integrated close to the antenna $(H = 0)$, only for the broadside scan. The dipole elements have been misoriented by about 5° $(\psi = 5°)$ with respect to the grating in order to deliberately introduce a cross-polar component. This is to emulate the cross-polar radiation from other circuit elements (the feed circuits, for example.) The design data of Fig.4 are also generally valid for gratings spaced above $(H \neq 0)$ the antenna, except at the resonance angle, θ_c given by (22.) As expected, better suppression of the cross-polar radiation can be achieved by using smaller D and/or larger w, because less space is available between the strips for the cross-polar field to escape through. About 20dB suppression can be easily obtained using a crude grating with fairly widely spaced strips, whereas more than 40dB suppression can be achieved from grating designs with narrow strip spacings.

3.2 Impedance Characteristics

Fig.5 shows the resonant frequency, f_r, resonant radiation resistance, R_r, and the Q-factor of a printed dipole array integrated with a strip grating on top $(H = 0)$, as a function of the separation, h_2, between the antenna and the grating layers. For these results, the grating strips are oriented perpendicular to the dipoles, as desirable. In this orientation, if the grating is spaced sufficiently above the antenna (large H,) it will significantly affect the cross-polarization characteristics but not the impedance characteristics of the antenna. However, when the grating is integrated close to the antenna $(H = 0)$, in addition to the polarization properties the impedance characteristics of the antenna can also be significantly affected. This loading effect for $H = 0$ is clearly seen in Fig.5, which is a result of the strong coupling between the evanescent fields of the antenna and the grating. Notice in Fig.5, in the limit when $h_2 = 0$ the resonant resistance of the dipole array tends to zero. This indicates the undesirable situation of having zero or negligible radiation if the grating is integrated too close to the antenna layer. Equivalently, the radiation bandwidth should be zero in this limiting situation, or the Q-factor is infinite, which is consistent with the results of Fig.5. In addition, the resonant frequency of the antenna in Fig.5 is also significantly reduced, or equivalently the resonant length is increased, due to the reactive loading effect.

3.3 Scan Characteristics

Fig.6 shows the scan characteristics of an infinite array of printed dipoles covered with a grating layer, when $\psi = 0$ and the grating is spaced far above (large H) the antenna layer. The corresponding scan characteristics of the array without the grating are

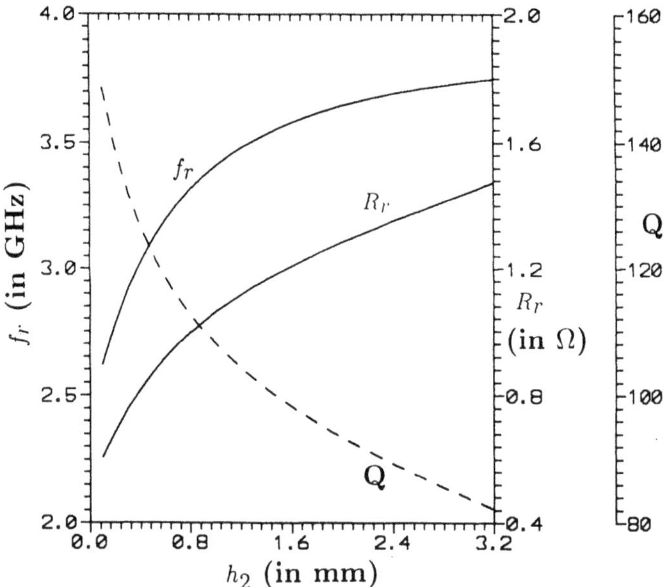

Fig.5 Dependence of resonant frequency, f_r, resonant radiation resistance, R_r, and quality factor, Q, on the distance, $h_2 + H = h_2$, between the grating and the array. Array unit cell: A = 35 mm, B = 35 mm; dipole parameters: a = 25 mm, b = 2 mm, $\psi = 0°$; substrate parameters: h_1 = 1.6 mm, H = 0, $\epsilon_{r_1} = 2.2$, $\epsilon_{r_2} = 2.2$; grating parameters: D = 2.5 mm, w = 1.0 mm.

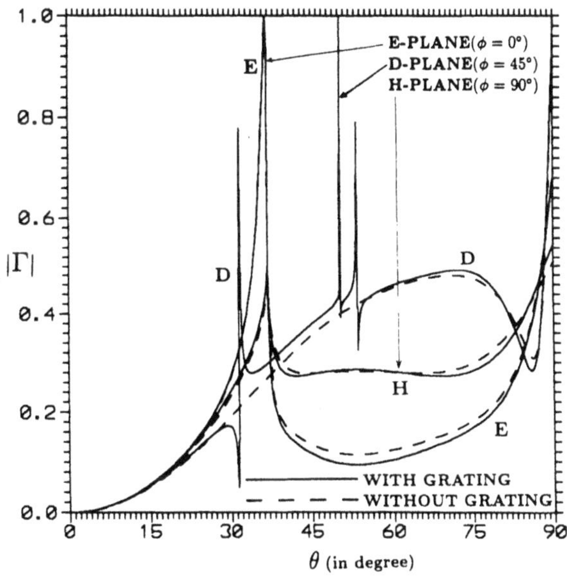

Fig.6 E-, H- and D-plane scan performance when grating is placed far from the antenna, showing resonance spikes due to a higher order parallel-plate-type mode and a grating mode. Frequency 3 GHz; array unit cell: A = 62.5 mm, B = 62.5 mm; dipole parameters: a = 37 mm, b = 1 mm, $\psi = 0°$; substrate parameters: h_1 = 1.6 mm, H = 55 mm, h_2 = 1.6 mm, $\epsilon_{r_1} = 2.2$, $\epsilon_{r_2} = 2.2$; grating parameters: D = 2.5 mm, w = 1.0 mm.

also shown for comparison. The grating introduces undesired scan-resonance effects (also called the scan-blindness effects [3]) in Fig.6 over narrow range of angles. With reference to the reduced spectral-plane diagram of Fig.7, each resonance angle in Fig.6 can be identified with a specific type of singularity (surface-wave, grating-mode, or higher-order parallel-plate type mode,) and is produced when one of the Floquet modes in Fig.7 encounters the particular singularity when scanned along a given plane. As discussed in previous sections, we know that a grating does not noticeably affect the impedance characteristics of the array, when placed sufficiently far away above the antenna. Interestingly, however, under the same above condition the grating in Fig.6 is seen to exhibit significant effects on the scan characteristics.

In contrast to Fig.6, Fig.8 shows the scan performance of the dipole array when the grating is integrated directly on top ($H = 0$) of the array. As can be seen, the E-plane scan performance remains almost unaffected due to the introduction of the grating strips. This is not the case for the H- or D-plane scanning, where the scan characteristics with the grating are significantly different from that without the grating over the entire scan plane, not just over narrow angles. This is due to the strong interaction, and the resulting loading effect, between the evanescent fields of the antenna and the grating in close proximity of each other. Due to varying order of coupling between the antenna and the grating strips, the loading is strongest in the H-plane ($\phi = 90°$), and the weakest in the E-plane ($\phi = 0°$). Further, scan resonance effects similar to that in Fig.6 are also seen in the D-plane scanning of Fig.8. However, these resonance angles occur when the array characteristics are already dominated by the loading effects, making them less meaninglful from practical considerations.

The scan resonance and loading effects demonstrated by Figs.(6,8) can be avoided by careful design. First, the resonance due to the higher order parallel-plate type mode can be eliminated by designing the spacing, H, between the antenna and the grating sufficiently less than $\lambda_0/2$. This type of scan resonance problems are absent for the designs when the grating strips are integrated on top of the antenna ($H = 0$). The scan resonance problems due to the grating mode effects can be solved by sufficiently reducing the H-plane spacing of the array, B, such that none of the Floquet modes in Fig.7 can reach the grating mode locus. The scan loading effect seen when $H = 0$ can also be minimized by reducing the H-plane spacing of the array. For demonstration, in Fig.9 we show the E-, H- and D-plane scan characteristics of an array design with $0.35\lambda_0 \times 0.35\lambda_0$ spacing, showing no scan resonance problems. The results for the array without the grating layer are also shown for comparison. It is seen that the scan performance with the grating is no worse than the scan performance without the grating. Due to the complex nature of the pole locus of the grating mode, one needs to carefully consider the specific spacing requirements on a case-by-case basis. The basic characteristics of the various pole loci we have discussed, together with the Floquet mode diagram similar to Fig.7, can be effectively used for design understanding.

4 Conclusions

A rigorous analysis of a new type of printed antenna configuration with the integrated strip gratings is presented. Our results demonstrate that such designs can be successfully used in order to achieve low or ultralow cross-polar radiation. A set of design data for the antenna impedance, bandwidth, resonant frequency, and polarization level are provided for selected antenna parameters. A successful design must account for the

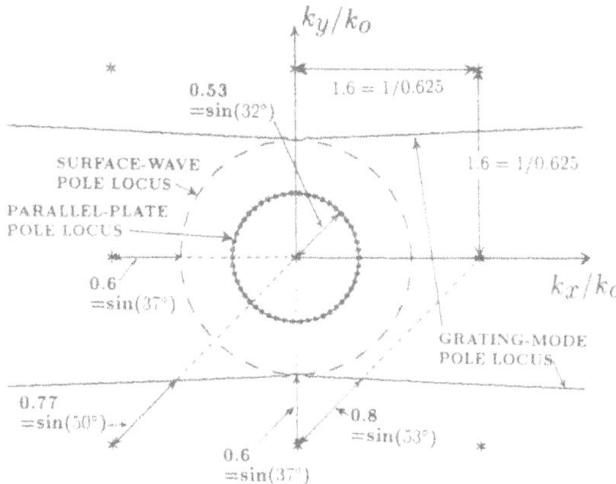

Fig.7 A Floquet mode diagram, drawn to scale for the specific parameters of Fig.6, showing the values of all critical scan-resonance angles. These angles can be identified with the respective resonance locations of Fig.6.

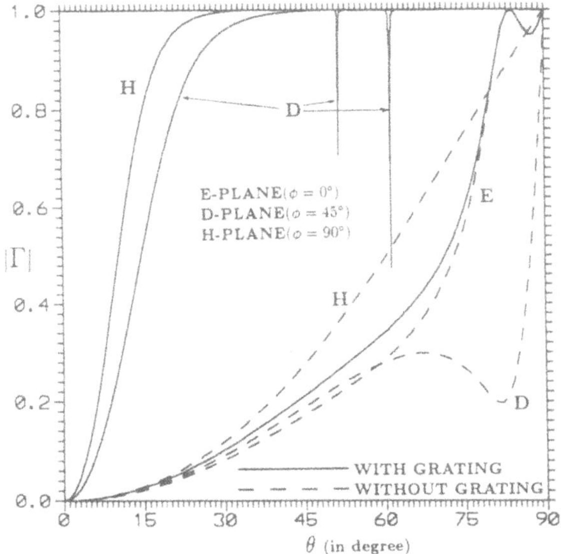

Fig.8 E-, H- and D-plane scan performance when grating is placed close to the antenna, showing the significant loading effect due to the grating. Frequency 3 GHz, $A = 50$ mm, $B = 50$ mm, $a = 30.35$ mm, $b = 1$ mm, $h_1 = 1.6$ mm, $H = 0$, $h_2 = 1.6$ mm, $\epsilon_{r_1} = 2.2$, $\epsilon_{r_2} = 2.2$, $D = 2.5$ mm, $w = 1.0$ mm, $\psi = 0°$.

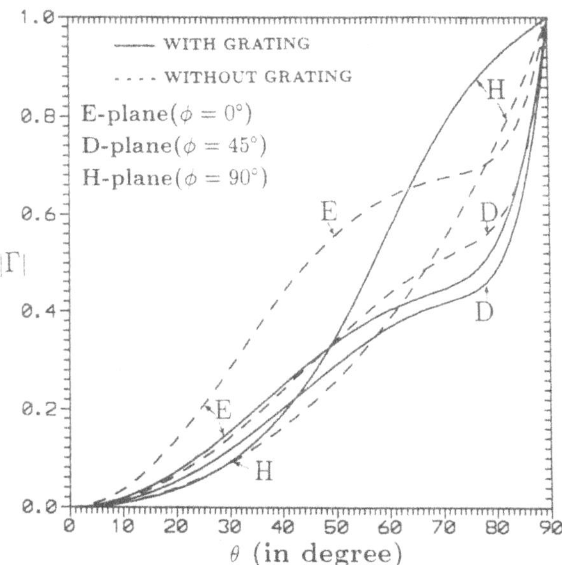

Fig.9 An example of a practical design with the grating layer integrated directly on top of the antenna, where the potential scan blindnesses due to the grating and surface-wave singularities, as well as the strong loading effect due to the grating layer have been avoided. Frequency 3 GHz; array unit cell: A = 35 mm, B = 35 mm; dipole parameters: a = 31 mm, b = 1 mm, $\psi = 0°$; substrate parameters: h_1 = 1.6 mm, $H = 0$, h_2 = 1.6 mm, $\epsilon_{r_1} = 2.2$, $\epsilon_{r_2} = 2.2$; grating parameters: D = 2.5 mm, w = 1.0 mm; array scanning: $\phi = 0°$ (E-plane), $\phi = 45°$ (D-plane), and $\phi = 90°$ (H-plane).

detrimental loading and resonance effects due to the grating layer. The coupling to the grating strips can have critical effects on the impedance, polarization, as well as the scan characteristics of the antenna. A fundamental understanding of the guided-wave effects in the multilayer grating structure is essential in order to help predict and avoid the possible resonance conditions.

References

[1] K. R. Carver and J. W. Mink. Microstrip Antenna Technology. *IEEE Transactions on Antennas and Propagation*, AP-29(1):2–24, January 1981.

[2] R. J. Mailloux, J. F. McIlvenna, and N. P. Kernweis. Microstrip Array Technology. *IEEE Transactions on Antennas and Propagation*, AP-29(1):25–37, January 1981.

[3] D. M. Pozar and D. H. Schaubert. Scan Blindness in Infinite Phased Arrays of Printed Dipoles. *IEEE Transactions on Antennas and Propagation*, AP-32(6):602–610, June 1984.

[4] N. K. Das and D. M. Pozar. Multiport Scattering Analysis of Multilayered Printed Antennas Fed by Multiple Feed Ports, Part I: Theory; Part II: Applications. *IEEE Transactions on Antennas and Propagation*, AP-40(5):469–491, May 1992.

[5] N.K. Das and D.M. Pozar. A Generalized Spectral-Domain Green's Function for Multilayer Dielectric Substrates with Applications to Multilayer Transmission Lines. *IEEE Transactions on Microwave Theory and Techniques*, MTT-35(3):326–335, March 1987.

[6] N. K. Das and D. M. Pozar. A Generalized CAD Model for Printed Antennas and Arrays with Arbitrary Multilayer Geometries. In L. Safai, editor, *Computer Physics Communication, Thematic Issue on Computational Electromagnetics*, North-Holland Publications, 1991.

[7] D. M. Pozar. Input Impedance and Mutual Coupling of Rectangular Microstrip Antennas. *IEEE Transactions on Antennas and Propagation*, AP-30(6):1991–1996, November 1982.

A NEW TYPE OF DUAL BAND MICROSTRIP ARRAY FED BY A PURELY TEM FEEDING NETWORK

Naftali Herscovici

anTeg, Inc.
52, Agnes Drive
Framingham, MA 01701
tuli@tiac.com

INTRODUCTION

The dual band microstrip arrays are of a particular interest in the satellite communication systems. Depending on the system, the upload/download channels can be quite apart, relatively to the typical bandwidth of a single patch antenna. Stacked patches can be used, however the achievable separation between the bands is still limited, if only one feeding point is to be used.

The new microstrip element proposed in this paper seems to overcome the above mentioned limitations. It consists of a rectangular patch, in which a slot was performed. The basic idea is that the patch resonates at the lower band and the slot resonates at the higher band. In the particular case described here, the patch and the feeding network are suspended over the ground-plane and thus the feeding network is based on purely TEM transmission lines . This eliminates the dispersion usually associated with microstrip lines which could hamper the performance of the dual band circuit.

In the past, slots were used on a patch for reducing the size of the patch. The size of the slot was determined so that the required (within limitations) reduction in size was achieved. Here the slot length is designed to be $\lambda/2$ at the higher frequency and the ground-to-patch separation is designed to be $\lambda/4$ at the higher frequency. At the lower band, since the slot is significantly under resonance, the patch radiation is dominant. At the higher band though, the radiation mechanisms are not ideally separated since the patch radiation could be quite significant, especially when the higher frequency is close to a higher mode resonance of the patch.

Directions for the Next Generation of MMIC Devices and Systems
Edited by N. K. Das and H. L. Bertoni, Plenum Press, New York, 1997

A variety of radiating elements based on the above mentioned principle were developed and successfully integrated in arrays for wireless communication applications.

THE 3D SUSPENDED MICROSTRIP CONCEPT

The dielectric losses associated with the microstrip arrays, are one of the factors which set the achievable gain limits of this type of antennas. An excellent study of these losses is given in [1-3]. In [3], also the losses associated with the feed network radiation and surface waves are incorporated in the overall efficiency computations. The implications of the dielectric presence, which are the very essence of the microstrip antennas for almost 50 years, were extensively treated in the literature.
In this paper, a new type of microstrip antenna is proposed. It consists of a circuit which includes only the metalization part of the 'printed circuit' which is suspended over the ground-plane. Some work was done in the past with suspended microstrip small *planar* arrays, usually supported by foam [4], however the technique described in this paper deals with a new class of microstrip antennas which as will be shown opens new possibilities in the design approach as well as types of applications. This new approach presents a number of advantages over the classical printed microstrip arrays or any other types of *planar* suspended microstrip:

a. it allows the use of *three-dimensional transitions*, from one level to another, so that the width of the lines can be controlled by changing the spacing between the line and the ground-plane; that also has an impact on the feed network radiation, and the coupling between the feed network and the radiating elements.
b. the feed network is purely TEM so there is no dispersion and no surface waves.
c. it supports greater changes in temperature, since there is no dielectric (with a temperature expansion coefficient different from that of the copper).
d. the efficiency is considerably higher, since there are no dielectric losses and no surface wave losses.
e. in some cases, the fabrication cost is lower than that of the conventional type.
f. can handle power levels (especially CW) significantly higher.
g. it allows for matching techniques 'inspired' by those used in the waveguide technologies.

Figure 1, shows the difference in the layouts of two patch sub-arrays:

a. a planar array where the whole circuit is 200 mils above the ground-plane.
b. a non-planar array where *only the radiating elements* are 200 mils above the ground-plane and the feeding network is only 100 mils above the ground-plane. The connection between the radiating elements and the feeding network is achieved through *Three Dimensional Transitions*.

The reduction in the size of the feeding network is evident. A smaller feeding network provides more space for the radiating elements (less grating lobes), radiates less and has less mutual coupling. Finally, the use of the *three dimensional transition* for the radiating element or the array feeding network allows for new degrees of freedom in the design:

a. the control of the feeding transmission line width. In the classical printed circuit type of microstrip antennas, the input impedance of the patch might be very high (or low) involving a very narrow (or wide) feeding line. Here, as shown in Figure 1, the width of the feeding line can change very rapidly and brought to a comfortable size by having it suspended at the appropriate height above the ground-plane.
b. the design of patches that are separated from the ground-plane by up to a <u>tenth</u> of a wavelength. In some applications, the bandwidth obtained with a such a patch would eliminate the need for a parasitic element. This kind of patches have a very high input impedance and using the classical approach (dielectric substrates) they are very hard to match. Again, this can be achieved with a *Three Dimensional Transition* which in some cases can be even perpendicular to the ground-plane.

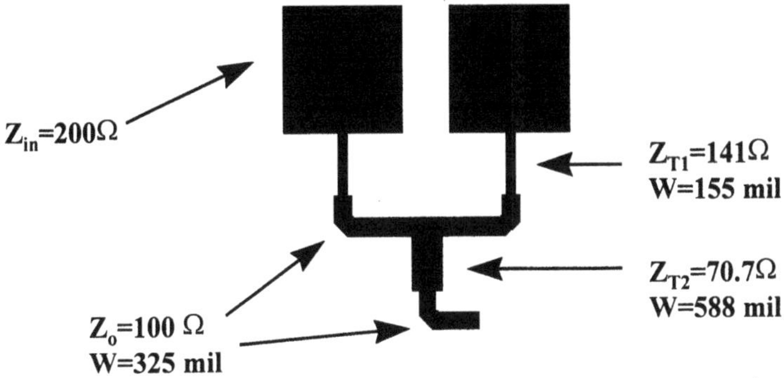

Planar Sub-Array; d=200 mil

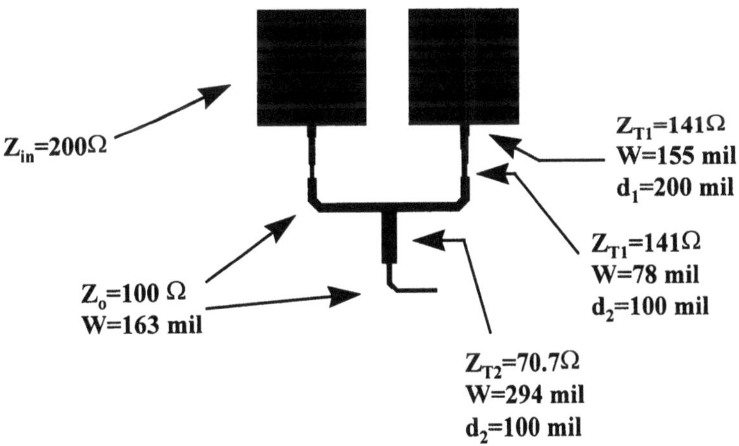

Non-Planar Sub-Array; d_1=200 mil, d_2=100 mil

Figure 1. Two microstrip sub-arrays.

THE MECHANICAL SUPPORT

Since there is no dielectric substrate or foam for support, the mechanical rigidity of the circuit is achieved through supporting posts which connect the center of the patch with the ground-plane. The supporting posts should be non-metallic for a number of reasons:

a. it would be very difficult to achieve good electrical contact between the post, the patch and the ground-plane.
b. it allows to use different metals for the circuit and the ground-plane. The bi-metal effect would have a negative impact on the circuit performance.

The effect of the perturbation by the post to the patch characteristics was calculated and found negligible when the ratio between the post diameter and the smaller dimension of the patch is less than 10%. The results of these calculations for a 2"×2" patch are shown in Figure 2. The patch is 250 mils above ground-plane, and the diameter of the post is 100 mils. Four cases were investigated:

a. the patch fed by a microstrip line.
b. the patch fed by a probe.
c. same as Case a., only the patch has an circular aperture in the center with a diameter of 100 mils
d. the patch is supported by a metallic post grounded at the other end. In both Case c. and d. the patch is fed by a microstrip line.

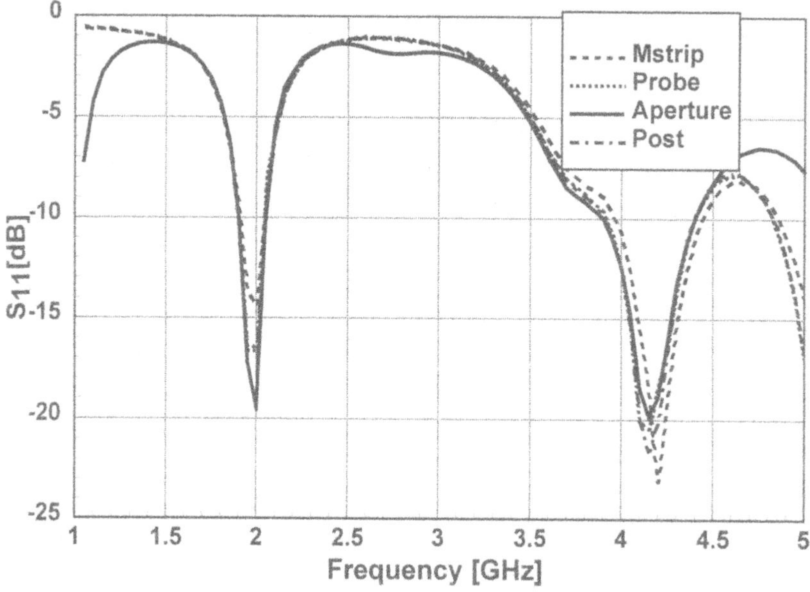

Figure 2. The impact of the supporting post on the patch performance.

THE DUAL-BAND SUSPENDEND PATCH

A number of different patch configurations were built and tested. Among them, the dual-band patch is of some interest for reasons mentioned above. The patch, shown in Figure 3 has a very high self impedance (about 500Ω) and was matched using a 3D transition. The measured return loss and its radiation patterns are shown in Figures 4 and 5 respectively.

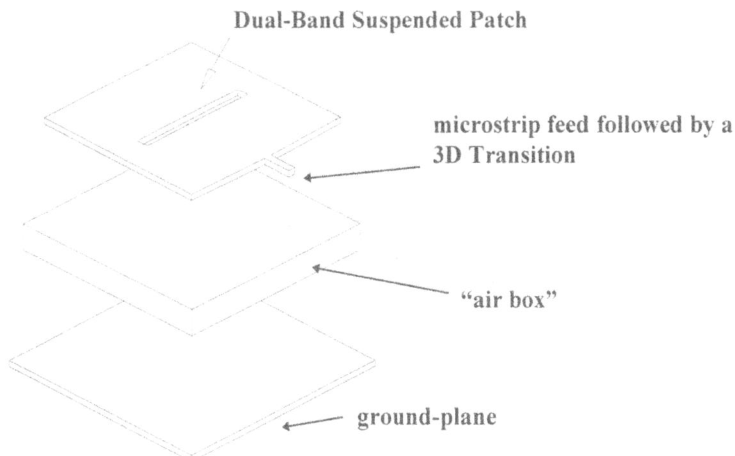

Figure 3. The geometry of the dual-band suspended patch.

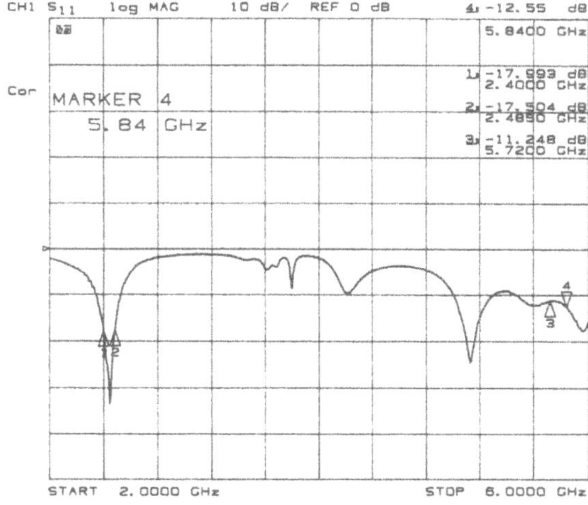

Figure 4. The return loss for the patch shown in Figure 3.

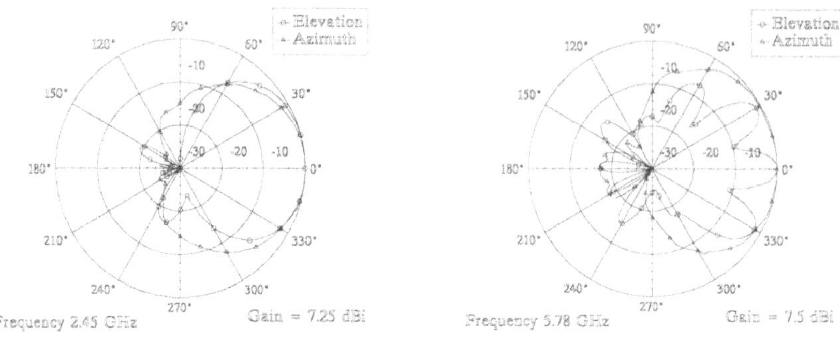

Figure 5. The radiation patterns of the dual-band suspended patch.

CONCLUSION

A new type of a dual-band microstrip patch was presented. The technology associated with it introduces a number of new degrees of freedom in the design and, in certain cases, allows a considerable reduce in the fabrication cost.
The author would like to thank **Dr. Jian Zhang** from **Zeland Software Inc**. for permitting the use of **IE3D** package in the analysis of the different configurations.

REFERENCES

1. E. J. Denlinger, 'Losses in of microstrip lines', *IEEE Trans. Microwave Theory Tech.*, vol. MTT-28, pp. 513-522, June 1980.
2. R. A. Pucel, J. D. Masse and C. P. Hartwig, 'Losses in microstrip' , *IEEE Trans. Microwave Theory Tech.*, vol. MTT-16, pp. 342-350, June 1968.
3. Ely Levine, Gabi Malamud, Shmuel Shrikman, 'A Study of Microstrip Array Antennas with the Feed Network', *IEEE Trans. Antennas and Propagation*, vol. 37, no. 4, pp. 426-434, April, 1989.
4. D.H.Schaubert, Private Communication.

THE INTEGRATED DIELECTRIC SLAB WAVEGUIDE-WEDGE ANTENNA

Gerald M. Whitman,[1] Felix Schwering,[2] Wan-Yu Chen,[1]
Anthony Triolo,[3] and Jitrayut Junnapart[4]

[1] New Jersey Institute of Technology, Newark, NJ
[2]CECOM, Fort Monmouth, NJ
[3]Lucent Technology, Whippany, NJ
[4]Ramkhumhung University, Thailand

ABSTRACT

Two methods are presented which yield radiation patterns for the dielectric wedge antenna fed by a dielectric slab waveguide of the same material. The first method of solution models the wedge as a sequence of step discontinuities and tracks forward and backward partial wave fields, expressed as modal expansions. The second method of solution uses a local mode theory together with the Schelkunoff equivalence principle. Both methods yield endfire radiation patterns for the dielectric slab/wedge antenna that are in excellent agreement over the important forward angular range from endfire to broadside. Radiation patterns are obtained for different dielectric constants and antenna lengths. The local mode method is seen to provide a computationally efficient and accurate method to study tapered dielectric antennas.

1. INTRODUCTION

Current interest in tapered dielectric antennas stems from the capability to integrate such antennas with dielectric waveguides for use in millimeter wave and integrated optical devices.[1–4] The structure under discussion is the dielectric slab waveguide feeding a dielectric wedge antenna of the same material (see Fig. 1). A fundamental even TE or TM surface wave mode of the dielectric slab waveguide is assumed to be incident on the dielectric wedge antenna. Rigorous theories, such as coupled-mode theory, can be used to find the fields.[5,6] However, these formulations are mathematically very complicated and have not yielded useful

Directions for the Next Generation of MMIC Devices and Systems
Edited by N. K. Das and H. L. Bertoni, Plenum Press, New York, 1997

solutions. Recently, two approximate approaches have been developed which have yielded highly accurate results for the radiation patterns of this integrated structure.[7–10] The first method, referred to as the mode matching method, is discussed in Section 2. In this method, the tapered region is approximated by a sequence of uniform slab segments of decreasing cross-sections. The analysis used is a modified version of the step-transition method that was first introduced by Marcuse[11,12] and improved upon by Suchoski, Jr. and Ramaswamy,[13] both of whom used the approach to study the effect of a continuous transition between dielectric slab waveguides of different cross-sections. This approach applies orthogonality relations to obtain a linear system of equations at each step discontinuity that is repeatable and numerically solved very efficiently. The second method, referred to as the local mode method, is discussed in Section 3. This approach uses a local mode theory together with the Schelkunoff equivalence principle to obtain the radiation pattern of the slab-wedge structure. The field expressions obtained using this method are much simpler since they are more analytical and require essentially only numerically evaluating an integral. Both methods yield very similar results and show that the radiation pattern is endfire with no minor lobes appearing in the forward angular range in the TE-case and very low sidelobes in the TM-case. Sidelobes, however, appear in experimentally determined radiation patterns of dielectric antennas fed by metal guides and, therefore, are due to scattering from the termination of the metal waveguide rather than from the dielectric antenna itself. In the following discussions of the two methods of solution only the TE-case will be presented since the TM-case is very similar. Data, however, will be presented for the two polarizations.

2. THE MODE-MATCHING METHOD - TE CASE

The dielectric slab waveguide feeding the dielectric wedge antenna is depicted in Fig. 1. Fig. 2 shows the staircase approximation of the wedge antenna. The fundamental even TE surface wave mode of the infinite slab waveguide of thickness $2D_1$ is assumed to be incident from $z = -\infty$ and to excite the wedge antenna. A time dependance of $exp(j\omega t)$ is assumed and suppressed. Geometry and excitation dictate that the field components are independent of "y" and are TE everywhere. The field components are E_{yi}, H_{xi} and H_{zi}. The subscript "i" denotes the region of space where the field exists; *i.e.*, $i = 1$ means the region $z < 0$ that is occupied by the semi-infinite dielectric slab of thickness $2D_1$, while $i = 2, 3, ..., M$ identifies regions occupied by slab segments of thickness $2D_i$ and equal lengths $\Delta z_i \equiv \Delta z$, where $D_i > D_{i+1} > D_M = 0$. Fig. 2 illustrates the staircase construction for $M = 5$.

In each region "i", the field components are represented as a superposition of both the finite discrete spectrum of surface wave modes and the infinite continuous spectrum of the radiation modes of the infinite dielectric slab of thickness $2D_i$.[12] The thickness D_i is chosen such that only the fundamental surface wave mode is above cutoff frequency. Since this mode does not experience cutoff, it can propagate along very thin slab waveguides. Hence, the transverse field components in region "i" are expressed as

$$E_{yi} = v_i(z)\Phi_i(x) + \int_0^{\infty} v_i(z,u)\Phi_i(x,u)du \tag{1a}$$

$$\eta_0 H_{xi} = i_i(z)\Phi_i(x) + \int_0^{\infty} i_i(z,u)\Phi_i(x,u)du, \tag{1b}$$

with expansion coefficients given by

$$v_i(z) = A_i e^{-j\beta_i z} + B_i e^{j\beta_i z} \tag{1c}$$

$$v_i(z,u) = A_i(u) e^{-j\beta_i(u) z} + B_i(u) e^{j\beta_i(u) z} \tag{1d}$$

$$i_i(z) = \frac{1}{jk_0}\frac{dv_i(z)}{dz}, \qquad i_i(z,u) = \frac{1}{jk_0}\frac{\partial v_i(z,u)}{\partial z}. \tag{1e}$$

In (1), the surface wave mode corresponds to β_i real in the range $k_0 \leq |\beta_i| \leq k_0 \epsilon_r^{1/2}$. The associate mode function is given by

$$\Phi_i(x) = \begin{cases} C_i \cos(k_{xi} D_i) e^{-\alpha_{xi}(|x| - D_i)}, & D_i \leq |x| < \infty \\ C_i \cos(k_{xi} x) \quad , & |x| \leq D_i \end{cases} \tag{2}$$

with

$$C_i = \left(\frac{\alpha_{xi}}{1 + \alpha_{xi} D_i}\right)^{1/2}, \qquad k_{xi} = \left(\epsilon_r k_0^2 - \beta_i^2\right)^{1/2}, \qquad \alpha_{xi} = \left(\beta_i^2 - k_0^2\right)^{1/2}, \tag{2a}$$

and dispersion and orthonormality relations, respectively,

$$k_{xi} \tan(k_{xi} D_i) = \alpha_{xi}, \qquad \int_{-\infty}^{\infty} \Phi_{mi}(x) \Phi_{ni}(x) dx = \delta_{mn}. \tag{2b}$$

Assuming the incident power to be unity gives $A_1 = \sqrt{\frac{2\eta_0 k_0}{\beta_1}}$. For the fundamental surface wave mode, the integer $n = 0$ is suppressed in the above equations. δ_{mn} is the Kronecker delta function. k_0 and η_0 are the free space wavenumber and intrinsic impedance, respectively.

The propagating radiation modes correspond to $\beta(u)$ real in the range $0 \leq |\beta(u)| \leq k_0$ and evanescent radiation modes occur when $\beta(u) = -j\,|\beta(u)|$ for $0 < |\beta(u)| < \infty$. The radiation mode functions are given by

$$\Phi_i(x,u) = \begin{cases} C(u)\left[\cos(vD_i)\cos(u(|x| - D_i)) - \frac{v}{u}\sin(vD_i)\sin(u(|x| - D_i))\right], \\ \qquad\qquad D_i \leq |x| < \infty \\ C(u)\cos(vx), \qquad |x| \leq D_i \end{cases} \tag{3}$$

with

$$C(u) = \left(\frac{\pi}{2}\left[\cos^2(vD_i) + (\frac{v}{u})^2 \sin^2(vD_i)\right]\right)^{-1/2} \tag{3a}$$

$$v = ((\epsilon_r - 1)k_0^2 + u^2)^{1/2}, \qquad \beta(u) = \begin{cases} (k_0^2 - u^2)^{1/2}, & 0 \leq u \leq k_0 \\ -j(u^2 - k_0^2)^{1/2}, & k_0 \leq u < \infty \end{cases} \tag{3b}$$

and orthonormality relation

$$\int_0^{\infty} \Phi_i(x, u)\Phi_i(x, u')dx = \delta(u - u'). \tag{3c}$$

$\delta(u - u')$ is the Dirac delta function. Surface wave and radiation modes are mutually orthogonal, *i.e.*,

$$\int_{-\infty}^{\infty} \Phi_{ni}(x)\Phi_i(x, u)dx = 0. \tag{4}$$

The longitudinal field component H_{zi} follows from Maxwell's equations.

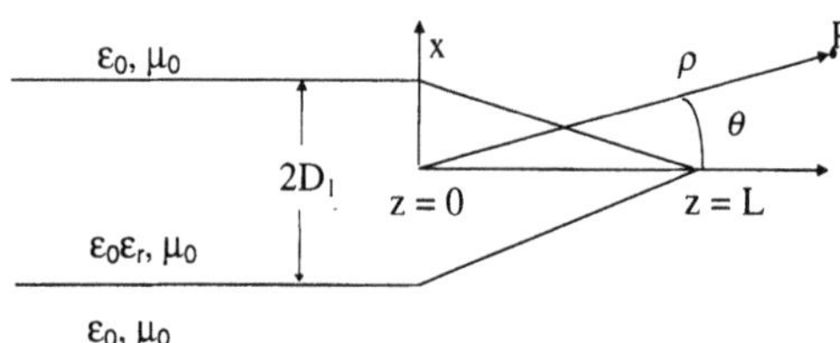

Figure 1. Dielectric slab waveguide feeding the dielectric wedge antenna of the same material.

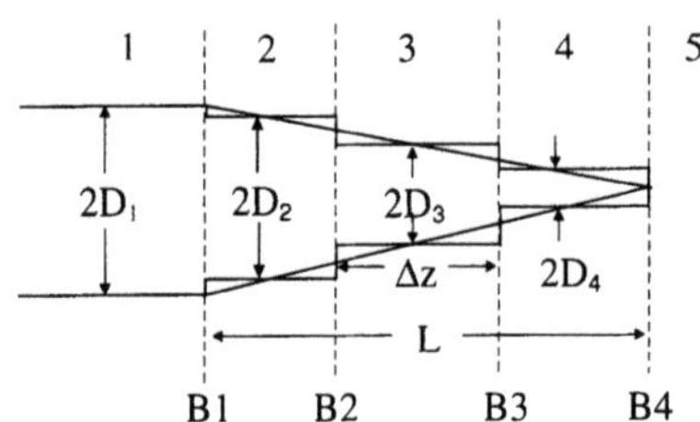

Figure 2. Staircase approximation of the dielectric wedge antenna.

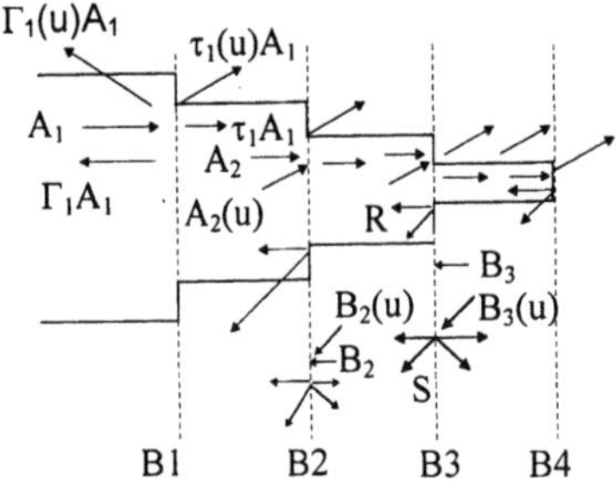

Figure 3. Wave progressions in the forward and backward directions.

Instead of applying boundary conditions to the total transverse field components in each region i, $i = 1, 2, ..., M$, and solving for all the unknown fields in each region simultaneously, it is more convenient to first find "partial" field solutions and then to construct the total field solution in each region as a superposition of both forward ($+z$) traveling and backward ($-z$) traveling partial fields. A partial wave field is one that is established due to scattering at a single step discontinuity while neglecting the effects of scattering at subsequent step discontinuities. The construction of partial wave field solutions is illustrated in Fig.[1] 3 for four slab segments. When a $+z$ propagating incident surface wave mode of amplitude A_1^{f1} first scatters at the step discontinuity located in the interface plane B_1, transmitted and reflected surface wave modes $\left(\tau_1^{f1} A_1^{f1}, \Gamma_1^{f1} A_1^{f1}\right)$ as well as transmitted and reflected radiation modes $(\tau_1^{f1}(u) A_1^{f1}, \Gamma_1^{f1}(u) A_1^{f1}, 0 \leq u < \infty)$ are established. All backscattered fields due to step discontinuities at $B_i, i = 2, 3, 4$, are momentarily neglected. Application of the boundary conditions of continuity of the tangential electric and magnetic fields at B_1 yield a system of equations which are then solved using a modified version of the procedure in [13]. Waves scattered in the forward direction that reach B_2 now involve both a single surface wave mode $(A_2^{f1} = \tau_1^{f1} A_1^{f1} e^{-j\beta_2 \Delta z})$ and the spectrum of radiation modes $(A_2^{f1}(u) = \tau_1^{f1}(u) A_1^{f1} e^{-j\beta_2(u)\Delta z}, 0 \leq u < \infty)$. These are scattered into transmitted and reflected surface wave modes $\left(\tau_2^{f1} A_2^{f1}, \Gamma_2^{f1} A_2^{f1}\right)$ as well as transmitted and reflected radiation modes $\left(\tau_2^{f1}(u) A_2^{f1}, \Gamma_2^{f1}(u) A_2^{f1}, 0 \leq u < \infty\right)$. Again, waves backscattered from step discontinuities at B_3 and B_4 are momentarily neglected. This process is repeated at each subsequent step discontinuity for waves traveling in the forward ($+z$) direction; the superscript $f1$ is used to designate the first forward progression of energy toward the tip. At the terminus (shown as B_4 in Fig. 3) waves traveling in the backward ($-z$) direction are established which now proceed toward the base of the wedge antenna. At B_3 in Fig. 3, the backward incident and scattered partial wave constituents are denoted by S. Boundary conditions of continuity of the tangential electric and magnetic field components at B_3 are then used to find the linear system of equations which is solved to find the expansion coefficients for the backward transmitted and reflected surface waves $(\tau_3^{b1}, \Gamma_3^{b1})$ and for the backward transmitted and reflected radiation modes $(\tau_3^{b1}(u), \Gamma_3^{b1}(u), 0 \leq u < \infty)$. The superscript $b1$ denotes the first backward going progression of energy toward the base of the wedge antenna. At B_2 in Fig. 3, incident backward traveling partial wave fields are taken to consist of two contributions; the first consists of the backscattered partial fields labeled R, that were neglected at B_2 but were generated in the scattering event which occurred during the first forward progression; the second consists of the scattered fields progressing in the $-z$ direction established at the scatter event labeled S. Hence, the backward traveling incident partial wave field that reach the step discontinuity in the plane B_2 have amplitudes given by $B_2^{b1} = (\tau_3^{b1} B_3^{b1} + \Gamma_3^{f1} A_3^{f1}) e^{j\beta_3 \Delta z}$ and $B_2^{b1}(u) = (\tau_3^{b1}(u) B_3^{b1} + \Gamma_3^{f1}(u) A_3^{f1}) e^{j\beta_3(u)\Delta z}$. In this manner, no partial fields are neglected in the final superposition of partial fields and progression in the forward and backward directions can be done as many times as is needed to achieve the desired degree of accuracy. Two forward and one backward progression were found to be sufficient to approximate the total field. Hence, the total electric field in region 5 and 1 of Fig. 2 are given by

$$E_{y5}^{total} \cong E_{y5}^{f1} + E_{y5}^{f2}, \quad z \geq L$$

[1] For convenience, the superscripts $f1$ and $b1$ which are defined later are suppressed in Fig. 3.

$$E_{y1}^{total} \cong E_{y1}^{f1} + E_{y1}^{b1}, \quad z \leq 0.$$

These fields are used to find the directive gain of the antenna.

As noted above, the unknown reflection and transmission coefficients at each step discontinuity are found by following a modified version of the procedure discussed in [13]. The modified procedure involves (1) using the modal representation of the partial field components to the left and to the right of a step discontinuity, (2) introducing normalized parameters, (3) applying the boundary conditions of continuity of the tangential electric and magnetic field components at each step discontinuity, (4) using mode orthogonality to obtain a simpler system of equations to solve, (5) truncating the infinite continuous spectrum such that evanescent modes are included,[2] (6) discretizing using Simpson's one-third rule and (7) correctly evaluating the double integrals over all u and all x, being careful not to interchange the order of integration unless permitted to do so. Thus at each step discontinuity and for the first forward progression, a linear system of equations is obtained for the unknown expansion coefficients $\tau_i^{f1}, \Gamma_i^{f1}, \tau_i^{f1}(u_j), \Gamma_i^{f1}(u_j), j = 1, 2, ..., N$, with the exception of $\Gamma_i^{f1}(k_0)$; the method does not yield values for $\tau_i^{f1}(0), \Gamma_i^{f1}(0)$ and $\Gamma_i^{f1}(k_0)$ and gives zero for $\tau_i^{f1}(k_0)$ which is non-physical. As mentioned before, [13] only applied their procedure to transitions between guides of different sizes whereas we have devised a scheme for handling the dielectric wedge.

THE LOCAL MODE METHOD - TE CASE

This method involves using a local mode theory with the Schelkunoff equivalence principle to determine the radiation pattern of the dielectric slab waveguide/wedge antenna depicted in Fig. 1. This is the same problem that is treated in Section 2 by using the mode-matching method. As in Section 2, the thickness of the dielectric slab waveguide is taken to support only the fundamental even TE surface wave mode. This mode is assumed to be incident in the $+z$ direction from $z = -\infty$ and to feed the dielectric wedge antenna. Utilizing the Schelkunoff equivalence principle, the radiation pattern of the antenna is derived from the polarization current in the dielectric region of the antenna structure. In other words, the polarization current in each volume element of the wedge and slab is regarded as a local Hertzian dipole moment, and the radiation pattern of the overall antenna is obtained by vector addition of the contributions of these individual radiating elements. For later use we note that the equivalence principle implies that the polarization current density is frozen in place while the dielectric material is removed so that each current element is radiating under free space conditions. If the polarization current density $j\omega P_y(x, z)\hat{y}$ is known exactly, the expression for the radiation pattern calculated in this way would be rigorous. In the present case, we are interested in a simple but acceptably accurate approximation, and $P_y(x, z)$ is determined using the "local mode" theory. This theory applies in the case of slender wedges of small apex angle and assumes that the polarization current density in the slab waveguide is associated with the unperturbed incident surface wave. If the slope of the boundary surfaces of the wedge is small, it can be expected that little energy will be reflected or scattered back towards the feed guide. Thus, using (1) and (2), the polarization current density in the slab waveguide is approximated by:

[2] In the numerical evaluation, the infinite continuous spectrum is truncated at $u = u_{\max}$ such that $u_j = j\Delta u$, $j = 1, 2, ..., N$, with $u_N = u_{\max} > k_0$ so as to include evanescent modes if needed.

$$\begin{aligned} j\omega P_y(x,z) &= j\omega\epsilon_0(\epsilon_r-1)E_y(x,z) \\ &= j\omega\epsilon_0(\epsilon_r-1)A_1\left(\frac{\alpha_{x1}}{1+\alpha_{x1}D_1}\right)^{1/2}\cos(k_{x1}x)e^{-j\beta_1 z}, |x|\le D_1, z<0. \end{aligned} \tag{5}$$

The propagation constant β_1 and the lateral wavenumbers k_{x1}, α_{x1} are determined by the dispersion relation via (2a) and (2b) with $D = D_1$. The local mode theory assumes, furthermore, that the incident surface wave continues into the wedge section in the sense that the cross sectional field distribution of the wedge field in any plane z = constant is that of the fundamental surface wave mode of a uniform guide having the local width $2D = 2D(z)$ of the wedge, while the complex amplitude of this mode is governed by phase accumulation and power conservation. In other words, as the mode travels down the wedge, phase increments are assumed to add while the total power transmitted through any z-plane is a constant independent of z. With (1) and (2), this results in the following expression for the polarization current density in the wedge:

$$\begin{aligned} j\omega P_y(x,z) &= j\omega\epsilon_0(\epsilon_r-1)E_y(x,z) \\ &= j\omega\epsilon_0(\epsilon_r-1)A_1\left(\frac{\beta_1}{\beta}\right)^{1/2}\left(\frac{\alpha_x}{1+\alpha_x D(z)}\right)^{1/2}\cos(k_x x)e^{-j\int_0^z \beta dz'}, \\ & \quad |x|\le D(z), 0\le z\le L, \end{aligned} \tag{6}$$

where β, k_x and α_x are functions of z as determined by the dispersion relation via (2a) and (2b).

From (6), the calculation of the contribution of the dielectric wedge to the field strength radiation pattern is straight forward. One obtains

$$R_W(\theta) = k_0^2\int_0^L Q(z')e^{-j\left[\int_0^{z'}\beta(z'')dz''-k_0 z'\cos\theta\right]}\left\{\frac{\sin w^+D(z')}{w^+}+\frac{\sin w^-D(z')}{w^-}\right\}dz' \tag{7}$$

with

$$u = k_0\sin\theta, \qquad w^{\pm} = u \pm k_x(z), \tag{7a}$$

$$Q(z) = \left[\frac{\beta_1}{\beta(z)}\frac{1+\alpha_{x1}D_1}{1+\alpha_x(z)D(z)}\frac{\alpha_x(z)}{\alpha_{x1}}\right]^{1/2}. \tag{7b}$$

The radiation angle θ is counted from the positive z-axis; see Fig. 1.

Calculation of the contribution to the radiated pattern of the slab waveguide is more of a problem since the slab extends to $z = -\infty$, and the asymptotic evaluation of the pattern integral, for $k_0 r \to \infty$, requires special care. The problem can be avoided by taking the approach indicated in Fig. 4. In the plane $z = 0 - \epsilon$ (termination plane of the slab), we introduce planar electric and magnetic current distributions, J_{sy} and M_{sx}, which are chosen such that they terminate the incident surface wave of the slab, reducing it to zero in the space range $z > 0$. Recall that in the local mode approximation, the incident surface wave is the only field present in the slab waveguide. With (1) and (2), we have

$$J_{sy}(x) = -H_x(x,0) = \begin{cases} -\frac{\beta_1}{\omega\mu_0} A_1 (\frac{\alpha_{x1}}{1+\alpha_{x1}D_1})^{1/2} \cos(k_{x1}x), & |x| \leq D_1 \\ -\frac{\beta_1}{\omega\mu_0} A_1 (\frac{\alpha_{x1}}{1+\alpha_{x1}D_1})^{1/2} \cos(k_{x1}D_1) e^{-\alpha_{x1}(|x|-D_1)}, & |x| \geq D_1 \end{cases} \quad (8)$$

and

$$M_{sx}(x) = -E_y(x,0) = \begin{cases} -A_1 (\frac{\alpha_{x1}}{1+\alpha_{x1}D_1})^{1/2} \cos(k_{x1}x), & |x| \leq D_1 \\ -A_1 (\frac{\alpha_{x1}}{1+\alpha_{x1}D_1})^{1/2} \cos(k_{x1}D_1) e^{-\alpha_{x1}(|x|-D_1)}, & |x| \geq D_1. \end{cases} \quad (9)$$

We compensate these fictitious current densities by assuming negative planar current distributions, $-J_{sy}$ and $-M_{sx}$, in the plane $z = 0 + \epsilon$ (base plane of the wedge). The problem may now be separated into two parts: the semi-infinite region terminated in the fictitious planar current density, shown on the left side of Fig. 4(b), and the wedge section with the negative fictitious current densities in its base plane on the right side.[3]

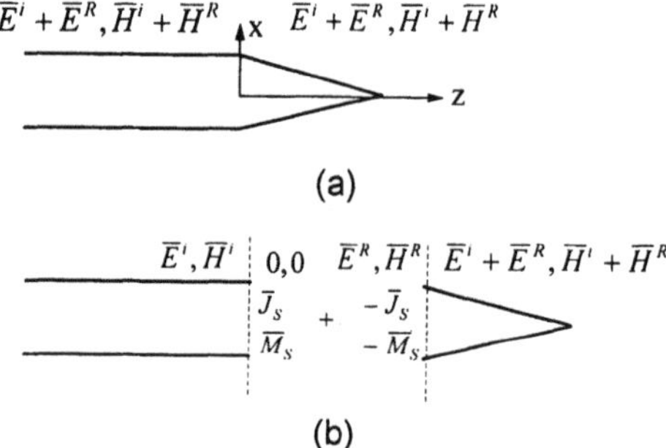

Figure 4. Equivalent current distributions; the shaded areas indicate the presence in free space of polarization current density, which replaces the dielectric material; the field $\overline{E}^i$, $\overline{H}^i$ is the incident surface wave mode determined from local mode theory while the field $\overline{E}^R$, $\overline{H}^R$ is the radiated field due to the polarization current density; the overbar means a vector quantity.

[3] Since the use of the equivalence principle implies that the dielectric material of the slab and wedge is removed (while the polarization currents stay in place as sources radiating in free space), the two sections of the antenna can be separated without affecting their individual radiation patterns.

The terminated slab region does not contribute to the radiation pattern. Its field is zero in the half space $z > 0$ and decreases exponentially with $|x|$ in the half space $z < 0$. Hence all radiation is provided by the structure on the right side of Fig. 4(b), which also means that the negative current distributions, $-J_{sy}$ and $-M_{sx}$, in the base plane of the wedge produce a radiation pattern equal to that produced by the polarization current density occupied by the semi-infinite slab region. Since these surface current distributions decrease exponentially with $|x|$, they are, in effect, restricted to a finite region of the plane $z = 0$ and the asymptotic form of their radiation field is easy to calculate, leading to the following closed form expression for the contribution of the semi-infinite dielectric slab waveguide to the field strength radiation pattern

$$R_G(\theta) = \frac{jk_0^2}{\beta_1 - k_0 \cos\theta}\left[\frac{\sin(u + k_{x1})D_1}{u + k_{x1}} + \frac{\sin(u - k_{x1})D_1}{u - k_{x1}}\right], \tag{10}$$

The total radiation pattern of the slab waveguide/wedge antenna is obtained by addition of R_W and R_G, and the power pattern of the antenna takes the form

$$G_D(\theta) = \frac{2\pi\rho S_{av}}{P_{rad}} = \frac{\pi\left|R_W(\theta) + R_G(\theta)\right|^2}{\int_0^\pi \left|R_W(\theta) + R_G(\theta)\right|^2 d\theta}. \tag{11}$$

The pattern (11) is normalized such that its value for the end-fire direction (main beam direction $\theta = 0°$) is the directivity gain of the antenna.

The local mode polarization current densities (5) and (6) are expected to be good approximations for narrow, slowly varying dielectric wedges, except near the tip. In this region, use of the local mode field yields polarization current that is highly inaccurate. However, only small errors in the far field are introduced because near the tip most of the field energy lies outside the dielectric wedge and the wedge cross-section is very small.[10] Justification for using the local mode field also follows from the fact that the local mode solution is the zero order solution to the four coupled integral differential equations that constitute a rigorous formulation of the dielectric wedge problem using the full wave theory.[10]

NUMERICAL RESULTS

The mode matching method, in principle, yields a rigorous determination of the radiation from the wedge provided that as more Δz_i slab segments are used in the staircase, the resultant radiation pattern converges. This requires that the field solution that is obtained at a single step discontinuity be sufficiently accurate. Hence, extensive comparisons were made with published data for scattering at a single step discontinuity, particularly, in the TE-case since published data is readily available for this polarization, but is very limited for the TM-case.[6–10] One such comparison is depicted in Fig. 5, which shows that the mode matching method presented here agrees with results for the radiation pattern at a single step discontinuity for the TE polarization case that were obtained using a Ritz-Galerkin variational procedure performed by T. Rozzi[16] and a finite element-boundary element approach used by Hirayama and Koshiba.[17] Reciprocity and energy conservation were found to be well satisfied for scattering at a single step using the mode matching method in the TE-case for all step sizes;[7,8]

in the TM-case, good satisfaction was obtained for small step sizes and for small dielectric constants, even when evanescent radiation modes are not used.

As mentioned in [7, 8], directive gain curves obtained using the mode matching method are not valid near $\theta = 90°$ where a discontinuous jump in the pattern occurs, and in the vicinity of $\theta = 180°$; the former is due to including in the asymptotic evaluation of the far field only the contribution from the first-order stationary phase point, whereas the latter is due to not considering the effect of the slab as θ approaches 180°. Additionally, in all plots of directive gain, the relative width of the semi-infinite slab is specified by

$$\frac{D_1}{\lambda_0} = \frac{1}{4}\frac{1}{\sqrt{\epsilon_r - 1}}. \tag{14}$$

This guarantees that only one surface wave mode is above cutoff and dictates, in the TE-case, that the normalized transverse wavenumbers $k_{x1}D_1$ and $\alpha_{x1}D_1$ in the slab and in free space, respectively, are independent of the dielectric constant. Physically this means that for different dielectric materials, the fundamental surface wave modes that propagate along the semi-infinite slab carry the same power within and outside the slab region and that the transverse field behavior of the surface waves remain independent of ϵ_r. In the TM-case since ϵ_r appears in the associated dispersion relation, these observations are no longer valid although the constraint equation (14) is used.

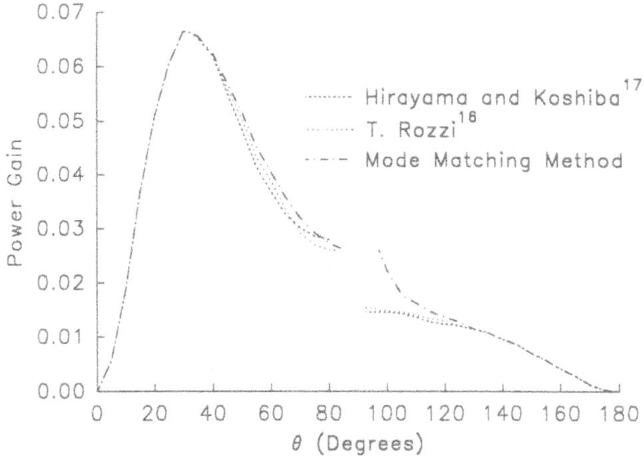

Figure 5. Comparison of the radiation patterns of power gain for a single step discontinuity for $k_0 D_1 = 1$, $\epsilon_r = 5$, $u_{\max} = 2k_0 D_1$, $N = 400$; TE case.

The directive gain plots in Figs. 6 and 7 show that as the number of steps is increased (so that the staircase more accurately models the wedge) the pattern converges for both the TE- and TM-polarizations. Shown in Figs. 8-10 are directive gain patterns for varying wedge lengths (L/λ_0 = 5, 10 and 15) and, in the TE-case, for two values of dielectric constants (ϵ_r = 2.56 and 12). Pattern plots were obtained using only propagating radiation modes since no change in the curves resulted when evanescent radiation modes were included. These directive gain patterns show that longer wedges are more directive with narrower beamwidths. Longer length wedges were anticipated to produce a more focused beam since the flow of energy is guided along the wedge in the forward direction over a longer distance. Fig. 11 shows for the TE-case that larger values of ϵ_r yield less directive patterns with broader beamwidths. The latter was to be expected because more energy gets scattered at the dielectric-air interface when ϵ_r is larger. The value of N specifies the number of elements used in discretizing the integral over the continuous transverse wavenumber "u" which is used in the representation of the radiation modes. Clearly, all patterns are endfire. In the TE-case, they exhibit no sidelobes in the forward direction, while in the TM-case, they display very small sidelobes. All patterns essentially decrease from a value of the order of 10dB in the endfire direction to very low dB values in the backscatter angular range, neglecting the vicinity of 180°.

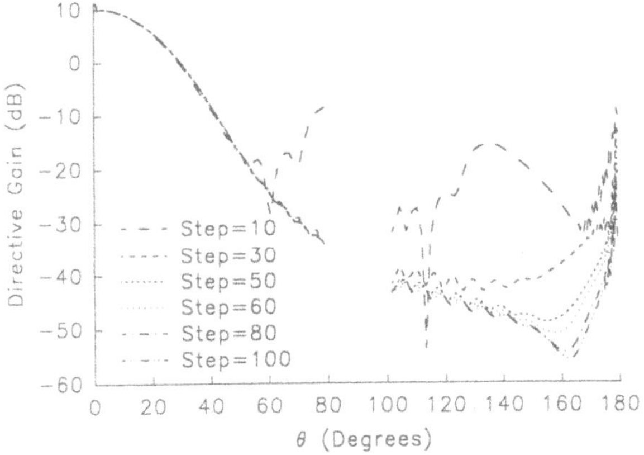

Figure 6. Radiation pattern of directive gain versus angle θ of the slab waveguide/wedge antenna for $\epsilon_r = 2.56(D_1 = 0.2\lambda_0)$, $L = 10\lambda_0$, $N = 100$, $u_{\max} = k_0 D_1$ for different number of steps; TE case.

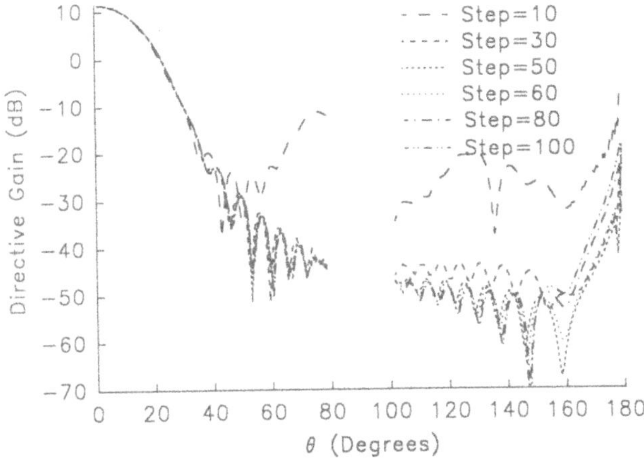

Figure 7. Radiation pattern of directive gain versus angle θ of the slab waveguide/wedge antenna for $\epsilon_r = 2.56(D_1 = 0.2\lambda_0)$, $L = 10\lambda_0$, $N = 100$, $u_{\max} = k_0 D_1$ for different number of steps; TM case.

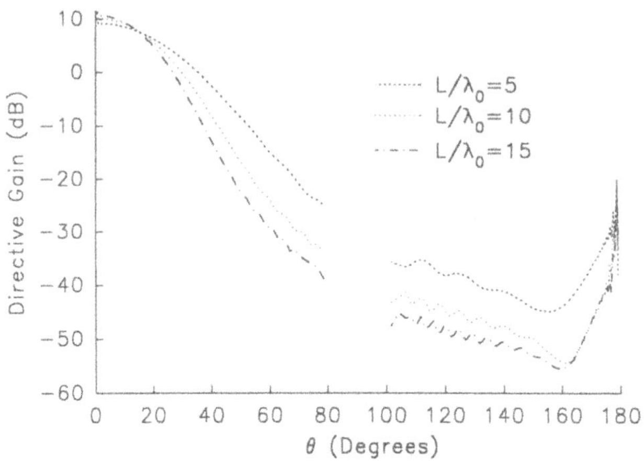

Figure 8. Radiation pattern of directive gain versus angle θ of the slab waveguide/wedge antenna for $\epsilon_r = 2.56 (D_1 = 0.2\lambda_0)$, 80 steps, $N = 100$, $u_{\max} = k_0 D_1$, and $L/\lambda_0 = 5, 10$ and 15; TE case.

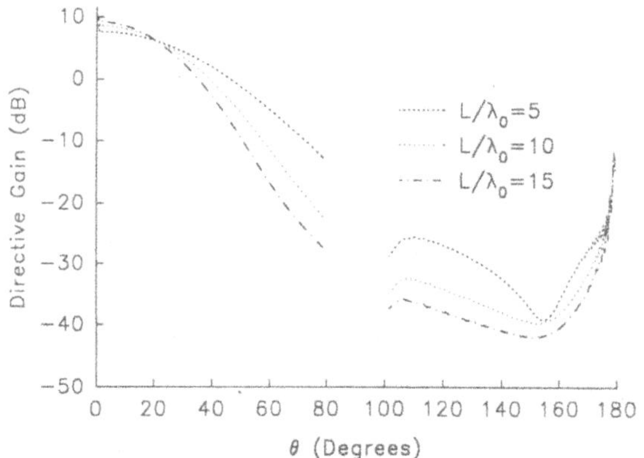

Figure 9. Radiation pattern of directive gain versus angle θ of the slab waveguide/wedge antenna for $\epsilon_r = 12 (D_1 = 0.075\lambda_0)$, 80 steps, $N = 100$, $u_{\max} = k_0 D_1$, and $L/\lambda_0 = 5, 10$ and 15; TE case.

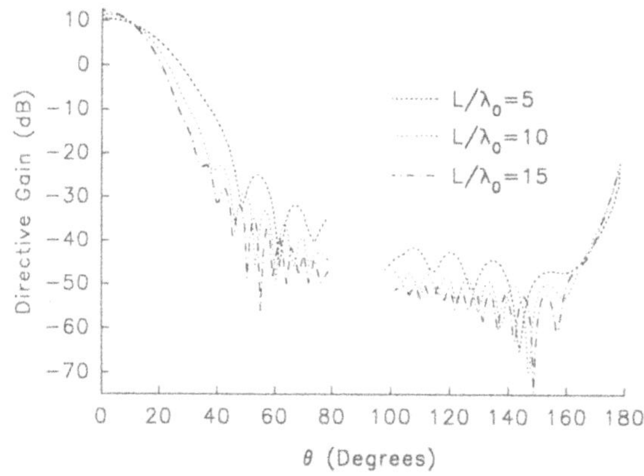

Figure 10. Radiation pattern of directive gain versus angle θ of the slab waveguide/wedge antenna for $\epsilon_r = 2.56 (D_1 = 0.2\lambda_0)$, 80 steps, $N = 100$, $u_{\max} = k_0 D_1$, and $L/\lambda_0 = 5, 10$ and 15; TM case.

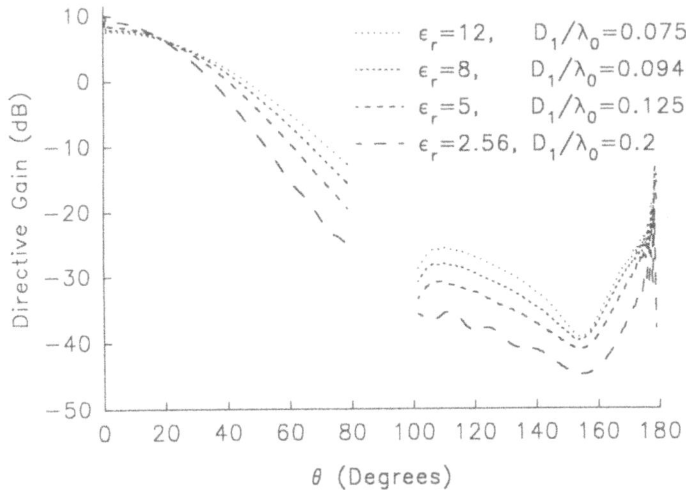

Figure 11. Radiation pattern of directive gain versus angle θ of the slab waveguide/wedge antenna for $L/\lambda_0 = 5$, 80 steps, $N = 100$, $u_{\max} = k_0 D_1$, and $\epsilon_r = 2.56, 5, 8$ and 12 (with $D_1/\lambda_0 = 0.2$, 0.125, 0.094, 0.075, respectively); TE case.

Figs. 12-14 compare directive gain patterns obtained via the local mode method and the mode matching method. They show that excellent agreement is obtained over the physically important forward range $0° < \theta < 90°$. In the backward range $90° < \theta < 150°$ differences do occur, but the dB levels are so low that for all practical purposes radiation is essentially negligible. Near 180°, as previously mentioned, the mode matching method is not valid. Only for $\epsilon_r = 2.56$ are results presented for the TM-case. A refinement in the theory is currently being implemented to obtain results for the TM-case for larger values of ϵ_r. This involves improving the satisfaction of the tangential electric field continuity boundary condition at each step discontinuity.

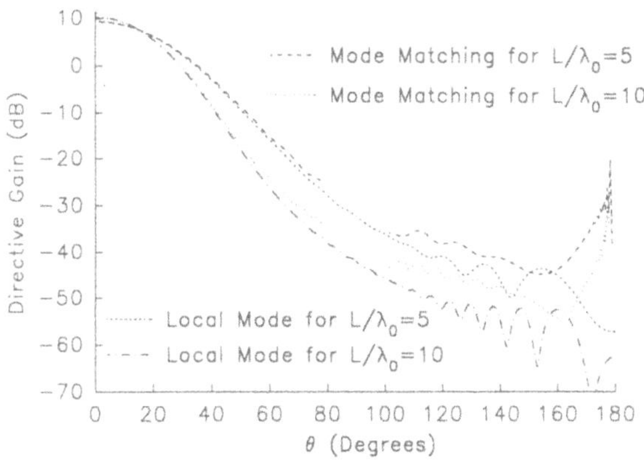

Figure 12. Radiation pattern of directive gain versus angle θ of the slab waveguide/wedge antenna for $\epsilon_r = 2.56$ ($D_1/\lambda_0 = 0.2$) and $L/\lambda_0 = 5$ and 10 using the local mode method and the mode matching method for 80 steps, $N = 100$, $u_{\max} = k_0 D_1$; TE case.

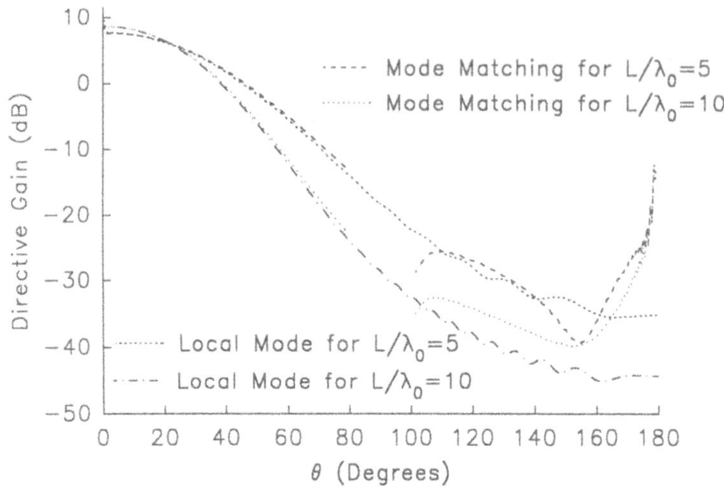

Figure 13. Radiation pattern of directive gain versus angle θ of the slab waveguide/wedge antenna for $\epsilon_r = 12$ $(D_1/\lambda_0 = 0.075)$ and $L/\lambda_0 = 5$ and 10 using the local mode method and the mode matching method for 80 steps, $N = 100$, $u_{\max} = k_0 D_1$; TE case.

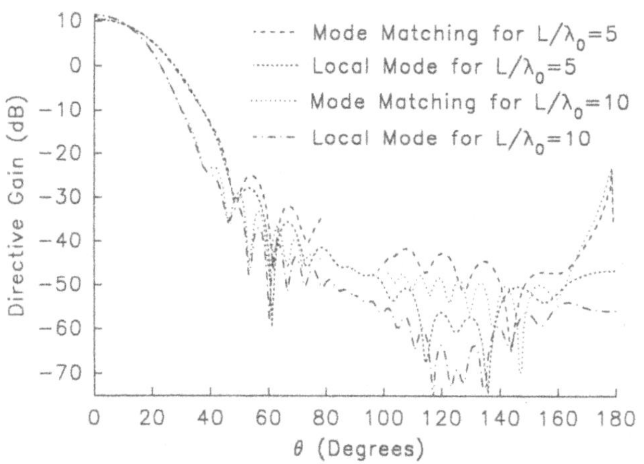

Figure 14. Radiation pattern of directive gain versus angle θ of the slab waveguide/wedge antenna for $\epsilon_r = 2.56$ $(D_1/\lambda_0 = 0.2)$ and $L/\lambda_0 = 5$ and 10 using the local mode method and the mode matching method for 80 steps, $N = 100$, $u_{\max} = k_0 D_1$; TM case.

5. CONCLUSION

For the dielectric wedge antenna excited by the fundamental TE or TM even surface wave mode of the dielectric slab waveguide of the same material, two independent methods of solution showed that the radiation pattern is endfire, that it smoothly decrease from a maximum of about 10dB in the forward direction, and that sidelobes are extremely low often below 25dB over an angular range that in the mode-match solution extends to about $\theta = 150°$. The mode matching method is not valid near 180° and data obtained via this method past 150° has to be viewed with caution.

While the mode matching method takes twenty-four minutes of CPU time on a DEC VAX4000 main frame for a staircase of eighty steps with each additional ten steps requiring

3 minutes more, the local mode method takes only 30 seconds on a 486, 66MHz personal computer using 1000 segments of length Δz_i to obtain convergence of the integral evaluation. Hence, the local mode method appears to be a very efficient method to study complicated geometries that are difficult to analyze mathematically and/or require intensive computational tools.

REFERENCES

1. Schwering, F. and Oliner, A.A., Millimeter-wave antennas in *Antenna Handbook*, Lo, Y.T. and Lee, S.W., Eds., Van Nostrand Reinhold Co., (1988).
2. Levin, B.J. and Keitzer, J.E., "Hybrid millimeter-wave integrated circuits," *Tech. Rept. ECOM-74-0577-F*, Contract DAAB07-74-C-0577, performed for the U.S. Army Electronics Command, Fort Monmouth, NJ by ITT Research Institute, Chicago, IL, October (1975).
3. Jacobs, H. and Chrepta, M.M., "Electronic phase shifter for millimeter-wave semiconductor dielectric integrated circuits," *IEEE Trans. Microwave Theory and Tech.*, Vol. MTT-22, No. 4, April (1974).
4. Baets, R. and Lagasse, P.E., "Calculation of radiation loss in integrated-optic tapers and Y-junctions," *Applied Optics*, Vol. 21, No. 11, June (1982).
5. Bahar, E., "Propagation of radio waves over a non-uniform layered medium," *Radio Science*, Vol. 5, July (1970).
6. Shevchenko, V.V., *Continuous Transitions in Open Waveguides*, Golem Press (1971).
7. Whitman, G.M., Schwering, F. and Chen, W-Y, "The tapered dielectric slab radiator," *International IEEE AP-S Symposium and URSI Radio Science Meeting*, Ann Arbor, MI, June 28–July 2 (1993).
8. Whitman, G.M., Schwering, F., and Chen, W-Y, "Rigorous TE solution to the staircase model of the dielectric wedge antenna fed by a slab waveguide," Submitted for publication to the *IEEE Trans. on Antennas and Propagation*, June (1995).
9. Schwering, F., Whitman, G.M., Chen, W-Y and Triolo, A. "The dielectric wedge antenna fed by a slab waveguide," *International IEEE AP-S Symposium and URSI Radio Science Meeting*, Newport Beach, California, June (1995).
10. Schwering, F., Whitman, G.M. and Triolo, A. "An approximation but accurate analysis of the dielectric wedge antenna fed by a slab waveguide using local mode theory and the Schelkunoff equivalence principle," submitted for publication to *IEEE Trans. on Antennas and Propagation*, November (1996).
11. Marcuse, D., "Radiation losses of tapered dielectric slab waveguides," *Bell System Tech. J.*, Vol. 49, No. 2, pp. 273-290, February (1970).
12. Marcuse, D., *Light Transmission Optics*, 2nd Edition, Van Nostrand Reinhold (1982).
13. Suchoski, P.G., Jr. and Ramaswamy, V., "Exact numerical technique for the analysis of step discontinuities and tapers in optical dielectric waveguides," *J. Opt. Soc. Am. A*, Vol. 3, No. 2, pp. 194-203, February (1986).
14. Snyder, A.W. and Love, J.D., *Optical Waveguide Theory*, NY, Chapman and Hall (1983).
15. Harrington, R.F., *Time-Harmonic Electromagnetic Fields*, NY, McGraw-Hill (1961).
16. Rozzi, T.E., "Rigorous analysis of the step discontinuity in a planar dielectric waveguide," *IEEE Trans. of Microwave Theory and Tech.*, Vol. MTT-26, No. 10, pp. 738-746, October (1978).
17. Hirayama, K. and Koshiba, K. "Numerical analysis of arbitrarily shaped discontinuities between planar dielectric waveguides with different thicknesses," *IEEE Trans. on Microwave Theory and Tech.*, Vol. MTT-38, No. 3, pp. 260-264, March (1990).

DUAL-POLARIZED RECTANGULAR DIELECTRIC RESONATOR ANTENNA

Z. Fan[1], Y.M.M. Antar[1], N. Sultan[2] and G. Séguin[2]

[1]Dept. of Electrical and Computer Engineering
Royal Military College of Canada
Kingston, ON K7K 5L0, Canada
[2]Space systems, Canadian Space Agency
6767 route de l'Aéroport
Saint-Hubert, Québec J3Y 8Y9, Canada

ABSTRACT

A theoretical analysis of a three-layer-microstrip / offset-slot coupled rectangular dielectric resonator antenna (DRA) is presented, and feasibility of using it to design a dual-polarized antenna is investigated. A combined method based on the modal expansion method and the spectral domain approach is developed. The accuracy of the theoretical analysis is verified by comparison with experimental results. Several dual-polarized DRAs with two offset slots are fabricated, and scattering parameters and radiation patterns are measured. It is found that by suitable designs isolation between two ports of -20 dB and cross-polarization level of -20 dB can be achieved.

INTRODUCTION

Recent studies of dielectric resonator (DR) antennas have shown that DRAs exhibit wider impedance bandwidth, smaller size, and no surface wave and less conductor losses when compared to microstrip patch antennas [1]. Most of theoretical and experimental work has been done on linearly polarized DRAs using a single feeding line. A few papers have also been published, dealing with circularly polarized DRAs using one or two feeding lines [2] [3].

In this paper we present a theoretical and efficient analysis of electrical properties of a rectangular DRA fed by a three-layer microstrip line through a slot on the ground plane. The method of analysis is based on the modal expansion method and the spectral domain approach, and used to compute Q factors, radiation patterns and input impedance. Also we experimentally investigate the feasibility of using the DRA to realize a dual-polarized antenna. In this dual-polarized DRA, two microstrip lines are used to excite two orthogonal fundamental modes of DRAs via two offset slots in

Directions for the Next Generation of MMIC Devices and Systems
Edited by N. K. Das and H. L. Bertoni, Plenum Press, New York, 1997

the ground plane in order to realize dual-polarization operation. Several dual-polarized slot-coupled antennas are designed and fabricated. Measured results for return losses, isolation between two ports, and radiation patterns for co- and cross-polarized fields are presented and discussed.

THEORETICAL ANALYSIS

Figure 1 shows the configuration of a three-layer-microstrip / offset-slot coupled rectangular dielectric resonator antenna. The slot in the ground plane, which is located symmetrically with respect to the microstrip line, is used to couple the energy from the microstrip line to the dielectric resonator. The slot has displacement x_c and y_c from the center of the DR in the x and y directions, respectively.

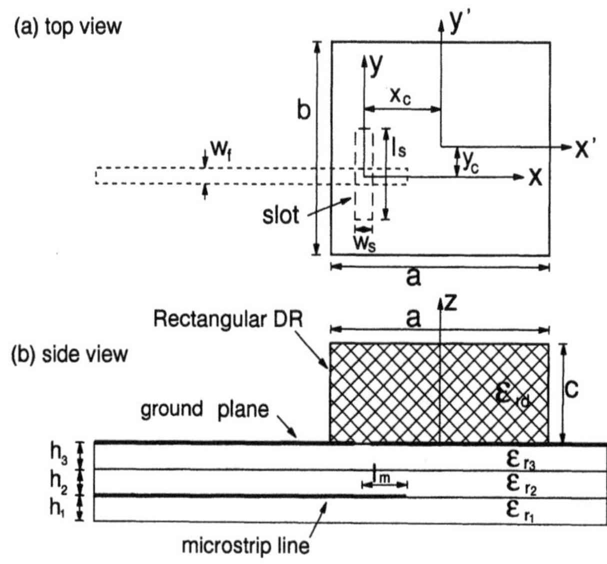

Figure 1: *A three-layer-microstrip / offset-slot coupled rectangular DRA*

Isolated DR on a Ground Plane. To compute natural resonant frequencies and electromagnetic fields of different modes a waveguide model can be used. In the model for TE to y' modes, the DR is treated as a dielectric body placed inside a rectangular waveguide with magnetic walls at $x' = \pm\frac{a}{2}$ and $z = c$ and with an electric wall at $z = 0$. The field components of the $TE^{y'}_{nlm}$ mode can be obtained from the electric scalar potential ϕ_{nlm}, for instance:

$$H^d_{y'_{nlm}} = \frac{j}{\omega\mu_0}\left(\frac{\partial^2\phi_{nlm}}{\partial x'^2} + \frac{\partial^2\phi_{nlm}}{\partial y'^2}\right) \tag{1}$$

After considering the boundary conditions on the walls of the waveguide, ϕ_{nlm} can be expressed as

$$\phi_{nlm}(x', y', z) = \sin[\kappa_{x'}(x' + \frac{a}{2})]\ \phi_l(y')\ \cos(\kappa_z z) \tag{2}$$

where $\kappa_{x'} = \frac{n\pi}{a}$ and $\kappa_{z'} = \frac{(2m-1)\pi}{2c}$, with $n, m = 1, 2, \ldots$. $\phi_l(y')$ satisfies

$$\frac{\partial^2\phi_l}{\partial y'^2} + \kappa^2_{y'}\phi_l = 0 \quad \text{for} \quad |y'| < \frac{b}{2}, \qquad \frac{\partial^2\phi_l}{\partial y'^2} - \gamma_0^2\phi_l = 0 \quad \text{for} \quad |y'| > \frac{b}{2} \tag{3}$$

in which $\kappa_{y'}^2 = \omega^2\mu_0\varepsilon_0\varepsilon_{rd} - \kappa_x^2 - \kappa_z^2$, $\gamma_0^2 = \kappa_x^2 + \kappa_z^2 - \omega^2\mu_0\varepsilon_0$. Applying the continuity conditions for electric and magnetic fields at $y' = \pm\frac{b}{2}$, we can obtain the solution to the above two equations and the characteristic equation for $\kappa_{y'}$

$$(\gamma_0 \sin\frac{\kappa_{y'}b}{2} + \kappa_{y'}\cos\frac{\kappa_{y'}b}{2})\ (\gamma_0\cos\frac{\kappa_{y'}b}{2} - \kappa_{y'}\sin\frac{\kappa_{y'}b}{2}) = 0 \tag{4}$$

Once finding $\kappa_{y'}$ ($= l\frac{\delta_l\pi}{b}$, where $0 < \delta_l < 1$, $l = 1, 2, \ldots$), the natural resonant angular frequency can be obtained as follows: $\omega_{nlm}^2 = (\kappa_{x'}^2 + \kappa_{y'}^2 + \kappa_z^2)/(\mu_0\varepsilon_0\varepsilon_{rd})$.

To compute radiation patterns of the $TE_{nlm}^{y'}$ mode, by using the image principle the ground plane can be equivalently replaced by the image of the DR. Applying the equivalence principle to the surfaces of the DR and its image of the same size, we can obtain equivalent magnetic currents at the $x' = \pm\frac{a}{2}$, $y' = \pm\frac{b}{2}$ and $z = \pm c$ surfaces and equivalent electric currents at the $y' = \pm\frac{b}{2}$ surfaces. From these equivalent currents, the electric and magnetic vector potentials and then the far-zone fields $(\vec{E}^r, \vec{H}^r)$ can be obtained. By integrating the Poynting vector, the radiated power P_r can be determined. Also by using a perturbation approach, power P_d dissipated in the dielectric and power P_c lost in the ground plane can be found. The total quality factor Q_{nlm} and radiation efficiency η_{nlm} can be determined from P_r, P_d, P_c and the stored magnetic energy W_H.

Three–Layer Microstrip Line with a Dielectric Cover. The spectral domain approach is used to obtain the field solution of the fundamental mode. The phase constant β is derived from the determinantal equation which is obtained by applying Galerkin's method to the boundary condistion on the conducting strip, and the characteristic impedance Z_0 is obtained based on the power–current definition. Electromagnetic field components are represented through their one–dimensional Fourier transforms: For convenience, the electromagnetic fields need to be normalized such that

$$\int_{y=-\infty}^{\infty}\int_{z=-\infty}^{0} \vec{e}\times\vec{h}\cdot\vec{x}\ dzdy = 1 \tag{5}$$

A Slot in the Ground Plane. The spectral domain approach is also used to find the fields generated by the slot in the ground plane of the microstrip line. The width of the slot is assumed to be electrically short, and hence only the x component of electric field E_x^s in the slot ($|x| < \frac{w_s}{2}, |y| < \frac{l_s}{2}$) need to be considered. $E_x^s(x,y)$ is expanded in terms of entire domain basis functions with expansion coefficients C_m

$$E_x^s(x,y) = \sum_{m=1}^{M} C_m\ E_m^s(x,y) = \sum_{m=1}^{M} C_m\ \frac{1}{w_s}\ \sin[\frac{m\pi}{l_s}(y + \frac{l_s}{2})] \tag{6}$$

Two–dimensional Fourier transforms can be used to represent the electromagnetic fields for the microstrip line side ($z < 0$), for example:

$$H_y^s(x,y,z) = \frac{1}{4\pi^2}\int_{-\infty}^{\infty}\int_{-\infty}^{\infty} \tilde{H}_y^s(\kappa_x,\kappa_y,z)e^{-j(\kappa_x x+\kappa_y y)}\ d\kappa_x\ d\kappa_y \tag{7}$$

Slot–Coupled Dielectric Resonator Antenna. A combined model, based on the modal expansion method and spectral domain approach, is developed to analyze the whole structure. By applying the equivalence principle and ensuring continuity of the tangential electric field across the slot, the coupling slot can be equivalently modelled as two magnetic currents $\vec{M}^s$ on the DR side and $-\vec{M}^s$ on the feed side, respectively,

For the DR side, the modal expansion method [4] is used to compute the field. The field can be expressed in terms of only resonant $TE^{y'}$ modes since the magnetic current $\vec{M}^s$ excites only $TE^{y'}$ modes. For example, the y component of the magnetic field can be obtained as

$$\vec{H}^d(\vec{M}^s) = \sum_{nlm} C_{nlm} \ \vec{H}^d_{nlm}(x,y,z) \tag{8}$$

where

$$C_{nlm} = \frac{j\omega \int_{S_s} \vec{M}^s \cdot \vec{H}^d_{nlm} \ dS}{[\omega^2 - \omega^2_{nlm}/(1 - j\frac{1}{Q_{nlm}})] \int_{V_g} \mu_0 |\vec{H}^d_{nlm}|^2 \ dV} \tag{9}$$

It should be noted that the total loss has been taken into account in terms of Q_{nlm} factor in the calculation of the field on the DR side, as seen from (9).

For the feed side, the reciprocity theory [5] can be used to obtain the voltage reflection coefficient R and the transmission coefficient T on the infinitely long microstrip line at the slot due to the slot discontinuity.

$$R = 1 - T = \frac{1}{4\pi} \int_{-\infty}^{\infty} \tilde{E}^s_x(-\kappa_y) \ \tilde{h}_y(\kappa_y, 0^-) \ d\kappa_y \tag{10}$$

The total magnetic field of the line can be expressed as

$$\vec{H}^f = \begin{cases} (\vec{h}_x + \vec{h}_{yz}) \ e^{-j\beta x} + R \ (\vec{h}_x - \vec{h}_{yz}) \ e^{j\beta x} & \text{for } x < 0 \\ T \ (\vec{h}_x + \vec{h}_{yz}) \ e^{-j\beta x} & \text{for } x > 0 \end{cases} \tag{11}$$

The y component of the total magnetic field in the vicinity of the slot on the feed side can be written as

$$H^f_y(x,y,0^-) = H^s_y(-\vec{M}^s) + (1 - R) \ h_y(\vec{J}^f) \tag{12}$$

By enforcing the continuity of the y component of the magnetic field in the slot, integral equations for the electric field in the slot E^s_x can be obtained. Galerkin's procedure is then used to solve these equations for expansion coefficients C_m. The relationship $1 - R = T$ in (10) implies that the equivalent circuit of the slot is a series impedance Z_s on the microstrip line $Z_s = Z_0 \ \frac{2R}{1-R}$.

EXPERIMENTAL VERIFICATION

To assess the accuracy of the analysis presented in the previous section, an offset–slot coupled DRA was fabricated and tested for the comparison with computed results for return losses and radiation patterns.

Computed and measured results for return losses are shown as a function of frequency in Figure 2. The agreement is seen to be reasonable. For example, measured resonant frequency, at which return loss is minimum, is 5.475 GHz; and computed resonant frequency is 5.625 GHz, so the difference is about 2.7%.

Figure 3 shows the comparison between computed and measured radiation patterns. There are some ripples in the measured radiation patterns, specially in the E plane, attributed to the effects of a finite ground plane. Given the fact that theory assumes the presence of an infinite ground plane, theory and experiment compare reasonably well.

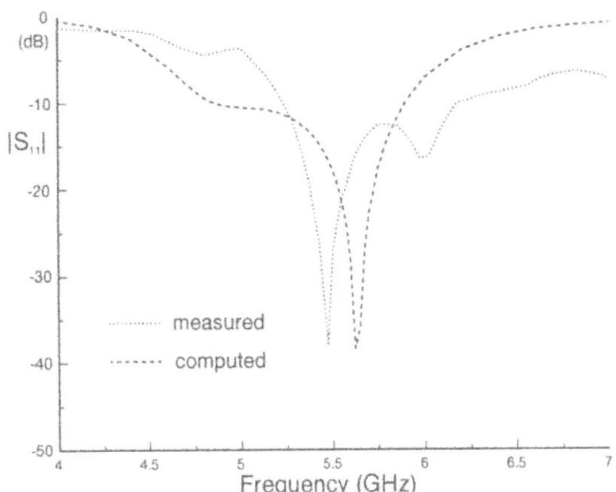

Figure 2: *Comparison between computed and measured return losses (ε_{rd} =10, a = b =20.3 mm, c =5.1 mm, l_s =7.5 mm, w_s =0.7 mm, ε_{r2} =10.2, w_f =0.6 mm, h_2 =0.635 mm, $h_1 = h_3$=0, l_m =1.804 mm, x_c =4.8 mm, y_c =0 mm)*

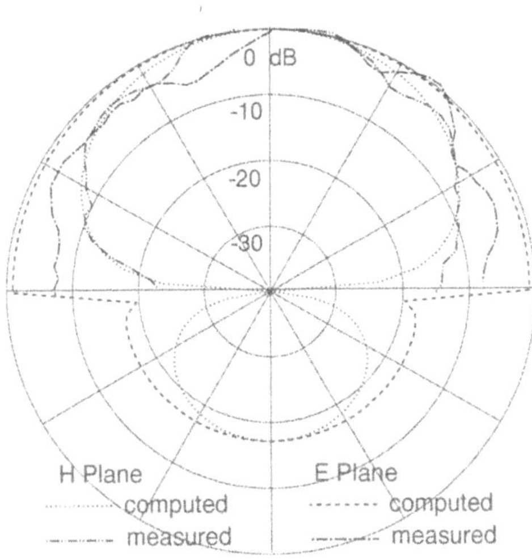

Figure 3: *Comparison between computed and measured radiation patterns at frequency of 5.5 GHz (The same structural and material parameters as in Figure 2)*

MEASURED RESULTS FOR DUAL–POLARIZED DRAs

Figure 4 shows the top view of a dual–polarized slot–coupled rectangular DRA with two offset slots. In a design procedure, the permittivity and dimensions of the DR are first found for a required central frequency and bandwidth. Second, the width of the feeding line is determined to obtain a specified value of characteristic impedance. Then, slot length and open stub length are optimised using an iterative method to achieve impedance matching. The above design procedure have been used to design several offset slot–coupled DRA to be used as dual–polarized antennas. In the following, measured results will be given for two samples.

Figure 5 shows measured scattering parameters of sample 1. Although curves for return losses of two input ports $|S_{11}|$ and $|S_{22}|$ do not have exactly the same shape due

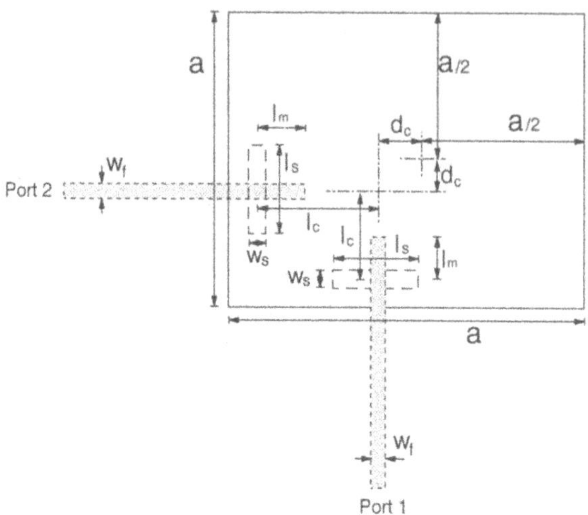

Figure 4: *Top view of a dual-polarized slot-coupled rectangular dielectric resonator antenna with two offset slots*

to the fabrication tolerance, the bandwidth for return losses less than -10 dB is about the same for the two ports. In fact, the bandwidth for port 1 is 19.8% from 5.125 to 6.25 GHz; the bandwidth for port 2 is 17.7% from 5.15 to 6.15 GHz. It is also seen that the isolation $|S_{12}|$ between two ports is about - 18 dB at the center of the band, but increases by about 3 and 4 dB at upper and lower ends of the band, respectively.

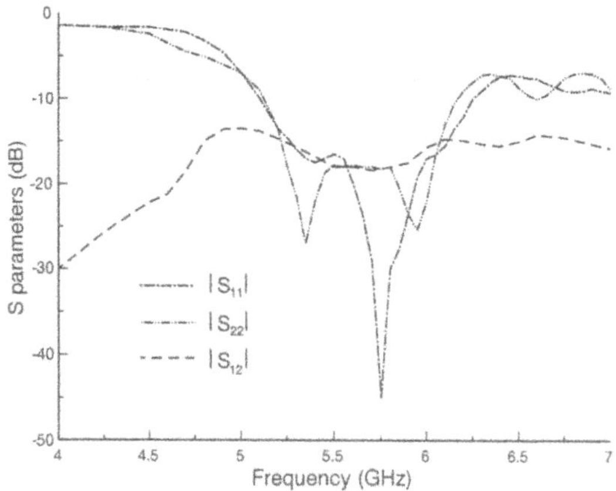

Figure 5: *Measured scattering parameters of sample 1 (ε_{rd} =10, a = b =20.3 mm, c =5.1 mm, l_s =7.5 mm, w_s =0.7 mm, ε_{r2} =10.2, w_f =0.6 mm, h_2 =0.635 mm, $h_1 = h_3 = 0$, l_m =1.804 mm, l_c =6.0 mm, d_c =-2.3 mm)*

Figure 6 shows measured H– and E–plane radiation patterns of sample 1 at frequency of 5.5 GHz. Both co–polarized and cross–polarized fields are given. The co–polarized field patterns are found to be broadside with some ripples especially in the E plane. These ripples are attributed to the edge diffraction of the finite ground plane. Compared with the co–polarization level, the cross–polarization level is about -20 dB below at broadside, but may be higher at other angles.

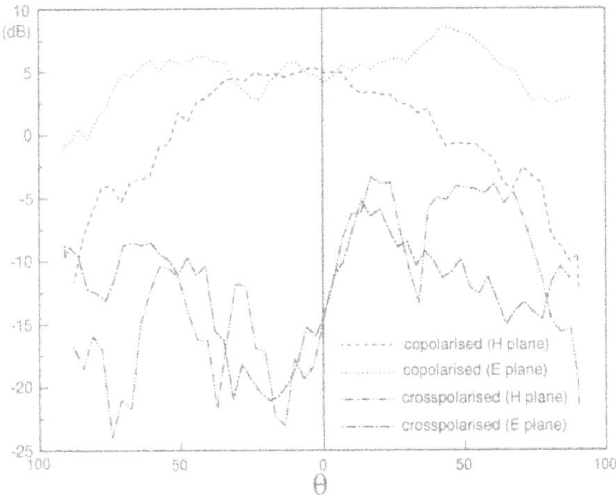

Figure 6: *Measured H– and E-plane radiation patterns of sample 1 at frequency of 5.5 GHz (The same structural and material parameters as in Figure 5)*

In Figures 7 and 8, measured scattering parameters of sample 2 as a function of frequency and H– and E-plane radiation patterns at the frequency of 5.5 GHz are shown. The bandwidths are found to be 18.5% from 5.275 to 6.35 GHz and 17.4% from 5.25 to 6.25 GHz for ports 1 and 2, respectively. The isolation is between -23 dB and -15 dB across the band. The cross-polarization level is about -20 dB at broadside. The cross-polarized fields are not symmetrical possibly because of offset slots, asymmetrical location of the DR ($d_c \neq 0$) and radiation from the connector. It has also been found that with decreasing slot length higher isolation and lower cross-polarization level can be obtained.

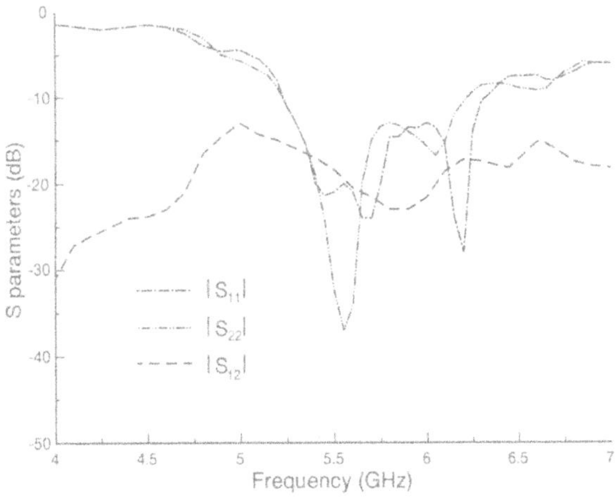

Figure 7: *Measured scattering parameters of sample 2 (ε_{rd} =10, a = b =17.8 mm, c =5.4 mm, l_s =7.5 mm, w_s =0.7 mm, ε_{r2} =10.2, w_f =0.6 mm, h_2 =0.635 mm, $h_1 = h_3 = 0$, l_m =1.784 mm, l_c =5.8 mm, d_c =-2.3 mm)*

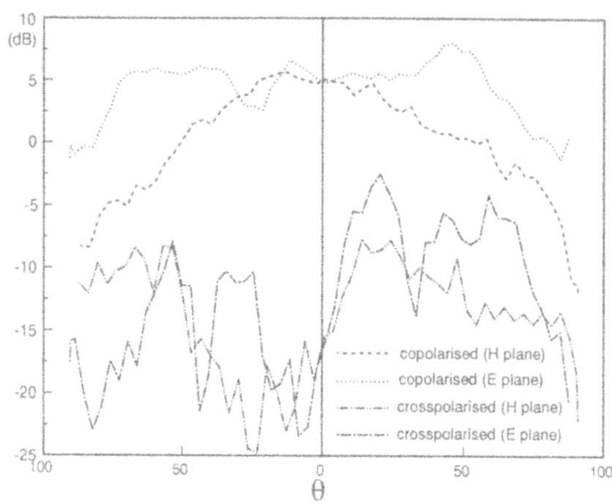

Figure 8: *Measured H– and E–plane radiation patterns of sample 2 at frequency of 5.5 GHz (The same structural and material parameters as in Figure 7)*

CONCLUSIONS

The approach combining the modal expansion method and the spectral domain approach has been described to analyze the slot coupled rectangular DRA with a three–layer microstrip line. The computed results for return losses and radiation patterns have been found to be in reasonably good agreement with measured data. Several dual–polarized DRAs with two offset slots have been fabricated and tested. Measured results for return losses, isolation between ports and radiation patterns for co– and cross–polarized fields have been presented. It has been found that by suitable designs isolation between two ports of -20 dB at the center frequency and cross–polarization level of -20 dB at broadside can be achieved. In order to further improve isolation and cross polarization, we are investigating other techniques, such as use of two–layer microstrip lines on different surfaces with two centered crossed slots.

References

[1] S.A. Long, M.W. McAllister and L.C. Shen, "The resonant cylindrical dielectric cavity antenna," *IEEE Trans. Antennas Propagat.*, vol.AP-31, pp.406–412, May 1983.

[2] M.B. Oliver, Y.M.M. Antar, R.K. Mongia and A. Ittipiboon, "Circularly polarized rectangular dielectric resonator antenna," *Electron. Lett.*, vol.31, pp.418–419, March 1995.

[3] R.K. Mongia, A. Ittipiboon, M. Cuhaci and D. Roscoe, "Circularly polarized dielectric resonator antenna," *Electron. Lett.*, vol.30, pp.1361–1362, Aug. 1994.

[4] R.F. Harrington, "Time harmonic electromagnetic fields," McGraw–Hill, New York, 1961.

[5] D.M. Pozar, "A reciprocity method of analysis for printed slot and slot-coupled microstrip antennas," *IEEE Trans. Antennas Propagat.*, vol.AP-34, pp.1439–1446, 1986.

SIGE MMIC'S - ON THE CURRENT STATE-OF-THE-ART

W. Heinrich

Ferdinand-Braun-Institut
D-12489 Berlin / Germany

INTRODUCTION

In recent years, the frequency limits of SiGe Heterostructure Bipolar Transistors (HBTs) have been extended to frequencies above 50 GHz. This opens new perspectives for Si-based monolithic circuits and raises questions up to which frequency range Si MMICs can be competitive with GaAs. This paper starts with an overview on the current activities in the field of SiGe HBT MMICs. In a second part we report on recent results in the 26 GHz and 38 GHz frequency range. While there are various activities in the 2 GHz range and for digital ICs, the scope of this paper is restricted to analog monolithically integrated circuits with SiGe HBTs for frequencies beyond 10 GHz.

So far, in this area mainly two research groups have built up a technological base: IBM, East Fishkill (New York), and Daimler Benz, Ulm (Germany). More recently, similar activities have been launched at the Institut für Halbleiterphysik, Frankfurt/Oder (Germany).

MMIC PROCESS

One fundamental choice in the development of Si-based MMICs is that of the substrate-resistivity type. High-resistivity Si substrates offer low-loss properties at microwave frequencies but are not fully compatible with standard Si process technology. The common low-resistivity substrates, on the other hand, lead to high losses and hence low quality factors of the passive elements such as transmission lines and spiral inductors. In order to avoid this the IBM SiGe MMIC approach[1] employs BCB thin-film circuit elements on top of the substrate. Since the IBM process is based on low-resistivity substrates the SiGe integrated circuits can be fabricated on a commercial CMOS line.

The Daimler Benz process, on the other hand, uses high-resistivity substrates. Recent advances[2] in the HBT fabrication applying thin MBE grown SiGe-alloy layers as the base yield cut-off frequencies above 100 GHz. Device design relies on the double heterojunction SiGe base approach. High base doping reduces the base sheet-resistance. In a collaboration with the Ferdinand-Braun Institut a coplanar MMIC process has been developed together with the CAD environment necessary for efficient circuit design[3]. Due to the high-resistivity substrate, special processing steps were necessary to prevent formation of a parasitic conducting layer on the wafer surface induced by surface charges in the passivation[4]. As a result, losses of the passive elements are dominated by the metallization. For spiral inductors, for instance, quality factors around 15 are obtained at 10 GHz, which is comparable to GaAs-based MMICs (see Fig. 1).

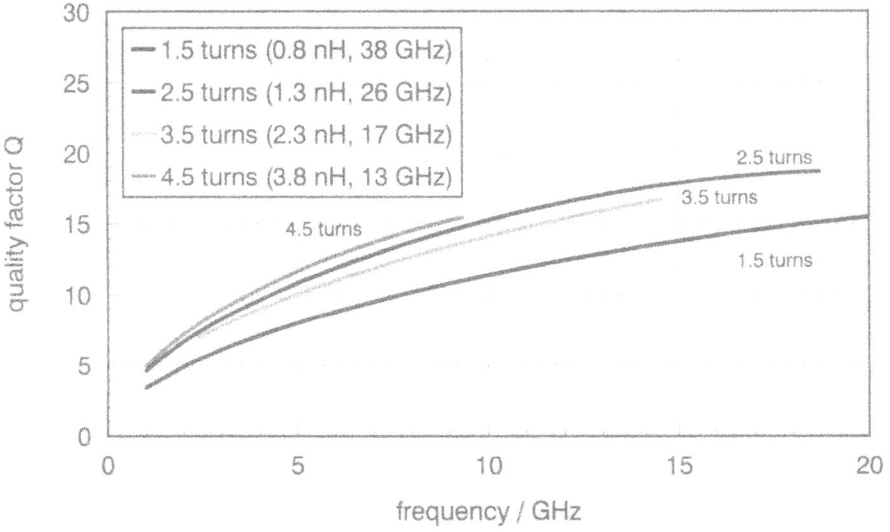

Fig. 1. Quality factors of spiral inductors on high-resistivity Si substrates (inductance and resonant frequency are given in the inset).

CIRCUIT PERFORMANCE

In the past, mainly current-gain and MAG (or unilateral gain) cut-off frequencies f_T and f_{max}, respectively, have been used as figures of merit when assessing SiGe HBT microwave capabilities. This data, however, accounts only for a small part of the high-frequency circuit performance of the HBT. Therefore, we will not elaborate on f_{max} or f_T record values in this paper but rather emphasize the SiGe HBT potential in MMIC circuit applications.

Table 1 provides an overview on analog MMICs fabricated recently by Daimler Benz and IBM that illustrate the state-of-the-art in SiGe HBT MMICs.

Table 1. Examples for analog SiGe-HBT MMICs fabricated by Daimler Benz[5] and IBM[6].

MMIC circuit type	performance	fabricated by
12 GHz VCO	19dBm, -80dBc/Hz phase noise	IBM
12 GHz power amp	> 6dB gain, 19dBm output	IBM
Broadband amplifier	17 GHz bandwidth, 8 dB gain	IBM
Gain block (2 stage)	18 GHz bandwidth, 9.5db gain	Daimler Benz
1-stage amplifier	4 dB gain @ 26 GHz	Daimler Benz
Oscillator	2 dBm @ 38 GHz	Daimler Benz

EXAMPLE: 38 GHZ OSCILLATOR

As an example to illustrate details of an HBT MMIC, a fixed-frequency oscillator in coplanar technique is presented in the following. The IC was fabricated by Daimler Benz while the Ferdinand-Braun Institut (FBH) was responsible for modelling, design and measurements.

As well known from GaAs MMICs, coplanar waveguides (CPW) with miniaturized dimensions lead to low-dispersive transmission-line elements. Based on field-level simulation (3D Finite-Difference in frequency domain), equivalent-circuit type models for the coplanar discontinuities and junctions were developed and implemented in a commercial software environment for MMIC design.

Fig. 2 shows the layout. The two emitter fingers are biased separately in order to be able to compensate for possible unsymmetries. Air bridges are necessary to suppress the parasitic slot-line mode on the CPW transmission lines. Chip size is 1.4 x 2.2 mm^2. The layout is not yet optimized with regard to chip size and packaging density. The oscillator delivers 2 dBm power at 38 GHz, with excellent design accuracy[7].

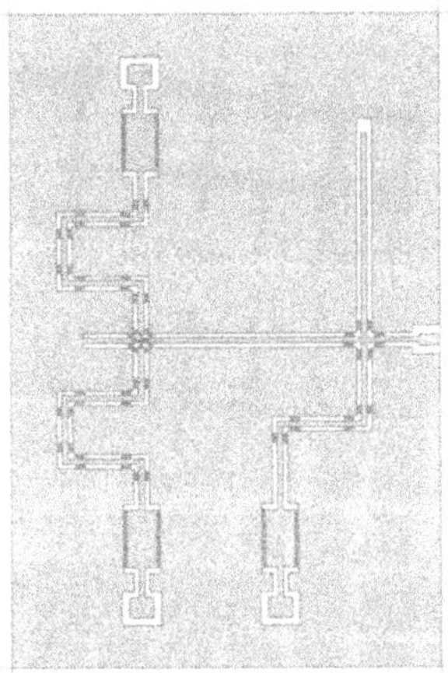

Fig. 2. Layout of the 38 GHz coplanar HBT oscillator.

CONCLUSIONS

MMICs with SiGe HBTs for operation beyond 10 GHz have become reality. As for Si-MMICs in general, substrate loss remains a critical issue. It can be resolved by employing either high-resistivity Si substrates, at the expense of process modifications, or by using dielectric film structures on top of the substrate. The present results demonstrate the potential of SiGe-HBT based MMICs for applications in the microwave and low mm-wave frequency range.

ACKOWLEDGMENTS

The author is grateful to J.-F. Luy from Daimler Benz and to B. Meyerson from IBM for providing information on the SiGe process. The work at the FBH is performed in collaboration with Daimler Benz, Ulm, and is funded by the German Ministry BMBF under contract 01 M 2938 B.

REFERENCES

1. M. Case, S. Maas, L. Larson, D. Rensch, D. Harame, and B. Meyerson, "An X-band monolithic active mixer in SiGe HBT technology," *1996 Int. Microwave Symposium Digest*, vol. 2, pp. 655-658.
2. A Schüppen, A. Gruhle, H. Kibbel, U. Erben, and U.König, "SiGe-HBTs with high f_T at moderate current densities," *Elect. Letters*, vol. 30, no. 14, pp. 1187-1188, 1994.
3. C. Rheinfelder, K. Strohm, F. Beißwanger, J. Gerdes, F.J. Schmückle, J.-F. Luy, W. Heinrich, "26 GHz coplanar MMICs," *1996 Int. Microwave Symposium Digest*, vol. 1, pp. 273-276.
4. R. Doerner, J. Gerdes, C. Rheinfelder, F.J. Schmückle, W. Heinrich, K. Strohm, F. Schäffler, and J.-F. Luy, "Modelling of passive elements for coplanar SiGe MMIC's," 1995 IEEE Int. Microwave Symposium Digest, Vol. III, pp. 1187-1190.
5. J.-F. Luy, *private communication.*
6. B. Meyerson, *private communication.*
7. C.N. Rheinfelder, F. Beißwanger, J. Gerdes, F.J. Schmückle, K.M.Strohm, J.-F. Luy, and W. Heinrich, "A coplanar 38 GHz SiGe-MMIC oscillator," IEEE Microwave and Guided-Wave Letters, vol. 6, no.11, to be published (1996).
6. B. Meyerson, *private communication.*

FM NOISE AND SYNCHRONIZATION BEHAVIOR

OF A SIMMWIC 76.5 GHZ FRONT-END

M. Singer[1], J.-F. Luy[2], A. Stiller[1,2], K.M. Strohm[2], E.M. Biebl[1]

[1]Lehrstuhl für Hochfrequenztechnik, Technische Universität München,
Arcisstraße 21, 80333 München, Germany
[2]Daimler-Benz Research,
Wilhelm-Runge Str. 11, 89081 Ulm, Germany

INTRODUCTION

SIMMWIC's (Silicon Monolithic Millimeter Wave Integrated Circuit) have gained increasing interest as receivers and transmitters in the millimeter wave region. They provide new solutions for near-range sensor and communication applications in the frequency range above 50 GHz. Particular advantages are the capability of very easy monolithic integration of a complete radiating oscillator, a high degree of technological reproducibility, small size and, thus, low cost[1]. A typical millimeter wave front-end consists of an IMPATT diode integrated in a planar resonator. This resonator acts simultaneously as an antenna resulting in an active antenna configuration. Due to the low q-factor of the resonator, for some applications, like FM-CW-radar, the frequency stability and the phase noise behavior of the front-end are insufficient. Using a phase-locked-loop is a practicable way to stabilize the oscillation frequency and suppress the phase noise. However, many millimeter wave components such as frequency dividers, voltage controlled oscillators (VCO's) and phase/frequency comparators are required resulting in an expensive and complex multichip system.

Due to the relatively short coherence length of the free running active antenna, the unstabilized front-end finds only short range applications[2]. For a larger operation range, frequency stabilization of the IMPATT oscillator is required, which is advantageously realized using subharmonic injection locking. In an integrated FM-CW system the signal of a tunable low noise reference oscillator, e.g. a Si/SiGe hetero bipolar transistor oscillator[3] (HBT), is injected at a subharmonic frequency into the front-end. In Figure 1 a low-cost integrated millimeter wave sensor system is depicted. The system consists of only three devices, the synchronizing oscillator at 25.5 GHz, the front-end at 76.5 GHz and the signal processing unit at IF band. The signal of the tunable low

Directions for the Next Generation of MMIC Devices and Systems
Edited by N. K. Das and H. L. Bertoni, Plenum Press, New York, 1997

and width 380 μm was found to match the impedance of the diode[5]. For best matching of active and passive part, the diode is located at 110 μm offset from the center of the resonator (Figure 2). The bias network consists of coupled microstrip lines. Via these coupled microstrip lines the injection power is fed symmetrically into the diode by means of a signal-ground prober. Dipole antenna and diode have been monolithically integrated on high resistive silicon with a thickness of 125 μm and backside metallization. This small substrate thickness results in low thermal resistance thus providing high thermal stability of the oscillator. The calculated radiation efficiency of the planar antenna is $\eta = 25$ %. The complete layer sequence of the diode was grown by silicon molecular beam epitaxy (Si-MBE). The device with a diameter of 22 μm is processed following a state-of-the-art SIMMWIC process[1].

RESULTS

At first we determined the oscillation frequency and the output power of the free running IMPATT oscillator depending on the DC bias current I. With $I = 25$ mA the oscillation starts at 76.67 GHz. By varying the diode current in the range from 25 to 45 mA the oscillation frequency may be tuned from 76.67 GHz to 76.31 GHz (Figure 3). The output power of the oscillator increases linearly by 0.33 dBm/mA for

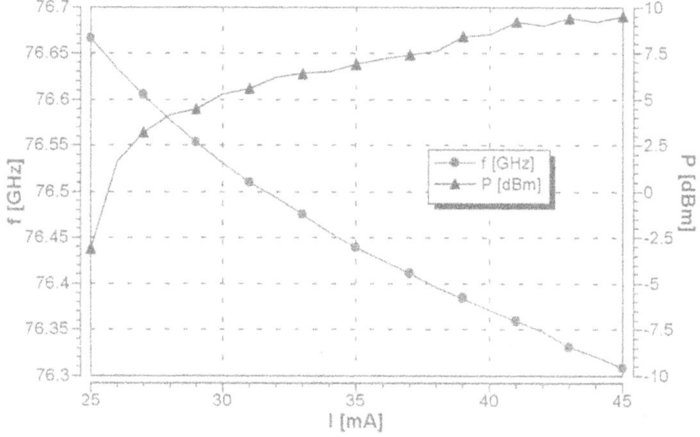

Figure 3. Output power P and oscillation frequency f depending on I

currents higher than 35 mA. At $I = 31.6$ mA the oscillation frequency adjusts at the desired value of 76.5 GHz with an output power of 6 dBm. Subsequently, we measured the phase noise of the free-running oscillator with the HP 3048 phase noise measurement system. For detecting the phase noise of the oscillator we used the frequency discriminator method. The radiated signal of the antenna under test (AuT) is detected by a horn antenna and mixed down to the baseband (10 MHz - 1.28 GHz) using a low noise SAW-oscillator at 640 MHz (Figure 4). At an offset carrier frequency of $\Omega = 100$ kHz we obtained a phase noise to carrier ratio $\mathcal{L}\ (\Omega) = -62$ dBc/Hz (Figure 5). In the next step, we measured the phase noise of the injection signal generated by a HP8340B synthesizer at 25.5 GHz. From these measured values we calculated the phase noise of an ideal noise free oscillator at 76.5 GHz subharmonically injection locked with this synthesizer. For $\Omega = 1,\ 10,\ 50,\ 100,\ 500$ kHz we predicted $\mathcal{L}_{min}\ (\Omega) = -60,\ -65,\ -77,\ -88,\ -103$ dBc/Hz, respectively. Now, the IMPATT oscillator was subharmonically injection locked. The tuning ranges and phase noises were

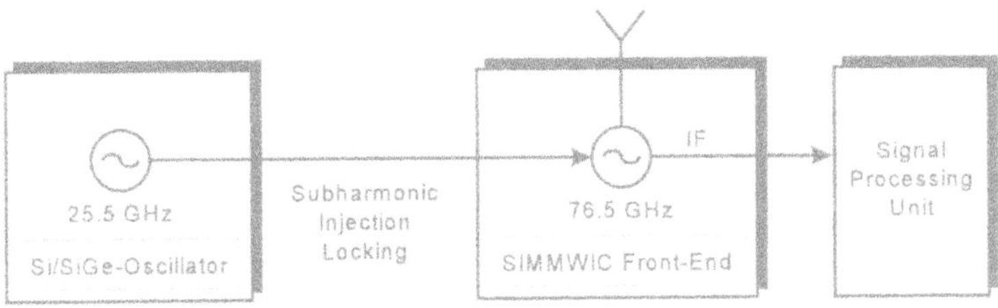

Figure 1. Millimeter wave near-range sensor system

noise Si/SiGe-HBT oscillator is injected into the SIMMWIC front-end via the bias lines (s. Figure 2). Even for a small power level of the injected signal the phase of the active antenna is locked to the synchronizing oscillator. This means, within a certain bandwidth around the carrier frequency, i.e. the synchronization bandwidth, the phase noise of the active antenna is dominated by the reference oscillator. Using the self-mixing properties of the active antenna due to the nonlinear behavior of the IMPATT diode, the IF is automatically generated. So the front-end is used as a transceiver. In the signal processing unit the IF signal is analyzed. To obtain good system performance, precise knowledgement of the stabilization and noise behavior of the front-end is required.

DESIGN AND FABRICATION

Matching of the low-impedance IMPATT diode in a planar configuration is critical and requires careful optimization of diode structure, geometry and resonator layout. For the design of the SIMMWIC front-end full wave analysis techniques based on electromagnetic field theoretical approaches have to be used for correct modelling of this strongly radiating structure. We used a rigorous full wave analysis taking into account all relevant loss mechanisms such as radiation losses, losses due to surface wave excitation, dielectric losses and ohmic losses. The analysis of the passive part of the oscillator is based on the electrical field integral equation (EFIE) in spectral domain, solved by means of method of moments[4]. The solution of the resulting complex system matrix delivers all interesting properties of the examined antenna, namely input impedance, gain, radiation pattern and radiation efficiency. A symmetric dipole of length 620 μm

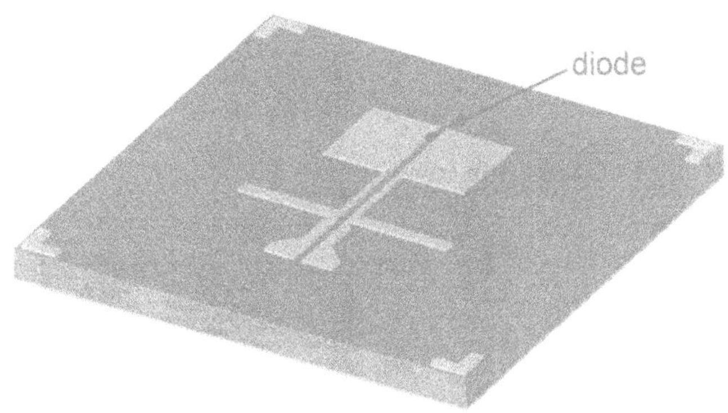

Figure 2. IMPATT oscillator on 125 μm Silicon

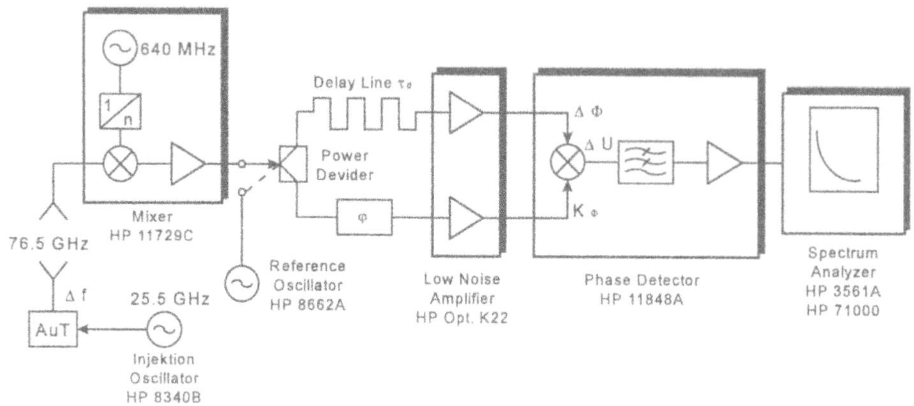

Figure 4. Phase noise measurement setup

measured for different input power levels and for different impressed diode currents.

The phase noise behavior of the injection locked oscillator depending on the injection power is depicted in Figure 6. The diode's bias current was adjusted to $I = 30$ mA. For small Ω (1 kHz, 10 kHz) the phase noise does not depend on the injection power level. Our results are in contrast to measurements by Zhang *et al.*[6] with subharmonically injection locked FET oscillators, where the phase noise increased with decreasing injection power near to the carrier. The measured phase noise is close to the predicted values for an ideal oscillator indicating that the phase noise behavior is dominated by the reference oscillator. For large offset Ω the phase noise decreases with increasing injection power level and remains almost constant above a certain level. This shows clearly that the synchronization bandwidth increases with increasing power of the injected signal. An injected power level of -3 dBm results in a synchronization bandwidth of about 1 MHz, which is sufficient for the planned applications.

In Figure 7 the 3-dB tuning range is depicted for different injection power levels and for different diode currents. We observed that frequency stabilization can only be achieved for frequencies higher than the oscillation frequency of the free-running oscillator. For an increase of the diode current the output power of the oscillator increases as well. The tuning range is mainly determined by the ratio of injected power level

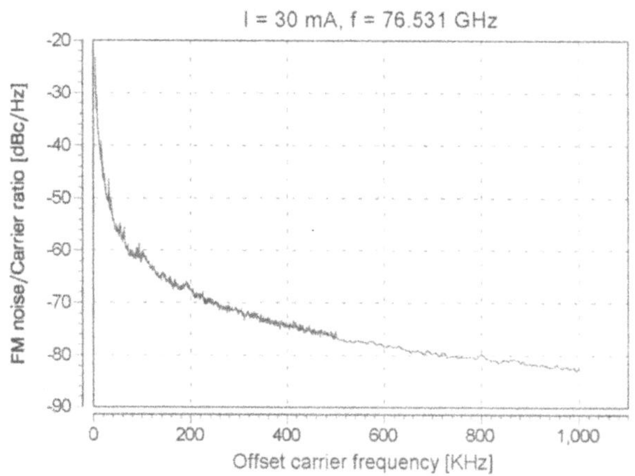

Figure 5. Phase noise of free-running IMPATT oscillator

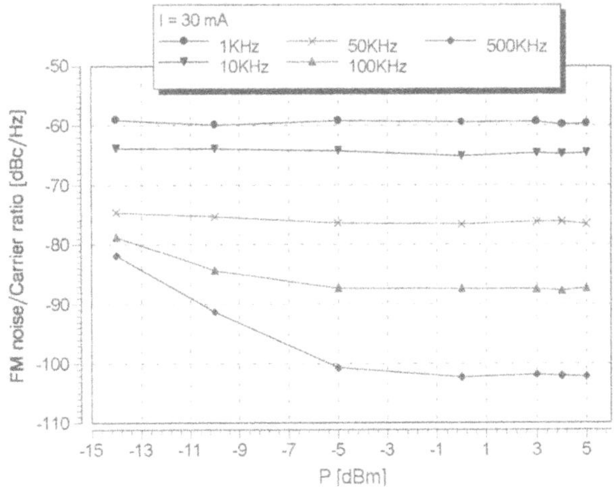

Figure 6. Phase noise behavior depending on injection power level

P_{inj} to the oscillator output power[7,8]. For a diode current of $I = 30$ mA we measured for $P_{inj} = 0$ dBm a 3-dB tuning range of 300 MHz and for $P_{inj} = 3$ dBm almost 600 MHz, respectively. For most applications at 76.5 GHz this bandwidth is sufficient. The smaller bandwidth at higher bias currents is a result from the increasing output power of the oscillator.

In another experiment we measured the phase noise behavior of the locked oscillator within the tuning range for $P_{inj} = 0$ dBm and $I = 30$ mA. The FM noise to carrier ratio $\mathcal{L}\ (\Omega)$ at different offset frequencies Ω is depicted in Figure 8. In the whole tuning range no increase of the phase noise is observed. The phase noise is only determined by the phase noise of the reference oscillator and not by the frequency offset to the free-running oscillator. Only at the edges of the tuning range the phase noise increases significantly.

Another crucial point for the application is the reliability of the employed circuit and the temperature stability. To avoid parasitic effects the front-end was measured in an electromagnetical shielded chamber. The still complete 4" wafer including the chip oscillator to be measured is mounted on a *WAFTHERM 'SP43A* thermal car-

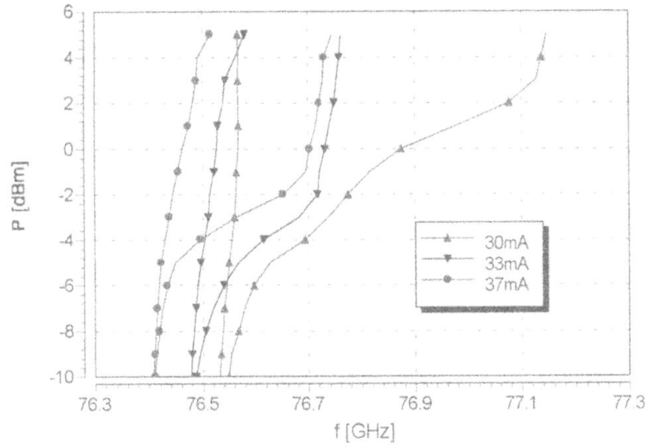

Figure 7. 3-dB tuning range of stabilized oscillator

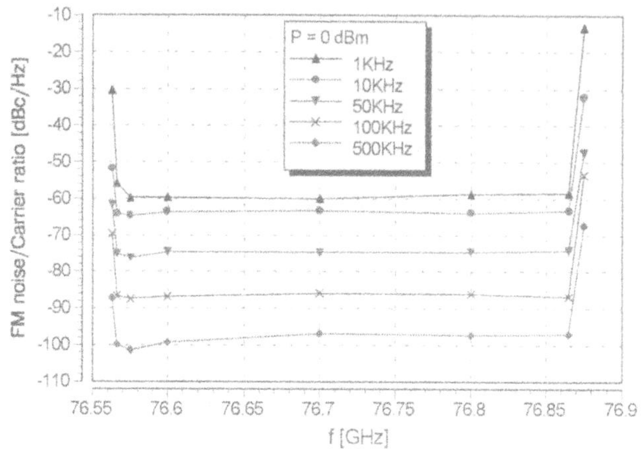

Figure 8. Phase noise behavior in the locking range

rier. The DC supply is accomplished via probing needles. At distance a above the oscillator a receiving horn antenna aligned to the dipole is mounted. By this measurement setup characteristic values as radiated power P_{rad}, oscillation frequency f_{osc} and DC voltage U_0 are measured. The measurement setup is depicted in Figure 9. The temperature ϑ is varied by 1 °C every 30 seconds, while the impressed DC current is kept at I_0 = 35 mA. The temperature sweep from −10 °C to + 90 °C is performed 3 times starting at 24 °C. U_0 and f_{osc} are normalized to the corresponding values at 24 °C (U_0 = 20.18 V, f_{osc} = 76.44 GHz). Both exhibit linear dependence versus temperature above ϑ = 10 °C. The temperature gradient for this temperature range is determined to $\Delta f/\Delta\vartheta = -0.006$ %/ °C respectively $\Delta U/\Delta\vartheta = 0.1$ %/ °C. Below the kink temperature of ϑ = 10 °C the temperature gradient of the oscillation frequency decreases to $\Delta f/\Delta\vartheta = -0.015$ %/ °C (Figure 10). At the same time degradation of the power spectrum as depicted in Figure 11 is observed. Nevertheless the radiated power P_{rad} remains nearly constant in the temperature range from −10 °C to

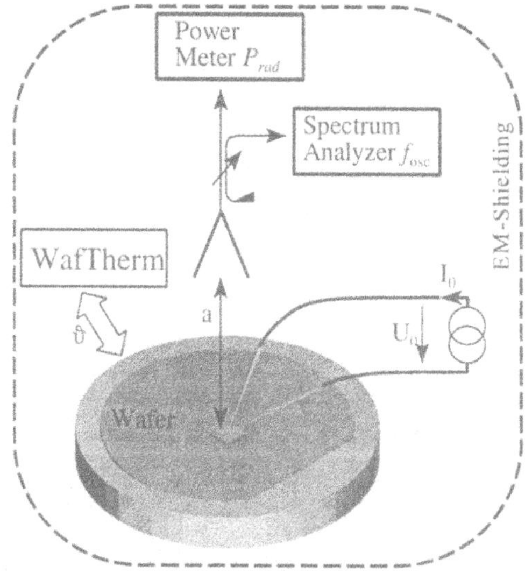

Figure 9. Temperature stability: measurement setup

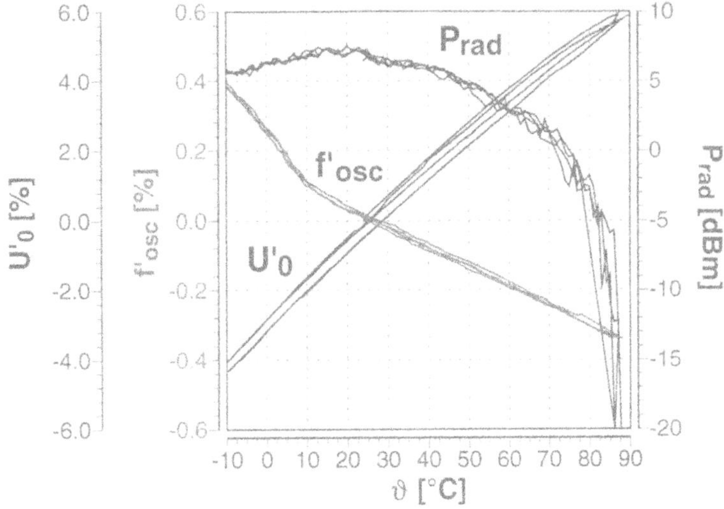

Figure 10. Temperature stability

+50 °C. Above 80 °C the radiated power rapidly declines to -20 dBm at 90 °C. Thus, for ambient temperatures up to 80 °C sufficient output power is achieved. This is very important for operation of the active antenna in adverse environmental conditions.

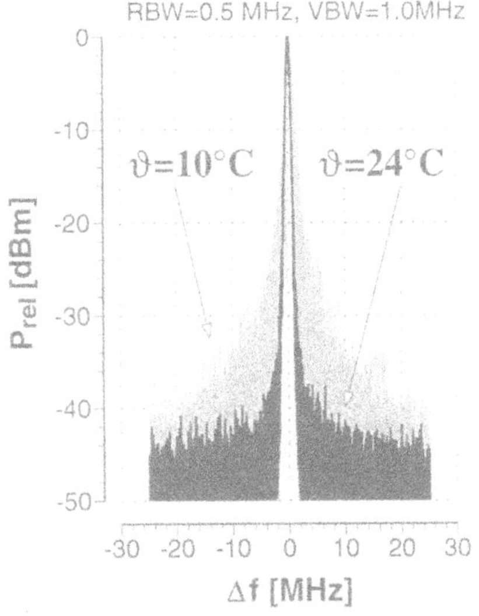

Figure 11. Degradation of power spectrum

CONCLUSIONS

In our contribution we presented the experimental results of investigations on subharmonically injection locked oscillators in the frequency range around 76.5 GHz. We gained important information about the required injected power level for a given tuning range in an integrated millimeter wave sensor system. For sufficient injected power level the phase noise behavior of the reference oscillator dominates the front-end. No increase of the phase noise is observed in the tuning range for sufficient injected power. To achieve a 3-dB tuning range of 600 MHz less than 3 dBm injection power is required. Our investigations of the temperature dependence showed clearly that the active antenna operates well below the thermal breakdown.

ACKNOWLEDGEMENT

This work has been supported by the German Bundesministerium für Bildung, Wissenschaft, Forschung und Technologie under project number 01 M 2938 B.

REFERENCES

1. J.-F. Luy, and P. Russer, *Silicon-Based Millimeter-Wave Devices*, Springer Verlag, Berlin Heidelberg (1994).
2. M. Gupta, R.J. Lomax, and G.I. Haddad, Noise considerations in self-mixing IMPATT-diode oscillators for short-range doppler radar applications, *IEEE Trans. on Microwave Theory Tech.* 22:34 (1974).
3. F. Beißwanger, U. Güttich, and C. Rheinfelder, Microstrip and coplanar SiGe M-MMIC oscillators, *Proc. of the 26th Europ. Microwave Conf.*, (1996).
4. A. Stiller, E.M. Biebl, J.-F. Luy, K.M. Strohm, and J. Büchler, A monolithic integrated millimeter wave transmitter for automotive applications, *IEEE Trans. on Microwave Theory Tech.* 43:1654 (1995).
5. M. Singer, A. Stiller, K.M. Strohm, J.-F. Luy, and E.M. Biebl, A SIMMWIC 76 GHz front-end with high polarization purity, *1996 IEEE MTT-S Int. Microwave Symp. Dig.* 2:1079 (1996).
6. X. Zhang, X. Zhou, and A.S. Daryoush, A theoretical and experimental study of the noise behavior of subharmonically injection locked local oscillators, *IEEE Trans. on Microwave Theory Tech.* 40:895 (1992).
7. R. Adler, A study of locking phenomena in oscillators, *Proc. IRE,* 34:351 (1946).
8. A.S. Daryoush, T. Berceli, R. Saedi, P.R. Herczfeld, and A. Rosen, Theory of subharmonic synchronization of nonlinear oscillators, *1989 IEEE MTT-S Int. Microwave Symp. Dig.* 2:735 (1989).

EXPERIMENTAL AND NUMERICAL RESULTS ON HIGH-FREQUENCY NOISE OF Si/SiGe HBTs

F. Herzel, B. Heinemann, P. Schley, U. Zillmann, D. Knoll, and D. Temmler

Institute for Semiconductor Physics
Walter-Korsing-Str. 2
15230 Frankfurt (Oder), Germany

INTRODUCTION

Silicon-Germanium heterojunction bipolar transistors (SiGe HBTs) are promising for wireless high-speed applications, based on the well-established silicon IC technologies. There are basically two concepts for the design of the vertical profile. The first concept[1,2] leaves the doping profiles of the Si bipolar junction transistor (BJT) essentially unchanged, but introduces a graded Ge profile with vanishing Ge concentration at the EB-junction. This gives rise to a drift field in the base. Advantages of this approach are the process latitude and the similarity to Si BJT processes. The second concept[3] uses a significantly higher Ge content, which is constant over the base. This allows a very high base doping resulting in low sheet resistivity. As a result, this so-called "true" HBT exhibits a low base resistance, high values of f_{max}, and low noise figures in the GHz region[4].

Our concept is similar to the latter but tries to combine the advantages of both concepts by using a Ge profile of medium concentration that is graded over the emitter side of the base only, and has a non-zero Ge concentration at the EB-junction. The vertical profile has been optimized numerically[5] and demonstrated experimentally[6].

The high-frequency noise figures (NF) of BJTs and HBTs are usually modeled by means of equivalent circuits. Recently, a thermodynamic approach to the high-frequency NF of BJTs and HBTs has been used to calculate the NF of SiGe HBTs by means of 2D device simulation[7,8]. It relates the NF to the small-signal parameters of the total transistor and the dc values. In contrast to the equivalent circuit approaches, the input for the simulation are device geometry and doping profiles instead of circuit element values. Thus, the device simulation approach to the NF is more closely related to the technological process, allowing one to predict the influence of technological quantities, such as the thermal budget, on noise figures.

In this paper, we show that measured and simulated minimum noise figures based on the thermodynamic approach agree well with conventionally measured ones. Based on 2D device simulation, we propose a noise optimized vertical profile. In order to

Directions for the Next Generation of MMIC Devices and Systems
Edited by N. K. Das and H. L. Bertoni, Plenum Press, New York, 1997

apply the thermodynamic approach to circuit simulation, a slight modification of the standard SPICE noise model is described, resulting in good agreement between NFs obtained by circuit and device simulations.

MEASUREMENTS

The tuned-out noise figure $\tilde{F}$ is defined as the noise figure with the input capacitance compensated by the imaginary part of the noise generator admittance. It is approximately given by[7,8]

$$\tilde{F} \approx 1 + \frac{1}{2kTg_G}\left(q|I_B| + 2kT\mathrm{Re}(y_{11}) + [q|I_C| + 2kT_C\mathrm{Re}(y_{22})]\left|\frac{\mathrm{Re}(y_{11}) + g_G}{y_{21}}\right|^2\right), \quad (1)$$

where q is the elementary charge, k the Boltzmann constant, g_G the real part of the generator admittance, T the ambient temperature, and T_C the effective collector temperature. T_C may significantly exceed the ambient temperature[7]. However, the term $4kT_C\mathrm{Re}(y_{22})$ may be neglected[7] for frequencies well below the transit frequency f_T, resulting in a simplification of Eq. (1). It is important to note that the y-parameters refer to the total transistor, including parasitics. The tuned-out noise figure $\tilde{F}$ can be calculated from the measured dc values and the measured s-parameters converted to y-parameters. The minimum noise figure F_{min} is then calculated from $\tilde{F}$ by minimizing $\tilde{F}$ numerically with respect to g_G.

In this paper, a SiGe HBT is investigated (Figure 1). Essential features are the epitaxy on a blanket substrate containing only subcollector wells, and the use of highly doped polysilicon for the emitter layer[6]. The emitter width of $W_E = 2.6\mu m$ is relatively large. Transistors with significantly narrower emitters are under development.

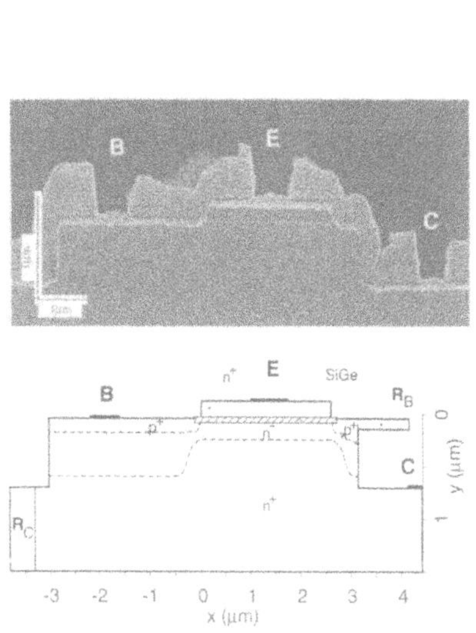

Figure 1. SEM cross-section of a SiGe HBT and region for device simulation.

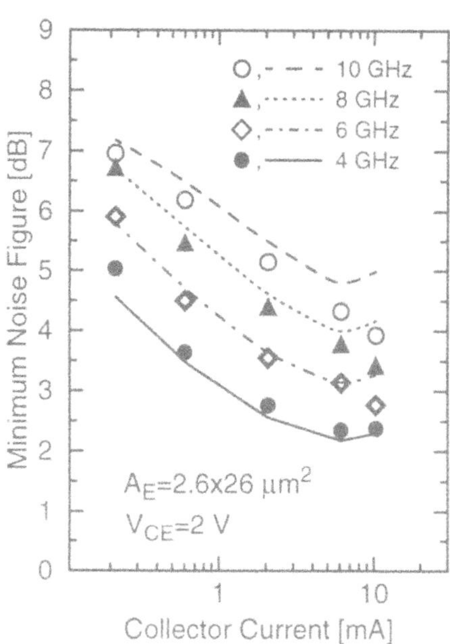

Figure 2. F_{min} versus collector current extracted from measured s-parameters (lines) and directly measured (symbols).

F_{min} obtained by the method described above is shown in Figure 2 (lines). For comparison, conventionally measured F_{min} are also plotted (symbols). A detailed discussion of these results will be published elsewhere[9]. Here we note that the good agreement between the two approaches illustrates that the thermodynamic approach to the NF can accurately describe the behavior of the noise figure in the GHz range, and allows the determination of the NF with acceptable accuracy with much reduced experimental effort.

DEVICE SIMULATION

Before we turn to design optimization by means of simulation, we demonstrate the accuracy of the 2D device simulation by comparing calculated NFs with measured ones. The region for the simulation is obtained from the SEM cross-section (Figure 1). The vertical profile was determined from SIMS measurements. To obtain the narrow B base profile, deconvolution[10] was required, since the high-frequency properties sensitively depend on the base profile. Using the measured B profiles without deconvolution would give, e.g., too small simulated transit frequencies and too large NFs. The simulated peak f_T of 56 GHz at V_{CE} = 2 V is close to the measured value of 54 GHz. In Figure 3, the NFs obtained by device simulation and measurement of the small-signal parameters, respectively, are shown. In order to estimate the accuracy of the approach, we assumed an error of 10 percent for the base width, boron dose and Ge concentration. As a result, we obtained an estimated error for the simulated minimum NF of ± 0.2 dB at 2 GHz and ± 0.5 dB at 10 GHz as shown in Figure 3.

Now we present the numerically optimized design of the doping profile and the Ge profile. A detailed description of this is given elsewhere[8]. To optimize the vertical profile, we used a device geometry with an emitter width of $W_E = 0.6\mu$m. Half the symmetrical device is shown in Figure 4.

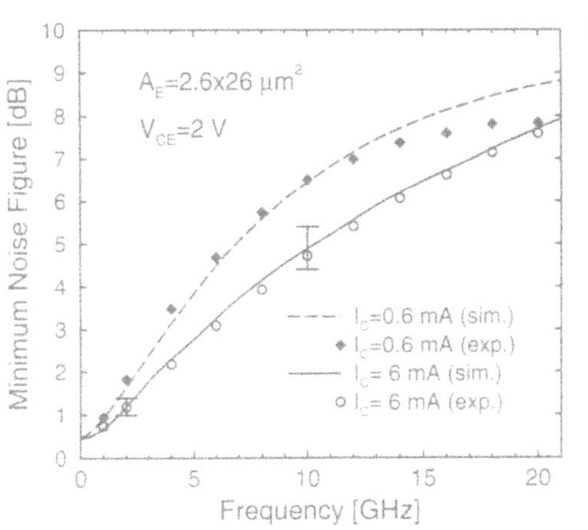

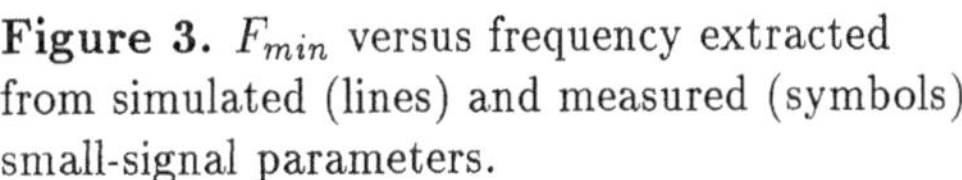
Figure 3. F_{min} versus frequency extracted from simulated (lines) and measured (symbols) small-signal parameters.

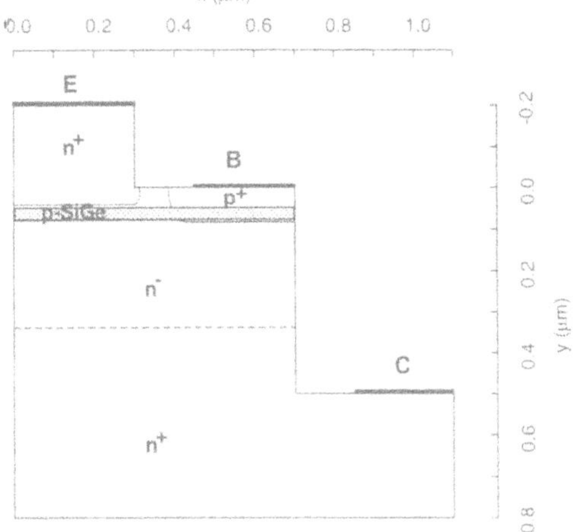

Figure 4. Advanced-technology transistor geometry for device simulation.

Three different Ge profiles (Figure 5) were investigated: a uniform Ge profile; a graded profile; and a half-graded profile with the grading on the emitter side of the base. The half-graded Ge profile has been shown to produce high values of f_T and f_{max} for our kind of B profiles resembling Gaussian distributions[5]. The B profiles shown in Figure 5 were calculated by process simulation starting from a rectangular profile of 10 nm width and $2 \times 10^{14}cm^{-2}$ dose. As can be seen in Figure 5, the basewidth decreases with increasing Ge fraction, since the strain in the SiGe layer reduces the B diffusion[10]. This is another advantage of using SiGe in the base of an npn-transistor, since a narrow base results in a small base transit time. As shown in Figure 6, the half-graded profile gives the best noise performance for the HBT. For comparison, measured noise figures[4] of a state-of-the-art SiGe HBT are also plotted in Figure 6.

In a previous publication[5], the f_{max}-optimized emitter and collector profile has been discussed. Here we note that the noise optimized collector is the same as the f_{max}-optimized one presented there. That means that the collector is best low doped. A collector width should be used that the BC-space charge region finishes at the transition between the low doped collector and the high doped subcollector for the desired collector-base voltage. The emitter-base capacitance was found to be of greater influence on the noise figure than on f_{max}. Therefore, the noise-optimized width of the low doped emitter is larger than for the f_{max}-optimized profile. For our transistor, the width of the low doped emitter that remains after the final thermal process should be about 40 nm.

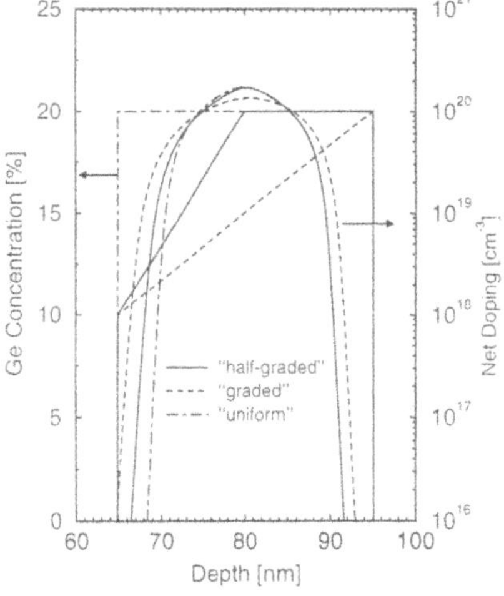

Figure 5. Ge and B profiles used for device simulation.

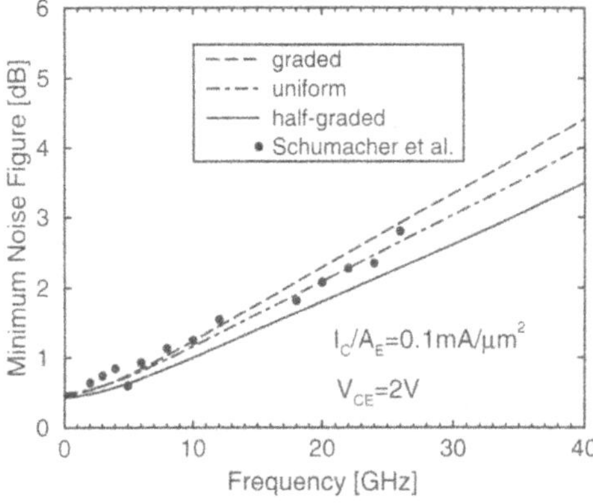

Figure 6. F_{min} versus frequency for the different Ge profiles.

CIRCUIT SIMULATION

Implementing the thermodynamic approach in circuit simulators requires the introduction of non-white noise sources, since the input noise current spectral density $\propto\mathrm{Re}(y_{11})$ strongly increases at high frequencies. This effect is not taken into account in most circuit simulators, such as SPICE3. However, only a slight modification of the thermal noise sources is required to include this effect[11]. Let $\tilde{y}_{11}$ be the input admittance for a short-circuited output with the emitter resistance R_E disregarded. (R_E is treated in the same fashion as in the standard SPICE3 noise model.) The input noise current spectral density may then be approximately expressed by internal SPICE parameters according to

$$4kT\mathrm{Re}(\tilde{y}_{11}) \approx 4kT\mathrm{Re}\left\{\left[(g_\pi + j\omega C_\pi)^{-1} + R_{BB}\right]^{-1}\right\}, \tag{2}$$

where g_π is the small-signal input conductance of the intrinsic transistor, C_π is the internal emitter-base capacitance, and R_{BB} is the series base resistance. An input noise current connected to the external base node and the internal emitter node with the spectral density given by Eq. (2) is used instead of the noise source associated with R_{BB}. A similar substitution is performed in the output circuit. An implementation of this improved SPICE noise model is presented elsewhere[11].

Here the SPICE parameters of a model transistor ($W_E = 1.5\mu$m) were extracted from the dc and ac characteristics obtained by 2D device simulation. Hence, the device simulation results serve as criteria for the quality of our improved compact model. To calculate the NF within SPICE3, we divided the total noise current spectral density transformed to the output, as obtained from the noise circuit analysis, by the corresponding contribution of the noise generator. As shown in Figure 7, there is a good agreement between the device simulation results and those of the improved SPICE model. We note that this agreement was achieved with model parameters determined from dc and ac characteristics, without an extra fitting of the SPICE parameters to the NF.

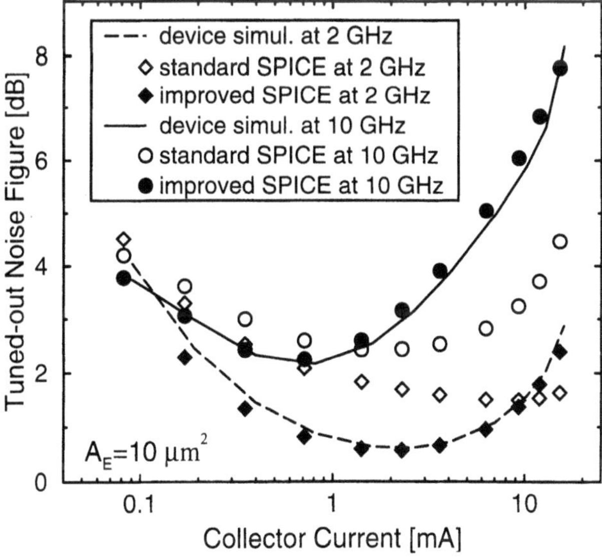

Figure 7. Tuned-out noise figure $\tilde{F}$ versus collector current obtained from device simulation and from circuit simulation (SPICE).

CONCLUSIONS

We have applied a thermodynamic approach to determine the high-frequency noise figure of SiGe HBTs from measurements and simulations of small-signal parameters. The NFs obtained by this method agree well with conventionally measured ones. 2D-device simulation is capable of describing the high-frequency noise in relatively good agreement with experiments. Based on extensive computer simulations, a noise optimized vertical profile has been proposed. By modifying the thermal noise sources in SPICE3, a good agreement between circuit simulation and device simulation was achieved.

REFERENCES

1. D.L. Harame, K. Schonenberg, M. Gilbert, D. Nguyen-Ngoc, J. Malinowski, S.-J. Jeng, B. Meyerson, J.D. Cressler, R. Groves, G. Berg, K. Tallmann, K. Stein, G. Hueckel, C. Kermarrec, T. Tice, G. Fitzgibbons, K. Walter, D. Colavito, T. Houghton, N. Greco, T. Kebede, B. Cunningham, S. Subbanna, J.H. Comfort, and E.F. Crabbé, A 200 mm SiGe-HBT technology for wireless and mixed-signal applications, *Techn. Dig. IEDM*: 437 (1994).
2. T.F. Meister, H. Schäfer, M. Franosch, W. Molzer, K. Aufinger, U. Scheler, C. Walz, M. Stolz, S. Boguth, and J. Böck, SiGe base bipolar technology with 74 GHz f_{max} and 11 ps gate delay, *Techn. Dig. IEDM*: 739 (1995).
3. E. Kasper, A. Gruhle and H. Kibbel, High speed SiGe-HBT with very low base sheet resistivity, *Techn. Dig. IEDM*: 79 (1993).
4. H. Schumacher, U. Erben, and A. Gruhle, Noise characterization of Si/SiGe heterojunction bipolar transistors at microwave frequencies, *Electronics Letters* 28:1167 (1992).
5. B. Heinemann, F. Herzel, and U. Zillmann, Influence of low doped emitter and collector regions on high-frequency performance of SiGe-base HBTs, *Solid-St. Electron.* 38:1183 (1995).
6. D. Knoll, B. Heinemann, K.-E. Ehwald, P. Schley, W. Röpke, D. Bolze, J. Schlote, F. Herzel, and G. Fischer, Comparison of P in situ spike doped with As implanted poly silicon emitters concerning Si/SiGe/Si HBT application, in *Proc. 25 th European Solid State Device Research Conference (ESSDERC'95)*, Edition Frontieres, Gif sur Yvette Cedex: 627 (1995).
7. F. Herzel and B. Heinemann, High-frequency noise of bipolar devices in consideration of carrier heating and low temperature effects", *Solid-St. Electron.* 38:1905 (1995).
8. F. Herzel and B. Heinemann, High-frequency noise analysis of Si/SiGe-heterojunction bipolar transistors, *Intern. Journal of Electronics* 81:37 (1996).
9. F. Herzel, P. Schley, B. Heinemann, U. Zillmann, D. Knoll, D. Temmler, and U. Erben, Experimental verification and numerical application of the thermodynamic approach to high-frequency noise in SiGe HBTs, *Solid-St. Electron.*, in press.
10. F. Herzel, K.-E. Ehwald, B. Heinemann, D. Krüger, R. Kurps, W. Röpke, and H.-P. Zeindl, Deconvolution of narrow boron SIMS depth profiles in Si and SiGe, *Surf. Interface Anal.* 23:764 (1995).
11. U. Zillmann and F. Herzel, An improved SPICE model for high-frequency noise of BJTs and HBTs", *IEEE Journal of Solid-State Circuits* 31:1344 (1996).

THE DEVELOPMENT OF Si AND SiGe TECHNOLOGIES FOR MICROWAVE AND MILLIMETER-WAVE INTEGRATED CIRCUITS

George E. Ponchak[1], Samuel A. Alterovitz[1], Linda P. B. Katehi[2], and Pallab K. Bhattacharya[2]

[1]NASA Lewis Research Center
Cleveland, OH 44135

[2]University of Michigan
Ann Arbor, MI 48109-2122

INTRODUCTION

Historically, microwave technology was developed by military and space agencies from around the world to satisfy their unique radar, communication, and science applications. Throughout this development phase, the sole goal was to improve the performance of the microwave circuits and components comprising the systems. For example, power amplifiers with output powers of several watts over broad bandwidths, low noise amplifiers with noise figures as low as 3 dB at 94 GHz, stable oscillators with low noise characteristics and high output power, and electronically steerable antennas were required. In addition, the reliability of the systems had to be increased because of the high monetary and human cost if a failure occurred.

To achieve these goals, industry, academia, and the government agencies supporting them chose to develop technologies with the greatest possibility of surpassing the state of the art performance. Thus, Si, which was already widely used for digital circuits but had material characteristics that were perceived to limit its high frequency performance, was bypassed for a progression of devices starting with GaAs Metal Semiconductor Field Effect Transistors (MESFETs) and ending with InP Psuedomorphic High Electron Mobility Transistors (PHEMTs). For each new material or device structure, the electron mobility increased, and therefore, the high frequency characteristics of the device were improved. In addition, ultra small geometry lithographic processes were developed to reduce the gate length to 0.1 μm which further increases the cutoff frequency. The resulting devices had excellent performance through the millimeter-wave spectrum.

Initially, the use of exotic materials with small geometry lithography resulted in poor yields. Therefore, early circuits were fabricated by bonding the transistors to ceramic substrates such as alumina upon which the matching circuits were built. As the processing technology matured and yield improved, Microwave Monolithic Integrated Circuits (MMICs) became possible. This higher level of integration reduced the circuit size and the

Directions for the Next Generation of MMIC Devices and Systems
Edited by N. K. Das and H. L. Bertoni, Plenum Press, New York, 1997

parasitic reactance created by the wire bonds which led to further improvements in the circuit characteristics. Today, GaAs and InP MESFET, PHEMT, and Heterojunction Bipolar Transistor (HBT) based MMICs have RF characteristics that are sufficient for nearly every application.

Unfortunately for the GaAs industry, the market for microwave circuits has changed dramatically over the past three years. While government agencies continue to buy circuits for their needs, the commercial market for personal communications, wireless communications, analog and digital mobile communications, intelligent vehicle highway systems, and industrial sensors has surpassed the military market. This has imposed a new requirement on the microwave industry; Once a circuit provides the needed service, cost is the most important parameter.

The exotic materials and small dimensions that yielded outstanding MMIC characteristics also resulted in high fabrication cost. In addition, since GaAs and InP circuit processing is not compatible with Si processing, they require separate and different equipment. Thus, the large capital investments made by the electronic industry and the corresponding low fabrication cost per silicon wafer area cannot be used by the GaAs MMIC industry, and these conditions are not expected to improve unless new markets develop for GaAs MMICs that would lower their fabrication cost. Lastly, higher levels of integration are not possible if MMICs are made from GaAs or InP and the data control and bias circuits are made from Si. Therefore, integration and packaging costs cannot be lowered and reliability improvements that are associated with integration cannot be realized.

To achieve lower cost and higher levels of integration, the microwave industry is reevaluating its reliance on GaAs and InP and the perceived disadvantages of Si which include poor high frequency performance due to low carrier mobility and high transmission line attenuation due to the low substrate resistivity. As silicon circuit manufacturers have reduced the gate size of their standard n-MOSFETs, Metal Oxide Field Effect Transistors, f_{max} has increased to 37 GHz [1], and self aligned Bipolar Junction Transistors (BJTs) have been fabricated with an f_{max} of 70 GHz [2]. More interesting though is the significant increase in microwave capabilities that arises from the introduction of Ge to create SiGe HBTs that have an f_{max} of 160 GHz [3]. These new devices should fulfill the requirements of the commercial industry through Ka-band, but further improvements are possible if the SiGe material structure and microwave component characteristics on Si are optimized.

This paper will present the efforts of research teams at NASA Lewis Research Center and The University of Michigan to grow higher quality SiGe on Si, characterize SiGe HBT material structures, and develop passive components on Si.

SiGe/Si HBT FABRICATION

High Ge content Si_xGe_{1-x} alloys ($1-x>0.4$) have a higher carrier velocity than Si that should translate into better microwave transistor characteristics[4], but there are practical limits to the Ge concentration. The most serious limitation comes from the fact that the Ge atomic lattice is 4.2 % larger than the Si lattice[5]. Therefore, strain is created within the lattice when SiGe is grown on a Si substrate, and the strain increases as the Ge content increases. If the strain reaches a critical value, misfit dislocations develop as is shown in the TEM microgragh in Figure 1. These thread type dislocation defects may degrade device performance, reliability, and fabrication yield if they terminate on or near the device channel. Thus, it is desirable to develop methods of reducing the defect density while maintaining a high Ge content.

Several authors have advocated the use of superlattices and graded structures that start with a very low Ge concentration at the SiGe/Si interface with increased Ge content as subsequent layers are added[6-11]. Although some reduction in misfit dislocations is obtained, the defect density is still too high[12] at approximately 10^9 cm^{-2} as seen in Figure 1. At the University of Michigan, an alternative approach that uses a low temperature silicon (LT-Si) buffer layer between the SiGe and the Si substrate has been developed[13]. This 0.1 μm thick layer is grown at 450 C by molecular beam epitaxy (MBE) followed by the SiGe HBT structure. It is seen in Figure 2 that the misfit dislocations are terminated in the LT-Si layer resulting in a defect density of $<10^4$ cm^{-2} at a thickness of 0.5 μm. The exact mechanism for the defect density reduction is currently unknown.

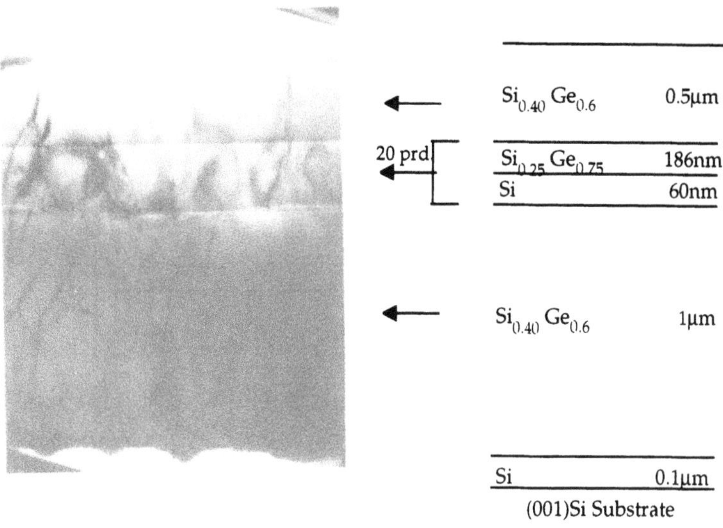

Figure 1. Cross-section TEM image of MBE grown SiGe heterostructure with 20 superlattices showing the dislocation propagation.

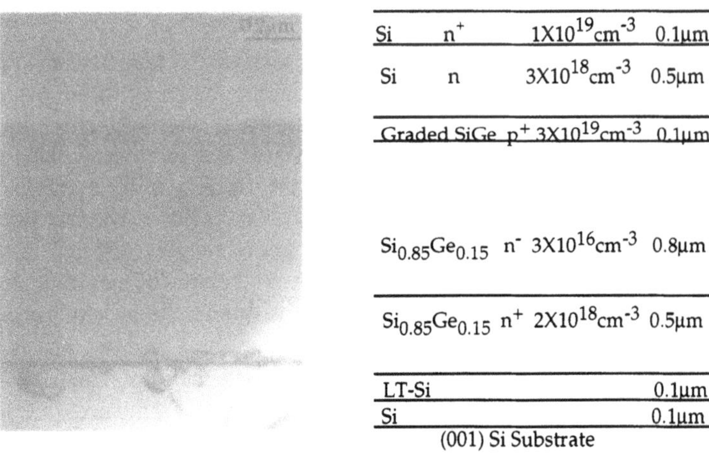

Figure 2. Cross-section TEM image of MBE grown SiGe HBT structure with LT-Si buffer layer showing the defect density reduction.

The structure shown in Figure 2 was used to fabricate self-aligned HBTs that incorporated Cr/Au contacts. As seen in the current-voltage curve for the 10 X 10 μm^2

device shown in Figure 3, the DC current gain is approximately 6, the breakdown voltage is greater than 10 V, and the Early voltage is approximately 1000 V. These results demonstrate that the LT-Si buffer does not degrade the DC characteristics of the HBT. Note that this device was not optimized for microwave performance.

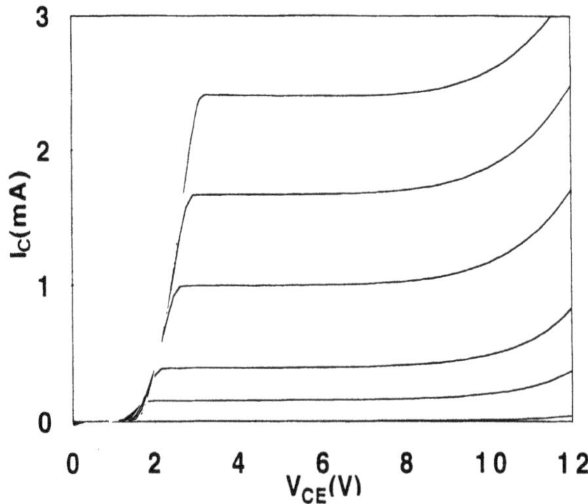

Figure 3. Measured current-voltage characteristics of 10X10 μm^2 emitter area HBT.

ELLIPSOMETRIC CHARACTERIZATION OF SiGe/Si HBT MATERIAL

Ellipsometry is an optical, non-destructive method to characterize semiconductor multilayers[14]. It uses dielectric calibration functions to interpret the experimental Ψ and Δ in terms of materials. Thus, accurate calibration functions are critical for materials studies. In this case, the dielectric functions of Si_xGe_{1-x} are known to high accuracy for the whole concentration range for relaxed material[15], but only very limited data is available for strained layers[16].

For a typical SiGe/Si HBT structure, ellipsometry is expected to estimate non-destructively the thickness of the emitter and base layers, the composition parameter x, and any dielectric layers on top of the emitter (thickness and dielectric constant). The ellipsometry work in this case is not straightforward, due to three problems: 1. The Si emitter layer is very thick; 2. There is strain in the Si_xGe_{1-x} base; 3. Usually there is a concentration gradient in the base layer. The first problem is due to the fact that the thick Si emitter layer is blocking the light below a wavelength of roughly 400 nm. However, many special features of the Si_xGe_{1-x} calibration functions are below 400 nm, thus decreasing the sensitivity for the x estimate. The second problem is due to the lack of accurate strained Si_xGe_{1-x} calibration functions, as mentioned above. The third problem makes the analysis rather complex, as the usual assumption of single phase layers is not obeyed.

We have performed studies on a variety of Si_xGe_{1-x} structures for devices and superlattices, both on Si and Ge[17-19]. First we tested an HBT structure with a thick emitter but a constant composition x in the base[17]. Thus we dealt with two of the three problems mentioned above.

A Si_xGe_{1-x} HBT structure was grown in a dual e-gun Si MBE system. The nominal structure obtained from the growth calibration is shown in Figure 4, left-hand side. The HBT material was characterized by x-ray diffraction (XRD) and secondary ion mass spectroscopy (SIMS) in addition to ellipsometry. The XRD result shows a composition x=0.83±0.002. Using the points where the Ge signal dropped to half the peak value, the SIMS depth profile gave an emitter thickness of 590±10 nm, and a base of 85±10 nm.

NOMINAL			ELLIPSOMETRY	
2 nm	Oxide		Oxide	4.55 +/- 0.07 nm
120 nm	Si n^{++}	$2X10^{19}$ Sb	Si	594.3 +/- 0.75 nm
480 nm	Si n^{+}	$2X10^{18}$ Sb		
92.3 nm	Si_xGe_{1-x}		Si_xGe_{1-x}	88.5 +/- 0.97 nm
	Si	$1X10^{17}$ Sb	Si substrate	
	Si substrate p			
x=0.8			x=0.862 +/- 0.008	
			MSE: $2.07X10^{-3}$	

Figure 4. Nominal structure (left-hand side) and ellipsometry results (right-hand side) of an HBT sample.

Ellipsometry measurements were performed at four angles of incidence, over spectrum range was 300-780 nm, in 5 nm steps. As usual in ellipsometry, a model was assumed with the nominal structure. As doping below the high 10^{19} cm^{-3} is not observable by ellipsometry, only four parameters were assumed as variables: oxide, emitter and base thickness, and the composition x. The final ellipsometry results are the values obtained for the four variable parameters using a non-linear least squares fit to the experimental results. We used the relaxed Si_xGe_{1-x} dielectric calibration functions. The model calculations using the best fitted four parameters shows good agreement with the experiment. The values of the four parameters are given in Figure 4, right-hand side, with the corresponding 90% confidence limits. MSE is a measure of the quality of the fit[17]. The ellipsometry determined thickness for the emitter and the base are in very good agreement with both the nominal and the SIMS values. The composition x is 0.03 higher than the extremely reliable value obtained by XRD. It turns out that a coherently strained Si_xGe_{1-x} layer under a Si layer measured by ellipsometry and using relaxed Si_xGe_{1-x} calibration functions will give a value of x which is higher by 0.03 vs. the XRD value[20]. The higher than expected oxide layer was due to previous heating to 150 °C for this particular sample.

As a second step to a complete HBT structure characterization, we tested several graded layers[19], with grading from x=1 to 0.7 and to 0.5. The graded layers were uncapped and nominally 1 μm thick. Ellipsometry measurements were done in the spectral range 300-800 nm, at three angles of incidence. Again, we used the relaxed dielectric calibration functions. The graded layer was assumed to be made of n thinner layers, each one with a constant composition x, where x varied linearly from one to the value on top, x_n (x_n(nom) is the nominal composition on the top surface). We found that for n>20 the results are independent of n. Thus, by using n=30 we got a good approximation to the continuously graded layer. Results for two of the samples are shown in Table 1. The main features of the results are: 1. The Si_xGe_{1-x} layer thickness is in excellent agreement with the nominal values; 2. The oxides are thicker than expected. We believe this is a result of the surface roughness, a common occurrence in Si_xGe_{1-x} films; 3. The composition parameter x_n is too high for all samples. We believe that the reason for the high x_n values is due to two factors: 1. The surface is probably rough; 2. There is residual strain in the samples. Sample B has a larger deviation vs. sample A, and it also has possibly more strain than A. The fact that the layers were thick does not mean that they were totally relaxed. XRD of a 0.61 thick Si_xGe_{1-x} layer with x=0.8 showed only a 65%-70% relaxation[21]. Thus it is unclear whether our samples were totally relaxed. It seems that more work on a variety of graded samples is required for an accurate determination of the composition.

Table 1. Ellipsometric results for graded Si_xGe_{1-x} on Si.

Sample	d(oxide) (Angstrom)	D (Si_xGe_{1-x}) (μm)	x_n	x_n (nom)
A	44.8	1.04	0.767	0.7
B	43.0	0.97	0.587	0.5

PASSIVE COMPONENTS ON HIGH RESISTIVITY SILICON

As early as 1965, it was recognized that microstrip transmission lines on standard CMOS grade Si wafers with a resistivity, ρ, of 0.1 to 10 Ohm—cm were too lossy, and High Resistivity Silicon (HRS) wafers with $\rho > 2500$ Ohm-cm are necessary for microwave and millimeter-wave microstrip[22]. Later experiments showed that $\rho > 2500$ ohm-cm was also required for Coplanar Waveguide (CPW) transmission lines[23]. A comparison of the attenuation of CPW on GaAs and a HRS wafer is shown in Figure 5. Although the attenuation of the Si line is greater the GaAs line by approximately 0.2 dB/cm, the effective dielectric constant of the Si line is lower than the GaAs line by approximately 7 %. Thus, if the attenuation is plotted in the more useful units of dB per wavelength, the attenuation of the Si line is only 0.1 dB/λ greater than the GaAs line as shown in Figure 6.

Besides microwave transmission lines on Si having a comparable attenuation to similar lines on GaAs, passive components and discontinuities also have similar characteristics. For example, consider the CPW short circuit terminated series stub that is often used as a series inductor or as a stop band filter element[24]. The characteristics of the stub fabricated on HRS and GaAs are shown in Figure 7. Note that except for a small shift in the resonant frequency due to the slightly lower relative permittivity of Si compared to GaAs, the characteristics are identical. Furthermore, if the loss factor, $1-|s_{11}|^2-|s_{21}|^2$, is calculated from Figure 7, it is seen that circuits fabricated on HRS do not suffer a degradation in performance when compared to circuits on GaAs.

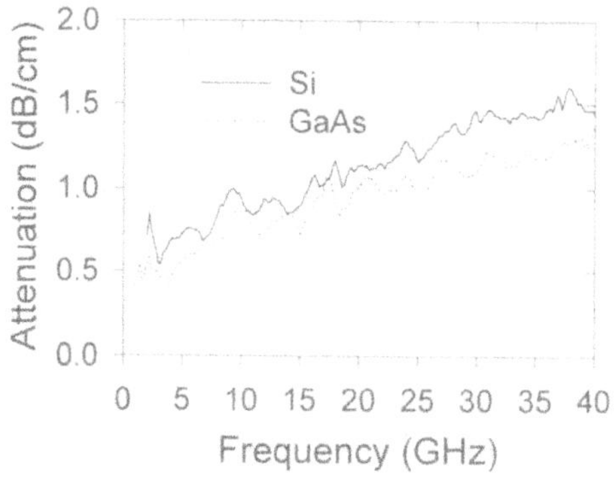

Figure 5. Attenuation per length of CPW line with S = W = 50 μm on GaAs and $\rho > 2500$ Ohm-cm Si.

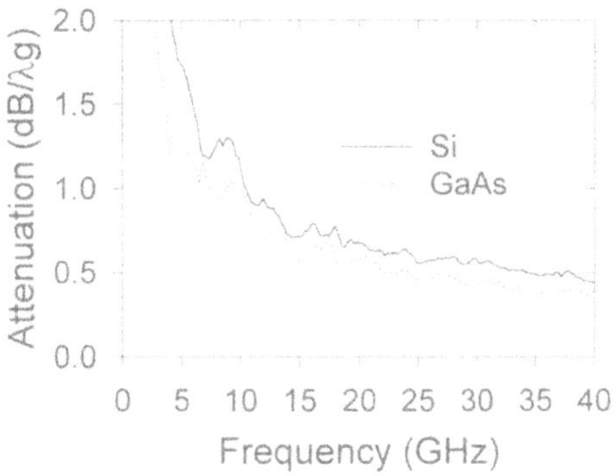

Figure 6. Attenuation per guided wavelength of CPW line with S = W = 50 μm on GaAs and ρ > 2500 Ohm-cm Si.

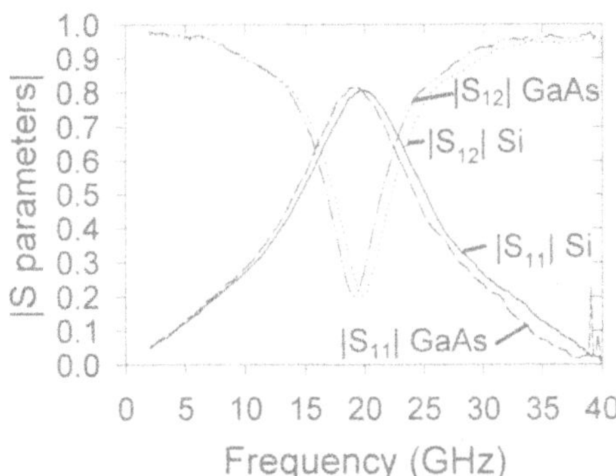

Figure 7. Magnitude of S-parameters of CPW short circuit series stub fabricated on GaAs and HRS.

DISCUSSION

The excellent results published in the literature and within this paper provide encouragement that Si MMICs are not only possible, but they can provide the performance required for the emerging commercial markets with further development. Although Si MMICs are used in nearly every RF system below 2 GHz, there has been very little development above 20 GHz. Furthermore, the integration of microwave and digital circuitry has not occurred.

To address these issues, the University of Michigan and NASA Lewis Research Center are working to develop SiGe HBTs for microwave applications utilizing the LT-Si buffer layer that was developed. In addition, full characterization of Si as a microwave substrate and methods to reduce the attenuation of transmission lines on Si by utilizing micromachining technology are being performed. These efforts will be demonstrated in a series of Ka-band Si MMICs.

REFERENCES

1. S. P. Voinigescu, S. W. Tarasewicz, T. MacElwee, and J. Ilowski, An assessment of the state-of-the-art 0.5 μm bulk CMOS technology for RF applications, *Proc. 1995 IEDM*: 721 (1995).
2. M. Ugajin, J. Kodate, Y. Kobayashi, S. Konaka, and T. Sakai, Very-high f_T and f_{max} silicon bipolar transistors using ultra-high-performance super self-aligned process technology for low-energy and ultra-high-speed LSI's, *Proc. 1995 IEDM*: 735 (1995).
3. A. Schuppen, U. Erben, A. Gruhle, H. Kibbel, H. Schumacher, and U. Konig, Enhanced SiGe heterojunction bipolar transistors with 160 GHz-f_{max}, *Proc. 1995 IEDM*: 743 (1995).
4. S. H. Li, J. M. Hinckley, J. Singh, and P. K. Bhattacharya, Carrier velocity-field characteristics and alloy scattering potential in $Si_{1-x}Ge_x$/Si, *Appl. Phys. Lett.* 63 (10): 1393 (1993).
5. J. C. Bean, Silicon-based semiconductor heterostructures: column IV bandgap engineering, *Proc. IEEE* Vol. 80, No. 4: 571 (1992).
6. P. M. Mooney, J. L. Jordan-Sweet, J. O. Chu, and F. K. LeGoues, Evolution of strain relaxation in step-graded SiGe/Si structures, *Appl. Phys. Lett.*, 66: 3642 (1995).
7. M. A. Lutz, R. M. Feenstra, F. K. LeGoues, P. M. Mooney, and J. O. Chu, Influence of misfit dislocations on the surface morphology of $Si_{1-x}Ge_x$ films, *Appl. Phys. Lett.*, 66: 724 (1995).
8. J. H. Li, V. Holy, G. Bauer, J. F. Nutzel, and G. Abstreiter, Strain relaxation of $Ge_{1-x}Si_x$ buffer systems grown on Ge (001), *Appl. Phys. Lett.*, 67: 789 (1995).
9. F. K. LeGoues, B. S. Meyerson, and J. F. Morar, Anomalous strain relaxation in SiGe thin films and superlattices, *Phys. Rev. Lett.*, 66: 2903 (1991).
10. G. Kissinger, T. Morgenstern, G. Morgenstern, and H. Richter, Stepwise equilibrated graded Ge_xSi_{1-x} buffer with very low threading dislocation density on Si (001), *Appl. Phys. Lett.*, 66: 2083 (1995).
11. F. K, LeGoues, Self-aligned sources for dislocation nucleation: the key to low threading dislocation densities in compositionally graded thin films grown at low temperature, *Phys. Rev. Lett.*, 72: 876 (1994).
12. K. Ismail, J. O. Chu, and B. S. Meyerson, High hole mobility in SiGe alloys for device applications, *Appl. Phys. Lett.*, 64: 3124 (1994).
13. K. K. Linder, F. C. Zhang, J.-S. Rieh, and P. Bhattacharya, Reduction of defect density in mismatched SiGe/Si by low temperature Si buffer layers, Submitted to *Appl. Phys. Lett.*: (1996).
14. P. G. Snyder, M. C. Rost, G. H. Bu-Abbud, J. A. Woollam and S. A. Alterovitz, Variable angle of incidence spectroscopic ellipsometry: application to GaAs-AlGaAs multiple heterostructures, *J. Appl. Phys.* 60: 3293 (1986).
15. G. E. Jellison Jr., T. E. Haynes and H. H. Burke, Optical functions of SiGe alloys determined using spectroscopic ellipsometry, *Opt. Mat.* 2: 105 (1993).
16. R. T. Carline, C. Pickering, D. J. Robbins, W. Y. Leong, A. D. Pitt and A. G. Cullis, Spectroscopic ellipsometry of SiGe epilayers of arbitrary composition $0<x<0.255$, *Appl. Phys. Lett.* 64: 1114 (1994).
17. R. M. Sieg, S. A. Alterovitz, E. T. Croke, M. J. Harrell, M. Tanner, K. L. Wang, R. A. Mena and P. G. Young, Characterization of SiGe/Si heterostructures for device applications using spectroscopic ellipsometry, *J. Appl. Phys.* 74: 586 (1993).
18. R. M. Sieg, S. A. Alterovitz, E. T. Croke and M. J. Harrell, Ellipsometric study of $Si_{0.5}Ge_{0.5}$/Si strained-layer superlattices, *Appl. Phys. Lett.* 62: 1626 (1993).
19. A. R. Heyd, S. A. Alterovitz and E. T. Croke, Characterization of high Ge content SiGe heterostructures and graded alloy layers using spectroscopic ellipsometry, *Mat. Res. Symp. Proc.* 358: 993 (1995).
20. C. Pickering, R. T. Carline, D. J. Robbins, W. Y. Leong, D. E. Gray and R. Greef, In-situ dual-wavelength and ex-situ spectroscopic ellipsometry studies of strained SiGe epitaxial layers and multi-quantum well structures, *Thin Solid Films* 233: 126 (1993).
21. C. Pickering, R. T. Carline, D. J. Robbins, W. Y. Leong, S. J. Barnett and A. G. Cullis, Spectrocopic ellipsometry characterization of strained and relaxed SiGe epitaxial layers, *J. Appl. Phys.* 73: 239 (1993).
22. T. M. Hyltin, Microstrip transmission on semiconductor dielectrics, *IEEE Trans. Microwave Theory Tech.*, Vol. 13, No. 6: 777 (1965).
23. S. R. Taub and P. G. Young, Attenuation and ε_{eff} of coplanar waveguide transmission lines on silicon substrates, *Dig. 11 th Benjamin Franklin Symp.*: 8 (1993).
24. N. I. Dib, L. P. B. Katehi, G. E. Ponchak, and R. N. Simons, Theoretical and experimental characterization of coplanar waveguide discontinuities for filter applications, *IEEE Trans. Microwave Theory Tech.*, Vol. 39, No. 5: 873 (1991).

94 GHZ SCHOTTKY DETECTOR WITH CMOS PREAMPLIFIER

Karl M. Strohm,[1] Josef Buechler,[1] Johann-Friedrich Luy,[1]
Michael Herrmann,[2] Dietmar Beck,[2] Erich Kasper [2]

[1]Daimler Benz Research Center
Wilhelm-Runge-Str. 11
D-89075 Ulm, Germany

[2]Institute of Semiconductor Engineering
University of Stuttgart
Breitscheidstr. 2
D-70174 Stuttgart, Germany

ABSTRACT

A multi chip module (MCM) method for combining high frequency circuits with low frequency readout electronics is investigated. The RF chip consists of a detector circuit with monolithically integrated zero bias Schottky barrier diodes and planar antenna structures on high resistivity silicon substrates. This rectifying antenna (rectenna) is manufactured by silicon monolithic millimeter wave integrated circuit (SIMMWIC) technology. The signal of the rectenna is amplified by a CMOS preamplifier mounted in a hybrid construction next to the rectenna. An amplification of 32 dB is measured. Maximum sensitivity of the detector circuit including preamplification is 1600 $mV/mWcm^{-2}$ at 94.6 GHz.

INTRODUCTION

SIMMWIC is the abbreviation for Silicon Millimeter Wave Integrated Circuits /1/. The potential of this technology is high, since silicon is worldwide the semiconductor material number one, it has good thermal and mechanical properties, it is available in large and low-cost substrates and enables the cointegration of RF-circuits with standard signal processing circuits. For the technological realization of SIMMWIC's high resistivity silicon substrates are necessary to keep RF substrate losses low. Also a technology has to be developed which preserves the high resistivity substrate characteristics during the fabrication process and which allows the realization of active silicon millimeter wave

Directions for the Next Generation of MMIC Devices and Systems
Edited by N. K. Das and H. L. Bertoni, Plenum Press, New York, 1997

devices. Up to now several SIMMWIC's have been realized /2/. With Transit Time Diodes oscillators and active antennas /3/, with Schottky diodes detector and mixer circuits /4/, with PIN diodes switches /5/ and with SiGe HBT's amplifiers, oscillators and frequency doublers have been fabricated /6/. The combination of SIMMWIC's with signal processing circuitry, especially with CMOS technology will be the next step for wider applications of SIMMWIC's. This combination may be done as multi chip modules (MCM) or in monolithic cointegration. As an early stage of this combination a SIMMWIC detector with a CMOS preamplifier is investigated.

DETECTOR CIRCUIT

Fig. 1 shows the layout of the detector circuit. It consists of an antenna structure combined with Schottky barrier diodes as rectifying elements. The whole circuit acts as a rectifying antenna (rectenna). The rectenna consists of two branches with ten $\lambda/2$ radiating elements which are placed at half wavelength distance along a microstrip line. The received millimeter radiation is guided by the microstrip line to the two monolithic integrated Schottky diodes both terminated by a common radial line sector. The rectified signal is taken via two connected filter networks.

Figure 1. Layout of the monolithic detector circuit

p-Schottky barrier diodes are used as rectifying elements. The diodes are realized in a coplanar configuration (Fig. 2). A cross section of the diode is shown in Fig. 3. As base material high resistivity substrates are used (ρ>4000 Ωcm). The fabrication process of the diodes is described elsewhere /7/. Standard buried layer processes, Si-MBE for the growth of the epitaxial layer and standard photolithographic patterning and lift-off processes for Schottky and ohmic metallization are used. Gold electroplating is applied for the definition of the final metallization layer.

DC-characteristics of the diodes are evaluated from I-V and C-V measurements. Series resistance, barrier height, ideality factor and junction capacitance strongly depend on doping parameters, diode geometry and anode metallization. By introducing a p^+ - spike on the epitaxial p-layer a barrier height reduction down to 0.4 V is achieved. These diodes may be used in zero bias detectors.

The detector chips are thinned to a thickness of 100 µm. The backside is metallized with electroplated gold.

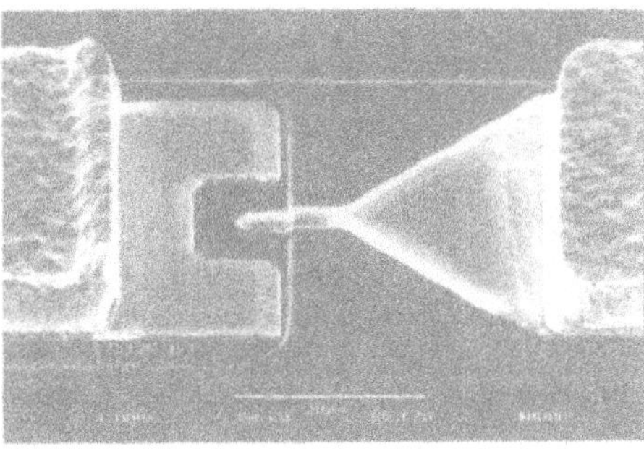

Figure 2. SEM of a monolithically integrated Schottky barrier diode

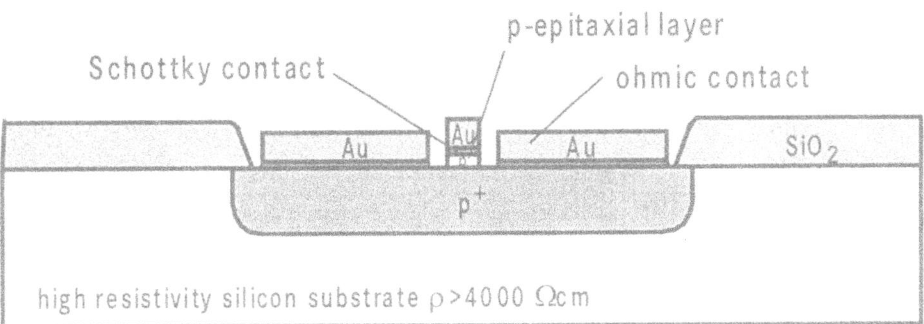

Figure 3. Cross section of a coplanar Schottky barrier diode

CMOS PREAMPLIFIER

Preamplification of the rectified RF signal is performed by a single stage inverting amplifier with a dc gain of 32 dB (cf. fig. 4) /8/. Amplifiers with differential inputs and two stages offering low frequency gains of around 70dB have also been realized on standard CMOS substrates (CZ, n-type, 4-6 Ωcm, <100> oriented) and on high resistivity substrates (FZ, p-type, 5000-7000 Ωcm, <100>). The float zone (FZ) substrate material is required to keep RF substrate losses low in the case of future monolithic integration of SIMMWIC RF circuits with analog and digital circuit functions realized in CMOS technology.

A comparison of these amplifiers and single MOS devices fabricated on both substrate materials showed a good similarity in electrical behaviour and a preservation of the high resistivity substrate characteristics outside the CMOS region.

Problems to be kept in mind regarding CMOS on FZ substrates are increased leakage due to lower substrate doping, facilitated punch-through effects and the non-existence of internal gettering as exhibited by CZ-substrates. An alternative solution to these problems may be the usage of SOI substrates on high resistivity silicon /9/. SOI substrates prevent leakage current, minmize parasitic capacitance, allow good isolation of the active devices and improve the overall speed of the CMOS circuits /10/.

The results so far offer a good prospect for monolithically integrated SIMMWIC/CMOS circuits in the near future.

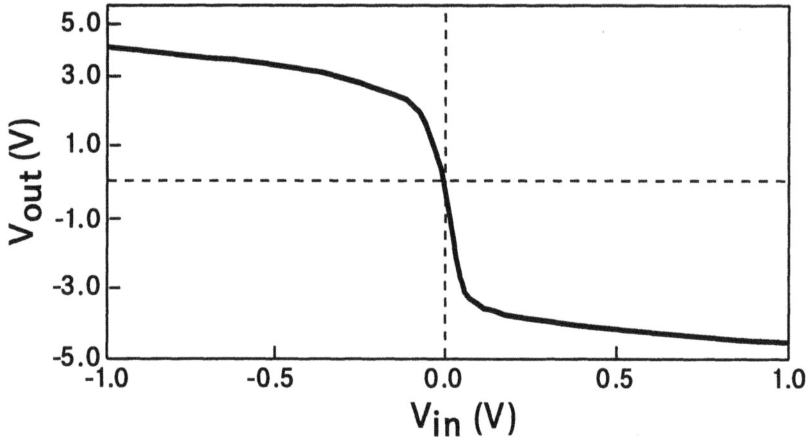

Figure 4: DC transfer characteristic of the single stage amplifier

HYBRID CONSTRUCTION

The detector chip is combined with a CMOS preamplifier as a multi chip module (MCM) in a hybrid construction. This was done as an early stage of the combination of SIMMWIC technology with standard CMOS technology. Likewise not in every case the monolithic integration is possible or useful. The high frequency chip for mm wave detection and the low frequency chip for amplification are mounted on a common substrate material, a silicon substrate (Fig. 5). This hybrid technology offers some advantages over monolithic integration, e. g. different technologies for the monolithic integration can be used, different backside processing of the different chips is possible and there is a free choice of substrate potential for the single chips. Either conductive or non-conductive adhesive is possible.

The SIMMWIC chip is mechanically thinned to a thickness of 100 μm due to the microstrip design of the antenna while the CMOS chip remains unthinned. As mechanical carrier a 7.5 x 20 mm^2 sized piece of an unthinned silicon wafer (<111>-oriented, high resistivity material with a thickness of 280 μm) is used.

The silicon substrate material is thermally oxidized to a thickness of 1 μm to avoid a conductive connection between the substrates of the CMOS and SIMMWIC chip. In this case, a conductive adhesive layer is used, but there is no problem in using a non-conductive adhesive. The signal connection is realized by one aluminium bond wire connecting the output of the rectenna to the input of the amplifier (see Fig. 5). The amplifier needs no special ground connection thus no further connection to the rectenna is needed.

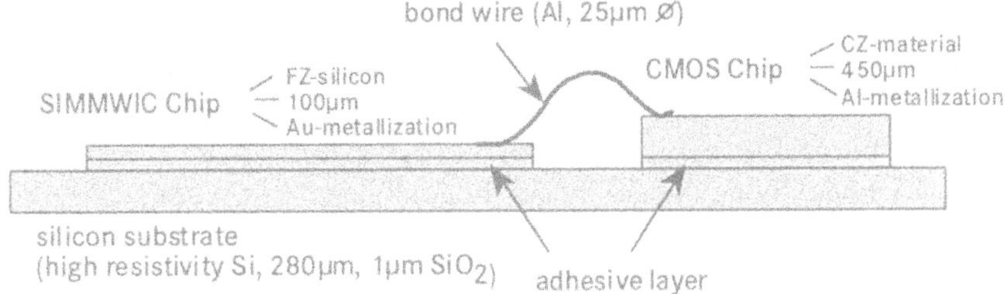

Figure 5. Hybrid construction

The RF/CMOS MCM is mounted in a 24 lead DIP ceramic package. The rectenna and CMOS amplifier connections are bonded with 25 µm diameter aluminum bond wires (Fig. 6).

Figure 6. Photo of the RF/CMOS MCM in 24 lead DIP ceramic package

RF MEASUREMENTS

The hybrid construction was measured under far field conditions. A W-Band transmitting horn antenna was placed vertical above the hybrid chip and the preamplifier output signal was measured. For the following measurements a hybrid construction with a single stage CMOS preamplifier operated with double supply voltages of ±6 V and an amplification of 42 was employed. Because zero bias diodes were used a forward current biasing of the diodes gives no significant increase of the observed output signal.

Figure 7 depicts the measured output signal of the CMOS amplifier as a function of the incident RF power density at a frequency of 94.6 GHz. As expected the measured output voltage is linearly increasing with incoming RF power. The sensitivity of the configuration is 1600 $mV/mWcm^{-2}$ including a preamplification gain (32 dB) of the detected LF signal. The sensitivity without preamplifier is 38 $mV/mWcm^{-2}$.

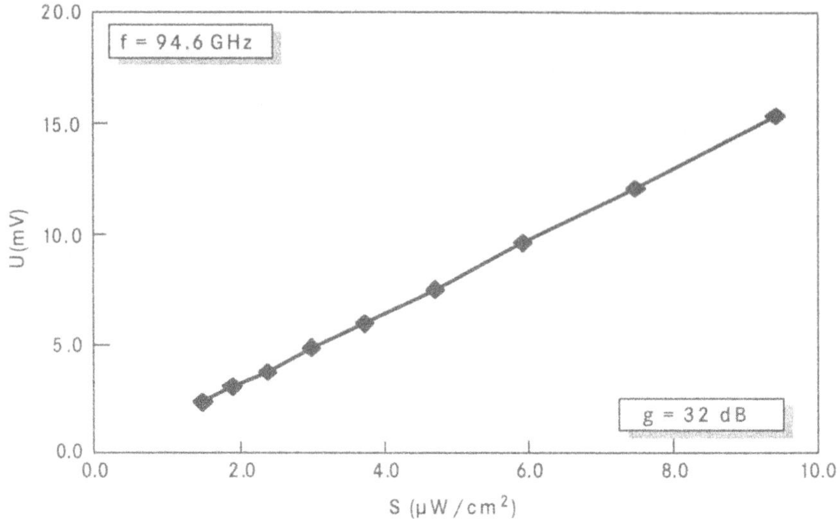

Figure 7. Output voltage as a function of incident power density S

The measured output signal (with preamplification) as a function of frequency is shown in figure 8.

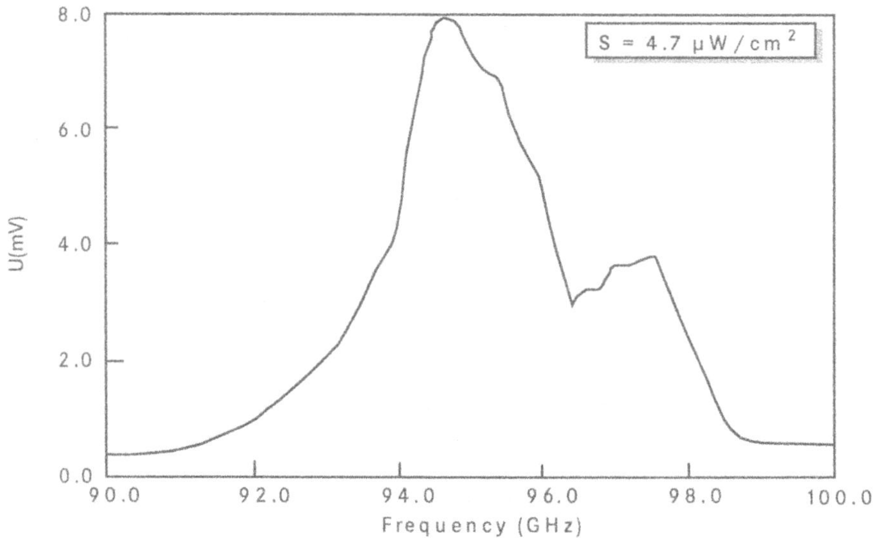

Figure 8: Output voltage as a function of frequency (S = 4.7 μW/cm^2)

The center frequency is 94.6 GHz with a 3 dB-bandwidth of 1.6 GHz according to the small-bandwidth layout of the antenna structure. Therefore this construction is well suited for selective detection of RF signals in the 95 GHz range.

The H-plane radiation pattern of the packaged module is shown in figure 9. The half power beam width is 17^0. The side lobe level is < -20 dB.

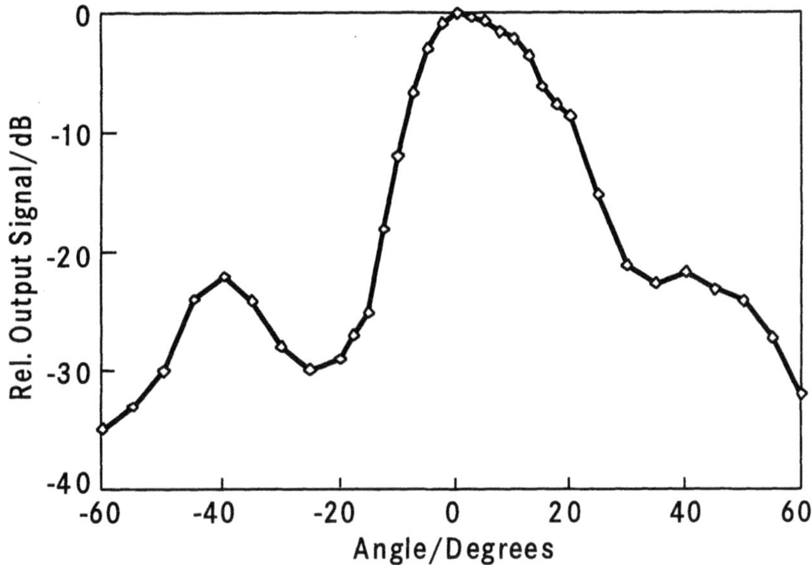

Figure 9. H-plane radiation pattern of the packaged module

CONCLUSION

A 94 GHz SIMMWIC detector was combined with a CMOS preamplifier. Both chips are mounted on a common silicon substrate and connected via aluminum bond wires. The sensitivity of the MCM configuration is 1600 $mV/mWcm^{-2}$ at 94.6 GHz including a preamplification gain (32 dB) of the detected LF signal instead of 38 $mV/mWcm^{-2}$ without preamplifier. The fabrication of CMOS circuits on high resistivity silicon substrates is promising and should allow the cointegration of SIMMWICs with CMOS signal processing circuits in near future.

ACKNOWLEDGEMENT

The authors are grateful to Prof. F. Schäffler (formerly Daimler-Benz) for growing the MBE-layers. This work is supported by the German Ministry of Research and Technology (BMBF) under contract No. 01 M 2938 A 8.

REFERENCES

1. P. Stabile, A. Rosen, A silicon technology for millimeter-wave monolithic circuits. *RCA Rev.* 45, 587-605 (1984)
2. J.-F. Luy and P.Russer (editors), *Silicon Based Millimeter-Wave Devices*, Springer, Berlin (1994)
3. M. Singer, A. Stiller, K. M. Strohm, J.-F. Luy, E. M. Biebl, A SIMMWIC 76 GHz front-end with high polarization purity, *1996 IEEE MTT-S Digest*, 1079-1082, (1996)

4. K. M. Strohm, J.-F. Luy, C. Rheinfelder, F. J. Schmückle, W. Heinrich, Monolithic integrated coplanar 77 Ghz SIMMWIC mixer, *26th. Europ. Microwave Conf.* (1996)
5. P. J. Stabile, A. Rosen, P. R. Herczfeld, Optically controlled lateral PIN diodes and microwave control circuit. *RCA Rev.* 47, 443-456, (1986)
6. W. Heinrich, SiGe MMICs - On the current state-of-the-art, *this issue* (1996)
7. K. M. Strohm, Silicon millimeter wave integrated circuit technology, in *Silicon-Based Millimeter-Wave Devices,* edited by J.-F. Luy and P. Russer, Springer-Verlag, p. 266, (1994)
8. P. R. Gray and R. G. Meyer, *Analysis and Design of Analog Integrated Circuits*, 3rd ed., Wiley, New York, (1993)
9. A.K. Agarwal, T. J. Smith, Maurice H. Hanes, R. L. Messham, M. C. Driver and H. C. Nathanson, MICROX™ - An affordable Silicon MMIC Technology, *Workshop - Silicon RF Technologies, 1995 IEEE MTT-S Int. Microw. Symp.* (1995)
10. T. Ohno, Y. Kado, M. Harada and T. Tsuchiya, *Experimental 0.25-μm-gate fully depleted CMOS/SIMOX process using a new two-step LOCOS isolation technique*, IEEE Trans. Electr. Dev., 42, 1481-1486, (1995)

SEMICONDUCTOR WAVEGUIDE COMPONENTS FOR ANALOG FIBER-OPTIC LINKS

R.B. Welstand[1], S.A. Pappert[2], J.T. Zhu[1], Y.Z. Liu[3], and P.K.L. Yu[1]

[1]University of California, San Diego; Department of ECE;
La Jolla, CA 92093-0407
[2]NCCOSC RDTE DIV 895; San Diego, CA 92152-5000
[3]Fermionics Corp.; 4555 Runway Dr.; Simi Valley, CA 93063

ABSTRACT

A Franz-Keldysh InGaAsP electroabsorption waveguide device is examined for dual-function operation in an externally modulated analog fiber-optic link. Either efficient optical modulator or photodetection is demonstrated with the same device by adjusting the device bias voltage. The dual function modulator/photodetector can be useful in compact transmit/receive front-end antenna architectures.

INTRODUCTION

Optical transmission of analog signals has potential use in high-speed antenna remoting and RF distribution systems. Depending on bandwidth, dynamic range, noise figure, and cost requirements, either direct modulation of a laser diode or external modulation using a waveguide modulator is used. External modulation is currently examined for the most demanding broadband link applications. However, two problems continue to plague the use of waveguide modulators, the limited spurious free dynamic range (SFDR) and the polarization sensitivity. Improvement in these two areas would provide attractive alternatives for several applications that could benefit from using high-fidelity, optical analog links. Improvement in SFDR and RF efficiency can be obtained by increasing the optical power used in the link. However, this demands large power-handling modulators and photodetectors. This paper addresses the power-handling and linearity performance of a waveguide component that serves both as a waveguide modulator and a waveguide photodiode.

WAVEGUIDE MODULATOR

The majority of studies on improving the linearity of optoelectronic modulators have been geared toward the lithium niobate modulators. Recent work make use of dual parallel modulation [1], mixed polarization [2], or other compensation schemes [3,4] that cause partial cancellation of the nonlinear components. Disadvantages for these approaches are the complexity of design and the tight tolerance requirements on the power splitters and recombining couplers.

The current approach employs an electroabsorbing (EA) InGaAsP layer, operating by the Franz-Keldysh effect (FKE) [5]. Except for slight modal disparity between the TE and TM polarizations in the waveguide, this modulation mechanism is much less polarization sensitive [6,7] than other electroopic effects. An additional feature of electroabsorption modulators is small drive voltage [8,9], typically on the order of a few (< 5) Volts.

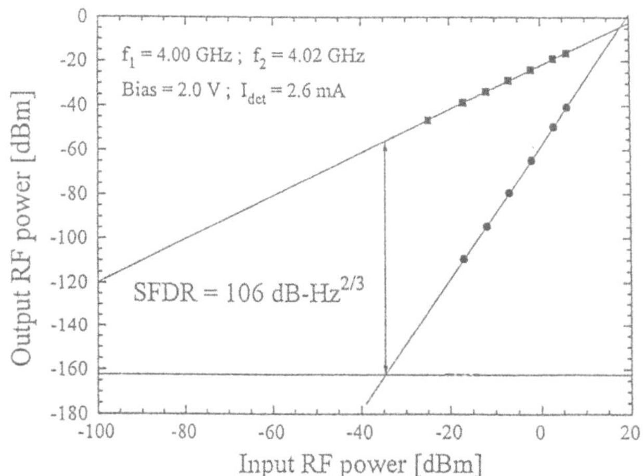

Figure 1. Two-tone linearity measurement of the optoelectronic devices a waveguide modulator showing fundamental (squares) and third-order intermodulation product (circles) signals. For the broadband measurement, with 43 mW (TM polarized) light incident onto the device, the extrapolated spurious-free dynamic range at 4 GHz is 106 dB-Hz$^{2/3}$ with 2.6 mA of detector photocurrent.

The structure of the device used in this letter is described in [10]. The active layer of the waveguide is a lattice-matched, 3500 Å thick, InGaAsP ($\lambda_Q = 1.24\ \mu m$) layer grown by OMVPE on n-type (S doped) InP substrate, sandwiched between p- and n-type InP layers, 1 μm and 0.375 μm thick, respectively. An additional p^+ InGaAs capping layer of 2000 Å is grown for ohmic contact. Rib-loaded waveguide structures have 5 μm mesa widths and are cleaved to 120 μm length. The waveguide facets are also anti-reflection coated and coupled to single mode fiber. The reverse breakdown voltage is in excess of 12 V.

As mentioned earlier, the primary concerns for the RF link are large SFDR and high RF efficiency. These are simultaneously maximized in broadband or multiple-octave applications by biasing the EA waveguide device at the maximum slope point on the DC transfer curve, where the second derivative is zero. This null point occurs at 2.0 V for the present device, which has a 3-dBe bandwidth of 11 GHz. The broadband linearity is investigated for the modulator biased at 2.0 V, and results of the two-tone measurement at 4 GHz, with 20 MHz tone separation, are shown in Fig. 1. For this measurement, with 43 mW incident optical power, the shot-noise-limited detector DC photocurrent is 2.6 mA. The dominant distortion is the third-order intermodulation term. The third-order intercept point (IP3) is -3.6 dBm (output referenced), thus the SFDR extrapolated to the calculated noise floor is 106 dB-$Hz^{2/3}$. The measured RF link loss is -17.4 dB at this incident power and modulator bias, as shown in Fig. 2.

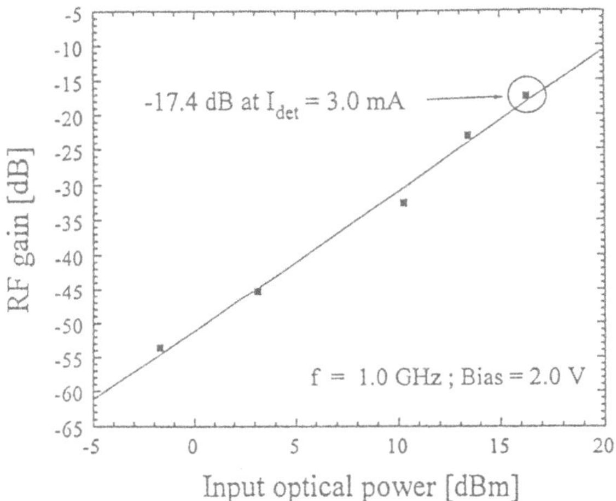

Figure 2. RF gain at 1 GHz versus input optical power of the optoelectronic device as a waveguide modulator.

It should be noted that biasing the modulator at the point on the transfer curve where the third derivative is zero, gives a reduction in the third-order intermodulation term [10]. This is desired in high center-frequency, narrow-bandwidth (sub-octave) applications, because the second-order distortions are out of the passband. For the present device, the null bias is 2.93 V with 37 mW incident optical power. Biasing at this null point results in a 0.66 mA DC photocurrent at the detector and an extrapolated SFDR of 124 dB-$Hz^{4/5}$. The dominant distortion is the third-order intermodulation term, which has a fifth-order power dependence. To maintain a high sub-octave and SFDR modulator performance over temperature, active modulator bias control is required. Simulation has shown that the modulator bias voltage must be maintained to better than ±3 mV to incur less than 5 dB SFDR degradation [10].

WAVEGUIDE PHOTODETECTOR

The EA device becomes a photodector when biased at 7.0 V to ensure a large absorption. Fig. 3 shows the detected RF signal power at 4 GHz plotted versus input optical power using two frequency beated YAG lasers. The measurement shows no RF saturation for power levels up to 4 mW, and the measured RF power agrees with values calculated from the measured DC responsivity of 0.45 A/W. Additionally, the DC responsivity is maintained to very large incident optical powers. The maximum photocurrent measured is 20 mA with the maximum incident power from the high power laser.

The nonlinearity of the waveguide detector is investigated at the second-harmonic frequency of the 4 GHz signal. The second-order intercept point (IP2) is extrapolated to +34.5 dBm (output referenced) from the measured second harmonics. The third-order intercept point limits the narrowband dynamic range of the link the same was as IP2 limits the broadband dynamic range. Determination of this nonlinear intercept-point is presently limited by the low optical power available from the two beated lasers in our measurement.

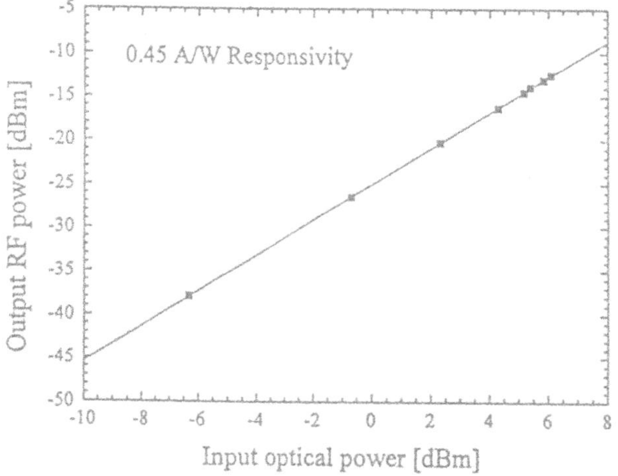

Figure 3. Detected RF signal power at 4 GHz plotted versus input optical power of the optoelectronic device as a waveguide detector.

DUAL-FUNCTION OPERATION

Previously, Giuliani, et. al. [11], have demonstrated a multifunctional amplifier-modulator-detector for digital photonic networks. Their two-electrode device acts as a modulator and power amplifier or as a pre-amplifier and detector. Our dual-function optoelectronic device has advantage in antenna applications where space is a premium. It eliminates the need for one optoelectronic device at the antenna front end, reducing component and optical fiber count which directly translates into space savings. Further packaging size benefit is obtainable if a reflection mode modulator is used resulting in a single-sided fiber pigtailed device. The reverse biased p-i-n diode modulator/detector requires an adjustable DC electrical bias to switch from modulator to detector operation. Microwave frequency conversion [12] or other signal processing may be done remotely at the site of the laser source making for a very simple antenna front end architecture. A schematic of a simplified antenna front end including the waveguide modulator/detector is shown in Fig. 4. The waveguide device works as a modulator in receive mode and a detector in transmit mode. The two modes can be remotely switched by control circuitry which can adjust the DC electrical bias with a switching time limited by the associated electronics. The modulation and detection characteristics of the semiconductor device are critical to this approach and our results show that little performance is sacrificed (5 dB added RF link loss due to reduced detector responsivity) in comparison to using a separate optical modulator and a dedicated 0.8 A/W responsivity detector at the antenna site.

CONCLUSION

In summary, we have demonstrated dual-function operation, as both a modulator and photodetector, of an InGaAsP/InP waveguide device. We have achieved the highest reported link gain of -17.4 dB for a semiconductor EA modulator with a broadband and narrowband SFDR of 106 dB $Hz^{2/3}$ and 124 dB-$Hz^{4/5}$. The electroabsorption detector, has a DC responsivity of 0.45 A/W and output 1P2 of 34.5 dBm. This detector can handle 20 mA of DC photocurrent.

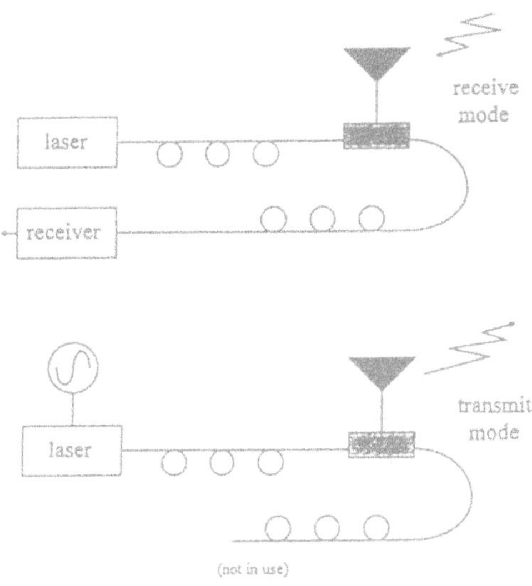

Figure 4. Schematic of antenna feed network used for transmit/receive applications. The network includes: antenna [black triangle], optoelectronic device [gray box], and remote laser and receiver.

ACKNOWLEDGMENT

The authors would like to acknowledge the funding support of DARPA, Rome Lab. and ONR.

REFERENCES

1. S. Korotky, R. De Ridder, "Dual Parallel Modulation Schemes for Low-Distortion Analog Optical Transmission," *IEEE J. on Selected Areas in Commun.*, Vol. 8, No. 7, pp. 1377-1381 (1990).
2. L Johnson, H. Rousell, "Linearization of an Interferometric Modulator at Microwave Frequencies by Polarization Mixing," *IEEE Photon. Technol. Lett.*, Vol. 2, No. 11, pp. 810-811 (1990).
3 L. Fock, R. Turner, "Reduction of Distortion in Analogue Modulated Semiconductor Lasers by Feedforward Compensation," *Electron. Lett.*, Vol. 27, No. 8, pp. 669-671 (1991).
4. G.E. Betts, "Linearized Modulator for Suboctave-bandpass Optical Analog Links," *IEEE, MTT Transaction*, Vol. 42, pp. 2642-2649 (1994).
5. W. Franz, *Z. Naturforsch.*, Vol. 13a, p. 484, and L Keldysh, *Soviet J. of Physics JETP*, Vol. 7, p. 788 (1958).
6. B. Knüpfer, P. Kiesel, M. Kneissl, S. Dankowski, N. Linder, G. Weimann, G.H. Döhler, "Polarization-Insensitive High-Contrast GaAs/AlGaAs Waveguide

Modulator Based on the Franz-Keldysh Effect," *IEEE Photon. Technol. Lett.*, Vol. 5, No. 12, pp. 1386-1388 (1993).
7. G. Mak, C. Rolland, K. Fox, C. Blaauw, "High-Speed Bulk InGaAsP-InP Electroabsorption Modulators with Bandwidth in Excess of 20 GHz," *IEEE Photon. Technol. Lett.*, Vol. 2, No. 10, pp. 730-733 (19900.
8. A. Crook, T. Cockerill, D. Forbes, C. Herzinger, T. DeTemple, J. Coleman, "Low Drive Voltage GaAs Quantum Well Electroabsorption Modulators Obtained with a Displaced Junction," *IEEE Photon. Technol. Lett.*, Vol. 6, No. 5, pp. 619-622 (1994).
9. F. Devaux, P. Bordes, A. Ougazzaden, M. Carre, F. Huet, "Experimental Optimisation of MQW Electroabsorption Modulators with up to 40 GHz bandwidths," *Electron. Lett.*, Vol. 30, no. 16, pp. 1347-1348 (1994).
10. R.B. Welstand, C.K. Sun, S.A. Pappert, Y.Z. Liu, J.M. Chen, J.T. Zhu, A.L. Kellner, P.K.L. Yu, "Enhanced linear dynamic range property of Franz-Keldysh effect waveguide modulator," *IEEE Photonics Technol. Lett.*, Vol. 7, pp. 751-753 (1995).
11. G. Giuliani, P. Cinguino, V. Seano, "Multifunctional characteristics of 1.5 μm two-section amplifier-modulator-detector SOA," *IEEE Photonics Technol. Lett.*, Vol. 8, pp. 367-369 (1996).
12. C.K. Sun, R.J. Orazi, S.A. Pappert, "Efficient microwave frequency conversion using photonic link signal mixing," *IEEE Photonics Technol. Lett.*, Vol. 8, pp. 154-156 (1996).

MICROWAVE PHOTONIC LINKS WITH VERY LOW (≈ 3 dB) AMPLIFIERLESS NOISE FIGURE

Charles H. Cox III, Edward I. Ackerman, and Gary E. Betts

Lincoln Laboratory, Massachusetts Institute of Technology
Lexington, MA 02173-9108

INTRODUCTION

Applications such as antenna remoting and CATV signal distribution require analog fiber optic links with very low noise figure, *NF*, and large dynamic range. Even a large link *NF* can be reduced by preceding the link with a sufficiently high-gain amplifier. However, the higher the *NF* of the amplifierless link the less feasible this approach is in practice, especially when one is simultaneously trying to satisfy a large dynamic range requirement. Consequently it is important to minimize the amplifierless link *NF*.

It has already been claimed that, under the constraint that the link input is passively and losslessly matched to the source impedance, 3 dB is the minimum amplifierless link *NF* that can be achieved [1]. This 3 dB limit, which we call the *lossless passive match limit*, is independent of the link gain, *G*. (We use "gain" here in the most general sense; *i.e.*, we call insertion loss "gain" of less than 0 dB.) Unfortunately, the 3 dB limit has proven hard to achieve in practice, partly because all realizable passive impedance-matching circuits have some loss and possibly some impedance mismatch as well.

In this paper we derive an expression for link *NF* based on an equivalent circuit model that takes into account two conditions not included in previously published models—namely, impedance matching circuit loss and impedance mismatch. In doing so we arrive at a *general passive limit* to fiber optic link *NF*, which can be less than 3 dB. The results predicted by this model correlate well with the measured performance of an experimental amplifierless external modulation link. When we had adjusted the modulator impedance-matching to maximize the gain of this amplifierless link we measured *G*=26.5 dB and *NF*=4.2 dB at 150 MHz, as reported earlier [2]. Adjusting the matching circuit in this link for lower *NF* instead of maximum *G* has yielded measured results of *NF*=3.5 dB and *G*=25.2 dB. To our knowledge 3.5 dB is the lowest noise figure reported for an analog photonic link in absence of any RF amplification. Efforts are under way to adjust the modulator interface circuit impedance in such a way as to further reduce the *NF*, possibly to below 3 dB, at the expense of reduced gain.

Directions for the Next Generation of MMIC Devices and Systems
Edited by N. K. Das and H. L. Bertoni, Plenum Press, New York, 1997

BACKGROUND

In a previous paper [2] we presented expressions for the available gain, G, and noise figure, NF, of an external modulation/direct-detection optical link in which both the modulator and photodetector are perfectly impedance-matched to the link input and output, respectively, using passive circuits. We repeat these expressions here in a form that is valid for both direct and external modulation links:

$$G = |g_1|^2 \frac{|Z_M|^2}{R_M} N_D^2 R_{out} ; \quad (1)$$

$$NF = 10 \log\left[2 + \frac{N_D^2 R_{out}}{kT\, G}\left(i_{RIN}^2 + i_{shot}^2\right) + \frac{1}{G} \right]. \quad (2)$$

In these equations Z_M is the impedance of the modulation device—*i.e.*, the external modulator or the directly modulated semiconductor laser. Additionally, R_M=Re{Z_M}, k is Boltzmann's constant, T=290 °K, N_D is the turns ratio of the lossless impedance transformer in the detector circuit, R_{out} is the link's output load impedance, and i^2_{RIN} and i^2_{shot} are the spectral densities of the *RIN* (relative intensity noise) and shot noise photocurrents. The term g_1 is the coefficient of the first-order term in the generalized expression for photocurrent as a function of the voltage across the modulation device, as will be discussed further on.

When equation (1) is substituted into the second term in the brackets of equation (2), it becomes clear that this term is inversely proportional to the quantity $|g_1|^2|Z_M|^2/R_M$. If this quantity could be made very large, then equation (1) would approach its lower limit of

$$NF = 10 \log\left[2 + \frac{1}{G} \right]. \quad (3)$$

We have named the $1/G$ term in equation (3) the *passive attenuation limit* because it is the same as the result one obtains for a passive attenuator. The equation also shows that the minimum attainable NF for a link with perfect, lossless passive input and output matching circuits is 2, or approximately 3 dB. This limit, which we call *lossless passive match limit*, arises for reasons that are not intuitively obvious. We therefore give a derivation of the 3 dB limit in this paper, followed by some brief intuitive arguments that explain its existence.

ANALYSIS OF LINK WITH LOSSLESS IMPEDANCE MATCHING CIRCUITS

Figure 1 shows an equivalent circuit for an intensity-modulation/direct-detection optical link with perfect and lossless input and output impedance matching circuits. At any frequency one can represent the modulation device and the detector (usually a reverse-biased photodiode) as impedances whose real parts are physical, ohmic resistances. In the detector, photocurrent i_D is generated in proportion to the intensity of optical illumination, which in turn depends upon the voltage v_M across the modulator or directly-modulated laser. We can express this relationship between i_D and v_M most generally as a Taylor series:

$$i_D = i_0 + g_1 v_M \sin \omega t + g_2 (v_M \sin \omega t)^2 + g_3 (v_M \sin \omega t)^3 + \ldots , \quad (4)$$

where ω is 2π times the modulation frequency, and where the magnitudes of coefficients i_0, g_1, g_2, *etc.* are dictated by the type of modulation device and its bias condition, as well as by the total optical insertion loss between the optical source and the photodetector. The coefficient g_n has units of $A \cdot V^{-n}$.

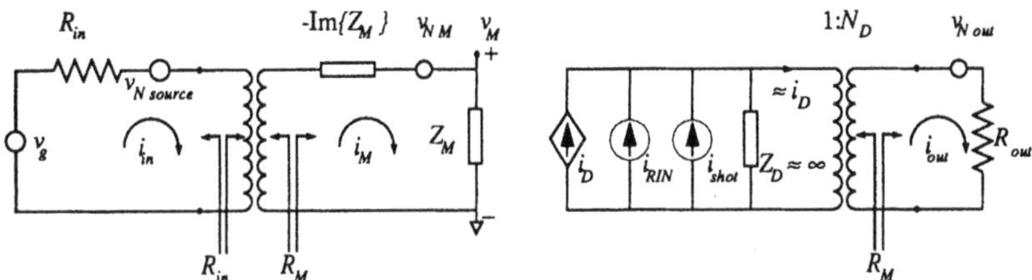

Figure 1. Equivalent circuit of an intensity-modulation/direct-detection optical link with perfect, lossless input and output impedance matching circuits.

The equivalent circuit also includes thermal noise voltages generated by the ohmic resistive portions of the input source, modulation device, and output load impedances. Reactive circuit elements transform the real parts of the device impedances so that from the input and output ports the link looks the same as, respectively, the source and load impedances. Using this equivalent circuit one derives the link's available gain G as follows:

$$G \equiv \frac{P_{out}}{P_{in,av}}, \tag{5}$$

where

$$P_{out} = |i_{out}|^2 R_{out} \tag{6}$$

$$= |i_D(\omega)|^2 N_D^2 R_{out} . \tag{7}$$

For small signals, $|i_D(\omega)| \approx |g_1| \cdot |v_M|$, so

$$P_{out} = |g_1|^2 |v_M|^2 N_D^2 R_{out} \quad = |g_1|^2 |i_M|^2 |Z_M|^2 N_D^2 R_{out} \tag{8}$$

$$= |g_1|^2 \frac{|i_{in}|^2 |Z_M|^2 R_{in}}{R_M} N_D^2 R_{out} . \tag{9}$$

Substituting equation (9) and the expression for available input power,

$$P_{in,av} \equiv |i_{in}|^2 R_{in} , \tag{10}$$

into equation (5) yields equation (1) for G.

Equations (9), (8), and (6) show how currents arising from the thermal noise voltages translate into total thermal noise power spectral density at the link output. That is,

$$\overline{N}_{out,th} = \left| \frac{v_{N,source}}{2R_{in}} \right|^2 |g_1|^2 \frac{|Z_M|^2 R_{in}}{R_M} N_D^2 R_{out} + \left| \frac{v_{N,M}}{2R_M} \right|^2 |g_1|^2 |Z_M|^2 N_D^2 R_{out} + \left| \frac{v_{N,out}}{2R_{out}} \right|^2 R_{out} . \tag{11}$$

Similarly, equation (7) shows how the photodetected laser relative intensity noise (*RIN*) current and shot noise current (see Figure 1) create additional noise at the link output:

$$\overline{N}_{out,RIN+shot} = \left(i_{RIN}^2 + i_{shot}^2\right) N_D^2 R_{out} \ . \tag{12}$$

The thermal noise voltage spectral densities in equation (11) are calculated as follows [15]:

$$\left|v_{N,source}\right|^2 = 4kT R_{in} \ ; \qquad \left|v_{N,M}\right|^2 = 4kT R_M \ ; \qquad \left|v_{N,out}\right|^2 = 4kT R_{out} \ ; \qquad (13), (14), (15)$$

and therefore the total noise spectral density at the output of the link is:

$$\overline{N}_{out} = \overline{N}_{out,th} + \overline{N}_{out,RIN+shot} = 2kT\,G + \left(i_{RIN}^2 + i_{shot}^2\right) N_D^2 R_{out} + kT \ . \tag{16}$$

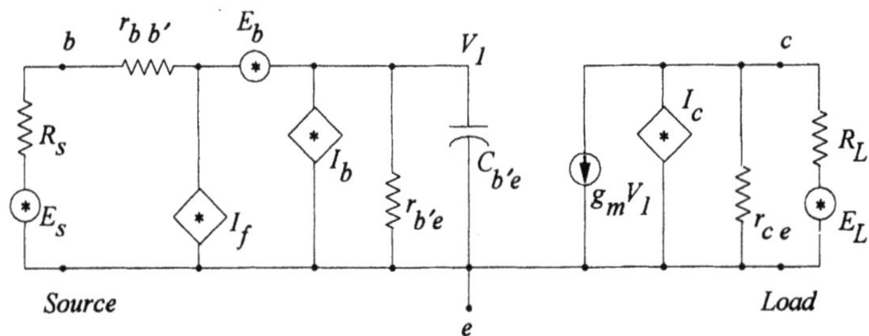

Figure 2. Hybrid-π small-signal equivalent circuit of a bipolar junction transistor (after [17]).

IEEE standards [16] define noise figure as "the ratio of the available output noise power per unit bandwidth to the portion of that noise caused by the actual source connected to the input terminals of the device, measured at a standard temperature of 290°K"—*i.e.*,

$$NF \equiv 10 \log\left[\frac{\overline{N}_{out}}{kT\,G}\right] . \tag{17}$$

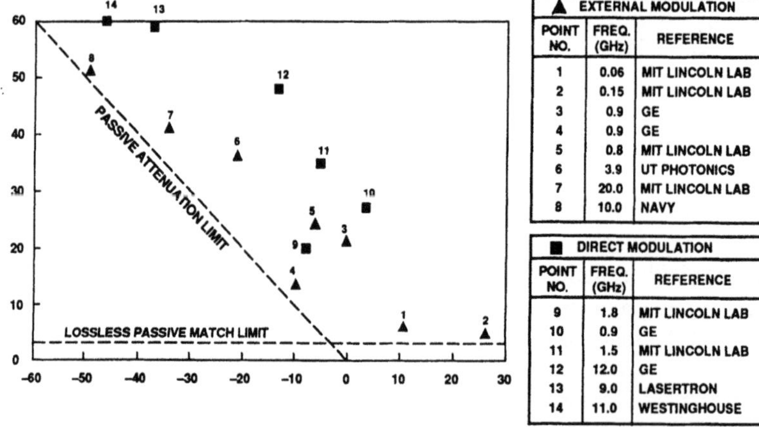

▲ EXTERNAL MODULATION		
POINT NO.	FREQ. (GHz)	REFERENCE
1	0.06	MIT LINCOLN LAB
2	0.15	MIT LINCOLN LAB
3	0.9	GE
4	0.9	GE
5	0.8	MIT LINCOLN LAB
6	3.9	UT PHOTONICS
7	20.0	MIT LINCOLN LAB
8	10.0	NAVY

■ DIRECT MODULATION		
POINT NO.	FREQ. (GHz)	REFERENCE
9	1.8	MIT LINCOLN LAB
10	0.9	GE
11	1.5	MIT LINCOLN LAB
12	12.0	GE
13	9.0	LASERTRON
14	11.0	WESTINGHOUSE

Figure 3. Published insertion gains and noise figures for experimental optical analog links.

Substituting equation (16) into this expression yields equation (2).

The above derivation has shown that the lossless passive match limit of NF=3 dB occurs for any two-port, three-terminal active device in which the entire real part of the input impedance is ohmic. Because intensity-modulation optical links fall into this category they differ from many other active two-ports. For instance most transistors, including BJTs and MESFETs, have input impedances with real parts arising not entirely from physical, ohmic resistances. In the hybrid-π small-signal equivalent circuit of a BJT shown in Figure 2, the base-spreading resistance $r_{bb'}$ contributes thermal noise due to its ohmic nature whereas the element $r_{b'e}$ (which together with $C_{b'e}$ usually dominates the input impedance of the transistor) does not. As a result, the minimum noise figure for a bipolar transistor has been shown [17] to be:

$$NF_{min} = 10 \log\left[1 + \sqrt{\frac{2 g_m r_{bb'} + 1}{g_m r_{b'e}}}\right], \tag{18}$$

where g_m is the device's transconductance. Equation (18) shows why a BJT's noise figure can be, and often is, less than 3 dB.

EFFORTS TO MINIMIZE NOISE FIGURE

Figure 3 shows the measured amplifierless optical link gains and noise figures reported by various researchers [1-14]. Most of these links have noise figures much greater than the lossless passive match limit of 3 dB. This situation can be attributed to the fact that, in practice, low link noise figure occurs only when several conditions are met simultaneously; we describe these conditions here and corroborate them using the equations above.

Notice in Figure 3 that the external modulation links generally have lower noise figures than the direct modulation links. This occurs because the only lasers that can be directly RF-modulated are semiconductor lasers, which generally have much greater RIN than the solid-state lasers that can be used in an external modulation link.

Equation (2) illustrates that, no matter how small the contribution of RIN, to achieve low noise figure also requires that the RF gain G be maximized. Previously [1, 3, 10-12] we have shown how narrowband impedance matching to the modulator and detector results in greater G. Equation (1) in this paper, which is valid for the case where impedance-matching has been applied to both devices using lossless passive circuit elements, shows that G is proportional to a sensitivity term $|g_1|^2|Z_M|^2/R_M$. Most external modulation links use Mach-Zehnder interferometric modulators, for which the sensitivity can be expressed as follows:

$$|g_1|^2 \frac{|Z_M|^2}{R_M} = \frac{\eta_D^2 G_{opt}^2 P_L^2}{4 V_\pi^2} \frac{|Z_M|^2}{R_M} \cos^2 \phi , \tag{19}$$

where η_D is the detector responsivity, G_{opt} is the laser-to-detector optical insertion gain (less than 1 for a link without an optical amplifier) including the insertion loss of the modulator, and P_L is the optical power output of the laser. The term V_π is the bias voltage necessary to impose an optical phase shift $\phi=\pi$ between the two arms of the interferometer. Maximum sensitivity occurs at $\phi=\pi/2$.

In [2] we described an external modulation link for which RIN was very low (less than −175 dB/Hz) and sensitivity was maximized by using a low-V_π (0.65 V) Mach-Zehnder interferometer to modulate a laser with substantial output power (P_L=400 mW). As was shown in Figure 3, this link achieved G=26.5 dB and NF=4.2 dB. What is not conveyed by Figure 3 is the fact that we had used an adjustable optical attenuator to step P_L gradually from about 1 mW up to 400 mW, and that the measured noise figures asymptotically approached a value near 4 dB instead of the 3 dB predicted by equation (3) for large G.

Equation (3) is based on a model of the link that assumes the modulator interface circuit is lossless and achieves a perfect match to the RF source. In the analysis that follows we reject this assumption and reach two important conclusions about noise figure:

I In the case of perfect impedance matching, loss in the modulator interface circuit increases the lower limit from 3 dB to a value equal to 3 dB plus the circuit loss in dB, and;

II The lower limit is also a function of the impedance mismatch between the RF source impedance and the link input impedance, and is not at its lowest in the case where these two impedances are equal (perfectly matched).

ANALYSIS OF LINK WITH CIRCUIT LOSS AND IMPEDANCE MISMATCH

Figure 4 shows the equivalent circuit for the intensity-modulation/direct-detection optical link already shown in Figure 1, but with some important changes. Instead of transforming R_M (the real part of the modulator impedance) to source impedance R_{in}, the lossless transformer makes the modulator appear to have resistance R'. Matching circuit loss is then simulated by using a combination of series and parallel resistances (R_1 and R_2, respectively) resulting in the purely real impedance R_{link}. To simulate any complex link input impedance we add a component with reactance X_{link} in series with this configuration.

It is desirable to express the resistances R_1 and R_2 as a function of the circuit loss they impose on the link. This can be done by looking at the perfect-match case where Z_{link}=R_{in}, in which case X_{link}=0 and R'=Z'', and:

$$R_{link} = R_1 + \left(R_2 \| R'\right), \quad \text{or equivalently,} \quad R' = \frac{R_2 R_{link} - R_1 R_2}{R_1 + R_2 - R_{link}}. \qquad (20), (21)$$

Given that

$$Z'' = R_2 \| (R_{link} + R_1) = \frac{R_{link} R_2 + R_1 R_2}{R_{link} + R_1 + R_2}, \qquad (22)$$

we arrive at

$$R_2 = \frac{R_{link}^2 - R_1^2}{R_1} \quad \text{and} \quad R' = \frac{R_{link}^2 - R_1^2}{R_{link}}. \qquad (23), (24)$$

At this point the derivation above which produced equation (2) for G of a link with lossless impedance-matching circuits (which we hereby call $G_{lossless}$) can be repeated line-for-line to yield an expression for G of the link in Figure 4 that includes the effects of circuit loss:

$$G = |g_1|^2 \frac{|Z_M|^2}{R_M} \frac{R_{link} - R_1}{R_{link} + R_1} N_D^2 R_{out} = \frac{R_{link} - R_1}{R_{link} + R_1} G_{lossless} . \tag{25}$$

Defining

$$G \equiv G_M G_{lossless} , \tag{26}$$

where G_M is the circuit's excess gain (less than 1 for a passive circuit), it is clear that

$$G_M = \frac{R_{link} - R_1}{R_{link} + R_1} , \tag{27}$$

and therefore

$$R_1 = \frac{1 - G_M}{1 + G_M} R_{link} , \quad R_2 = \frac{4 G_M}{1 - G_M^2} R_{link} , \quad \text{and } R' = \frac{4 G_M}{(1 + G_M)^2} R_{link} . \tag{28), (29), (30}$$

These equations make sense because removing the lossy elements from Figure 4 should yield G_M=1 (so that G=$G_{lossless}$), which is equivalent to setting R_1=0 and R_2=∞. Repeating the gain derivation yet again for $Z_{link} \neq R_{in}$ yields:

$$G = \frac{4 R_{link} R_{in}}{|Z_{link} + R_{in}|^2} |g_1|^2 \frac{|Z_M|^2}{R_M} G_M N_D^2 R_{out} , \tag{31}$$

which is valid for any link input impedance and for any amount of modulator interface circuit loss. In the case where Z_{link}=R_{link}=R_{in} and G_M=1 this gain reduces to the expression that was valid for the perfect lossless match case [equation (2)].

With the resistances R_1 and R_2 now expressed as functions of the loss they represent, it is possible to derive the noise figure as a function of that loss as well. Figure 4 shows the thermal noise voltages generated by these two impedances. Including these two additional noise voltages in the derivation yields the following expression for noise figure:

$$NF = 10 \log \left[1 + \frac{1 - G_M}{1 + G_M} \frac{R_{link}}{R_{in}} + \frac{1 - G_M}{1 + G_M} \frac{\left[(1 + G_M) R_{in} + (1 - G_M) R_{link} \right]^2}{4 G_M R_{link} R_{in}} \right.$$
$$\left. + \frac{R_M^2 (R_{link} + R_{in})^2}{G_M R_{link} R_{in} |R_M + Z_M'|^2} + \frac{N_D^2 R_{out}}{kT G} \left(i_{RIN}^2 + i_{shot}^2 \right) + \frac{1}{G} \right] . \tag{32}$$

Equation (32) was derived for the case where X_{link}=0 in order to simplify the expression and more clearly illustrate the points **I** and **II** that we mentioned above and repeat here—namely:

I In the case of perfect impedance matching (*i.e.*, when R_{link}=R_{in} and Z_M'=R_M), loss in the modulator interface circuit (G_M < 1) increases the noise figure's lower limit from 3 dB to a value equal to 3 dB plus the circuit loss in dB; *i.e.* for very large G,

$$NF = 10 \log \left[1 + \frac{1 - G_M}{1 + G_M} + \frac{1 - G_M}{1 + G_M} \frac{1}{G_M} + \frac{1}{G_M} \right] = 10 \log \left[\frac{2}{G_M} \right] = 3\text{dB} + 10 \log \left[\frac{1}{G_M} \right] . \tag{33}$$

This is what we have named the *general passive match limit* [2]. The above analysis has helped to explain why the lowest noise figure that we achieved previously was about 4 dB. As we reported earlier [2], we separately measured the insertion loss of the circuit we used to impedance-match the modulator in our link, and had found it to be 0.7 dB. Additionally:

II From equation (32) we see there is a *general passive limit* that is a function of the impedance mismatch between the RF source impedance and the link input impedance. This limit is <u>not</u> minimized when these two impedances are equal, as is most clearly illustrated when the circuit loss is ignored (*i.e,* when $G_M = 1$), in which case for very large G,

$$NF = 10\ \log\left[1 + \frac{R_M^2\left(R_{link} + R_{in}\right)^2}{R_{link}\, R_{in}\left|R_M + Z'_M\right|^2}\right]. \tag{34}$$

Note that for perfect impedance matching equation (34) reduces to the lossless passive match limit of 3 dB. It appears from this expression that the key to circumventing the lossless passive match limit is a modulator interface circuit that would make the second term less than 1, using high-Q components in an effort to also minimize the interface circuit loss.

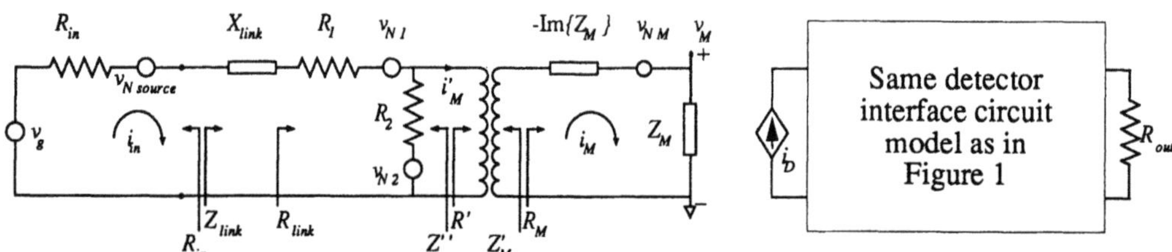

Figure 4. Equivalent circuit for an intensity-modulation/direct-detection optical link with arbitrary input impedance Z_{link}, and resistors R_1 and R_2 to simulate loss in the modulator interface circuit.

EXPERIMENTAL RESULTS

Using the same external modulation link components described in [2], we adjusted the capacitance and inductance in the modulator interface circuit while monitoring the 150 MHz performance on a noise figure meter. The lowest noise figure we were able to obtain was 3.5 dB, at the expense of some gain (G=25.2 dB compared to 26.5 dB for the perfect match case). Although to our knowledge this is the lowest noise figure that has been achieved for an amplifierless optical link, it is still in excess of the 3 dB lossless passive match limit we were hoping to circumvent by varying Z_{link}. The interface circuit in the link had been originally designed to perfectly match the modulator impedance to 50 Ω, and not for maximum adjustability. Currently under way is the assembly of a circuit that should allow us to achieve almost any value of Z_{link}. We plan to report measurements of link noise figure for a broad range of input impedances, and also to compare these results with those predicted by the model we have presented here.

CONCLUSIONS

3 dB is the lowest *NF* that can be achieved for an amplifierless optical link with perfect lossless impedance matching to the RF source. However, investigation of circuit loss and impedance mismatch effects has shown that circuit loss increases the minimum link *NF*, whereas impedance mismatch should enable a reduction of *NF* to below 3 dB.

ACKNOWLEDGMENTS

The authors thank Roger Helkey, Fred O'Donnell, Joelle Prince, Kevin Ray, Gil Rezendes, Harold Roussell, Mike Taylor, and Allen Yee for technical assistance, and Colin Aitchison of Brunel University, Uxbridge, UK, for helpful discussions. This work was supported under Air Force contract #F19628-95-C-0002. Opinions, interpretations, conclusions and recommendations are those of the authors and are not necessarily endorsed by the United States Air Force.

REFERENCES

[1] C. H. Cox III, G. E. Betts, and L. M. Johnson, "An analytical and experimental comparison of direct and external modulation in analog fiber-optic links," *IEEE Trans. Microwave Theory Techn.*, 38: 501-509 (1990).

[2] C. H. Cox III, E. I. Ackerman, and G. E. Betts, "Relationship between gain and noise figure of an optical analog link," *1996 IEEE MTT-S Symp. Digest*, 1551-1554 (1996).

[3] E. Ackerman, D. Kasemset, and S. Wanuga, "An external modulation L-band link with a 117-dB spurious-free dynamic range," *Microwave J.*, 34:158 (1991).

[4] E. Ackerman, S. Wanuga, J. MacDonald, and J. Prince, "Balanced receiver external modulation fiber-optic link architecture with reduced noise figure," *1993 IEEE MTT-S Symp. Digest*, 615-618 (1993).

[5] C. H. Cox III, G. E. Betts, and A. C. Yee, "Incrementally lossless, broad-bandwidth analog fiber-optic link," *IEEE/LEOS Summer Topical Meeting on Broadband Analog Optoelectronics: Devices and Systems*, Monterey, California (1990).

[6] S. W. Merritt, W. L. Glomb, and J. D. Farina, "Minimum noise figure microwave transmission system," *Second Annual DARPA/Rome Laboratory Symp.*, Monterey, California (1991).

[7] G. E. Betts, C. H. Cox III, and K. G. Ray, "20-GHz optical analog link using an external modulator," *Broadband Analog Optoelectronics—Devices and Systems*, Monterey, California (1990).

[8] S. A. Pappert, M. H. Berry, S. M. Hart, R. J. Orazi, and S. T. Li, "Remote multioctave electromagnetic field measurements using analog fiber optic links," *IEEE Antennas and Propagation Society International Symp.*, Chicago, IL (1992).

[9] C. H. Cox III, "Intrinsic gain in analog fiber-optic links and its effect on noise figure," *XXIV General Assembly of The International Union of Radio Science* (1993).

[10] E. Ackerman, D. Kasemset, S. Wanuga, D. Hogue, and J. Komiak, "A high-gain directly modulated L-band microwave optical link," *1990 IEEE MTT-S Symp. Digest*, 153-155 (1990).

[11] C. H. Cox III, D. Z. Tsang, L. M. Johnson, and G. E. Betts, "Low-loss analog fiber-optic links," *1990 IEEE MTT-S Symp. Digest*, 157-160 (1990).

[12] E. Ackerman, D. Kasemset, S. Wanuga, R. Boudreau, J. Schlafer, and R. Lauer, "A low-loss Ku-band directly modulated fiber optic link," *IEEE Photonics Techn. Lett.*, 3:185-187 (1991).

[13] D. Atlas, "A 20 GHz bandwidth InGaAsP/InP MTBH laser module," *IEEE Photonics Techn. Lett.*, 5:123-126 (1993).

[14] D. K. Davies and A. P. Goutzoulis, "Wavelength-multiplexed analog fiber optic link for wideband radio-frequency and local oscillator signal transmission," *Optical Engineering*, 31:2323 (1992).

[15] H. Nyquist, "Thermal agitation of electric charge in conductors," *Phys. Rev.*, 32:110-113.

[16] H. Haus, *et al.*, "IRE standards on methods of measuring noise in linear twoports, 1959," *Proc. IRE*, 48:60-68 (1959).

[17] C. D. Motchenbacher and F. C. Fitchen. *Low-Noise Electronic Design*, John Wiley & Sons, New York (1973).

POLYMER OPTICAL WAVEGUIDES FOR MULTI-CHIP MODULES

Tsang-Der Ni

U.S. Army Research Laboratory
Physical Sciences Directorate
Fort Monmouth, NJ 07703

INTRODUCTION

Polymer optical waveguides are being used in next generation mixed signal Transmitter/Receiver multi-chip module combining photonic and microwave monolithic integrated circuits (MMICs) technologies. The optical interconnections for the multi-chip module is becoming an important issue due to the development of MSM photodetector on the MMICs. However, optical chip-to-chip interconnects require low-loss, wide-angle, and high accuracy in a short propagation length presenting restrictions to waveguide design. The use of polymer optical waveguides have a number of advantages over optical fiber such as flexible chip-to-chip interconnection, low cost, and easy fabrication. Additionally, polymer waveguide processing technique is compatible with a variety of semiconductor materials (Si, GaAs, InP), and can be fabricated with low loss (<1 dB/cm)[1] as compared with semiconductor slab waveguide. These characteristics are important for applications in high speed signal distribution in multi-chip module incorporating both MMICs and hybrid optoelectronic integrated circuits (OEICs).

Optical beam splitters are important components for signal distribution in optoelectronic T/R modules. The conventional Y-junction beam splitters suffer severe radiation losses when the branching angle is larger than 2 degrees[2,3]. The main reason for the radiation loss is due to the optical phase mismatch between the input waveguide and two output waveguides. Many efforts have been proposed to reduce the radiation loss. Ulrich *et al.* demonstrated self-imaging properties of multimode waveguides to avoid the branching waveguide and to reduce the scattering loss of waveguide[4]. Heaton *et al.* demonstrated a GaAs/AlGaAs integrated optical 1 to N beam divider which uses symmetric mode mixing in center-fed multimode waveguide[5]. This technique has the advantage of making the device shorter, and less sensitive to fabrication tolerances. However, the overall fabrication process is still complicated.

Another method called phase front accelerator exhibits a lower refraction index at central of the junction[6,7]. More, integrated microprisms were adopted by many research groups[8,9]. However, all the methods reported before only show the theoretical analysis and simulation results. The difficulties in fabricating this kind of beam splitter are controlling of the

Directions for the Next Generation of MMIC Devices and Systems
Edited by N. K. Das and H. L. Bertoni, Plenum Press, New York, 1997

refraction index for the waveguides and the prism at branching area, and the high complexity fabrication processes of using materials with different refraction indices.

In this paper, we introduce the processes to fabricate the polymer waveguides, and then take advantage of the ease of fabrication of polymer waveguides to demonstrate two type of beam splitter: center-fed mode mixing beam splitter and hybrid Y-junction beam splitter. The comparison of this two types of beam splitters is made. Both types of beam splitters can be fabricated on low temperature co-fired ceramic incorporating Si, GaAs components as found in standard multi-chip modules. The applications of the polymer waveguide for the multi-chip module is briefly introduced.

WAVEGUIDE FABRICATIONS

The requirements for material used in optical waveguide fabrication are fast cured response, ability to form hard film, and adhesion to substrate. In addition, they have some other requirements including refraction index control, and the need to maintain a consistent physical shape and optical performance in environments specific to high-speed electronic application.

The staring material is a commercially available spin-on polymer (Norland 81, 68)[10]. This polymer, originally designed as a single part adhesive for optical elements, is a clear, colorless, liquid photopolymer that is cured when exposed to ultraviolet (UV) light within range of 325-375 nm. Norland 81 and 68 has an index of refraction of 1.56 and 1.54, viscosity of 300 and 5000, respectively.

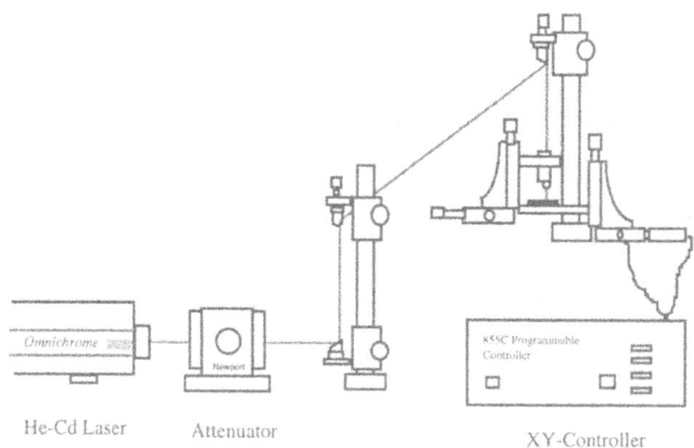

Figure 1. Experimental setup for laser direct writing technique.

GaAs wafer were used as the substrate material to emphasize the applicability to GaAs based MMIC system. Because of the index of refraction of GaAs is about 3.7, a cladding layer is needed to act as a buffer layer. After finishing the processing for the cladding layer, the polymer for the core is spread on the top of the cladding layer. After spinning, the polymer is patterned by exposing it with the laser writing system as shown in Fig. 1. The laser writing system consisted of a He-Cd laser with wavelength 325 nm and 442 nm and imaged via a series lenses to project onto a set of computer-controlled XY stages. After exposing, the pattern is developed by rinsing the polymer in acetone. The unexposed regions rinse away, leaving a patterned channel waveguide on the cladding layer.

The measured minimun physical dimensions of thickness and width are about 1 μm and 5 μm, respectively. The measured results suggested that the lower incident energy with high scan speed produce the minimum dimension of waveguides.

DESIGN, FABRICATION AND TEST OF CENTRAL MODE-MIXING COUPLER

The proposed polymer optical beam splitter based on the self-imaging effect in center-fed multimode waveguide is shown in Fig. 2. The GaAs wafer was used as the substrate material to emphasize the applicability to OEICs' system. The core of the waveguide has a higher refraction index than the cladding to guide the light in the waveguide at the core-cladding interface. This device has one single-mode input waveguide, one multimode waveguide, and two single-mode output waveguides. The 1-to-2 beam splitter is designed by using symmetric mode mixing in center-fed multimode planar waveguide. The electric field of the light in the input guide is assumed to be symmetric. The field amplitude E(x,z) in the multimode waveguide can be written as a sum of symmetric modes[5],

$$E(x,z) = \sum_{n=1,3,5,..}^{N} E_n \cos\left(\frac{n\pi}{w} x\right) \exp jk_{zn}z \qquad (1)$$

where N is the total number of modes in the waveguide, E_n is the complex amplitude of each mode, w is the width of the multimode waveguide, k_{zn} is the propagation constant of n-th mode and is defined as $k_{zn} = k - n^2\pi^2/(2w^2k)$, where $k = 2n_c\pi/\lambda_o$, n_c is the refractive index of the core material, and λ_o is the free space wavelength of light.

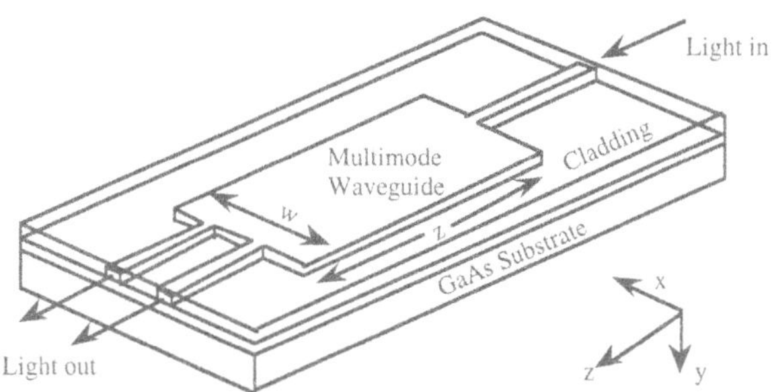

Figure 2. Schematic diagram of 1 to 2 polymer optical beam splitter.

The simulated field intensity distribution from Eq. (1) in the z-direction for the multimode polymer optical waveguide is shown in Fig. 3. A value of $n_c = 1.56$, $w = 80$ µm, and $\lambda_o = 630$ nm is assumed in this simulation. In Fig. 3(a), two sharp resonant modes are observed at the location of z = 7.9 mm with space of 40 µm. The different output number of waveguides is shown in Fig 3(b). It is obvious that the more output waveguides, the shorter the device under the same cavity width. The device fabrication is based on the simulated data for output number of two.

A He-Ne laser is used as the source to test this beam splitter. The output is monitored by a laser beam analyzer and power meter. As shown in Fig. 4, two output beams with similar intensity are clearly observed. The higher order modes can also be observed in Fig. 4 and are due to the larger dimensions of the waveguide. The measured power coupling is about 5 dB rather than the expected value of 3 dB. This is because of the coupling loss from a fiber to the input waveguide.

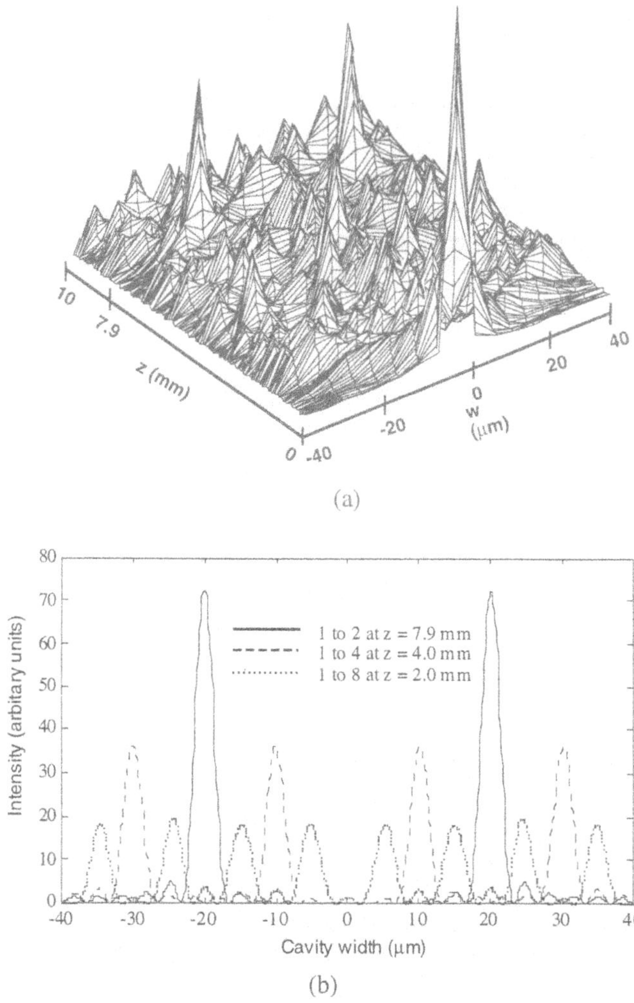

Figure 3. The simulation results of the center-fed mode mixing waveguide; (a) 3-D profile, and (b) the 2-D intensity profile for N = 2, 4, and 8.

DESIGN, FABRICATION AND TEST OF LOW-LOSS, LARGE-ANGLE Y-JUNCTION COUPLER

The conventional Y-junction beam splitter and the new proposed hybrid polymer beam splitter are shown in Fig. 5. The refractive index of the input, output waveguides and microprism are n_1, n_2, and n_p, respectively. The microprism has a lower refractive index than the optical waveguide ($n_p<n_1$, n_2). The input light incident to the polymer waveguide passes through the microprism directly coupling to the two branches. The microprism causes additional phase delay to compensate the optical phase at the junction, therefore the phase-front of the optical wave can be bent with minimum radiation loss. The light path can be traced by using simple geometric optics, with the branching angle θ calculated by using Snell's law. In our design, a value of $n_1 = n_2 = 1.56$, and $n_p = 1.5$ was chosen based on the selected material, resulting in a calculated branching angle θ of about 4.5°. A conventional Y-junction beam splitter was also designed with the same branching angle, and was used to compare with the new optical beam splitter.

A He-Ne laser at wavelength of 632 nm was coupled through a single mode fiber and used as the input light to the polymer waveguide. Fig. 6 shows the 3-D output light intensity

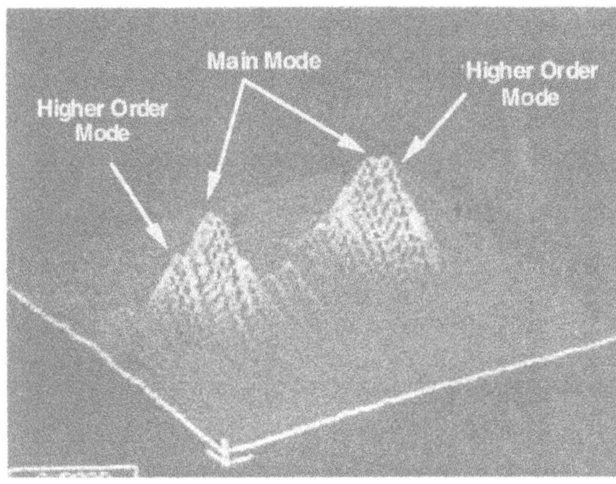

Figure 4. Measured 3-D output beam profile of the multimode waveguide.

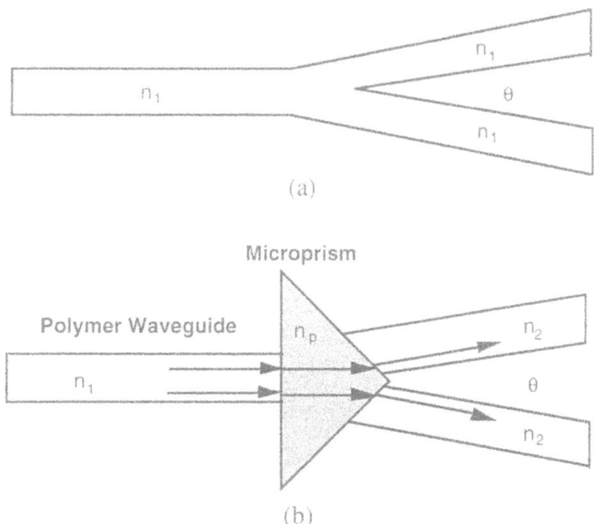

Figure 5. The Y-junction beam splitters, (a) conventional structure, (b) proposed hybrid beam splitter incorporating a microprism.

profile at the near-field for the same beam splitter. Two similar output beams are observed. However, the radiation intensity profile is also observed beside these beams. This radiation loss is due to the optical phase mismatch between the input waveguide and two output waveguides.

Fig. 7 shows the 3-D output light intensity profile of the newly designed hybrid beam splitter with a microprism. It is clear that the two output beams are more distinctly separated than in the conventional Y-junction beam splitter. The radiation loss is minimized due to the phase compensation through the microprism. The measured insertion loss is about 4 dB rather than the theoretical expected value of 3 dB. This is because of the coupling loss from the fiber to the input waveguide. Meanwhile, the insertion loss for the conventional Y-junction polymer beam splitter with same branching angle is about 6 dB which is 2 dB higher than the hybrid beam splitter.

In addition, another advantage of using polymer waveguide rather than the semiconductor waveguide is the refractive index of polymer and microprism are compatible (in the range of 1.5 to 1.7). This advantage will reduce the light reflection at the interface which is caused by the index mismatch between the two media.

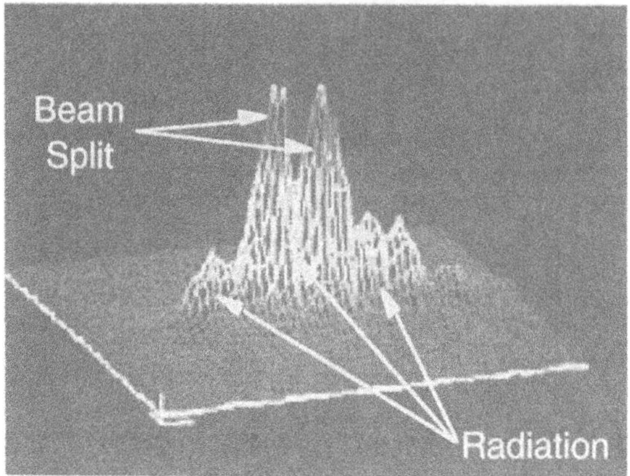

Figure 6. The 3-D output light intensity profile of the Y-junction beam splitter.

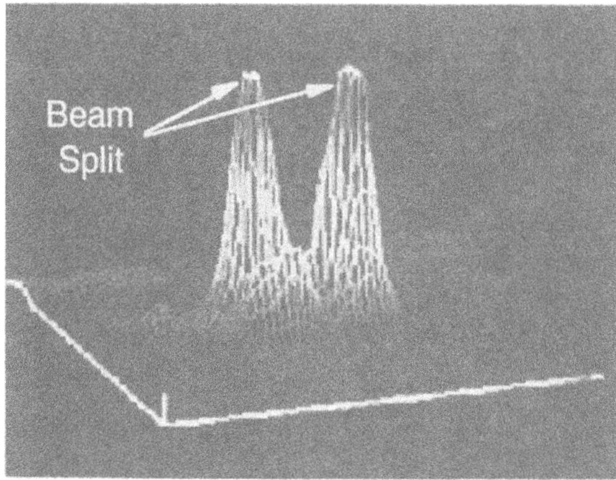

Figure 7. The 3-D output light intensity profile of the hybrid Y-junction beam splitter incorporating a microprism.

APPLICATIONS

One of the important applications of using the polymer waveguides is for MMIC multi-chip module. The low temperature co-fired ceramic (LTCC) is used as the substrate and package material. The LTCC is a recrystallizable ceramic tape process, which can be stacked and fired to fabricate custom packages, circuits, and modules. The use of LTCC allows multi-layers for RF and DC electrical distribution, and buried capacitors and resistors reducing the circuit size, weight, and parasitic effects.

The designed modules include multiple layers, and each layer contains a portion of circuits, transmission lines. The polymer waveguide can also be fabricated on one of the layers and then stacked each layer together, and co-fired to a single piece as shown in Fig. 8. After these process, the MMIC chips are then mounted on the substrate.

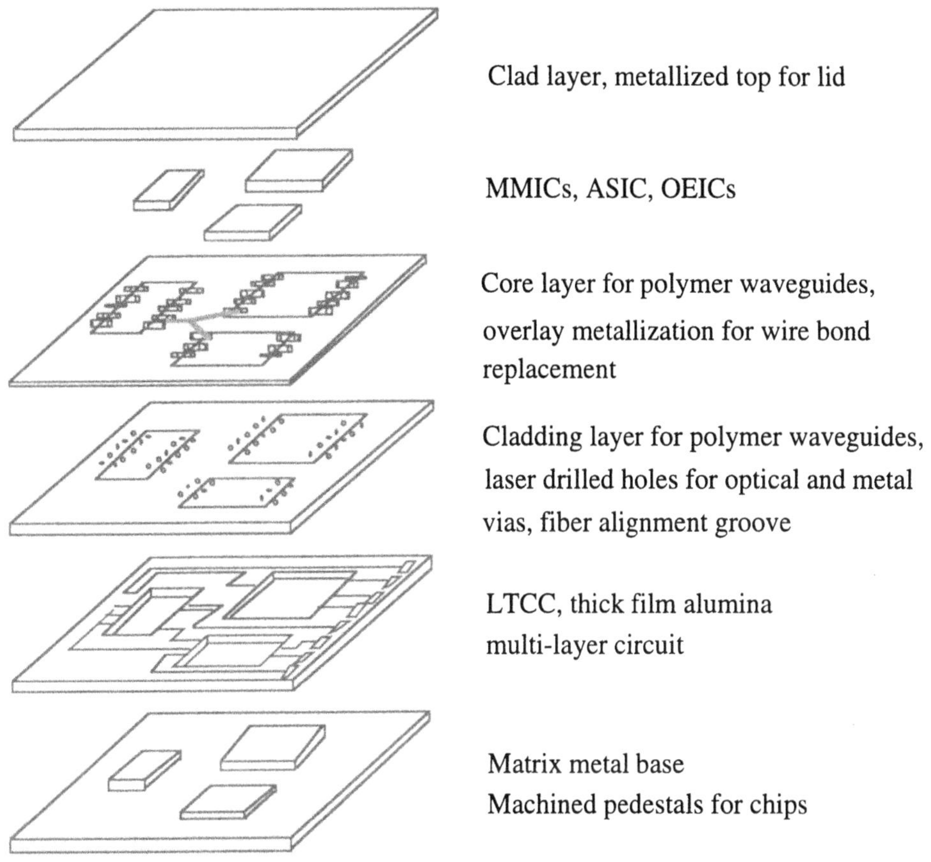

Figure 8. Multiple layers of low temperature cofired ceramic.

CONCLUSIONS

Polymer waveguides are envisioned to perform chip-to-chip, as well as chip to package interconnects in MCM. In this paper, we demonstrate polymer optical beam splitters based on the self-imaging effect in center-fed multimode waveguide. This type of beam splitter can be extended to contain more output ports (1 x N) which would be useful for optical signal distributions networks (i.e. greater distribution capacity). This type of beam splitter will allow for more efficient use of space in high density packaging.

We also have demonstrated a wide-angle hybrid polymer optical waveguide beam splitter incorporating microprism for bending the optical light. It has been shown that the radiation loss is small with relatively wide branching angle as compared to the conventional Y-junction beam splitter. This kind of beam splitter also shows a great flexibility to change the branching angle of the light. For example, a branching angle of 13° can be achieved by selecting the material with refractive index of 1.7 from commercially available polymers or polyimides. Further increase the branching angle is attainable by rotating the microprism[11], or using the multiple microprisms[12,13].

ACKNOWLEDGMENT

The authors wish to thank M. Popick, J. DeMarco and D. Brocking for the technical assistance, and T. Higgins for the computer program.

REFERENCES

1. Krchnavek, R.R., Lalk,G.R., and Hartman, D.H., "Laser Direct Writing of Waveguides Using Spin-on Polymers," *Journal of Applied Physics,* **66**(11), pp. 5156-5160, Dec. 1989.
2. Sasaki, H., and Mikoshiba, N., "Normalized Power Transmission in Single Mode Optical Branching Waveguide," *Electronics Letter,* Vol.17, pp.136-138, Feb. 1981.
3. Weissman, Z., Hardy, A., and Marom, E., "Mode-Dependent Radiation Loss in Y-Junction and Directional Couplers," *IEEE Journal of Quantum Electronics*, Vol.25, No.6, pp.1200-1208, June 1989.
4. Ulrich, U., and Ankele, G., "Self-Imaging in Homogeneous Planar Optical Waveguides," *Applied Physics Letter,* Vol. **27**(6), pp.337-339, 1975.
5. Heaton, J.M.,Jenkins, R.M., Wight, D.R., Parker, J.T., Birbeck, J.C.H., and Hilton, K.P., "Novel 1-to-N Way Integrated Optical Beam Splitters Using Symmetric Mode Mixing in GaAs/AlGaAs Multimode Waveguides," *Applied Physics Letter*, Vol.**61**(15), pp.1754-1756, 1992.
6. Hanaizumi, O., Miyagi, M., and Kawakami, S., "Low Radition loss Y-Junctions in Planery Dielectric Optical Waveguide," *Optic Communications*, Vol.51, pp.236-238, Sept. 1984.
7. Hung, W.Y.,Chan, H.P., and Chung, P.S, "Novel Design of Wide-Angle Single-Mode Symmetric Y-Junction," *Electronics Letter,* Vol.24, pp.1184-1185, Sept. 1989.
8. Korotky, S.K., Marcatlli, E.A.J., Veselka, J.J., and Bosworth, R.H., "Greatly Reduced Losses for Small-Raduis Bend in Ti:liNbO3 Waveguides," *Applied Physics Letter,* Vol.**48**, pp.92-94, Jan. 1986.
9. Hirayama, K., and Koshiba, M., "A New Low-loss Structure of Abrupt Bends in Dielectric Waveguide," *Journal of Lightwave Technology*, Vol.10, pp.563-569, May 1992.
10. Norland Products Inc., New Brunswick, NJ 08902.
11. Lin, H.B., Cheng, R.S., and Wang, W.S., "Wide-Angle Low-Loss Single-Mode Symmetric Y-Junctions," *IEEE Photonics Technology Letters*, Vol.6, No.7, pp.825-827, July 1994.
12. H. Hatami-Hanza, Chu, P.L., and Lederer, M.J., "A New Low-Loss Wide-Angle Y-Branch Configuration for Optical Dielectric Slab Waveguides," *IEEE Photonics Technology Letters,* Vol.6, No.4, pp.528-530, April 1994.
13. Lin, H.B., Chen, W.L., Wei, P.K., and Wang, W.S., "A Novel Type Power Divider with Large-Angle Low-Loss Characteristics," *IEEE Journal of Quantum Electronics,* Vol.31, No.4, pp.735-742, April 1995.

LOW-VOLTAGE MMIC HBT VCO FOR MILLIMETER-WAVE COMMUNICATION SYSTEMS

Jenshan Lin[1], Young-Kai Chen[1], Dexter A. Humphrey[2], Robert A. Hamm[1], Roger J. Malik[1], Alaric Tate[1], Rose F. Kopf[1], and Robert W. Ryan[1]

[1]Bell Laboratories, Lucent Technologies, Murray Hill, New Jersey 07974
[2]Jet Propulsion Laboratory, Pasadena, California

ABSTRACT

HBT VCO is used as an example to demonstrate the low voltage operation of MMIC devices at millimeter-wave frequencies. The VCO is operated at 1.5 V DC supply and consumes less than 10 mA. DC-to-RF conversion efficiency is above 10% at 26.5 GHz and 5% at 33.5 GHz, respectively. Its phase noise levels at 100 kHz and 1 MHz offset frequencies are -80 dBc/Hz and -110 dBc/Hz, respectively.

INTRODUCTION

Low voltage operation is not only a trend in digital circuits but also in analog circuits[1]. It became a challenge for RF circuit design, especially for those circuits requiring linearity, power output, and efficiency[2]. Typical examples are power amplifiers and oscillators. In these circuits, performance usually degrades as bias voltage decreases because the knee voltage of transistor limits the voltage or current swing and start to affect the linearity and output power at its fundamental frequency.

On the other hand, as the wireless communication of voice, video, and data grows, the increasing demand of channels and bandwidth is driving the transceiver systems toward millimeter-wave frequencies[3]. A typical example is Local Multipoint Distribution Service (LMDS)[4]. The Federal Communications Commission (FCC) has proposed the 31-31.3 GHz band for two-way LMDS operations. Since the maximum available gain (*Gmax*) has a 20 dB roll-off per decade approaching its cutoff frequency, devices with high cutoff frequencies are critical in developing millimeter-wave circuits.

In this paper, we use Voltage Controlled Oscillator (VCO) as an example. Low-phase noise VCO is a key component in radio front end. At millimeter-wave frequencies,

Directions for the Next Generation of MMIC Devices and Systems
Edited by N. K. Das and H. L. Bertoni, Plenum Press, New York, 1997

VCO's built by HBT devices deliver lowest phase noise comparing to other high-frequency devices[5,6]. We have built MMIC VCO's in Ka-band using InGaAs/InP HBT technology and demonstrated the lowest phase noise published so far[7]. Its phase noise levels at 100 kHz and 1 MHz offset frequencies are -80 dBc/Hz and -110 dBc/Hz, respectively. Particularly, these VCO's are operated at only 1.5 V DC supply voltage and current consumption is only 6.8 mA at 26.5 GHz and 9.5 mA at 33.5 GHz, respectively. Their DC-to-RF conversion efficiency is above 10% and 5% throughout the tuning ranges of two VCO's at 26.5 GHz and 33.5 GHz, respectively. In addition, Coplanar Waveguide (CPW) structure is implemented in the design to eliminate via hole processing and thus reduce the cost.

DEVICE TECHNOLOGY AND FABRICATION

The device used for this VCO is InGaAs/InP HBT. The InGaAs/InP HBT wafer was grown in the Metal-Organic Molecular Beam Epitaxy (MOMBE) system. The HBT device used in the VCO has an emitter size of 3 μm x 10 μm. The collector doping and thickness were 1 x 10^{16} cm^{-3} and 500 nm, respectively. The short-circuited current gain cutoff frequency f_T is 100 GHz and the maximum oscillation frequency f_{max} is greater than 60 GHz (Fig. 1). The maximum available gain at 30 GHz is 7 dB with a collector current of 15 mA.

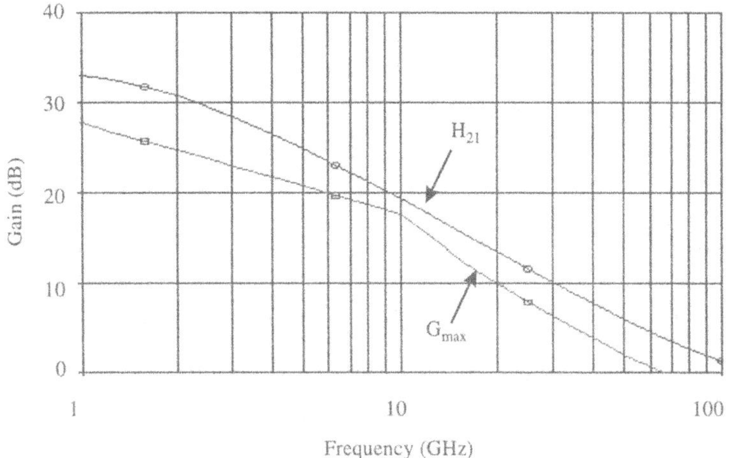

Figure 1. Short-circuited current gain and maximum available gain of InGaAs/InP HBT at V_{CE} = 1.5 V and I_E = 15 mA.

SiO_2 was used for passivation as well as the dielectric layer of MIM capacitor. To reduce the parasitic capacitance and support the cross-over bridges used in CPW structure, polyimide (ε_r = 2.0) with thickness of 1 μm was used. Since the ground plane is on the same side of the circuitry in CPW structure, via hole back side processing is not required.

CIRCUIT DESIGN

The VCO and Phase-Lock Loop (PLL) together provide a stable reference signal source for up/down conversion in communication systems (Fig. 2). The oscillators were designed in a CPW structure using common base configuration. The schematic is shown in Fig. 3. The negative resistance of oscillator is realized by a transistor in common base configuration whereas the varactor is realized by a base-collector junction of transistor. Though the common base configuration is used in low power oscillator design, it provides an advantage of broadband tuning[8].

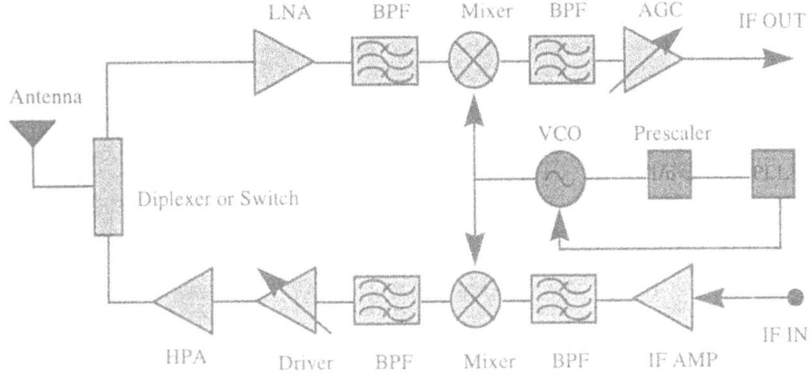

Figure 2. System diagram of a wireless transceiver.

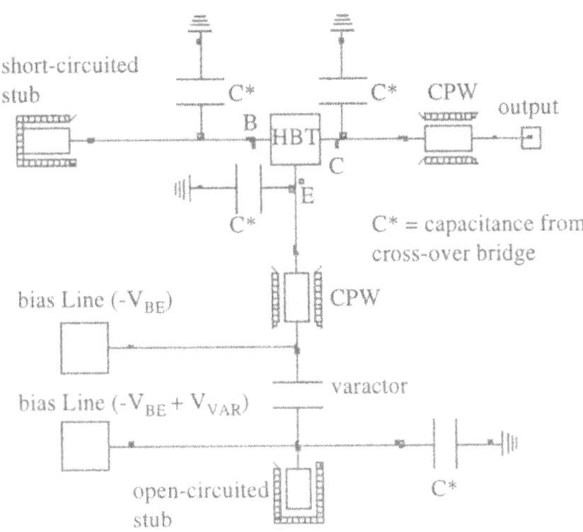

Figure 3. Schematic of VCO circuit.

The base of the HBT is connected to the ground through a CPW short-circuited stub functioning as an inductor to increase the negative resistance. The resonant tank determining the oscillation frequency is connected to the emitter. The varactor is inserted in the resonant tank to provide frequency tuning. The collector is connected to the VCO output port through a section of CPW line for impedance matching. Since the small-signal maximum available gain of the transistor at 33.5 GHz is not high enough, the small-signal negative resistance is not large enough to initialize the oscillation. Therefore, in the 33.5 GHz oscillator design, a low-impedance quarter-wavelength transformer is essential to reduce the load impedance from 50 Ω to a much lower impedance. For the 26.5 GHz oscillator design, the negative resistance at collector is high enough to start the oscillation at this frequency. No impedance matching is required and a simple 50 Ω line is used.

A typical problem associated with CPW circuit design is the excitation of coupled slot mode(even mode). This usually happens near the discontinuities in a circuit. To avoid this problem, cross-over bridges were used at discontinuities to suppress even mode and support the odd mode(CPW mode).

DC bias lines were designed by using quarter-wavelength high impedance lines and Metal-Insulator-Metal (MIM) capacitors to create RF chokes at oscillation frequencies.

Two oscillators were designed. One has center frequency at 26.5 GHz and the other has center frequency at 33.5 GHz. The layout of the chip size is 1.75 mm x 1.2 mm for both oscillators (Fig. 4).

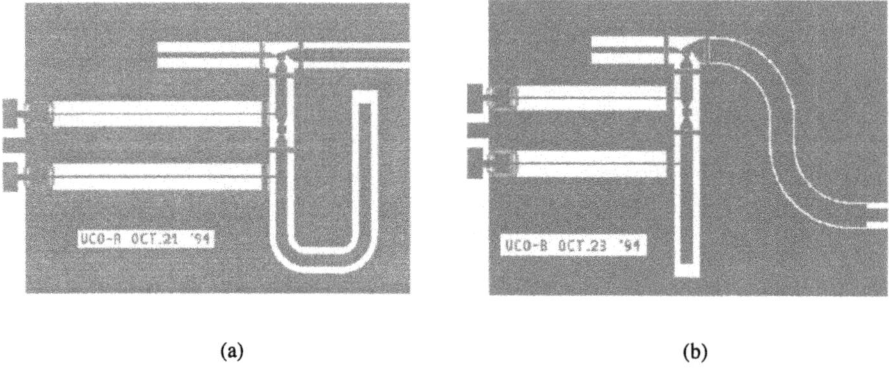

(a) (b)

Figure 4. Layout of the VCO. (a) 26.5 GHz, (b)33.5 GHz.

RESULTS

Frequency tuning and output power were measured at the bias condition of V_{CE} = 1.5 V (V_{BE} ~ 1.0 V and V_{CB} ~ 0.5 V). The emitter current I_E was 6.8 mA for VCO at 26.5 GHz and 9.5 mA for VCO at 33.5 GHz, respectively (Fig. 5). With varactor tuning voltage from 0 V to 1.5 V, the frequency tuning range is 1.42 GHz for VCO at 26.5 GHz and 0.75 GHz for VCO at 33.5 GHz, respectively. The DC-to-RF conversion efficiency within the tuning range is above 10 % for VCO at 26.5 GHz and 5% for VCO at 33.5 GHz, respectively. The measured phase noise levels at 100 kHz and 1 MHz offset frequencies are -81 and -112 dBc/Hz for VCO at 26.5 GHz and -79 and -107 dBc/Hz for VCO at 33.5 GHz.

The capacitance tuning performance of varactor was also measured at 18 GHz (Fig. 6). Capacitance tuning ratio is 1.47:1 from 0 V to 1.5 V. Quality factor increases from 12 at 0 V to 56 at 1.5 V. Series RC model is assumed and the data was extracted from one-port S-parameter measurement on Network Analyzer. Though the quality factor of varactor at 0 V bias is not high, it does not affect the output power and the overall quality factor of the oscillator remains about the same level.

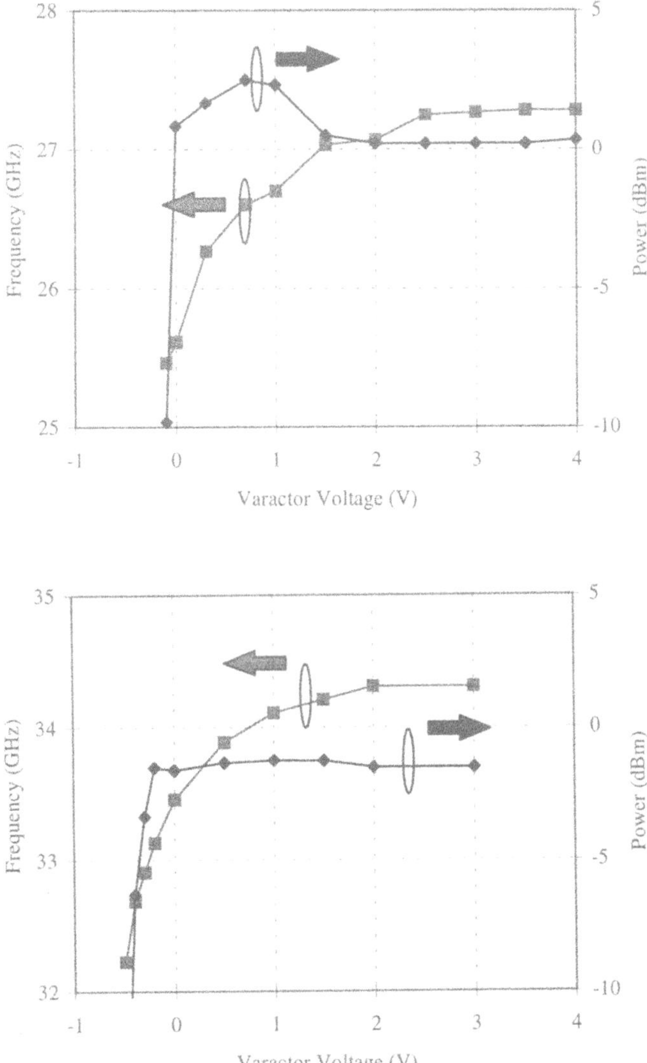

Figure 5. Frequency tuning and output power of VCO.

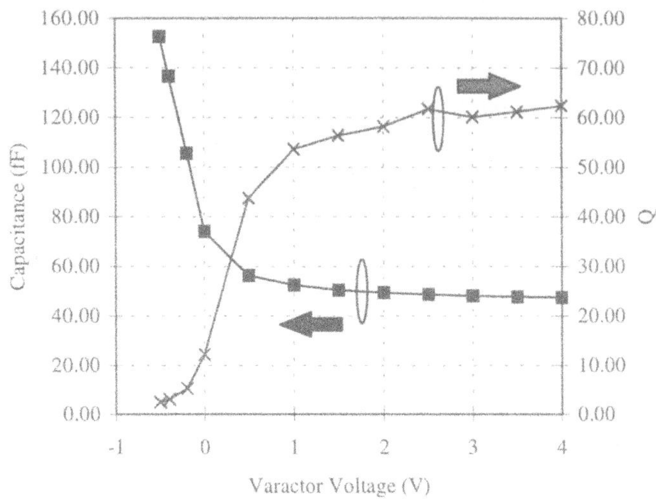

Figure 6. Tuning performance of varactor diode.

CONCLUSION

The features of low DC power consumption, high efficiency, and low phase noise enable this HBT VCO technology to be applied to low-voltage millimeter-wave communication systems. Coplanar waveguide circuit design approach was discussed in this paper. With CPW topology, the VCO circuit can be easily integrated with other digital circuits without additional via hole back side processing as in microstrip circuits. This simplifies processing steps in fabrication and increases the yield in production. It also demonstrates the feasibility of integrating low voltage VCO with prescaler, PLL, and other low voltage digital circuits to have a millimeter-wave synthesizer on a single chip.

ACKNOWLEDGMENTS

The authors would like to thank Dr. W. F. Brinkman, Dr. A. Y. Cho, Dr. D. V. Lang, and Dr. R. E. Howard for their continuous support on this project.

REFERENCES

1. A. A. Abidi, Low power radio frequency IC's for portable communications, *Proc. IEEE,* 83:544 (1995).
2. K. Inosako, N. Iwata, and M. Kuzuhara, 1.2V operation 1.1W heterojunction FET for portable radio applications, *IEEE IEDM Technical Digest,* 185 (1995).
3. G. Bechtel, Millimeter-wave communications: current use and future trends, *Proc. Third Ann. Wireless Symp.* 455 (1995).
4. Millimeter Wave Report, ISSN 1087-8386, Vol. 2, No. 2, August 15, (1996).
5. P. M. Asbeck, M.-C. F. Chang, J. A. Higgins, N. H. Sheng, G. J. Sullivan, and K.-C. Wang, GaAlAs/GaAs heterojunction bipolar transistors: issues and prospects for applications, *IEEE Trans. Electron Dev.*, 36:2032 (1989).

6. M. Madihian and H. Takahashi, A low-noise K-Ka band oscillator using AlGaAs/GaAs heterojunction bipolar transistors, *IEEE Trans. Microwave Theory Tech.*, 39:133 (1991).
7. J. Lin, Y. K. Chen, D. A. Humphrey, R. A. Hamm, R. J. Malik, A. Tate, R. F. Kopf, and R. W. Ryan, Ka-band monolithic InGaAs/InP HBT VCO's in CPW structure, IEEE Microwave and Guided Wave Lett., 5:379 (1995).
8. G. Gonzales, Microwave Transistor Amplifier, Prentice Hall (1984).

A TRIPLE-GATE MESFET VOLTAGE VARIABLE ATTENUATOR FOR MIILIMETER-WAVE APPLICATIONS

Maureen Daly, Ravi Khanna, and Dylan Bartle

Alpha Industries
Woburn, Massachusetts

ABSTRACT

A GaAs MMIC Voltage Variable Attenuator (VVA) has been developed for use at millimeter wave (MMW) frequencies incorporating triple-gate 0.25 μm power MESFET devices. This circuit exhibits exceptional power handling capabilities at all attenuation states, handling a minimum of 10 dBm input power, which occurs at it's 3 dB attenuation state, and greater than 26 dBm while operating in it's minimum attenuation state. The triple-gate VVA measures 1.7 dB in the low attenuation state and greater than 32 dB in the high attenuation state at 30 GHz. Additional on-wafer small signal and large signal measurement data is presented and compared to a similar single-gate MESFET VVA.

INTRODUCTION

This paper presents a MMW VVA that has been developed utilizing triple-gate MESFET devices. Millimeter wave VVAs are used in digital microwave radio transmitters, wireless cable, as well as other applications requiring gain control functions. Input power constraints can limit the versatility of placement of a VVA in the transmit chain. The power handling of a switch is typically limited by the voltage breakdown mechanism in the high impedance state and maximum current handling in the low impedance states. VVAs, which use MESFETs as voltage variable resistors, also use mid-range levels, which typically exhibit power handling which is lower than at either of the two extreme attenuation states. The power handling capability decreases significantly when the circuit is biased to low attenuation states, i.e. at biases close to pinchoff, and increases as attenuation increases.

The current limited power handling of the VVA may be increased by simply increasing the device periphery. However, substantial increases in the shunt device peripheries is not a viable option at millimeter wave frequencies due to the corresponding increase in off capacitance. Increasing the RF voltage handling, thereby the power handling, of a switch has been accomplished at L-Band by combining multiple FETs in series.[1] The observed improvement in power handling was attributed to capacitive voltage division among three cascoded FETs, allowing in theory three times the voltage handling of a single FET. However, high frequency performance limitations are expected due to the increase in the parasitic contributions of the cascoded FETs.

To significantly increase the power handling capabilities while still maintaining the high frequency performance with minimal effects on insertion loss, triple-gate MESFET devices, as opposed to three cascoded devices, have been investigated.[2,3] Consolidating the three gates into one channel minimizes the parasitic contributions of three cascoded devices and reduces the overall chip size. In addition, it was proposed that the RF voltage

Directions for the Next Generation of MMIC Devices and Systems
Edited by N. K. Das and H. L. Bertoni, Plenum Press, New York, 1997

divides more evenly across the triple-gate device's depletion region thereby gaining the *full* theoretical increase in power handling capability. [2] It was predicted that a triple-gate MESFET can handle three times the RF voltage of a single-gate device before becoming nonlinear. The results showed that the triple-gate MESFET did increase power handling in a switching application [2] as well as in a low frequency VVA circuit [3]. In this paper we investigate the use of a triple-gate MESFET device in a higher frequency VVA application. A triple-gate VVA and a single-gate VVA, were designed, fabricated, and evaluated to determine the power handling advantages of a triple-gate MESFET at MMW frequencies.

VOLTAGE VARIABLE ATTENUATOR DESIGN

A standard T-attenuator configuration that uses MESFETs as voltage variable resistors was adopted in the design of the triple-gate VVA. The T-attenuator requires two negative control voltages, VC1 to bias the series devices and VC2 to bias the shunt devices. The negative control voltages, supplied to the gate terminals, are biased through a 2K resistor to present a RF open at the gate terminal of the MESFET. The presented VVAs are composed of two single-gate MESFET's, each of which have 10 gate fingers of 80 μm unit length, as the series components and four MESFET's, each of which have 4 gate fingers of 50 μm unit length (4x50), combined distributedly as the shunt element, as illustrated in Figure 1. The only difference between the two circuits fabricated is the use of 4x50 single-gate devices or 4x50 triple-gate devices as the shunt components. This variation allows direct examination of power handling increases due to implementation of the triple-gate approach described above.

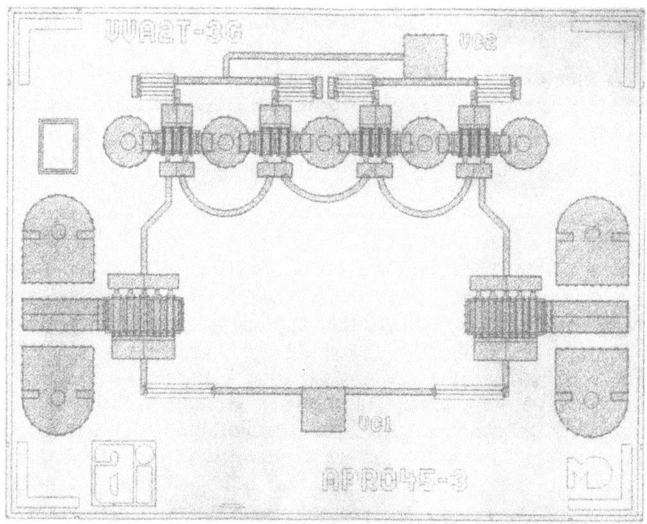

Figure 1. Voltage Variable Attenuator Circuit Layout

The primary factors limiting low loss performance in the minimum attenuation state at MMW frequencies is the off-capacitance, Coff, and off-resistance, Roff, of the shunt devices and the on-resistance, Ron, of the series devices. The series device is made with a large periphery which minimizes it's loss contribution to the circuit performance. The shunt elements introduce a frequency dependent effect on attenuation [4]. By using optimally long, high impedance transmission lines, the minimum attenuation may be optimized for high frequency operation. Inductive transmission lines are used in conjunction with the off-capacitance of the FET to create an artificial transmission line. The high impedance lines, in effect, tune out the shunt off-capacitance. The maximum obtainable attenuation is limited by the on-resistance of the shunt devices. The VVA presented consists of four shunt MESFETs in order to obtain the desired attenuation range.

CIRCUIT FABRICATION

The MMIC circuits were fabricated using the Alpha Industries standard 0.25 μm power MESFET process, allowing for easy integration with other power circuits. The circuits were fabricated from epitaxially grown wafers with a uniform channel doping and a highly doped cap layer. Devices were first mesa isolated, followed by the formation of ohmic contacts with alloyed AuGeNi Au. Gates were defined by direct write e-beam lithography and metallized with Ti/Pd/Au. The spacing between each gate of the triple-gate FET is 2.5 μm, and the source-to-drain spacing is 7.0 μm. The gates lie in individual channel recesses that are etched prior to gate formation. The purpose of the channel recess is to increase breakdown voltage, and are commonly used in MESFET power processes. The single-gate FETs have a source-to-drain spacing of 2.0 μm. The devices were passivated with PECVD silicon nitride. The circuits were plated with 3 μm of gold using an airbridge technology. The plating reduces the resistance of the transmission lines and the airbridges reduce parasitic capacitances at all crossovers. After frontside processing, the wafers were ground to a 4 mil thickness, after which via holes were formed by reactive ion etching. A schematic cross-section comparing the completed single-gate and triple-gate FETs is shown in Figure 2.

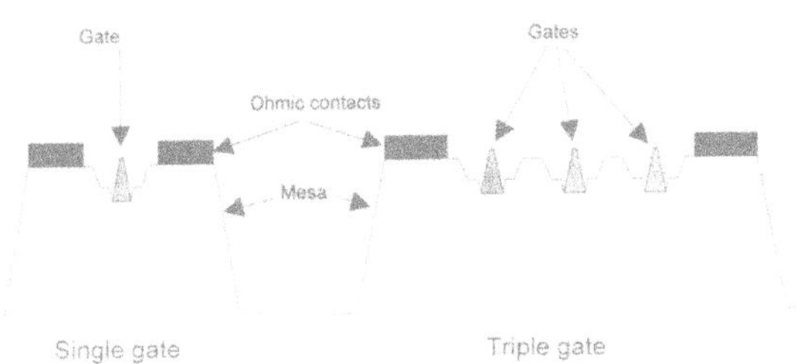

Figure 2. Single-gate MESFET and Triple-Gate MESFET Cross Section

MEASUREMENT RESULTS

On-wafer small signal and large signal measurements of the triple-gate and single-gate VVA circuits were performed. In addition to circuit measurements, two-port RF measurements and DC measurements of discrete FETs, which were used in the circuits, were also taken. In order to simplify the on-wafer measurements of the many circuits VC1 was set to 0V. All of the measurements discussed are with a single control voltage controlling the shunt devices.

Small Signal Measurements

On-wafer small signal measurements were performed as VC2 was varied between 0V and -3V with 0.2 V steps. Figure 3. depicts the attenuation versus frequency characteristic of the triple-gate VVA from 20 GHz to 50 GHz. It's minimum attenuation state varies from 1.3 dB at 20 GHz to 1.7 dB at 30 GHz and further degrades to 3 dB at 40 GHz. Measurements were performed up to 50 GHz and show a useful VVA that has a maximum low attenuation state of 3.4 dB at 45 GHz. The VVA provides better than 30 dB of attenuation at frequencies above 30 GHz. A monatomic degradation to 24 dB at 20 GHz is observed for the given bias conditions. The return loss at 30 GHz degrades from greater than 20 dB to 8 dB as atteunation increases. The VVA performs very well with one control voltage at frequencies up to 50 GHz if a slight degradation in return loss at the highest attenuation state (40 dB) can be tolerated. When the VVA is used at frequencies below K-band, biasing the series component will improve the return loss of the circuit and will be necessary at RF frequencies to obtain adequate return losses. The control voltage curves versus attenuation characteristics will be different at RF frequencies than at microwave frequencies.

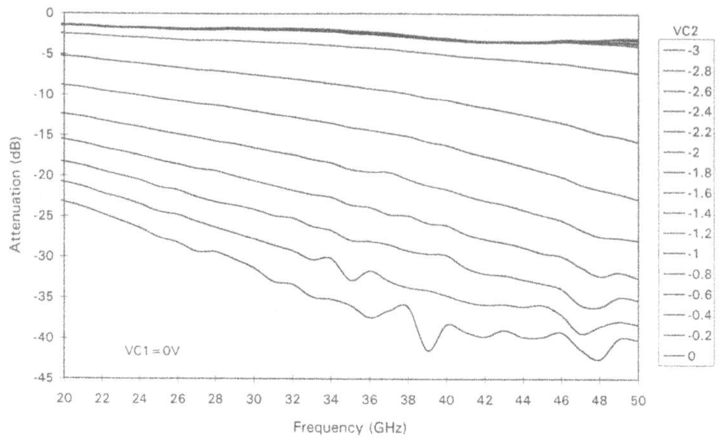

Figure 3. Attenuation vs. Frequency of the Triple-Gate Voltage Variable Attenuator

It can be observed in Figure 4. that the attenuation versus control voltage relationship is fairly linear from minimum to maximum attenuation levels at the MMW frequencies. This makes design of control circuitry less complicated. While operated at the same bias condition, VC1=0V, the attenuation for the single version exhibits greater sensitivity to control voltage when compared to the triple-gate VVA. Note that the triple-gate VVA will be less sensitive to control voltage fluctuations as it uses almost twice the control voltage range as the single-gate VVA for a similar attenuation range.

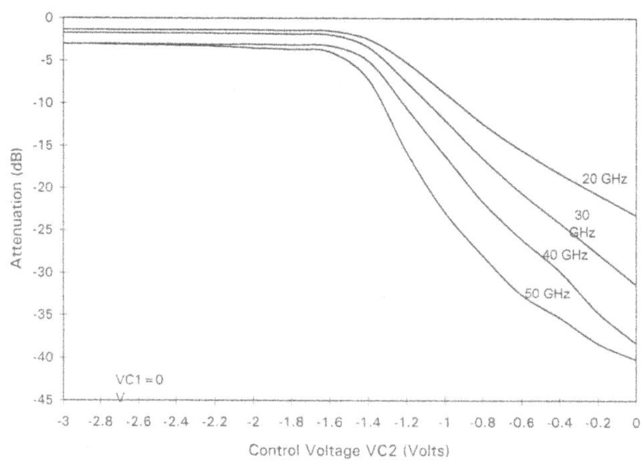

Figure 4. Attenuation vs. Control Voltage of the Triple-Gate Voltage Variable Attenuator

Power Measurements

On-wafer power measurements have been performed at 30 GHz on both the single-gate and triple-gate VVA for a variety of control voltages. The 1 dB compression point, (P1dB) defined as the input power that changes the attenuation by 1 dB will be used as a figure of merit. The primary focus is on the region of operation where a VVA's power handling is at it's lowest value. This significant drop in P1dB occurs as the shunt devices are biased less negatively and brought out of pinchoff, corresponding to the low attenuation states. The power handling then increases as the attenuation is increased to it's maximum attenuation state.

The triple-gate VVA shows exceptional performance in all of it's measured attenuation states. This circuit will handle greater than 26 dBm while operating in it's minimum attenuation state then degrades to +10 dBm before it begins to increase again. In comparison, the single-gate VVA degrades to below 4 dBm in it's low attenuation states. Figure 5. illustrates this significant difference between the power handling of the triple-gate VVA and the single-gate VVA. Measured data of four single-gate circuits and four triple-gate circuits is depicted. The input power level at which the attenuation level changes by 1 dB is plotted. The arrows indicate that the maximum available input power from the measurement equipment was reached before the circuits began to compress. The P1dB in this area is greater than 15 dBm, though there are no measurements of how much higher.

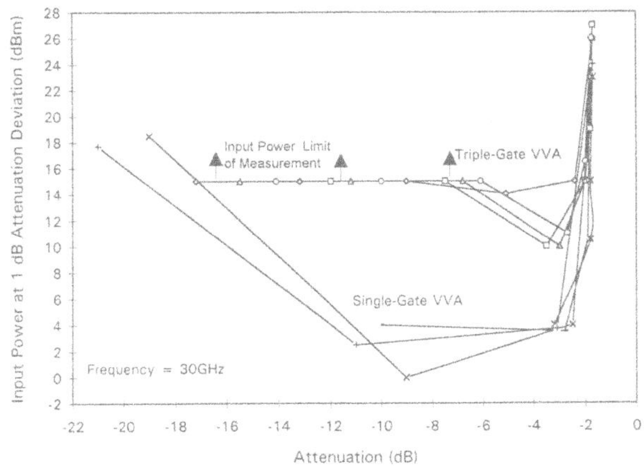

Figure 5. Triple-Gate and Single-Gate Voltage Variable Attenuator Power Handling

DISCUSSION

The triple-gate VVA exhibits improvements over the single-gate VVA in both power handling and small signal attenuation versus control voltage characteristics. Since the two circuits differ only by the number of gates within each of the shunt FETs, the differences in RF performance are due to a difference in the performance between the triple and single gate FET. The small signal performance difference may be attributed to a difference in the on-resistance versus gate (control) voltage characteristics of the FET, as shown in Figure 6(a). Small signal models from the literature typically model triple-gate FETs as three single-gate FETs in series, with some of the parasitic elements eliminated.[2,3] It follows that Ron should be approximately three times larger for the triple-gate FET as compared to a lone single-gate FET, this can be observed in the plot shown in Figure 6(a), in the region from 0 to -1.1V.

The S11 of a discrete triple versus single gate FET in an LRL structure is shown in Figure 6(b). In this structure, the gate is connected to a DC control pad, while the source and drain are connected to input and output transmission lines. The shift in S11 clearly shows the effect of this resistive shift. In addition, the small signal model of the FET accurately predicts the S11 change when the channel resistance is increased by a factor of three. The increase in Ron of the triple-gate device also accounts for the decrease in maximum obtainable attenuation compared to the single-gate circuit.

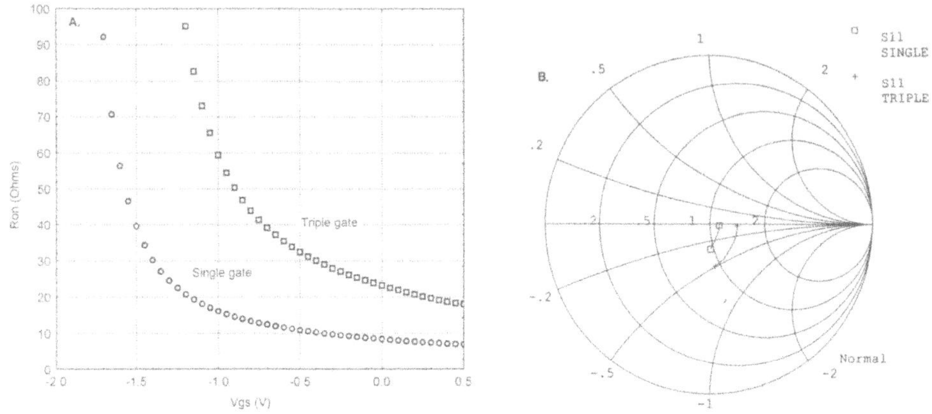

Figure 6. On-Resistance vs. Vgs of Triple and Single Gate MESFETs

The improvement in power handling can also be attributed to differences in the DC characteristics of the FETs. A comparison of the Id-Vd curves of the triple and single gate FET is shown in Figure 7. In this plot, the drain current versus drain voltage characteristic of a triple and single gate FET is shown, with the control biases of the FETs set to bring them into the same on-resistance, corresponding to a similar attenuation state. It can be observed that the triple-gate Id-Vd curve has a larger effective knee voltage than the single-gate FET. Since larger swings in current and voltage can be obtained before the onset of current saturation, the P1dB point is increased. Thus, at a given attenuation, power handling increases because the effective knee voltage is increased. In Figure 8, the control bias was set to bring the FETs into a state which would result in low attenuation in the VVA. It can be observed that the triple-gate curve has higher breakdown voltage than the single-gate FET. Larger voltage swings can be obtained before the onset of breakdown in the triple gate structure, again resulting in better power handling.

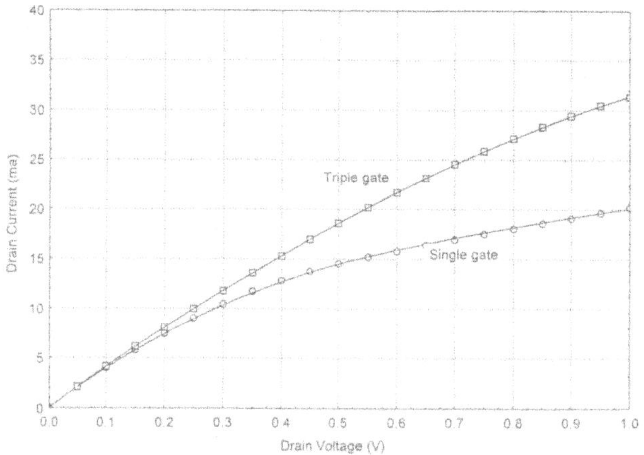

Figure 7. Comparison of Drain Current vs. Drain Voltage Between Triple and Single Gate FETs at Similar (High) Attenuation States. Note the extended linear region in the triple-gate FET.

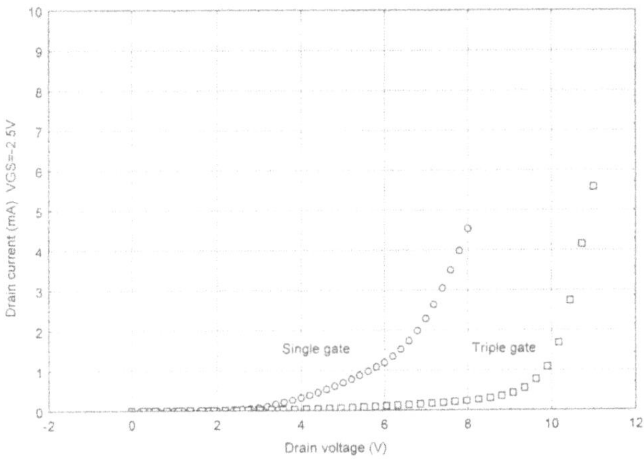

Figure 8. Comparison of Drain Current vs. Drain Voltage Between Triple and Single Gate FETs at Similar (Low) Attenuation States. Note the increased breakdown of the triple-gate FET.

CONCLUSIONS

Triple-gate devices have been used successfully in a MMW VVA and have been shown to outperform their single-gate counterparts in two areas. The triple-gate VVA exhibits a more desirable attenuation versus gate voltage characteristic, and most importantly has significantly better power handling than the single-gate circuit with compatible low loss performance. It can handle a minimum of +10 dBm input power. The improvement of both small signal attenuation and power handling is due to the improvements in the DC characteristics of the triple-gate FETs. The effective knee voltage as well as breakdown voltage are both higher in the triple-gate FET.

ACKNOWLEDGMENTS

The authors would like to thank Dr. Zev Bogan for his encouragement and technical support, and Hong Pham for all of the on-wafer testing. We would also like to thank Dave Whitefield for insightful discussion in analyzing the DC and RF data. We are grateful to Don Mitchell for the process development and fabrication of the triple gates.

REFERENCES

1. Mitchell B. Shifrin, Peter J. Katzin, Yalcin Ayasli, "Monolithic FET Structures for High-Power Control Component Applications," IEEE Transactions on Microwave Theory and Techniques, Vol. 37, No. 12, December 1989

2. F. McGrath, C. Varmazis, C. Kermarrec, R. Pratt, "Multi Gate FET Power Switches," Applied Microwave, pp. 77-88, Summer 1991

3. Horng Jye Sun, and James Ewan, "A 2-18GHz Monolithic Variable Attenuator Using Novel Triple-Gate MESFETs," IEEE MTT-S Digest, pp. 777-780, 1990

4. Manfred J. Schindler, Annamarie Morris, "DC-40 GHz and 20-40 GHZ MMIC SPDT Switches," IEEE Transactions on Microwave Theory and Techniques, Vol. MTT-37, No. 12, December 1987

MODELING AND CAD

TYPES OF LEAKY DOMINANT MODES AND SPECTRAL GAPS ON PRINTED-CIRCUIT LINES

Arthur A. Oliner

Department of Electrical Engineering
Polytechnic University
Brooklyn, NY 11201

I. INTRODUCTION

Not long ago it was generally believed that the dominant mode on open printed-circuit transmission lines would be purely bound at all frequencies. Now we know that, for most such lines, the dominant mode is purely bound at ***low*** frequencies, in agreement with common perception, but becomes leaky ***above*** a critical frequency. For certain lines, however, the behavior is different; for example, the dominant mode on conductor-backed slot line or coplanar waveguide of infinite width is leaky at ***all*** frequencies, whereas for microstrip line there is an ***additional*** dominant mode, which is leaky at high frequencies but nonphysical at low frequencies. It is important to know when leakage may occur because the presence of leakage can lead to power loss, crosstalk, and package effects which would deteriorate circuit performance.

The paper first explains why leakage occurs at all, and then summarizes the nature of the leakage fields. Several examples are presented to show the leakage behavior on such lines as CPW, slot line and coplanar strips, for which leakage occurs only above a certain frequency. It is then explained why conductor-backed slot line and CPW of infinite width are leaky at ***all*** frequencies, and how to modify the substrate to avoid such leakage.

The transition region in frequency between the bound and leaky portions of the dominant mode is called a ***spectral gap,*** because the guided-mode solution is ***nonphysical*** over at least part of that frequency range. The paper will describe the interesting fine structure within this gap, explain why the nonphysical behavior is necessary, and indicate what is happening to the total field when the guided mode becomes nonphysical.

For many structures, the spectral gap has the common form mentioned above, whereas for others the spectral gap is quite different. In these latter cases, we find unusual and unexpected modal behavior, two different basic examples of which are presented here.

Directions for the Next Generation of MMIC Devices and Systems
Edited by N. K. Das and H. L. Bertoni, Plenum Press, New York, 1997

The first example of unusual modal behavior involves microstrip line. The characteristic feature for this line is the presence of an ***additional*** dominant mode. Of the two dominant modes, one (the conventional mode) remains bound at all frequencies, whereas the other is leaky at high frequencies but takes the form of a pair of nonphysical improper real solutions at all lower frequencies, effectively producing a ***one-sided*** spectral gap.

The second example arises when certain lines, which have standard spectral gaps, have their ***relative cross-sectional dimensions*** changed, for example, by increasing the slot width in slot line, or the strip width in coplanar strips. For those cases, it is found that the spectral gaps disappear, and the previously separate bound and leaky portions of the dominant mode overlap, producing a range of ***simultaneous propagation,*** which can in fact be quite wide in frequency.

The topic of leaky modes on printed-circuit transmission lines has captured the interest of many individuals, for two rather different reasons. One reason, a very practical one, is that leakage effects can produce serious performance difficulties in microwave and millimeter-wave integrated circuits, such as power loss, crosstalk between neighboring portions of the circuit, and coupling to modes of the package.

The second reason is that continuing studies of these leakage properties are constantly producing new surprises. The complicated leakage behavior, examples of which are mentioned above, arise because the cross section of the printed-circuit line is ***inhomogeneous,*** and the new surprises referred to above serve to further stimulate the imagination and offer new challenges. Fortunately, an increasing number of researchers from all parts of the world have been contributing new ideas and new results. The material presented in this paper relies heavily on the excellent studies conducted by two independent groups: Profs. H. Shigesawa and M. Tsuji of Doshisha University, Kyoto, Japan, and Prof. D. R. Jackson of the University of Houston in Texas, together with his colleagues, Prof. J. T. Williams and Dr. D. Nghiem. Although their results, as well as those of others, are credited in the references, special acknowledgment should be given to their pioneering, as well as continuing, contributions.

II. LEAKAGE FROM PRINTED-CIRCUIT TRANSMISSION LINES

A. Background

Experience with microwave integrated circuits tells us that the dominant mode on the printed-circuit transmission lines used in those circuits is purely bound at lower frequencies. When the frequency is raised sufficiently, however, recent research informs us that in almost all cases the dominant mode becomes leaky. But there are interesting and important exceptions which we will consider later, such as certain conductor-backed lines which are leaky at all frequencies, and microstrip line, which supports an additional leaky mode at higher frequencies.

When the dominant mode is purely bound, it possesses a real propagation wavenumber β, which is the phase constant of the mode, and a field variation which decays transversely. It is customary to plot the dispersion behavior of the mode as β/k_0, where k_0 ($= 2\pi/\lambda_0$) is the free-space wavenumber, as a function of frequency f or normalized frequency h/λ_0, where h is the substrate height. If the mode is completely dispersionless, the curve on such a plot will become a straight horizontal line. The dominant modes on printed-circuit lines, such as microstrip line, slot line or coplanar waveguide (CPW), are not dispersionless but are mildly dispersive, with the dispersive behavior increasing slightly at higher frequencies. As a result, a plot of β/k_0 vs. f will yield a dispersion curve that is quite flat at low frequencies, but which increases somewhat as the frequency increases. Basically, we should view the dispersion curve as being relatively flat, but increasing a bit as the frequency is raised.

Away from the central region of the line, the structure consists of a dielectric layer which may or may not be placed on a metal ground plane, depending on the specific transmission line. For example, microstrip line is placed on a grounded dielectric layer, whereas coplanar strips (slot line with finite-width plates) are located on a dielectric layer without a ground plane. The lowest surface wave that can be supported by a grounded dielectric layer is the TM_0 surface wave, and it is the only mode that can propagate down to zero frequency. For an ungrounded dielectric layer, there are two surface waves, the TE_0 and the TM_0, that can propagate down to zero frequency, but the lowest surface wave is the TE_0 one. It is important to note that the lowest surface waves on the grounded and ungrounded dielectric layers have opposite polarizations, with the transverse electric field orientations being vertical and horizontal, respectively.

Let us designate the propagation wavenumber of a surface wave as k_s, where the subscript indicates "surface". On a plot of k_s/k_0 vs. f, we find that surface waves are very dispersive, going from $k_s/k_0 = 1$ at zero frequency to $k_s/k_0 = \sqrt{\varepsilon_r}$ asymptotically at infinite frequency. This dispersion behavior corresponds physically to the field spreading far out into the air region at lower frequencies and being drawn into the dielectric substrate at higher frequencies. If there is a metal top cover over the circuit, the limit at zero frequency will no longer be unity but becomes some value between unity and $\sqrt{\varepsilon_r}$, depending on the filling factor of the dielectric layer. The main point to remember here is that the surface waves are very dispersive whereas the dispersion curves for the dominant modes on the printed-circuit lines are relatively flat.

B. Why and When Leakage Occurs

When a dominant mode becomes leaky, it does so by changing into a leaky mode, with the leakage occurring in the form of a surface wave on the surrounding substrate. Let us assume that the leakage power is leaking away at angle θ with respect to the transmission-line axis z, as seen in Fig. 1, which shows a top view of a strip or slot. The mode has the phase constant β, and the surface-wave wavenumber is k_s. From the figure we can see that

$$\cos\theta \approx \beta/k_s \tag{1}$$

where the approximate sign is used because the wavenumber β is actually complex when leakage occurs. An exact expresssion can be readily derived, and it is found to be

$$\cos\theta = \beta/[k_s^2 + \alpha^2 + \alpha_x^2]^{1/2} \tag{2}$$

where α and α_x are the leakage constant and the value of the transverse field increase, respectively, of the leaky dominant mode, and all quantities in (2) are real. The simpler

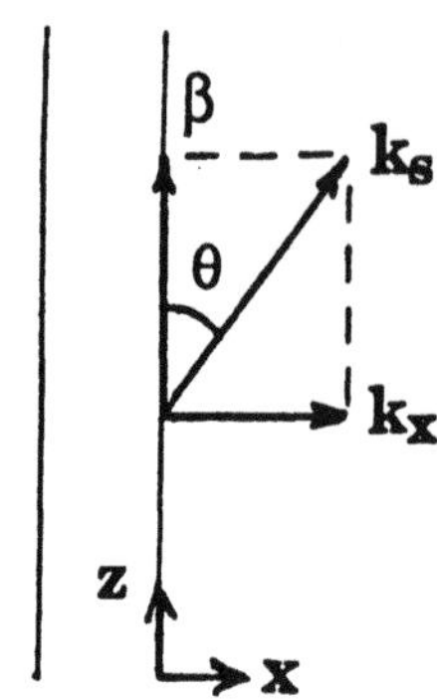

Fig. 1 Sketch showing the angle θ of leakage when the dominant mode becomes leaky. A top view of a transmission-line strip or slot is shown.

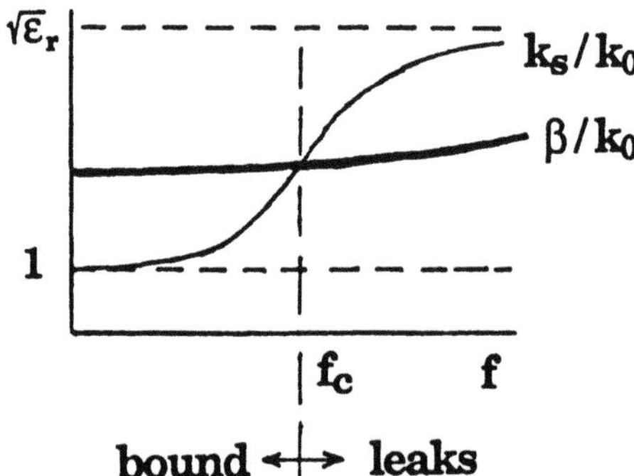

Fig. 2 A typical dispersion plot for a printed-circuit line. The curve for the dominant mode (labeled β/k_0) is seen to be relatively flat, whereas that for the surface wave (labeled k_s/k_0) exhibits strong dispersion. For frequencies greater than f_c the mode becomes leaky, in accordance with eq. (3).

expression (1) is a very good approximation, however, and is helpful in understanding the basic physics.

From (1) we can see that, for real power to leak away at angle θ, we must have $\beta < k_s$, or, dividing by k_0,

$$\beta/k_0 < k_s/k_0 \tag{3}$$

Relation (3) is the simple condition that must be satisfied for leakage to occur [1].

In Sec. A above, we explained that if we obtain plots of β/k_0 vs. frequency f for the dominant mode on the printed-circuit line, and of k_s/k_0 vs. f for the surface wave on the surrounding substrate, we would find that the plot for β/k_0 would be relatively flat, with a small increase as f increases, but that the plot for k_s/k_0 would exhibit strongly dispersive behavior. Such plots for a typical printed-circuit structure are shown in Fig. 2, where the expected behaviors are illustrated. The two dispersion curves are seen to cross at a critical frequency f_c. When the frequency f is greater than f_c, condition (3) is satisfied, and the dominant mode becomes leaky. When $f < f_c$ the mode remains purely bound, in agreement with experience.

When $f = f_c$, and therefore $\beta = k_s$, we can see from relation (1) that $\theta = 0$, meaning that at the onset of leakage the leakage direction is parallel to the axis of the transmission line. As the frequency is increased, the two dispersion curves are seen to separate, so that angle θ increases and the leakage direction swings out away from the axis of the line.

The ***polarization*** of the leakage power depends on whether the substrate on which the line is placed has or does not have a ground plane. If it has a ground plane, the lowest surface wave is the TM_0 surface wave, and the electric field polarization is basically vertical. If the dielectric layer of the substrate is ungrounded, the lowest surface wave is the TE_0 surface wave so that the electric field polarization of the leakage power is horizontal. For most printed-circuit lines the polarization of the leakage field is vertical; examples of such lines are conventional coplanar waveguide (CPW) and slot line with side strips of infinite width, and conductor-backed coplanar strips and CPW with side strips of finite width. Examples of lines that would leak with horizontal polarization are coplanar strips and CPW with side strips of finite width but without conductor backing.

All of the examples listed just above leak only at higher frequencies, consistent with condition (3) and the dispersion behavior shown in Fig. 2. Microstrip line furnishes a special case, and will be discussed in Sec. IV, A.

A different type of special case is furnished by ***conductor-backed*** slot line or CPW with side plates of infinite width. Those lines leak at ***all*** frequencies, even down to zero frequency [2]. An examination of the substrate structure away from the central portion of these lines reveals that it consists of parallel-plate waveguide completely filled with the substrate dielectric material. The dispersion curve for that parallel-plate structure on the plot in Fig. 2 is simply a straight horizontal line for which $k_s/k_0 = \sqrt{\varepsilon_r}$. Since that line always lies above the β/k_0 dispersion curve for the transmission line, condition (3) is satisfied at all frequencies. Such conductor-backed slot lines or CPWs are not useful for integrated-circuit applications because the leakage that is always present will produce crosstalk and package effects that will deteriorate circuit performance.

On the other hand, conductor backing offers various practical advantages, such as a heat sink, greater mechanical strength, a better range of Z_0 values, etc. Therefore, after it was realized that leakage will occur even at low frequencies, and after crosstalk problems were encountered in practice, several modifications in the structure were proposed which would permit the advantages of conductor backing to be retained but would eliminate the leakage. The best of these modifications involve fabricating the substrate in two layers [3,4]; the upper layer, on which the circuit would be placed, retains its high value of dielectric constant, and the lower layer would have a much lower value of dielectric constant. The resulting substrate structure, composed of parallel-plate guide with two dielectric layers, would have a dispersion curve in Fig. 2 that would be much lower than the $k_s/k_0 = \sqrt{\varepsilon_r}$ line appropriate for the earlier single-layer version. By choosing the right values for the height and for ε_r for the lower layer, it is easy to obtain a dispersion curve for the two-layer substrate structure that lies below the dispersion curve for the slot line or the CPW, so that leakage is eliminated over the frequency range of operation. At higher frequencies, the dispersion curve for the two-layer substrate bends upward and will cross the slot line or CPW dispersion curve, so that leakage will then occur for such higher frequencies.

C. Examples of Leaky-Mode Behavior

A typical example of the dispersion behavior for a printed-circuit transmission line whose dominant mode is purely bound at lower frequencies but leaky above some critical frequency is shown in Fig. 3, for conventional coplanar waveguide (CPW) with side plates of infinite width. The structure is shown in the inset in that figure, and the dispersion behavior follows the form shown in Fig. 2. In addition, we have in Fig. 3 a plot of the normalized leakage constant α/k_0 as a function of normalized frequency h/λ_0, where h is the substrate height. The leakage is seen to begin when the curve for the surface wave exceeds the curve for the CPW mode, in agreement with condition (3).

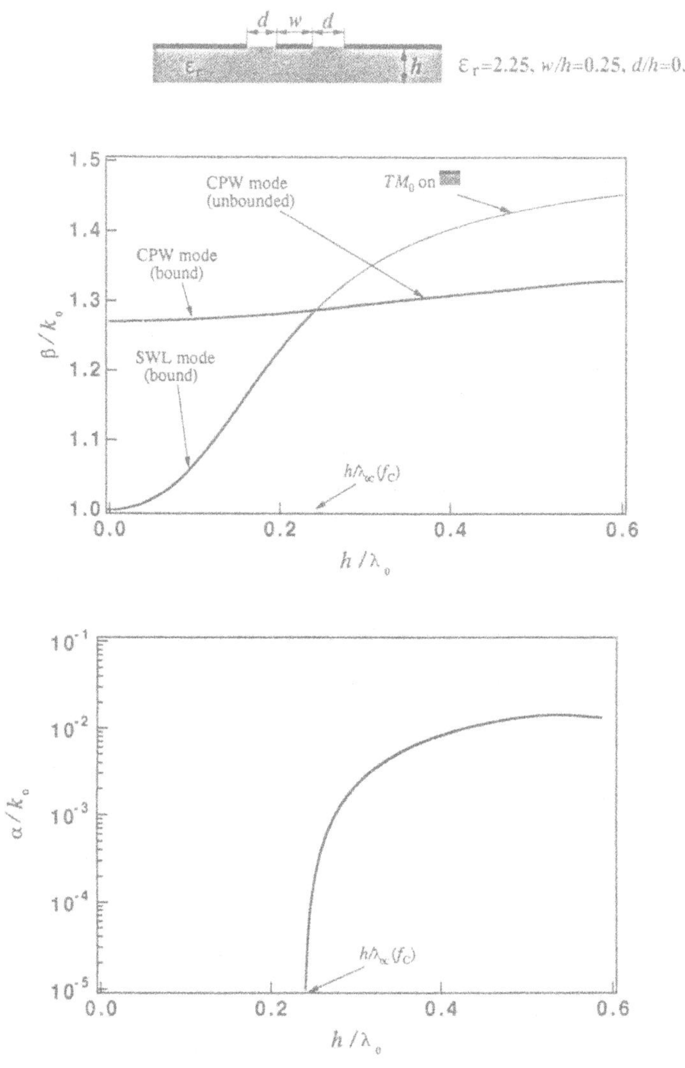

Fig. 3 Dispersion behavior, showing β/k_0 and α/k_0 vs. h/λ_0, for conventional coplanar waveguide. The structure is shown in the inset. The mode becomes leaky at higher frequencies, in agreement with the sketch in Fig. 2.

It should be noted that for leakage the substrate needs to be electrically relatively thick, about 0.2 for h/λ_0. For the lower frequencies in the microwave range it is easy to achieve electrically thin substrates; as a result this type of leakage was not encountered. (As mentioned under B above, this statement does not apply for conductor-backed slot line and CPW with infinite-width side plates, which leak at all frequencies unless appropriately modified.) When the frequency is increased into the millimeter-wave range, however, and the attempt is made to scale the whole structure, including all circuit dimensions, to be consistent with the much smaller wavelengths, it is found that it is not easy to commensurately reduce the substrate height h, for mechanical reasons. Such structures therefore require values of h/λ_0 that may be significantly higher than the direct scaling would indicate. For the millimeter-wave range, therefore, it is quite possible to have the dominant mode become leaky, and one must be careful to insure that this does not happen.

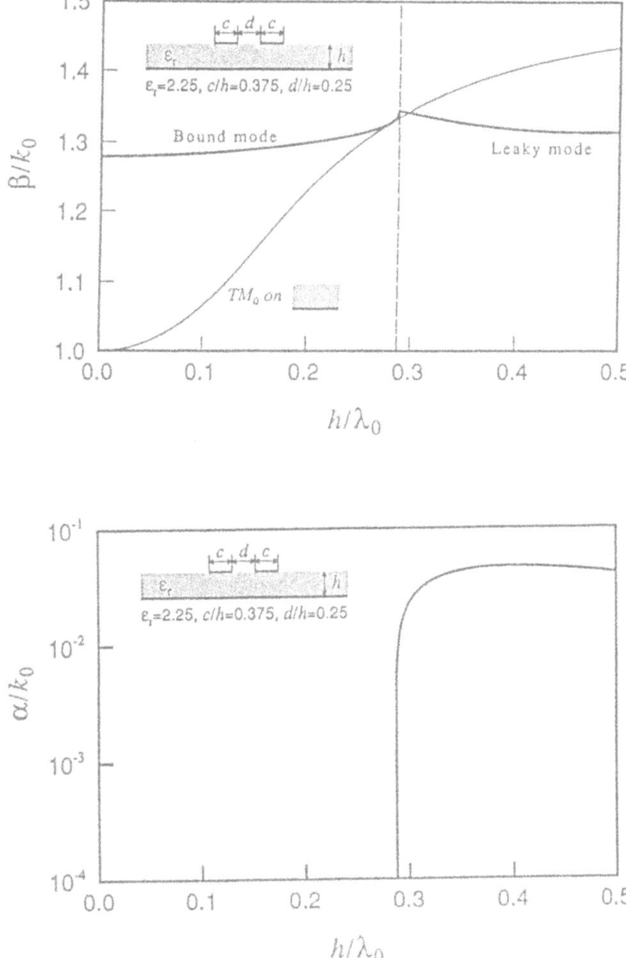

Fig. 4 Behavior of β/k_0 and α/k_0 vs. h/λ_0 for conductor-backed coplanar strips, which has a much higher leakage constant α than the structure in Fig. 3 does. As a result, the curve for β/k_0 exhibits a noticeable peak at the transition region between the bound and leaky portions.

When the transmission line has a ***top cover***, the dispersion curve for the surface wave can shift upwards substantially, particularly if the top cover is not far above the circuit. The terminal value at $h/\lambda_0 = 0$ (zero frequency) then jumps from unity to a value determined by the "filling factor" of the dielectric material. As a result, the whole surface-wave curve will rise, and the onset of leakage will occur at a lower frequency. The amount by which the frequency is lowered will depend on how close the top cover is chosen to be. This consideration could be one reason why a circuit at millimeter wavelengths works well before it is placed in a package but suffers from crosstalk after being placed in it.

When the leakage constant is significantly greater, the ***transition region*** between the bound and leaky portions of the dominant mode exhibits an interesting peak, as we may observe in Fig. 4 for conductor-backed coplanar strips. We may note that the peak value for α/k_0 in Fig. 4 is about six times that in Fig. 3. Otherwise, the curves in Figs. 3 and 4 are reasonably similar. When examined further, the transition region actually shows a very interesting fine structure, including a small frequency range over which the guided-mode solution is ***nonphysical***. Because of the nonphysical aspect, this transition region has been termed a ***spectral gap.*** Such behavior raises a variety of interesting questions, which are addressed in the next section, Sec. III.

III. SPECTRAL GAPS

A. Wavenumber Behavior Inside the Spectral Gap

When one views the transition between the bound and leaky portions of the dominant-mode dispersion curve in Fig. 3, it appears that the transition is a smooth one. In Fig. 4, however, which corresponds to a structure with a much higher leakage rate, one sees that the transition is not so smooth, but in fact represents some complicated behavior. That transition region (for the same structure but with slightly different relative dimensions) is presented in Fig. 5 (from Fig. 3 in [5]) on a greatly expanded scale. If the transition region in Fig. 3 were replotted on an even more greatly expanded scale, one would obtain behavior that would be qualitatively exactly the same as that seen in Fig. 5. The first detailed treatment of a transition region of this type was presented for a leaky-wave antenna [6]; the curve in [6] looks different from that in Fig. 5 in two ways, because the transition occurs at $\beta = k_0$ instead of $\beta = k_s$ and because the frequency dependence is reversed, but the principles are identical.

As we begin in Fig. 5 with the solid line at the lower frequencies, we are following the purely bound portion of the dominant mode. As indicated, the solution is real and is proper, or spectral, those two terms being exactly equivalent to each other. A ***proper*** (or spectral) solution possesses fields that ***decay*** in the transverse (x) direction (along the air-dielectric interface perpendicular to the axis of the transmission line), so that the solution satisfies the boundary condition at infinity in the transverse direction. An ***improper*** (or nonspectral) solution is one whose fields ***increase*** transversely; that solution therefore violates the boundary condition at transverse infinity. The fields of a ***leaky*** mode increase transversely away from the central region of the transmission line, but the leakage occurs at an angle, so that, as is well known, the solution remains physical.

As we follow the ***bound*** portion of the dominant mode in the direction of increasing frequency, we note that the solution approaches the dot-dashed curve representing the TM_0

surface wave, and that it finally touches that curve at point 1 (the symbol 1 will be used in the text to refer to the symbol "1 with a circle around it" in Fig. 5). As those two curves approach each other, the transverse decay rate of the bound mode decreases, until at point 1, for which $\beta = k_S$, the decay rate has become zero and the field density has thinned out to zero. The limit point 1 is therefore in actuality already nonphysical.

For frequencies slightly higher than point 1, the solution becomes ***improper*** (or nonspectral) ***real,*** and is represented by a dashed curve. There are in fact two improper real solutions, which meet at point 2. The second one is then seen to continue in the direction of lower frequencies.

These improper real solutions maintain a constant amplitude in the longitudinal (z) direction and they increase in the transverse (x) direction. Since they do not decay as they progress, and they increase transversely without limit, they fill all of space and would require infinite power. It is therefore clear that those improper real solutions must be ***nonphysical.*** The rate of transverse increase becomes greater as the dashed curve moves from point 1, where the rate of increase is zero, toward point 2. The portion of the dashed curve that doubles back to lower frequencies after point 2 represents a solution with even greater rates of transverse increase.

For frequencies higher than point 2, the solution remains improper but becomes ***complex***, representing a ***leaky*** mode. The leakage begins at point 2, but the leakage rate and the rate of transverse increase there are zero, and the angle of leakage with respect to the axis of the transmission line is likewise zero. As frequency increases further, and we follow the solid curve from point 2 to point 3, the leakage rate α becomes finite, and the leakage angle θ increases from zero in accordance with (2). The values of β and the rate of

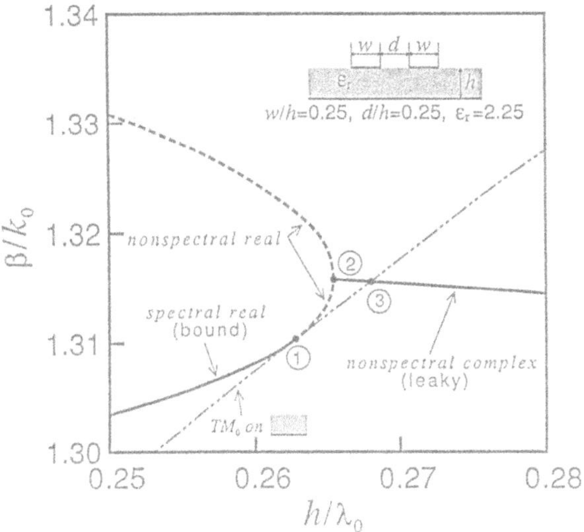

Fig. 5 Greatly expanded plot of the transition region (spectral gap) between the bound and leaky portions of the β/k_0 curve seen in Fig. 4, illustrating the complicated behavior in that region. (The structure is identical to the one for which Fig. 4 is valid except for the strip width, which is called w here but c there.) The different portions of the spectral gap are discussed in detail in the text.

transverse increase do not change much. As the leaky-mode solution continues past point 3 in the direction of increasing frequency, the value of α/k_0 continues to increase rapidly, and the leakage direction continues to swing away from the line axis. Since this improper complex solution decreases in the propagation direction, and the leakage occurs at an angle, the leaky mode is ***physical***, as is well known.

The spectral gap is defined as the frequency range between points 1 and 3. Within this range, the improper real solutions between points 1 and 2 are without question nonphysical, but some significant questions remain regarding the extent of physical meaning for the leaky mode portion between points 2 and 3. For example, if the data on Fig. 5 were replotted on the steepest-descent plane, the values between 2 and 3 would not be captured in the deformation of the original path to the steepest-descent path, whereas the portion for frequencies greater than point 3 would be captured. That result signifies that, to first order, the portion between points 2 and 3 do not contribute to the total field due to a source placed near the transmission line. Furthermore, $\beta > k_s$ for the portion between points 2 and 3, so that the solution there corresponds to a slow wave, rather than a fast wave, representing a contradiction.

The way to resolve the apparent contradiction above, and to assess the extent to which the portion between points 2 and 3 is physical, is to actually solve for the ***total*** fields due to a source located near the guiding structure, and to then compare the fields with those due to the leaky-mode solution alone. Such an analysis has recently between performed for two quite different structures [7,8] but the results are qualitatively the same. Those results indicate that the values of the ***wavenumber***, $\beta - j\alpha$, obtained between points 2 and 3 are reasonably correct, with their accuracy decreasing as one approaches point 2. However, when one attempts to calculate the relative ***amplitudes*** of those solutions by taking the residues of the leaky-mode poles (due to some specific source), the results are much less accurate, but again with the accuracy improving as point 3 is approached. In one study [9] the radiation patterns derived from the leaky-mode wavenumbers were calculated and compared with those from an exact analysis. It was found that the pattern ***shapes*** agreed with each other extremely well over the whole range from 2 to 3, but that the amplitudes of the radiation fields did not, particularly as point 2 was approached. We should note, however, that the patterns are calculated from an integration that utilizes the wavenumber values, and therefore represent a less-sensitive test of the validity of the wavenumber values than the wavenumbers would by themselves.

From the above-mentioned studies, one may conclude that the leaky-mode solutions within the spectral gap (between points 2 and 3) do have ***some*** physical validity, but that it is limited. Furthermore, the solutions become less meaningful as one penetrates further into the spectral gap (approaching point 2). We may summarize the nonphysical nature of the solutions within the spectral gap by stating that the improper real solutions (between points 1 and 2) are entirely nonphysical, but that the leaky-mode solutions (between points 2 and 3) are partially valid physically and that they gradually lose physical meaning as one enters further into the spectral gap.

B. The Physical Meaning of the Nonphysical Behavior

The properties described above for the spectral gap lead us to two basic questions: Why should the transition region have a nonphysical portion? What does it mean physically when the guided-mode solution is nonphysical? Let us consider these questions sequentially.

Why should the transition region have a nonphysical portion? On the low-frequency side of the spectral gap the dominant mode is bound, and its fields ***decay*** in the transverse direction. On the high-frequency side, where the dominant mode is a leaky mode, the fields ***increase*** transversely. The transition region between the two must provide a mechanism that allows a decaying field to be transformed into an increasing field; of necessity, there must be at least a point at which the field's transverse variation is zero. That point, as explained in Sec. III A, occurs at point 1 in Fig. 5. But, in addition, the solution on the high-frequency side is complex and leaks power at an angle. This added requirement calls for some additional fine structure within the spectral gap, which takes the form, first, of the improper real solutions between points 1 and 2 to provide the transition to a transverse field increase, and, second, of the improper complex solutions between points 2 and 3 to provide the power leakage at an angle while maintaining the transverse field increase so that the final leaky mode is physical. (There are actually two improper complex solutions, the leaky mode and its complex conjugate. The conjugate solution has the same value of β, so that it cannot be plotted independently on Fig. 5, but its value of α has the opposite sign, corresponding to a longitudinal field increase which, of course, is nonphysical for a passive structure.)

The nonphysical nature of the spectral gap is therefore necessary in order to allow for the transition from a transversely decaying bound-mode field to a transversely increasing but physical leaky-mode field.

What does it mean physically when the guided-mode solution is nonphysical? Within the spectral gap, we encounter a frequency range over which the guided-mode (eigenvalue) solution is nonphysical. To understand what this implies physically, we must examine the ***total*** field due a source appropriate to the circuit.

For a given printed-circuit transmission line we would have a feed structure designed to excite the desired dominant mode efficiently. This feed structure will also excite a small amount of surface-wave power which will radiate approximately radially away from the feed (plus some radiation into space waves if there is no top cover). For a properly designed feed, this surface-wave contribution should be negligibly small, and the guided-mode field should dominate the total field. When the guided-mode field is nonphysical, however, as is the case in much of the spectral gap, it can no longer contribute to the total field. What happens then?

Two studies [7,8] have examined this question, at least in part. The first one [7] considers the case of the spectral gap for a higher-order guided mode on a grounded two-layer dielectric structure. For higher frequencies, this guided mode is a surface wave; at frequencies below the spectral gap, the mode becomes leaky. The space-wave field was examined both within the spectral gap and outside of it for a line-current source. Outside of the spectral gap, the guided mode was the dominant contributor to the total field, and the space-wave contribution was relatively small, as expected. For frequencies within the spectral gap, however, where the guided mode is nonphysical, it was found that the space-wave field increased very substantially, and in effect "replaced" the guided mode in the total field.

In the second study [8] the guiding structure was an unusual form of microstrip line composed of a dielectric layer on a ground plane with a top metal cover very close to the dielectric-air interface. That structure supports two "dominant" modes, a bound mode and a leaky mode ("dominant" means that the form of the strip current is the same for both and is that of the usual microstrip dominant mode). At very low frequencies, for this choice of dimensions, the dispersion curves for the leaky mode and the surface wave cross each other and a spectral gap is created. The bound mode remains physical down to zero frequency. The total strip current, consisting of the currents for the bound mode, the leaky mode, and the surface wave from the feed, was examined for frequencies far from, and very near to, the beginning of the spectral gap. It was found that, far from the spectral gap, both the bound and the leaky modes were strongly excited, but the surface-wave contribution was very small. As the spectral gap was approached, it was found that the leaky-mode amplitude did decrease, but the surface-wave contribution remained very small, and instead the bound-mode amplitude began to increase significantly. Unfortunately, the investigation did not continue into the spectral gap, but the authors plan to do so. From the limited data so far, however, it appears that the "replacement" of the leaky mode within the spectral gap will turn out to be the bound mode, rather than the surface-wave contribution.

The answer, then, to what happens physically when a guided mode becomes nonphysical (when it is in a spectral gap) is that some other contribution to the total field then becomes much larger, to "replace" the no-longer-contributing guided mode.

IV. UNUSUAL AND UNEXPECTED MODAL BEHAVIORS

For most printed-circuit transmission lines, the spectral gap has the "standard" form described in Sec. III, but for other lines the spectral gap can be quite different. Furthermore, by simply changing the relative cross-section dimensions of many lines, the spectral gap can change form and, in the process, produce unexpected modal behavior. The above comments are illustrated below by two very different basic examples.

A. A New Leaky Mode on Microstrip Line

Surprisingly, at higher frequencies the dominant mode on microstrip line behaves differently from the dominant mode on most other printed-circuit lines. First of all, the dispersion curve for the conventional dominant mode on microstrip line never crosses the dispersion curve for the TM_0 surface wave, which is the relevant surface wave on the surrounding substrate. The curve for the surface wave always lies under the one for the dominant mode, and approaches it asymptotically at very high frequencies [10]. As a result, the conventional dominant mode remains ***bound*** (proper real) at ***all*** frequencies. These remarks apply when the substrate is isotropic.

Second, when the substrate of microstrip line is ***anisotropic*** in the right way, it is found that the dominant mode will leak at the higher frequencies, but that the leakage is in the form of the TE_1 surface wave, not the TM_0 surface wave [11]. The spectral gap between the bound mode at lower frequencies and the leaky mode at higher frequencies follows the "standard" form discussed in Sec. III. The behavior at higher frequencies with respect to leakage is therefore completely different for isotropic substrates and for those with anisotropy in the right way. This response to anisotropy is completely different from what one finds for slot line or CPW, whether conductor-backed or not, and whether or not the widths of the side plates (strips) are finite or infinite. For those lines the introduction of anisotropy produces a quantitative difference but not a qualitative one.

Since microstrip line already appears to offer some modal behavior different from that of other lines, the next discovery may be less surprising than otherwise. Further examination of the properties of microstrip line on an isotropic substrate revealed that there is an ***additional*** dominant mode, and that it is leaky at the higher frequencies [12,13]. This additional mode has a current distribution and a transverse field behavior near the strip that are exactly of the same form as those of the conventional bound mode, which is present simultaneously. As the frequency is decreased, this new mode encounters a spectral gap, which is highly unusual because it is ***one-sided***. This behavior is summarized in Fig. 6.

In Fig. 6 let us first observe the curves for the bound mode (shown with short dashes) and the TM_0 surface wave (shown dotted). Those curves are well known, and are what everyone would expect. The new curves are the solid lines, representing the new leaky mode, and the dash-dot lines, which are the extension of the new leaky mode to lower frequencies. All of the curves, except for the solid curve which points to the scale on the right side, present the normalized phase-constant values β/k_0; the solid curve that rises rapidly represents the normalized leakage constant α/k_0. If we compare the spectral gap here with that shown in Fig. 5, we can easily identify the locations here of points 2 and 3. There is no point that corresponds to point 1, however, because both the upper and lower improper real solutions continue down to zero frequency, forming a one-sided spectral gap.

Since the improper real solutions are nonphysical, this new mode cannot be detected at lower frequencies; there we would find only the conventional bound mode, in agreement with experience. Above a sufficiently high frequency, however, corresponding in Fig. 6 to about 5.5 GHz or $h/\lambda_0 = 0.08$, the presence of the new leaky mode would be felt. In Fig. 7 we present the calculated field distributions for both the bound mode and the new leaky

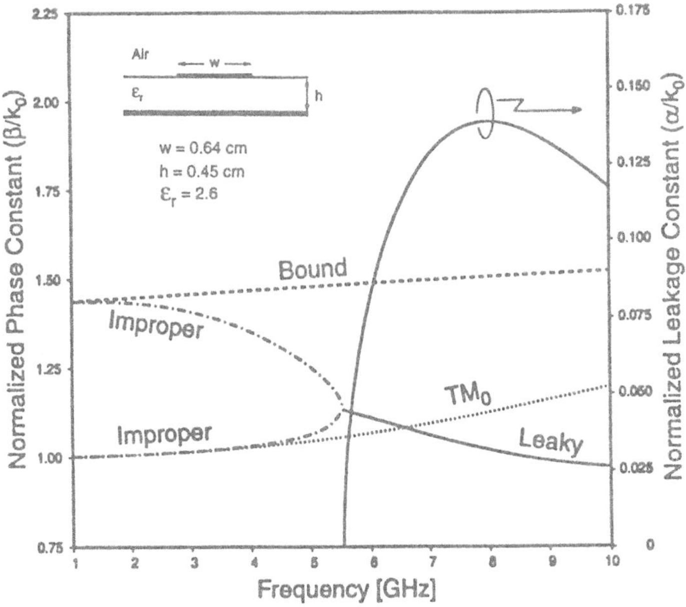

Fig. 6 Dispersion behavior in the form of β/k_0 and α/k_0 vs. frequency for microstrip line on an isotropic substrate. For the higher frequencies a new leaky mode is present, but that mode cannot be detected at low frequencies because it turns into a pair of improper real solutions.

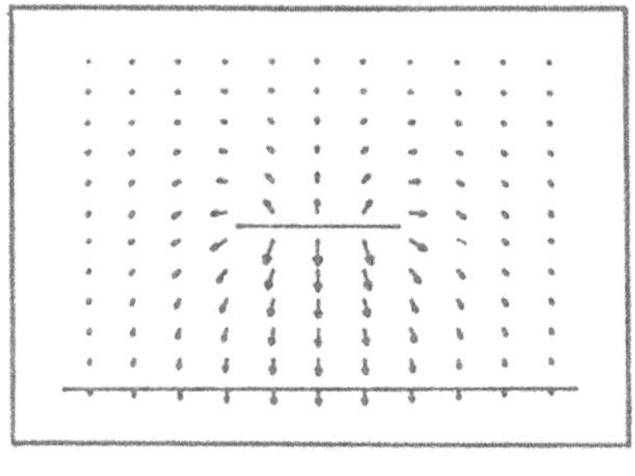

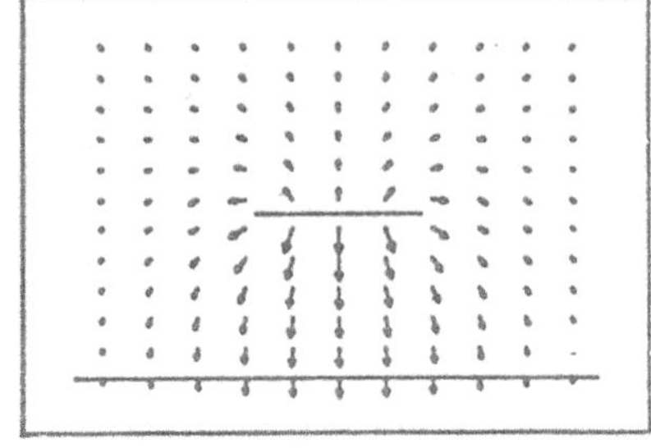

Fig. 7 Calculated vector electric-field distributions for both the bound mode and the real part of the new leaky mode on microstrip line. Because they are so similar, a feed designed to excite the bound mode would also excite the leaky mode.

mode for $h/\lambda_0 = 0.12$. It is seen that the field patterns are not identical but are very similar to each other. That implies that a feed structure designed to excite the bound mode would inevitably excite the new leaky mode also, and with comparable amplitude. Although no measurements were made directly of the wavenumber values of the new leaky mode, indirect measurements are completely consistent with all of the statements made above[13].

B. Simultaneous Propagation of Bound and Leaky Modes: A New General Effect

Although the new leaky mode and the conventional bound mode of microstrip line, discussed just above in Sec. A, are present simultaneously, the new effect to be described here is different in nature. In addition, it applies to a large variety of printed-circuit lines, not to a specific one. What is most astounding, however, is that the simultaneous-propagation effect, to be described below, is obtained by varying ***only*** the relative cross-section dimensions of the printed-circuit line.

We saw in Secs. II, B and C, that in most cases the dominant mode is bound at lower frequencies but becomes leaky above a critical frequency, and that, at the transition between the two mode types, there exists a spectral gap, whose properties are discussed in Sec. III. In the usual situation, therefore, we find that the bound dominant mode can be observed only ***below*** the frequency at which the spectral gap begins, and the leaky dominant mode propagates only ***above*** the frequency at which the spectral gap ends, so that the two solutions are ***completely separated*** from each other.

Until recently, everyone thought that when the ***relative dimensions*** of the cross section are changed, the only effects are to modify the values of the propagation wavenumber and the characteristic impedance. We now know that this assumption is incorrect.

It was found [5] that, when the strip widths of conductor-backed coplanar strips are modified, there can exist a frequency range within which the bound and leaky modes can propagate ***simultaneously***, and that the spectral gap then ***disappears***. It was also found that the frequency range over which the simultaneous propagation occurs can in fact become quite large. Further studies [14] showed that the effect was not restricted to that particular transmission line, but that this effect is actually quite ***general***, applying to many different lines. A detailed paper [14] shows such behavior for conventional slot line, and another paper [15] demonstrated similar behavior for conductor-backed coupled slot lines.

The usual behavior, where the bound and leaky solutions are completely separated from each other in frequency, is found when, for coplanar strips, the strip widths are relatively narrow, and, for slot lines, the slot spacing is relatively narrow. When the strip widths and the slot spacing, respectively, are ***increased*** sufficiently, the propagation behavior changes systematically into the regime in which the bound and leaky ranges overlap. The theoretical explanation for this unexpected physical effect is associated with the presence of a new improper real solution. This new solution is clearly unphysical, but its evolution as a

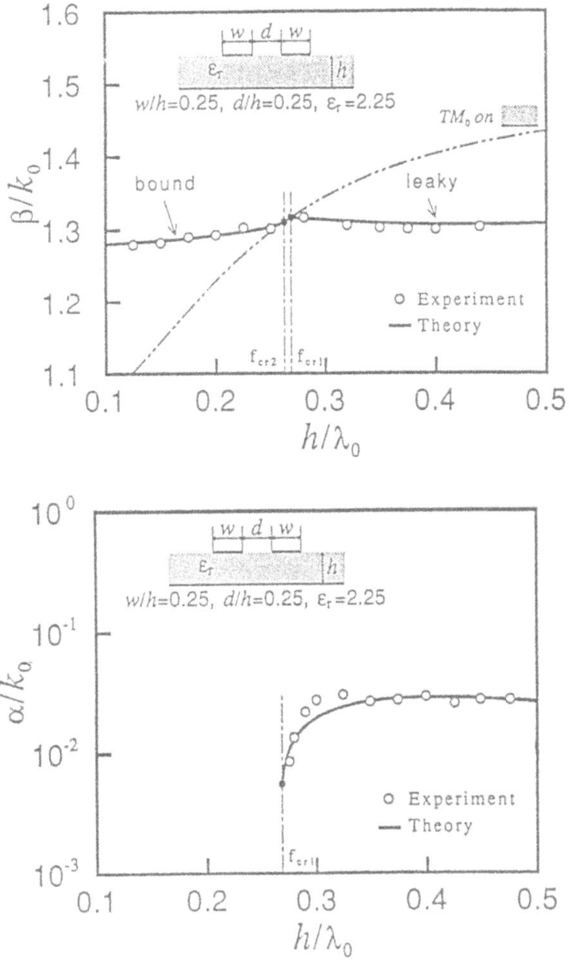

Fig. 8 The variations of β/k_0 and α/k_0 vs. h/λ_0 for conductor-backed coplanar strips with w/h = 0.25. With these relatively narrow strips, the dispersion behavior is what we would normally expect. The solid lines represent the theoretical calculations, and the open circles are the results of measurements.

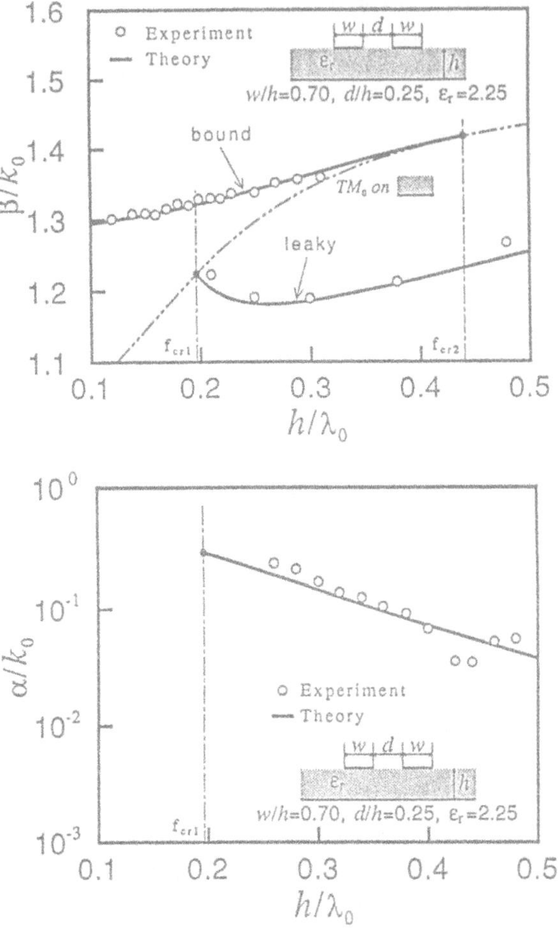

Fig. 9 Same as in Fig. 8 but for wider strips, w/h = 0.70 instead of 0.25. The dispersion behavior is now seen to be dramatically different from that in Fig. 8, with the bound mode and the leaky mode propagating simultaneously over a two-to-one frequency range.

function of the strip width or slot spacing mentioned above, and its interaction with the other solutions, lead to the phenomenon of simultaneous propagation. A detailed description of the evolution and role of this new improper real solution is contained in [15], as well as measurements of both α/k_0 and β/k_0 which verify the results of the calculations.

In the present paper numerical results are presented for two specific strip widths for conductor-backed coplanar strips. The first case, shown in Fig. 8, holds for w/h = 0.25, where w is the strip width and h is the substrate height. The relative dimensions for this case correspond to values commonly employed, and Fig. 8 presents the variations of β/k_0 and α/k_0 vs. normalized frequency. The behavior is what one would expect, in the light of the discussion in Sec. II; the bound and leaky portions of the dominant mode are completely separated from each other in frequency, and there is a small spectral gap between them. The measured values, represented by the open circles, show quite good agreement with the theoretical values.

As the strip width w is increased sufficiently, to about $w/h = 0.4$, the bound solution rises somewhat and moves to higher frequencies; correspondingly, the leaky solution moves down and extends to lower frequencies. When w/h increases to 0.50, the spectral gap has disappeared completely, and the end points of the bound and leaky solutions have moved to about $h/\lambda_0 = 0.32$ and 0.27, respectively, producing a range of simultaneous propagation of about 15% in frequency.

When the strip width is increased still further, but not that much further, to $w/h = 0.70$, the result becomes rather dramatic, as seen in Fig. 9. The bound solution continues on to greater than $h/\lambda_0 = 0.4$ and the leaky solution drops down in frequency to below $h/\lambda_0 = 0.2$. We thus see that the leaky and bound portions of the dominant mode now propagate simultaneously over a greater than ***two-to-one*** frequency range. Since the strip currents and the fields near to the strips for the bound and leaky solutions are quite similar to each other, a feed structure designed to excite the bound mode in this overlap region would surely also excite the leaky mode. The simultaneous-propagation phenomenon has effectively reduced the upper frequency limit for the bound mode alone by about 25% as compared to the behavior shown in Fig. 8, and a designer may need to keep this point in mind.

V. CONCLUSIONS

The presentations in this paper are in part a review and in part new material. They are intended to provide a synthesis in summary form of much of the present state of knowledge, with stress on physical descriptions of the various phenomena.

The paper first points out that the dominant modes on the various printed-circuit transmission lines used in microwave and millimeter-wave integrated circuits are not purely bound at all frequencies, as was initially believed, but can become leaky under the proper circumstances. For most lines, the dominant mode is bound at low frequencies and becomes leaky only above some critical frequency. For other lines, the behavior is quite different; leakage can be present at all frequencies or there can be a leaky dominant mode at some frequencies in addition to the bound mode. It is of great practical importance to know when leakage can arise because its presence can cause power loss, crosstalk, and undesired package effects, which in turn can seriously disrupt the performance of the circuit.

Section II explains why and when leakage can occur and describes the nature of the leakage. It then discusses the various types of leaky dominant modes and how they appear on different transmission lines. Numerical examples are also presented. In Sec. IV two recent examples of unusual and initially unexpected modal behavior are considered, first in qualitative terms and then in numerical detail. As seen there, the numerical results have been verified by measurements.

The transition between the bound and leaky frequency regions contains a frequency range, usually small, within which the modal solution is nonphysical, and is therefore called a "spectral gap." The spectral gap is treated in some detail in Sec. III, where issues are discussed such as why a nonphysical region is necessary and what it means physically to have a range in frequency over which the mode is nonphysical.

The complicated modal phenomena, including the leakage and spectral-gap behavior, which we find in connection with the printed-circuit transmission lines, arise because of the dielectric substrate. Furthermore, the substrate is employed asymmetrically in the guide cross section, thereby making it both inhomogeneous and asymmetrical. The surprisingly rich variety of phenomena that result have provided a continuing challenge for us, and we have found much gratification in our successively deeper levels of understanding.

REFERENCES

1. H. Shigesawa, M. Tsuji, and A. A. Oliner, Dominant mode power leakage from printed-circuit waveguides, *Radio Science* 26:559 (1991).
2. H. Shigesawa, M. Tsuji, and A. A. Oliner, Conductor-backed slot line and coplanar waveguide: dangers and full-wave analyses, *IEEE MTT-S Int. Microwave Sympos. Digest* p. 199 (1988).
3. N. Das, Characteristics of modified slotline configurations, *IEEE Int. Microwave Sympos. Digest* p. 777 (1991). Also, N. Das, Methods of suppression or avoidance of parallel-plate power leakage from conductor-backed transmisssion lines, *IEEE Trans. Microwave Theory Tech.* 44:169 (1996).
4. Y. Liu and T. Itoh, Leakage phenomena in multilayered conductor-backed coplanar waveguide, *IEEE Microwave and Guided Wave Lett.* 39:426 (1993).
5. M. Tsuji, H. Shigesawa, and A. A. Oliner, Simultaneous propagation of both bound and leaky dominant modes on conductor-backed coplanar strips, *IEEE Int. Microwave Sympos. Digest* p. 1295 (1993).
6. P. Lampariello, F. Frezza, and A. A. Oliner, The transition region between bound-wave and leaky-wave ranges for a partially dielectric-loaded open guiding structure, *IEEE Trans. Microwave Theory Tech.* 38:1831 (1990).
7. H. Ostner, J. Detlefsen, and D. R. Jackson, Radiation from one-dimensional dielectric leaky-wave antennas, *IEEE Trans. Antennas Propagat.* 41:344 (1995).
8. C. Di Nallo, F. Mesa, and D. R. Jackson, Excitation of leaky modes on multilayer stripline structures, submitted to *IEEE Trans. Microwave Theory Tech.*
9. A. A. Oliner, D. R. Jackson, and H. Ostner, Puzzles relating to radiation fields within the spectral gap between surface waves and leaky waves, *URSI Radio Science Meeting Digest* p. 247 (1995).
10. H. Shigesawa, private communication (1995).
11. M. Tsuji, H. Shigesawa, and A. A. Oliner, Printed circuit waveguide with anisotropic waveguides: a new leakage effect, *IEEE Int. Microwave Sympos. Digest* p. 783 (1989).
12. D. Nghiem. J. T. Williams, D. R. Jackson, and A. A. Oliner, Existence of a leaky dominant mode on microstrip line with an isotropic substrate: theory and measurement, *IEEE Int. Microwave Sympos. Digest* p. 1291 (1993). Also, accepted for publication in *IEEE Trans. Microwave Theory Tech.* (1997).
13. D. Nghiem, J. T. Williams, D. R. Jackson, and A. A. Oliner, Suppression of leakage on stripline and microstrip structures, *IEEE Int. Microwave Sympos. Digest* p. 145 (1994).
14. H. Shigesawa, M. Tsuji, and A. A. Oliner, Simultaneous propagation of bound and leaky modes on printed-circuit lines: a new general effect, *IEEE Trans. Microwave Theory Tech.* 43:3007 (1995).
15. Y.-D. Lin and Y.-B. Tsai, Surface wave leakage phenomena in coupled slot lines, *IEEE Microwave and Guided Wave Lett.* 4:338 (1994).

2-D INTEGRAL SPECTRAL DOMAIN ANALYSIS OF LEAKY MODES IN COVERED AND UNCOVERED MICROSTRIP LINES

R.Marqués and F.Mesa

Microwave Group
Fac. of Physics, University of Seville
Av. Reina Mercedes s/n. 41012 Seville, SPAIN
e-mail: *marques@cica.es, mesa@cica.es*

INTRODUCTION

Complex leaky modes on printed circuit lines have been analyzed using approximate techniques [1], Wiener–Hopf techniques [2], Mode–matching [3] and spectral domain integral techniques [4] –[9]. The method of moments (MoM) in the spectral domain is a very general formulation, providing a straightforward technique for obtaining the dispersion relation of the line. This relation is obtained as an implict equation whose solutions are the zeros of a given function of the spectral variable. In this paper, we will analyze both the MoM technique for obtaining this equation in covered and uncovered microstrip lines and the equation itself. Special attention will be devoted to study the possible location and evolution of the solutions in the complex plane. This analysis will provide guidelines for the search of the leaky modes of printed lines. It would be also useful as a previous step towards a rigorous 3-D spectral domain analysis of the practical excitation of leaky modes by physical sources.

THE IMPLICIT DISPERSION EQUATION

The structure under analysis is a microstrip line on a dielectric substrate with an optional upper ground plane (see Fig.1). A wave propagating along the positive z-direction will be assumed, with fields of the kind $\mathbf{E} = \mathbf{E}_0(x,y)e^{-jk_z z}e^{j\omega t}$ (the factor $e^{j\omega t}$ will be suppressed in the following) where k_z is a complex propagation constant having $\mathrm{Re}(k_z) \geq 0$; $\mathrm{Im}(k_z) \leq 0$. The integral equation for the tangential electric fields on the strip interface is given by:

$$\mathbf{E}_t = \int_{-w/2}^{w/2} \overline{\mathbf{G}}(x - x'; k_z) \cdot \mathbf{J}_s(x')dx' = 0 \;\; ; \;\; -w/2 < x < w/2 \tag{1}$$

Directions for the Next Generation of MMIC Devices and Systems
Edited by N. K. Das and H. L. Bertoni, Plenum Press, New York, 1997

where $\overline{\mathbf{G}}(x-x';k_z)$ is the transverse Green's dyad in the spatial domain [6], and $\mathbf{J}_s$ the surface current on the strip. As is well known, leaky modes are nonspectral solutions of the Maxwell equations, and therefore the definition of the Green's dyad is not unique. The different choices for the Green's dyad are given by the different inversion contours C used in the expression:

$$\overline{\mathbf{G}}(x-x';k_z) = \frac{1}{2\pi}\int_C \tilde{\overline{\mathbf{G}}}(k_x;k_z)e^{-jk_x(x-x')}dk_x \tag{2}$$

that relates the spatial-domain Green's dyad to the dyadic spectral domain Green's function $\tilde{\overline{\mathbf{G}}}(k_x;k_z)$. Some apparent restrictions are imposed to the inversion contour C:

- C must be symmetric by inversion with respect to the origin in the k_z plane, owing to the symmetry of the substrate
- C must run along the real axis as $k_x \Longrightarrow \infty$, to ensure that the fields remain finite for any finite value of x [6].

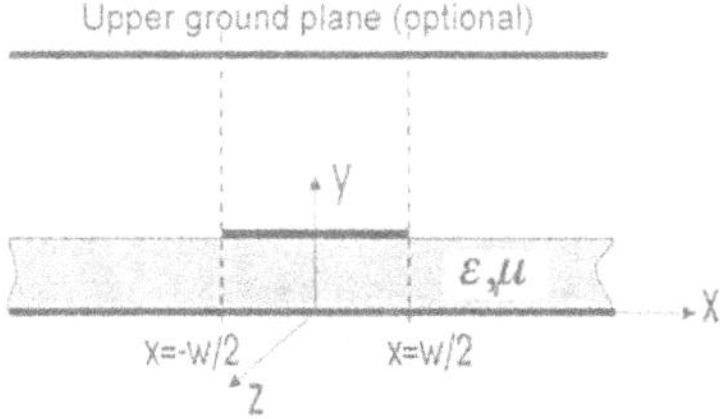

Figure 1: Covered/uncovered microstrip line on a dielectric substrate

Once the inversion contour C has been chosen, (1) can be expressed in the spectral domain as:

$$\mathbf{E}_t = \frac{1}{2\pi}\int_C \tilde{\overline{\mathbf{G}}}(k_x;k_z)\cdot\tilde{\mathbf{J}}_s(k_x)e^{-jk_x x}dk_x = 0 \;\; ; \;\; -w/2 < x < w/2 \;, \tag{3}$$

where $\tilde{\mathbf{J}}_s$ is the Fourier transform of the surface current density on the strip. Then the MoM is applied, with suitable basis, $\mathbf{J}_n$, and weight functions, $\mathbf{w}_n(x)$, giving the implicit equation of dispersion which can be expressed as:

$$f(k_z) = \mathrm{Det}\left(< \tilde{\mathbf{w}}_m^* | \tilde{\overline{\mathbf{G}}} | \tilde{\mathbf{J}}_n >\right)_C = 0 \tag{4}$$

where $\tilde{\mathbf{w}}_n$ are the Fourier transforms of the weight functions $\mathbf{w}_n$, and the inner products can be evaluated in the spectral domain using:

$$< \tilde{\mathbf{w}}_m^* | \tilde{\overline{\mathbf{G}}} | \tilde{\mathbf{J}}_n >_C = \int_C \tilde{\mathbf{w}}_m^*(k_x^*)\cdot\tilde{\overline{\mathbf{G}}}(k_x;k_z)\cdot\tilde{\mathbf{J}}_n(k_x)\,dk_x \;. \tag{5}$$

It should be emphasized the double conjugation $\tilde{\mathbf{w}}_m^*(k_x^*)$ in the integrand of (5). This double conjugation arises from the generalization of the Parseval theorem to complex integration contours, and guarantees that $f(k_z)$ is an analytical function, except for some point singularities that will be discussed below [6]. It is also worthwhile to note that complex solutions of (5) should appear in pairs of complex conjugate roots k_z and k_z^*, being those solutions with $\mathrm{Re}(k_z) > 0$, $\mathrm{Im}(k_z) > 0$ physically meaningless.

Different leaky mode solutions to (4) will appear when different inversion contours C are used in the definition of the inner products (5). In covered lines the spectral Green's function is a meromorphic function whose poles $k_{x,n}$ are given by:

$$\gamma_n^2 = k_{x,n}^2 + k_z^2 \, , \tag{6}$$

where k_z is the unknown propagation constant of the leaky mode and γ_n are the wavenumbers of the different uniform plane waves that can propagate in the background waveguide [6]. For complex propagation constants k_z, these poles are located along the hyperbolae defined by:

$$\mathrm{Re}(k_x)\mathrm{Im}(k_x) = -\mathrm{Re}(k_z)\mathrm{Im}(k_z) \tag{7}$$

in the complex k_x-plane [7]. Owing to the Cauchy theorem, the inner products (5) for two different inversion contours C and C' will differ only if the poles $k_{x,n}$ have different relative locations with respect to C and C'. This fact is illustrated in Fig.2, where the inversion contour C surrounds only the pole corresponding to the first waveguide mode $k_{x,1}$, and the contour C' surrounds the two first waveguide mode poles. For a given inversion contour, the waveguide modes participating in the leakage are those enclosed btween the inversion contour and the real axis. As the far field for $|x| \to \infty$ must be a superposition of outgoing waves, all the enclosed waveguide modes must be above cutoff for physical leaky modes [7].

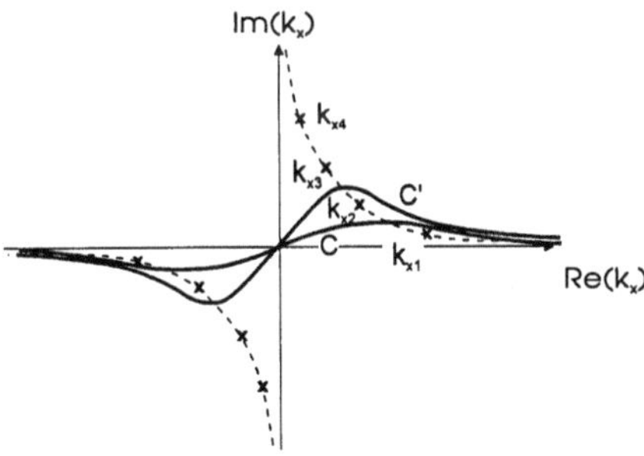

Figure 2: Inversion contours for covered microstrip line leaky mode analysis

The situation is somewhat different for uncovered lines, because in this case the spectral Green's dyad has branch cuts in $k_x = \sqrt{k_0^2 - k_z^2}$, where k_0 is the free-space wavenumber. The usual choice of the branch cuts (see Fig.3) implies the definition of a proper and an improper Riemann sheet, defined by $\mathrm{Im}(k_y) < 0$ ($k_y \equiv \sqrt{k_0^2 - k_x^2 - k_z^2}$)

and $\mathrm{Im}(k_y) > 0$ respectively. For these definitions and for complex values of k_z, the branch cuts in the complex k_x-plane are located along the hyperbolae defined in (7). The spectral Green's dyad has also one or more surface-wave poles on the proper Riemann sheet, given by (5), where γ_n are now the eigenvalues of the surface waves that can propagate in the background waveguide. It can be noticed that inversion contours lying entirely on the proper Riemann sheet can not provide leaky modes with volume-wave radiation, for the far field should vanish as $y \rightarrow \infty$. To obtain leaky modes exhibiting volume-wave radiation, the inversion contour must surround the branch points, and then lies partially on the improper Riemann sheet [1] [8] [9]. All these facts are illustrated in Fig.3. The inversion contour C corresponds to a leaky mode whose far field is composed only by the first surface-wave mode. The contour C' instead corresponds to a leaky mode that radiates in both volume and surface waves.

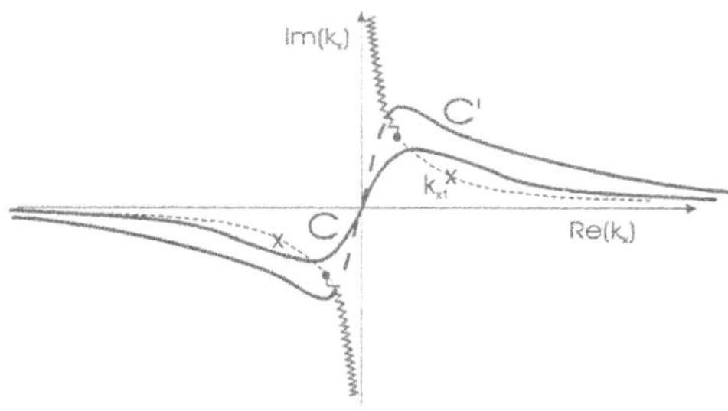

Figure 3: Inversion contours for uncovered microstrip line leaky mode analysis

ANALYSIS OF THE IMPLICIT DISPERSION EQUATION

The solution of the implicit dispersion equation (4) implies to compute the function $f(k_z)$ (4) and search for its zeros. In this section we will analyze the behavior of $f(k_z)$ in the k_z complex plane. The basis and weight functions, $\mathbf{w}_n$ and $\mathbf{J}_n$, are defined in a finite interval of the x variable and therefore its Fourier transforms are analytic, making that the singularities of $f(k_z)$ are related to those of the spectral Green's function $\tilde{\overline{\mathbf{G}}}$. For covered striplines it has been reported the presence of square–root type branch points of $f(k_z)$ at $k_z = \gamma_n$ [6]. If the branch cuts are chosen as in Fig.4, the upper Riemann sheet can be chosen as the so called *proper* Riemann sheet, i.e. the sheet that corresponds to the choice of the inversion contour C of (5) along the real axis of the complex k_x plane. Only real bound modes of the line will appear on this sheet; complex leaky modes as well as unbounded real modes will appear on the lower sheets. We will define the remaining Riemann sheets in Fig.4 giving the set of waveguide modes $\{\gamma_n\}$ which participate in the leakage of power (i.e. giving the set of waveguide modes between the inversion contour and the real axis). Thus, the Riemann sheet defined by the set (γ_1, γ_2) corresponds to a choice of the inversion contour as C' in Fig.2.

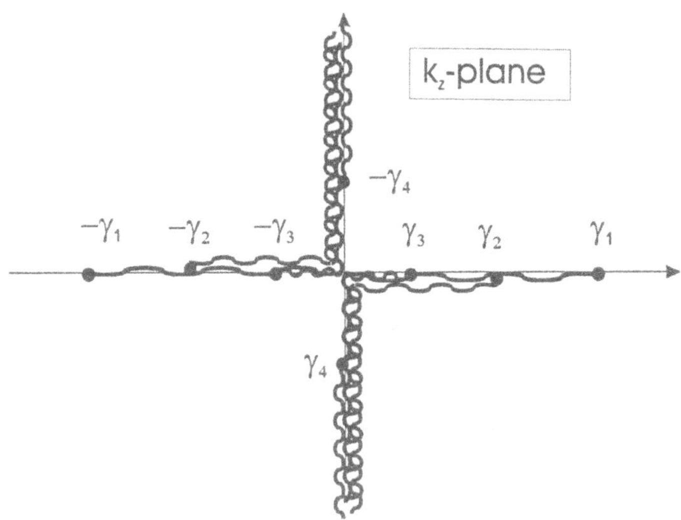

Figure 4: Branch points and branch cuts of $f(k_z)$. The upper Riemann sheet is the *proper* sheet. The lower sheets are defined by the set of waveguide modes $\{\gamma_n\}$ participating in the leakage

There is another interesting choice of the branch cuts. They can be chosen along the real axis, finishing in $k_z \rightarrow \infty$. In Fig.5 we have plotted this possibility for the branch cut corresponding to the first waveguide mode γ_1. In this case, the second and the fourth quadrants of the upper Riemann sheet does not correspond to the same quadrants of the *proper* Riemann sheet (see Fig.4). The fourth quadrant in particular corresponds to a choice of the inversion contour in (5) like the path C in Fig.2, with a symmetrical choice for the second quadrant. If the branch cut corresponding to the second waveguide mode γ_2 is also chosen in the same way, the upper fourth and second quadrants will now correspond to the (γ_1, γ_2) sheet of Fig.4, and so on.

There are some interesting conclusions that can be obtained from Figs.4 and 5:

- First of all, we can see in Fig.5 that the *proper* Riemann sheet and the sheet defined by a choice of the inversion contour like C in Fig.2 are only connected through the segments of the real axis defined by $\gamma_1 > |k_z| > \gamma_2$. This conclusion can be generalized saying that the proper Riemann sheet is only connected to the sheets of Fig.2 defined by sets of consecutive poles, i.e. by sets of the $(\gamma_1, ...\gamma_i, \gamma_{i+1}, ...\gamma_n)$ kind, through the $\gamma_n > k_z > \gamma_{n+1}$ zones of the real axis.

- A second conclusion affects to the onset of complex leaky modes. They can only appear in pairs of complex conjugate roots of (4). Therefore, two real modes (unbounded and/or bound) must join at the onset of a leaky mode, giving rise to the pair of complex conjugate roots. Nevertheless, all the real modes (bound and unbounded) are restricted to the $|k_z| > \gamma_1$ zone of the real axis, since for other real values of k_z there will be a non-compensated lateral power flux from or towards the line, violating thus the Poynting theorem[1]. In consequence, all the complex leaky modes must exhibit a *spectral gap* region with $\mathrm{Re}(k_z) > \gamma_1$ at the onset (of course, they can be also leaky modes that remain complex for all frequencies).

[1]For the same reason there will not be possible an hypothetical purely imaginary root of (4)

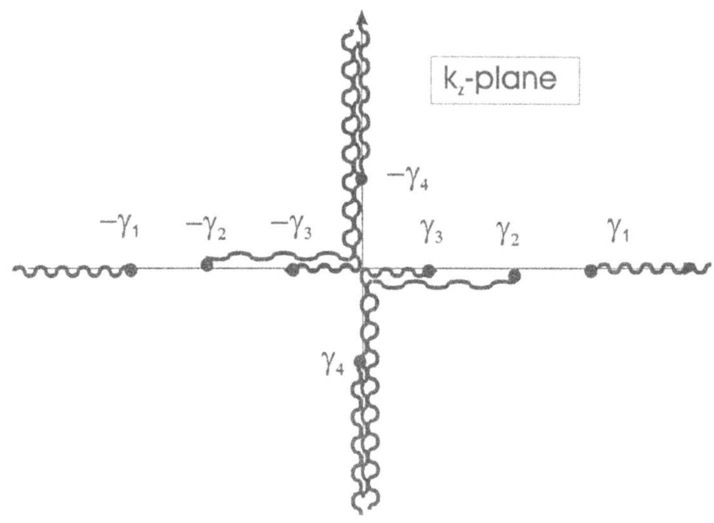

Figure 5: Another choice of the branch cuts for $f(k_z)$. The branch cut associated with the first waveguide mode, γ_1, has been chosen in order to include in the upper Riemann sheet (second and fourth quadrants) the leaky modes with the first waveguide mode participating in the leakage

- A third conclusion also affects to the onset of leaky modes. It is apparent from Figs.4 and 5 that real modes on different Riemann sheets can never join together, except for the case of a real bound mode and an unbounded mode of the (γ_1) sheet of Fig.4. Therefore, complex leaky mode on the (γ_1) sheet of Fig.4 can be originated by the combination of either a bound and an unbounded real modes, or two unbounded real modes. But leaky modes on any other sheet only appear after the joining of a pair of unbounded modes *on the same* Riemann sheet of Fig.4.

- Finally, a complex leaky mode on a given Riemann sheet of Fig.4 can never go through the branch cuts from a sheet to another one. In fact, this would imply its transformation into a real mode on the real axis in the $|k_z| < \gamma_1$ zone, which has been found to be impossible [2]. Therefore all the solutions of (5) must remain always on the same Riemann sheet of Fig.4, except in the case of the proper and the (γ_1) Riemann sheets.

For uncovered lines, the picture of the singularities of $f(k_z)$ is more complicated because another branch point appear in $k_z = k_0$. This new branch point is not of the square–root type, a fact that substantially affect to the sheet structure of $f(k_z)$. However, in this kind of structures, there is always at least a surface–wave branch point $\gamma_1 > k_0$, and some of the aforementioned properties of the roots of (5) remain still valid. In particular, it must be true that bound real modes can not be connected with any other improper modes than those lying on the (γ_1) Riemann sheet. This fact can be seen in Fig.6, where the dispersion characteristics of some microstrip higher order modes are shown. At low frequencies, the higher order bound modes (EH_1 and EH_2) migrates to the (γ_1) Riemann sheet, becoming real unbounded modes (which are not shown in the figure). Moreover, there also appear two volume–wave leaky modes (LM1

[2]Or its transformation into a purely immaginary root of (5), which is also impossible

and LM2), which are obtained following an inversion contour similar to those labeled as C' in Fig.3 . These leaky modes are not connected with the bound modes, a fact that is more apparent for the EH_2 mode. The closeness of the first TM_0 waveguide mode to the free space wavenumber, k_0, makes this fact less apparent for the EH_1 mode, but it has been carefully tested by numerical computations.

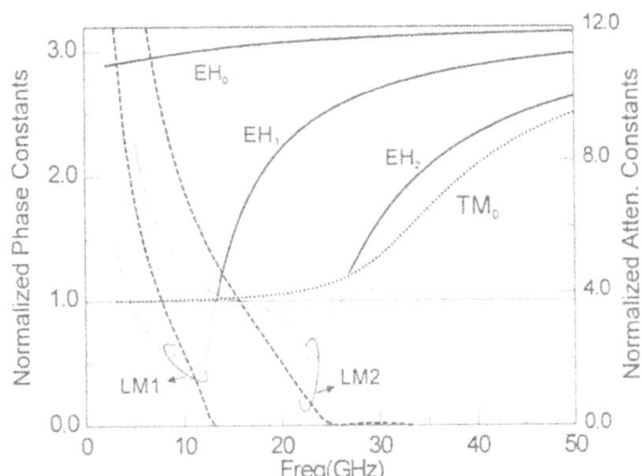

Figure 6: Normalized propagation constants of an uncovered microstrip on an uniaxial dielectric substrate ($\epsilon_x = \epsilon_z = 9.8\epsilon_0$, $\epsilon_y = 10.3\epsilon_0$, $w/h = 1$). The EH_n modes have are bound modes lying on the proper Riemann sheet of $f(k_z)$. The LMn modes are on the (γ_1, k_0) Riemann shhet, i.e., they have been computed using an inversion contour as C' in Fig.3

Finally, it is interesting to note that the aforementioned analysis of the function $f(k_z)$ has been fruitfully used in the spectral domain analysis of the excitation of leaky modes by finite sources made in [10]. In this work it is shown that the branch points and Riemann sheets of the 3-D spectral domain Green's function of the stripline are the same that those of $f(k_z)$ and that the pole singularities of the 3-D spectral domain Green's function are the zeroes of $f(k_z)$. Using these properties a *necessary* condition of excitation was shown in [10]. This *necessary* condition of leakage states that a leaky mode lying on the $(\gamma_1, \gamma_2, ..., \gamma_n)$ Riemann sheet must satisfy the condition $\gamma_n > \text{Re}(k_z) > \gamma_{n+1}$, in order to be significantly excited by a physical source. This statement is justified because only leaky modes satisfying this condition are close enough to the *proper* Riemann sheet to affect the field calculations in any way [10] (see also the first aforementioned conclusion on the properties of $f(k_z)$)

CONCLUSION

The method of moments in the spectral domain has been applied to the analysis of leaky modes in covered and uncovered microstrip lines. The dispersion relation has been obtained as an implicit equation in the complex plane whose solutions are the zeros of a function $f(k_z)$ of the spectral variable. Some properties of this function and of its zeros have been deduced and tested by numerical computations. It has been shown that $f(k_z)$ has square–root type branch points at the locations of the background waveguide modes. The sheet structure of $f(k_z)$ has been also studied. The evolution and properties of the zeros of $f(k_z)$ have been analyzed, showing that they must remain inside the same Riemann sheet when frequency varies, except for the solutions on the *proper* Riemann sheet (bound modes). These solutions can transform into unbounded modes and, eventually, into complex leaky modes *on a particular* Riemann sheet of $f(k_z)$. The spectral gap and other known properties of leaky modes can be also deduced from the features of $f(k_z)$. This study can be also fruitfully used in the 3-D spectral domain analysis of the physical excitation of leaky modes.

ACKNOWLEDGMENT

This work was supported by the DGICYT, Spain (Proj. n. TIC95-0447)

References

[1] T.T. Wu. Theory of the microstrip. *Journal of Applied Physics*, 28(3):299–302, March 1957.

[2] A.A. Oliner. Leakage from higher modes on microstrip line with apllication to antennas. *Radio Science*, 22:907–912, November 1987.

[3] M. Tsuji, H. Shigesawa, and A.A. Oliner. Printed–circuit waveguides with anisotropic substrates: a new leakage effect. In *Proc. of IEEE MTT-S Digest*, pages 783–786, (USA), 1989.

[4] J. Boukamp and R.H. Jansen. Spectral domain investigation of surface wave excitation and radiation by microstrip lines and microstrip disk resonators. In *Proc. 13rd European Microwave Conference*, pages 721–726, September 1983.

[5] D. Phatak, N.D. Das, and A.P. Defonzo. Dispersion characteristics of optically excited coplanar striplines: Comprehensive full–wave analysis. *IEEE Transactions on Microwave Theory & Technics*, MTT-38:1719–1730, November 1990.

[6] F. Mesa and R. Marqués. Integral representation of spatial Green's function and spectral domain analysis of leaky covered strip–like lines. *IEEE Transactions on Microwave Theory & Technics*, MTT-43:828–837, April 1995.

[7] F. Mesa and R. Marqués. Power based considerations on the spectal domain analysis of leaky waves in covered strip-like transmission lines. *Proc. of IEE Microwaves, Optics & Antennas Pt.H*, 143(1):25–30, January 1996.

[8] K.A. Michalski and D. Zheng. Rigorous analysis of open microstrip lines of arbitrary cross section in bound and leaky regimes. *IEEE Transactions on Microwave Theory & Technics*, MTT-37:2005–2010, December 1989.

[9] J.M. Grimm and D.P. Nyquist. Spectral analysis considerations relevant to radiation and leaky modes of open-boundary microstrip transmission line. *IEEE Transactions on Microwave Theory & Technics*, MTT-41(1):150–153, January 1993.

[10] C. Di Nallo, F. Mesa, and D.R. Jackson. Excitation of leaky modes on multilayer striplines. In *Proc. of IEEE MTT-S Digest*, San Francisco (USA), June 1996.

EXCITATION OF LEAKY MODES ON PRINTED CIRCUIT STRUCTURES BY PRACTICAL FEEDS: AN INVESTIGATION OF PHYSICAL MEANING

F. Mesa[1], C. Di Nallo[2], and D. R. Jackson[3]

[1] Microwave Group, Dept. of Electronics and Electromagnetism, University of Seville, Avda. Reina Mercedes s/n, 41012 Seville, Spain

[2] Department of Electronic Engineering, University of Rome "La Sapienza", Via Eudossiana 18, 00184 Rome, Italy

[3] Dept. of Electrical and Computer Engineering, University of Houston, Houston, TX 77204-4793

INTRODUCTION

The existence of leaky modes on printed circuit transmission lines has been the subject of considerable interest recently[1]. These modes are usually undesirable since they result in increased attenuation of the signal, and may result in crosstalk with adjacent circuit components and other spurious effects, including interference with bound modes that also propagate on the line[2]. Of particular interest is the existence of leaky *dominant* modes on the structure[2,3]. A dominant leaky mode (as opposed to a leaky *higher-order* mode), is one that has a current distribution on the conducting strip that closely resembles that of a quasi-TEM mode of propagation. Therefore, such a leaky mode will typically be excited quite strongly by a customary feed. Leaky dominant modes have been found on multilayer stripline structures[2], coplanar waveguide and slotline[4], coplanar strips[5], microstrip line with an anisotropic substrate[6], and recently on microstrip line with an isotropic substrate[7].

Although the propagation properties of leaky modes on printed-circuit lines has been studied quite thoroughly in recent years, much less attention has been devoted to the interesting and practical issue of excitation of these modes by a practical source, such as a delta-gap feed on the conducting strip or a probe feed. The issue of excitation by a *finite-size feed* is an important one, since it can be used to define the degree of physical meaning of the leaky mode[8]. For printed circuit structures, the excitation of the current on the conducting strip by a finite source provides a convenient way to investigate the physical meaning of the leaky modes, and this is the subject of the present paper. Results will be presented for the two-layer stripline structure shown in Fig. 1. One advantage of this structure is that it allows considerable flexibility in controlling the phase constants of the leaky, bound, and parallel-plate modes, by selection of the dimensions and permittivities. The source consists of a delta-gap feed located on the strip.

Directions for the Next Generation of MMIC Devices and Systems
Edited by N. K. Das and H. L. Bertoni, Plenum Press, New York, 1997

In this paper the calculation of the strip current due to the source excitation is performed by constructing a numerical Green's function. The calculated strip current is numerically exact, under the assumption that the strip width is small (since a fixed transverse dependence of the longitudinal current is assumed, and the transverse current is neglected). The numerical Green's function is obtained by Fourier transforming the source in the longitudinal (z) direction. The problem is thus essentially reduced to one of calculating the strip current due to an infinite set of phased line source excitations (a one-dimensional Green's function problem). The one-dimensional Green's function is in turn calculated from a spectral integration in the transverse wavenumber (k_x) plane, which is the same type of integration used to solve for the modal solutions on the guiding structure[9]. One of the properties of the k_x integration is that different choices are possible for the path of integration[2]. A real-axis path defines a modal solution that is bound in the transverse ($\pm x$) directions, while a path that detours around the poles of the background structure results in a solution that is improper in the transverse directions. Different choices of path in the transverse wavenumber plane give rise to *branch cuts* in the longitudinal wavenumber plane for the integration in k_z that determines the numerical Green's function. A careful consideration of these branch cuts provides much insight into the physical meaning of the leaky modes that are excited by the source, corresponding to the poles in the k_z plane.

The numerically exact strip current is compared to the current of the leaky mode alone, defined from the residue contribution of the leaky-wave pole in the k_z integration of the numerical Green's function. The degree of physical meaning for the leaky mode is defined by the correlation between the exact and leaky-mode currents. The generalized pencil of functions (GPOF) method[10], which resolves the exact current into a set of exponential waves, is used to help quantify the correlation.

ANALYSIS

A. Formulation for Strip Current

The conducting strip is assumed to be infinite in the $\pm z$ directions, and perfectly conducting. Another assumption is that the strip width is sufficiently small that the transverse (x-directed) current may be neglected. A delta-gap voltage feed is assumed to exist at $z=0$. For the delta-gap feed, there is an impressed electric field on the surface of the conducting strip, of the form

$$E_z(x,z)=T(x)\,L_g(z). \tag{1}$$

The function $T(x)$ is taken as unity over the width of the strip. The function $L_g(z)$ is $\delta(z)$ for an idealized delta-gap feed. However, in order to make the transform converge faster, $L_g(z)$ is taken as

$$L_g(z)=\frac{1}{d}\frac{e^{-\frac{1}{2}\left(\frac{z}{d}\right)^2}}{\sqrt{2\pi}}, \tag{2}$$

which has the transform

$$\tilde{L}_g(k_z) = e^{-\frac{1}{2}(k_z d)^2}. \tag{3}$$

The current on the strip is represented as

$$J_{sz}(x,z) = \eta(x) I(z), \tag{4}$$

where

$$\eta(x) = \begin{cases} \dfrac{1}{\pi\sqrt{\left(\dfrac{w}{2}\right)^2 - x^2}} & w/h_{\min} < 2 \\ \\ \dfrac{1}{w} & w/h_{\min} \ge 2, \end{cases} \tag{5}$$

with $h_{\min} = \min(h_1, h_2)$. A spectral-domain method is used to solve for the current on the strip, by requiring that the electric field produced by the strip current equal the impressed field of the delta-gap source. A Galerkin testing procedure results in the solution

$$J_{sz}(x,z) = \eta(x) \cdot \frac{1}{2\pi} \int_{-\infty}^{\infty} \tilde{L}_g(k_z) R_g(k_z) e^{-jk_z z}\, dk_z, \tag{6}$$

where

$$R_g(k_z) = \frac{2\pi}{\int_{-\infty}^{\infty} \tilde{\eta}^2(k_x) \tilde{G}_{zz}(k_x, k_z)\, dk_x}, \tag{7}$$

with $\tilde{G}_{zz}(k_x, k_z)$ a component of the spectral-domain Green's function (the Fourier transform of the field $E_z(x,z)$ due to a unit-amplitude z-directed electric dipole). Equation (6) is the numerical Green's function for the current on the strip when excited by a delta-gap feed.

The function $R_g(k_z)$ has poles in the k_z plane at the values of the propagation constants of the guided modes on the structure, either k_z^b for the bound mode or k_z^{lw} for a leaky mode. This is because the denominator in Eq. (7) is precisely the same integral that appears in the solution of the propagation constant for the guided-mode on the structure. The residue contribution to the integral (6) at a pole gives, by definition, the current amplitude of the guided mode (either bound or leaky), and defines the excitation coefficient of the guided mode.

B. Discussion of Integration Paths

An important consideration in the evaluation of the numerical Green's function is the choice of path in the k_x plane for the evaluation of the function $R_g(k_z)$ in Eq. (6). As mentioned previously, there are poles in the k_x plane, corresponding to the parallel-plate modes of the background structure, located at

$$k_x = k_{xp} = \left(k_{pp}^2 - k_z^2\right)^{1/2}, \tag{8}$$

where k_{pp} is the propagation wavenumber of a parallel-plate mode. In most practical cases only the fundamental TM_0 mode ($k_{pp} = k_{TM0}$) is above cutoff. The path of integration in the k_x plane may either be chosen to detour around the poles or not. For example, if k_z is chosen in the fourth quadrant of the complex plane, the poles in the k_x plane will be in the first and third quadrants, as shown in Fig. 2 (illustrated for a single pair of poles $k_{pp} = k_{TM0}$). There are two possible paths shown: the real axis path (which will yield a bound solution) and the one that detours around the poles (which will yield an improper solution). Therefore, the function $R_g(k_z)$ is a multi-valued function, which implies the existence of branch cuts in the k_z plane, in order to restrict the function $R_g(k_z)$ to being single-valued. By circling the branch point in the k_z plane, and examining the corresponding path of integration in the k_x plane, it is concluded that the branch cut corresponds to a *two-sheeted* Riemann surface, as does a square-root type of branch point (two round trips around the branch point result in the same path of integration in the k_x plane). The important observation that *poles of the background structure* give rise to *branch points in the longitudinal wavenumber plane,* which is key to the discussion below on the physical meaning of the leaky modes, was recognized originally by Nyquist[11]. The branch cuts play a crucial role in providing insight into the physical meaning of the leaky mode excited on the structure.

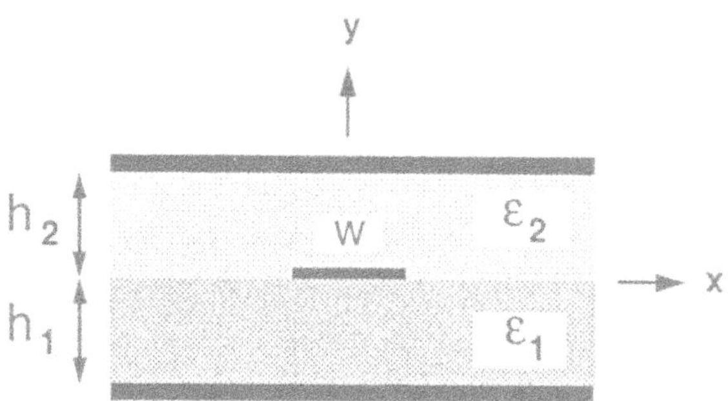

Fig. 1. Two-layered stripline structure.

The exact shape of the branch cuts is arbitrary. However, a convenient choice is the Sommerfeld branch cut, in analogy with the same shape of branch cut that is commonly used when dealing with the wavenumber mapping shown in Eq. (8). The Sommerfeld choice of branch cuts is shown in Fig. 3 for the case of two modes above cutoff. A convenient property of the Sommerfeld branch cut is that all points on one of the sheets (denoted as the *top sheet*) of the k_z plane correspond to paths in the k_x plane that do not detour around the poles in the k_x plane - that is, the path is the *real axis*. Points on the *bottom sheet* correspond to paths that *detour around the poles* in the k_x plane (as shown in Fig. 2).

The branch cuts in the k_z plane provide insight into the physical meaning of a leaky mode that is excited by the source. The path of integration in the k_z plane is along the real axis, except that the path detours around the bound-mode poles that lie on the real axis (above the pole on the positive real axis, below the one on the negative real axis). If a leaky-mode pole is close to the integration path in the k_z plane, and the residue of the pole is not too small, then the pole will make a strong contribution to the path integration. This contribution will result in a field behavior that closely resembles the field of the leaky-mode, i.e., the field from the residue of the pole. If the pole is further from the path, its contribution will be blurred out, and the integrand will not have a sharply peaked well-defined component; consequently, the field calculated from the path integration will not resemble that of the leaky mode alone. Therefore, a *necessary* condition for a leaky mode to have physical meaning is that the leaky-mode pole be close to the path of integration.

The term "close" means close in the *Riemann surface sense*, not in the geometrical sense. To illustrate this, consider several possible pole locations in the k_z plane, shown in Fig. 3 as points A, B, C, and D (only poles in the right-half of the complex plane are shown for simplicity). The (real axis) path stays on the top sheet of all branch points. Point A is on the top sheet of all branch points and corresponds to the location of the bound-mode pole, which is on the real axis and has $k_{zr} > k_{TM0}$, while points B, C, and D are possible locations

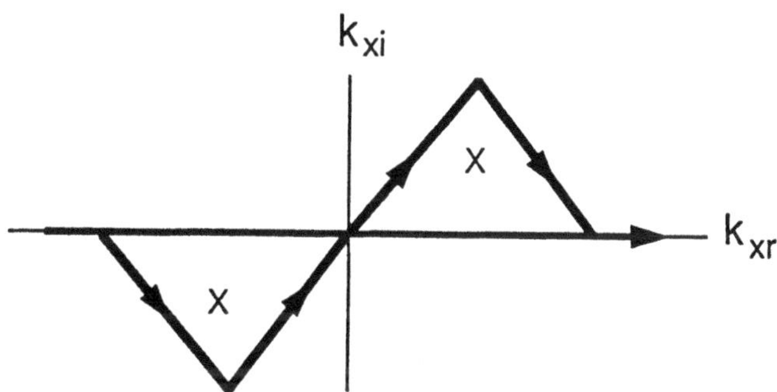

Fig. 2. Two possible paths of integration in the k_x plane. The real-axis path yields the bound-mode solution, while the path that detours around the poles yields the leaky-mode solution.

of the leaky-mode. Clearly, point A is close to the path (and this agrees with the well-known fact that the bound mode is always physically meaningful).

Points B and C are assumed to reside on the bottom sheet of the TM_0 branch point, and the top sheet of all others (corresponding to the path shown in Fig. 2, that detours around only the TM_0 parallel-plate poles). Because point B has $k_{zr} > k_{TM0}$, but is on the opposite sheet to the one the path is on, it is "far" from the path, even though it may have coordinates (k_{zr}, k_{zi}) that are *geometrically* close to the path. Point C has $k_{TE1} < k_{zr} < k_{TM0}$, and is close to the path.

Point D has $k_z < k_{TE1}$, and is located on the bottom sheet of both the TM_0 and TE_1 branch points. This point is also close to the path.

For the leaky-mode poles, any other location on different sheets, other than points C and D, would not be close to the path. A concise way to summarize this necessary condition for physical meaning is that provided by the "condition of leakage" [2]. This criterion states that, in order for a leaky mode to have well-established physical meaning, the value of the phase constant $\beta = \mathrm{Re}(k_z^{lw})$ must be *consistent* with the choice of path used to obtain the leaky mode. The word consistent means that the path must detour around (capture the residues from) only those poles for which $\beta < k_{pp}$, and none others.

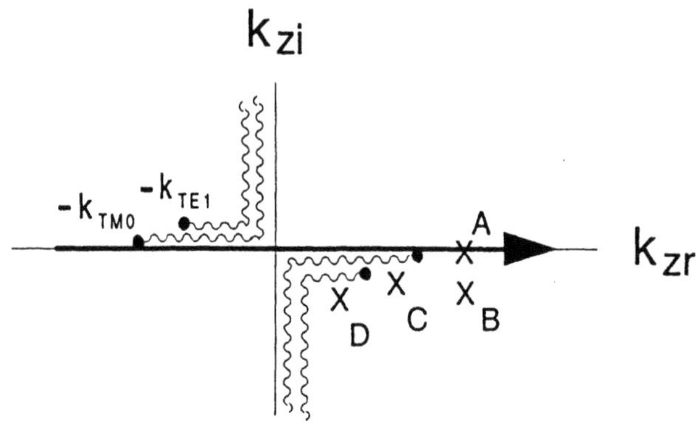

Fig. 3. The k_z plane when two modes (TM_0 and TE_1) are above cutoff, showing various possible pole locations.

RESULTS

The numerically exact current on the conducting strip (from Eq. (6)) is compared with the current of the leaky mode, in order to illustrate the determination of physical meaning. The current of the leaky mode is obtained by calculating the residue contribution from integral (6), which defines the excitation coefficient of the leaky mode. (The excitation coefficient of the bound mode is also calculated in this way, although the focus will be on the leaky mode). To help quantify the comparison, the generalized pencil of functions (GPOF) method[10] is used to approximate the exact strip current with a set of exponential waves.

The dispersion curves showing the normalized phase constants for the bound mode, leaky mode, and the TM_0 mode of the two-layered structure in Fig. 1 are shown in Fig. 4. The structure has a wide strip, and the permittivity of the bottom layer much larger than that of the top layer (which is air). This results in a large separation between the dispersion curves for the three different solution in Fig. 4, which makes the results easier to interpret.

Table 1 shows a comparison between the theoretical excitation coefficients and the GPOF fit versus frequency, to study how the physical meaning of the leaky mode changes with frequency. The GPOF waves with the largest amplitudes are shown for each frequency. For all frequencies, the agreement between the theoretical values and the GPOF results is excellent for the bound mode, as expected (since this mode always has physical meaning). For the leaky mode, the agreement is quite good at 8 GHz, and becomes progressively worse as the frequency decreases. This is because the leaky mode is approaching the "spectral gap" [12], which is the region below approximately 1.5 GHz, for which $\beta > k_{TM0}$.

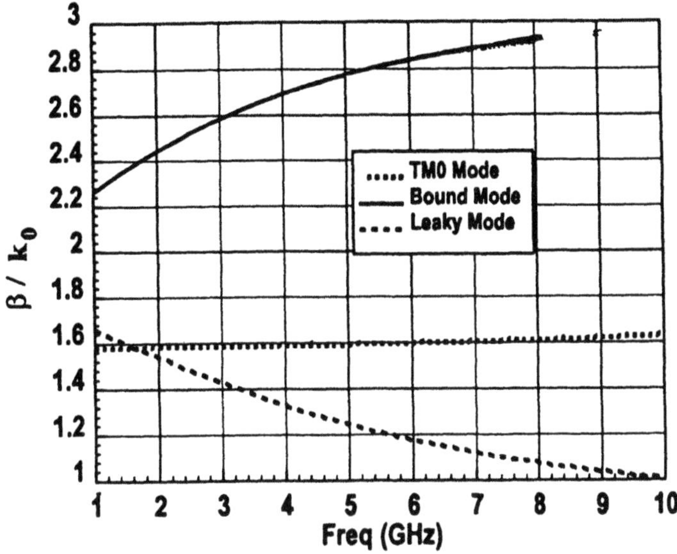

Fig. 4 Dispersion plot showing the wavenumbers of the bound and leaky modes for the structure of Fig. 1. Also shown is the dispersion curve for the propagation wavenumber of the TM_0 parallel plate mode. ε_{r1} = 10.0, ε_{r2} = 1.0, h_1 = 1.0 mm, h_2 = 0.5 mm, w = 7.0 mm.

TABLE 1

Theoretical propagation constants and amplitudes (real, imaginary) determined from the pole locations and the residues of the poles in the k_z plane, and those determined numerically by the GPOF method, for a delta-gap source.

f (GHz)			k_{z0} / k_0	AMPLITUDE
8	Theory	bound leaky TM_0	(2.9277, 0) (1.0800, -0.32283) (1.6119, 0)	(0.02483, 0) (0.02286, -0.00452)
8	GPOF		(2.9274, 0.000015) (1.0828, -0.32201) (1.5974, -0.23874) (1.6110, -0.04738)	(0.02486, 0.000031) (0.0234, -0.00780) (0.00078, 0.00147) (-0.00038, 0.00016)
4	Theory	bound leaky TM_0	(2.6936, 0) (1.3336, -0.36879) (1.5885, 0)	(0.0306, 0) (0.0275, -0.0.0004)
4	GPOF		(2.6935, -0.000055) (1.3520, -0.32484) (1.5938, -0.08322)	(0.0306, 0.00001) (0.0213, 0.0125) (-0.00121, 0.00143)
2	Theory	bound leaky TM_0	(2.4453, 0) (1.5441, -0.28270) (1.5830, 0)	(0.03645, 0) (0.0287, 0.00871)

ACKNOWLEDGEMENTS

Prof. Mesa wishes to acknowledge the financial support provided by NATO and by CICYT, Spain (project number TIC95-0447) during the course of this research.

REFERENCES

[1] H. Shigesawa, M. Tsuji, and A. A. Oliner, "Dominant mode power leakage from printed-circuit waveguides," *Radio Science*, Vol. 26, no. 2, pp. 559-564, March-April 1991.

[2] D. Nghiem, J. T. Williams, D. R. Jackson, and A. A. Oliner, "Proper and improper dominant mode solutions for stripline with an air gap," *Radio Science*, Vol. 28, no. 6, pp. 1163-1180, Nov.-Dec. 1993.

[3] F. Mesa and R. Marques, "Low-frequency leaky regime in covered multilayer striplines," *IEEE Trans. Microwave Theory Tech.*, Vol. 43, pp. 2831-2838, Dec. 1995.

[4] H. Shigesawa, M. Tsuji, and A. A. Oliner, "Conductor-backed slotline and coplanar waveguide: dangers and full-wave analysis," *IEEE Intl. Microwave Symp. Digest*, pp. 199-202, New York, NY, May 1988.

[5] M. Tsuji, H. Shigesawa, and A. A. Oliner, "Simultaneous propagation of both bound and leaky dominant modes on conductor-backed coplanar strips," *IEEE Intl. Microwave Symp. Digest*, pp. 1295-1298, Atlanta, GA, June 1993.

[6] M. Tsuji, H. Shigesawa, and A. A. Oliner, "Printed circuit waveguides with anisotropic substrates: a new leakage effect," *IEEE Intl. Microwave Symp. Digest*, pp. 783-786, Long Beach, CA, June 1989.

[7] D. Nghiem, J. T. Williams, D. R. Jackson, and A. A. Oliner, "Leakage of the dominant mode on microstrip with an isotropic substrate: theory and measurements," *IEEE Trans. Microwave Theory Tech.*, Oct. 1996 (to appear).

[8] H. Ostner, J. Detlefsen, and D. R. Jackson, "Radiation from one-dimensional dielectric leaky-wave antennas," *IEEE Trans. Antennas and Propagation*, Vol. 41, pp. 344-348, April 1995.

[9] N. K. Das and D. M. Pozar, "Full-wave spectral-domain computation of material, radiation, and guided wave losses in infinite multilayered printed transmission lines," *IEEE Trans. Microwave Theory Tech.*, Vol. 39, no. 1, pp. 54-63, Jan. 1991.

[10] T. P. Sarkar and O. Pereira, "Using the matrix pencil method to estimate the parameters of a sum of complex exponents," *IEEE Antennas and Propagation Magazine*, Vol. 37, no. 1, pp. 48-55, Feb. 1995.

[11] D. P. Nyquist and D. J. Infante, "Discrete higher-order leaky-wave modes and the continuous spectrum of stripline," *IEICE Trans.*, Vol. E78-C, no. 10, pp. 1331-1338, Oct. 1995.

[12] H. Shigesawa, M. Tsuji, and A. A. Oliner, "The nature of the spectral gap between bound and leaky solutions when dielectric loss is present in printed-circuit lines," *Radio Science*, Vol. 28, no. 6, pp. 1235-1243, Nov.-Dec. 1993.

Complex Characteristic Impedance of a Leaky Conductor-Backed Slotline: Alternate Analysis Methods

Nirod K. Das

Weber Research Institute
Department of Electrical Engineering
Polytechnic University, Route 110
Farmingdale, NY 11735, USA

ABSTRACT:

We present in this paper alternate definitions of "equivalent complex characteristic impedance" for a conductor-backed slotline, that can be useful for circuit modeling purposes. Standard methods for computing the characteristic impedance of non-leaky transmission lines would not apply when leakage exists. Two definitions are discussed, that are compared with each other, as well as validated by comparing with rigorous 3D simulations of a practical circuit geometry. The basic definitions can also be extended to other general leaky printed lines.

1 Introduction

Though the leakage mechanism in certain types of printed transmission lines have been recognized for sometime [1] - [4], only the propagation behavior of such leaky lines have been investigated. The impedance characteristics of leaky lines have not been studied. It is possible to design novel practical circuits such as couplers, antenna feeds, signal transitions, etc., that make novel use of the power leakage concept in printed lines. For such circuit applications, it is important to define a characteristic impedance, Z_c, for an ideal infinite-length leaky line, that can be useful for simple circuit modeling of practical finite-length sections. Unfortunately, the standard definitions of characteristic impedance commonly used for non-leaky transmission lines would not apply when leakage exists. For a leaky conductor-backed slotline (CBS), for example, defining an "equivalent characteristic impedance" is made complicated due to its non-conventional growing fields. Due to indefinite growing nature of the transverse fields of a CBS, which increase to infinity at large distances, the total cross-sectional power for a given slot voltage would become infinitely large. Therefore, a standard power-voltage definition [Z_c = (voltage)^2/(cross-sectional power),] of the characteristic impedance, commonly used for regular slotlines, can not be used here. This will result in a trivial

Directions for the Next Generation of MMIC Devices and Systems
Edited by N. K. Das and H. L. Bertoni, Plenum Press, New York, 1997

zero value of the characteristic impedance, which is not practically meaningful. The voltage-current definition (Z_c = voltage between the two conductors / the total current) will also fail to work, because the total current on the groundplane of a CBS can not be properly defined.

2 Theory

In this paper we will present two definitions for the characteristic impedance of a conductor-backed slotline (CBS.) The geometry of a CBS, which is known to be leaky at all frequencies [1,2], is shown in Fig.1. A simple quasi-static distributed model for a general leaky transmission line is shown in Fig.2, that includes material as well as leakage loss. For a CBS, particularly, the leakage loss is a distributed radiation process that can be modeled as distributed shunt conductance, G'. Similarly, for a leaky strip-type transmission line the leakage loss can be incorporated via distributed series resistance, R'. The G' or R' can be viewed as equivalent to radiating antenna elements connected in parallel or series with the transmission line. The total fields of the transmission line of Fig.2 consists of two parts: (1) the radiation fields produced due to the distributed radiating elements, that are exponentially growing in transverse directions, and (2) the quasistatic transmission line fields.

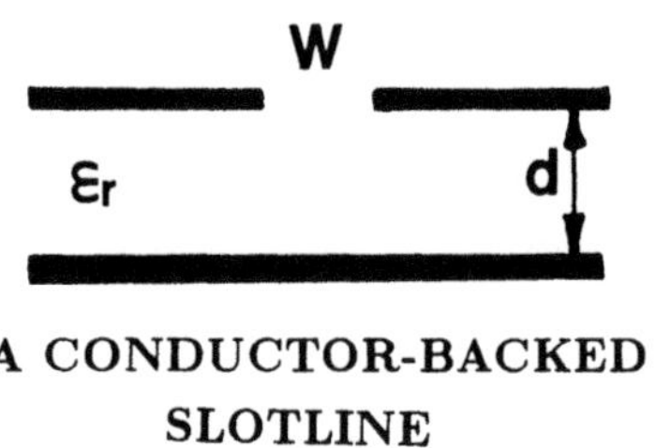

Fig.1: Geometry of a conductor-backed slotline (always leaky) that is analyzed for the characteristic impedance study. The propagating direction is assumed along $\hat{x}$, the transverse direction is $\hat{y}$, and the normal direction is $\hat{z}$ with zero reference on the plane of the conductor backing.

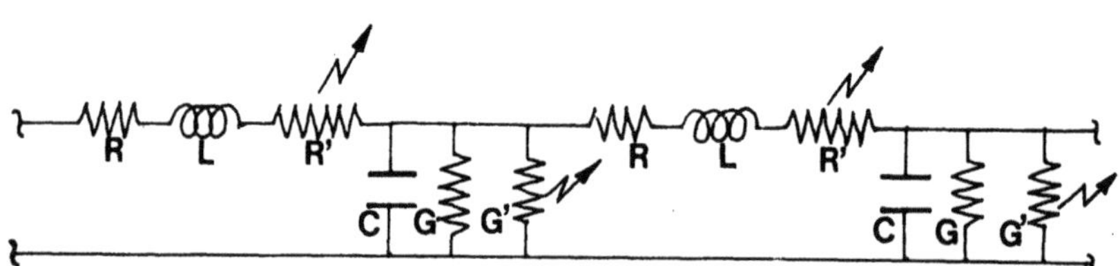

Fig.2: A generalized equivalent circuit of a transmission line with distributed radiation. R and G constitute material loss (metal and dielectric loss,) which are neglected for the present study. R' and G' are respectively the series radiation resistance and the shunt radiation conductance per unit length. R' is assumed zero for a conductor-backed slotline (the leakage is modeled as shunt type radiation.) L and C are quasi-TEM parameters.

The quasistatic transmission line fields are bound to the vicinity of the transmission line, and dictate the characteristic impedance of the line. Whereas, the exponentially growing radiation fields from the distributed sources are a part of the loss process that do not directly contribute to the power transmission along the transmission line (the effect of radiation is indirectly accounted for through the distributed radiation resistance R' or conductance G'.) We analytically extract out the exponentially growing leakage fields from the total fields of a CBS. This is implemented using a residue theory on the complex spectral plane [2]. The rest of the fields are called the "bound-mode fields." The transverse power associated with these bound-mode fields is then calculated by integrating over the transverse plane. We now define a characteristic impedance using the bound-mode power, P_b:

$$Z_c = \frac{P_b^*}{|V|^2}. \tag{1}$$

This characteristic impedance, referred to as the "bound-mode characteristic impedance," was proposed by us in [5], and is expected to be valid for circuit modeling purposes when leakage exists. It may be noted, the characteristic impedance obtained above is a complex number, unlike the purely real number for a non-leaky, lossless transmission line.

We propose here a second method based on a simple transmission line theory of Fig.2. First, we find the radiation conductance, G', of Fig.2 by analytically calculating the power radiated away by the leakage fields. The following procedure is used:

The spectral domain Green's functions due to an $\hat{x}$ directed equivalent magnetic current source (or, $\hat{y}$ directed slot field), can be used in order to express the $\hat{x}$ component of the magnetic field, H_x, of the CBS of Fig.1 [6,7,8]. We are only interested in the field at the $z = d_-$ plane (slightly below the plane of the slot).

$$H_x(x,y,z=d_-) = \frac{e^{-jk_e x}}{2\pi} \int_C \frac{j\varphi_p^2}{\varphi_1 \omega \mu_0} \cot(\varphi_1 d) F(k_y;W) e^{jk_y y} dk_y, \tag{2}$$

$$F(k_y;W) = \frac{\sin(k_y W/2)}{k_y W/2}, \tag{3}$$

$$\varphi_p = \sqrt{k_0^2 \epsilon_r - k_e^2}; \;\; k_e = \beta - j\alpha; \;\; \mathrm{Im}(\varphi_p) > 0, \tag{4}$$

$$\varphi_1 = \sqrt{\varphi_p^2 - k_y^2}. \tag{5}$$

Here $F(k_y;W)$ is the Fourier transform of the transverse variation, $f(y)$, of slot field of the CBS, which is assumed uniform over the width W of the slot. k_e is the complex propagation constant of the leaky CBS obtained separately using [2], and ω is the frequency of operation in rad/s. The contour of integration C in (2) is properly deformed on the complex k_y plane respectively below and above the poles of the integrand at $-\varphi_p$ and $+\varphi_p$ [2].

Now, the singular part, H_{sx}, in (2) produces the exponentially growing leakage fields that carries away power from the central transmission line. Contribution due this singular part can be analytically extracted from (2) using residue theory.

$$H_{sx}(x,y>0,z=d_-) = -e^{-jk_e x} F(-\varphi_p;W) \frac{\varphi_p}{2d\omega\mu_0} e^{-j\varphi_p y}; \tag{6}$$

$$H_{sx}(x,y>0,z=d_-) = -e^{-jk_e x} F(\varphi_p;W) \frac{\varphi_p}{2d\omega\mu_0} e^{j\varphi_p y}. \tag{7}$$

The circuit parameter G' can be expressed using reaction per unit length of the above H_{sx} with the equivalent magnetic current (or slot field) distribution of the slotline.

$$G' = \text{Real}\left(\int_{-W/2}^{W/2} H_{sx}(x, y, z = d_-) f(y) e^{jk_e x} dy\right) = \text{Real}\left(\frac{\varphi_p}{2d\omega\mu_0}\left(\frac{\sin(\varphi_p W/2)}{\varphi_p W/2}\right)^2\right), \tag{8}$$

From the known values of G' and k_e, we can find the values of the circuit parameters L, C, and Z_c of Fig.1 using transmission line principle.

$$Z_c = \frac{|\text{Im}(k_e^2))|}{k_e G'}. \tag{9}$$

We refer to this method as the circuit extraction method. Though the method is based on a quasistatic model of the transmission line, it should provide a first order approximation for comparison with the more rigorous definition obtained earlier in (1). Also, in order for the method to be implemented, it is assumed that the transverse variation of the slot fields are known, which in (3) is assumed to be uniform for simplicity.

In addition to the above two methods, we have also performed 3D electromagnetic simulations of a realistic circuit geometry in order to validate the characteristic impedance definitions proposed. A finite-length section of a short-circuited CBS stub of length $L = 20\lambda_0$, excited at the center by a shunt current source (see Fig.3 inset,), is used for the simulation. The numerical simulation of the above circuit geometry is separately performed using a full-wave method of moments. The characteristic impedance of the line is derived from the input impedance, Z_{in}, seen by the excitation source:

$$Z_c = \frac{2Z_{in}}{j\tan(k_e L/2)} \tag{10}$$

When the length of the stub, L, or the attenuation constant, α, of the transmission line is sufficiently large (in most cases we have simulaed,) the stub approximates two infinite-length transmission lines connected in parallel across the input current source (see Fig.3 inset for the stub geometry). Under this consitoin, $Z_c \simeq 2Z_{in}$, which should also be evident from (10) because now $j\tan(k_e L/2) \simeq 1.0$.

Because (10) models a real circuit component, it should provide a reliable measure of the characteristic impedance of the leaky line. However, due to the significant complexity of a 3-D simulation, it is always desirable to use a simpler definition based on a 2-D analysis such as the two definitions discussed earlier. It may be noted, that the definition of (10) do not account for the discontinuity effects at the short-circuit ends, or at the feeding location of the stub, or interactions between the two discontinuity effects. Because these effects are usually small, the definition of (10) should at least provide a reliable first-order estimate for comparison with the other definitions.

3 Results

Fig.3 and Fig.4 show comparison of characteristic impedance values obtained using (1,9,10), for different substrate thicknesses, d, and substrate dielectric constant, ϵ_r, respectively. More 3D simulations have been done with different stub lengths and source positioning, showing consistent results. It may be noted here, that the imaginary parts of the characteristic impedance, which have much smaller magnitudes compared to the

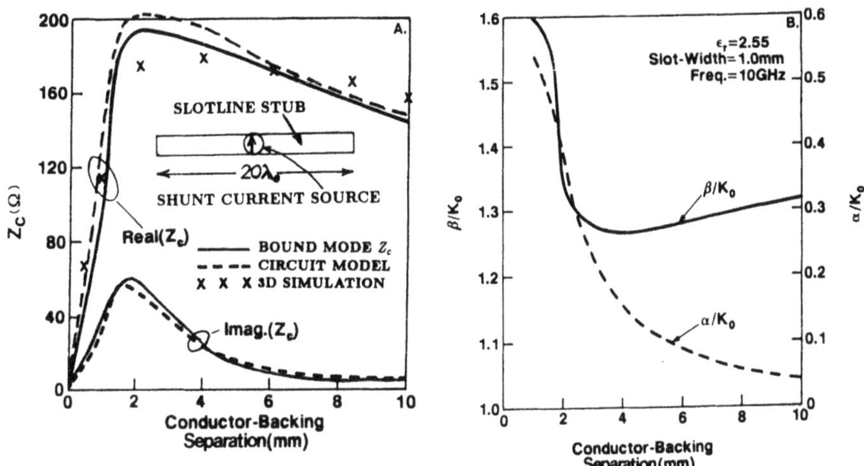

Fig.3: (a) Comparison of the equivalent characteristic impedance for different values of conductor-backing separation, d, obtained using (i) the "bound-mode Z_c" definition, (ii) the circuit model, and (iii) the 3D simulation. The 3D full-wave electromagnetic simulation was performed first to find the input impedance seen by a delta-gap shunt current source at the center of a $20\lambda_0$ long slotline (see the inset,) from which the characteristic impedance was calculated. Slot width = 1mm, ϵ_r = 2.55, frequency = 10GHz. (b) The corresponding values of β and α used for the circuit model derivation.

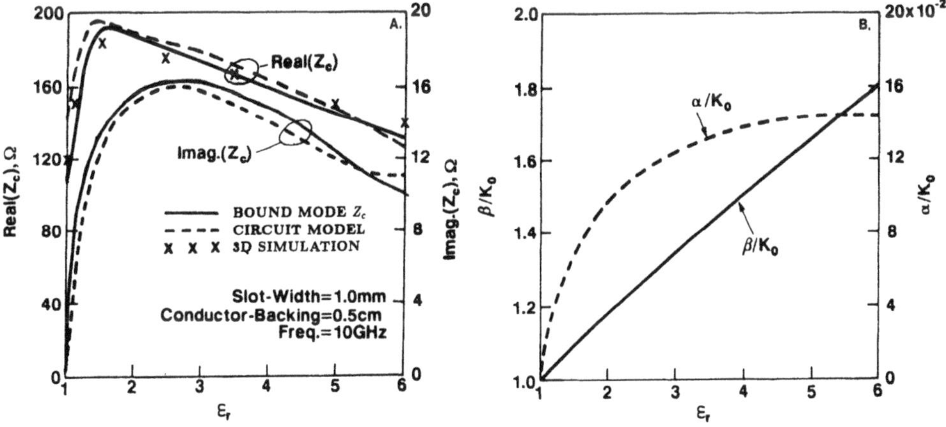

Fig.4: Results similar to Fig.3, but for different values of ϵ_r, with d = 0.5cm, frequency = 10GHz, slot width = 0.1cm.

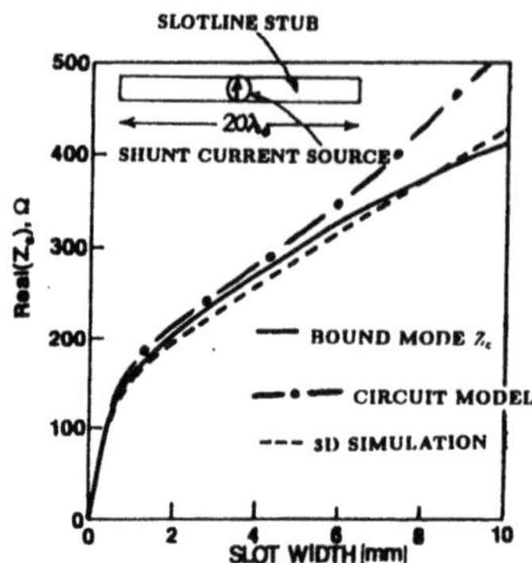

Fig.5: Results similar to Fig.3a (only real part of Z_c are shown,) but for different values of slot width, W, with $\epsilon_r = 2.55$, $d = 0.8$cm, frequency = 10GHz. The corresponding values of α and β can be obtained from [2], having the same set of physical parameters. Poorer results from the circuit definition for larger values of W can be explained due to inaccurate modeling of the slot fields in (5) using a simple uniform distribution.

real parts, could not be derived properly from the 3D simulations. This is because, the additional reactive impedance usually seen at the input discontinuity masks the much smaller reactive values contributed by the transmission line characteristic impedance. The independent results of Fig.3 and Fig.4 are seen to agree well, which should validate the proposed definitions. In Fig.3, the characteristic impedance is zero for $d = 0$, which is due to the short circuiting effect of the bottom ground plane, and increases with d. For large values of d, the parallel plate mode excitation is reduced, which explains the decrease of the imaginary part of Z_c and the leakage constant α with d. In Fig.4, when $\epsilon_r = 1.0$, the geometry turns into a TEM structure with no leakage ($\alpha = 0$), and hence a purely real value of Z_c.

Fig.5 shows data similar to Figs.3,4, but for different values of slot width, W (only the real part of characteristic impedance has been shown.) The general trend of the characteristic impedance in Fig.5 obtained using the circuit definition is similar to that from the "bound-mode characteristic impedance" definition and the 3-D simulation. The actual values compare well for small slot widths, which further validates our proposed models. However, the comparison of actual values in Fig.5 is poorer for larger W. This could be possible due to poor modeling of the slot electric field for large slot widths. In (5) the slot field has been assumed to be uniform over the width of the slot. This is a valid approximation only for small W, and the approximation can be poor for larger values of W.

4 Conclusion

As we conclude from our investigation, the new definitions of characteristic impedance should be useful for circuit designs of transmission lines that are predominantly leaky. The basic theory can be similarly extended to other slot- and strip-type transmission lines. The bound-mode impedance definition is inherently more rigorous than the circuit extraction method. However, the circuit extraction method is analytically much simpler than the bound-mode power method, and can be useful at least as a good first order approximation in most practical situations.

Acknowledgement

This work was funded by the US Army Research Office, Research Triangle Park, NC, under grant DAAH04-95-1-0254.

References

[1] H. Shigesawa, M. Tsuji, and A. A. Oliner. Conductor Backed Slotline and Coplanar Waveguide: Dangers and Fullwave Analyses. *IEEE Microwave Theory and Techniques Symposium Digest*, 199–202, 1988.

[2] N. K. Das and D. M. Pozar. Full-Wave Spectral-Domain Computation of Material, Radiation and Guided Wave Losses in Infinite Multilayered Printed Transmission Lines. *IEEE Transactions on Microwave Theory and Techniques*, MTT-39(1):54–63, January 1991.

[3] L. Carin and N. K. Das. Leaky Waves in Broadside-Coupled Microstrips. *IEEE Transactions on Microwave Theory and Techniques*, MTT-40(1):58–66, January 1992.

[4] M. Tsuji, H. Shigesawa, and A. Oliner. New Interesting Leakage Behavior on Coplanar Waveguides of Finite and Infinite Widths. *IEEE Transactions on Microwave Theory and Techniques*, MTT-39(12):2130–2137, December 1991.

[5] N. K. Das. Power Leakage, Characteristic Impedance and Leakage Transition Behavior of Finite-Length Stub Sections of Leaky Printed Transmission Lines. *IEEE Transactions on Microwave Theory and Techniques*, MTT-44(4):526–536, April 1996.

[6] N. K. Das and D. M. Pozar. Multiport Scattering Analysis of Multilayered Printed Antennas Fed by Multiple Feed Ports, Part I: Theory; Part II: Applications. *IEEE Transactions on Antennas and Propagation*, AP-40(5):469–491, May 1992.

[7] N. K. Das. *A Study of Multilayered Printed Antenna Structures.* PhD thesis, Department of Electrical and Computer Engineering, University of Massachusetts, Amherst, September 1987.

[8] N. K. Das and D. M. Pozar. A Generalized CAD Model for Printed Antennas and Arrays with Arbitrary Multilayer Geometries. In L. Safai, editor, *Computer Physics Communication, Thematic Issue on Computational Electromagnetics*, North-Holland Publications, 1991.

SUPPRESSION OF LEAKAGE ON PRINTED TRANSMISSION LINES USING FINITE-WIDTH DIELECTRIC LOADING

George W. Hanson and Alexander B. Yakovlev

Department of Electrical Engineering and Computer Science
University of Wisconsin-Milwaukee
Milwaukee, Wisconsin 53211

INTRODUCTION

The suppression and avoidance of undesirable leakage effects has been recently investigated for stripline and microstrip structures [1] as well as for some other conductor-backed printed transmission lines [2],[3]. It has been found that through the use of suitable dielectric loading, leakage in many printed structures can be suppressed over all frequencies, or over some frequency range. The general idea is either to decrease the velocity of the transmission line modes, increase the velocity of the background modes into which energy is leaked, or a combination of both effects. Other techniques to suppress and/or avoid leakage from conductor-backed transmission line configurations have been presented in [2], including the use of shorting pins and incorporation of a transversely inhomogeneous substrate layer.

In this paper the use of a finite-width dielectric ridge superstrate is investigated for suppression of leakage in microstrip transmission line structures. Positioning a superstrate ridge in an area of electromagnetic field concentration leads to a decrease in the velocities of the transmission line modes, while having no effect on the background surface wave modes. Leakage is suppressed as in previously proposed geometries which use an infinite loading superstrate layer, while confining the superstrate to the immediate vicinity of the conducting strips. In this way, other lines or devices do not need to be covered with the superstrate, which may be beneficial when integrating several circuit functions. Other benefits also exist, such as eliminating the possibility that the superstrate itself will contribute to surface wave propagation and leakage, which, for example, has been shown to result in the resumption of leakage at high frequencies for conductor-backed coplanar waveguide.

Directions for the Next Generation of MMIC Devices and Systems
Edited by N. K. Das and H. L. Bertoni, Plenum Press, New York, 1997

MATHEMATICAL FORMULATION

The structures to be analyzed are shown in Fig. 1. If the finite-width dielectric ridge region is removed, Fig.'s 1(a) and 1(b) represent a single microstrip transmission line and two coupled microstrip lines, respectively. The mathematical formulation of the integral equation model utilized in this work is briefly described in the following. Mathematical details may be found in [4],[5], where the incorporation of electromagnetically complex media is also considered.

The fundamental relationship upon which to build the integral equation model is between an electric-type line source immersed in an inhomogeneous background environment, and the field it maintains. The background environment consists of the infinite dielectric layering with the material ridge and conducting strips removed. This leads to the electric-type Green's function, and to the integral relationship between a general source and the resulting field.

The finite-width material ridge is removed from the problem by replacing it with equivalent volume polarization currents. Equivalent volume currents, rather than surface ones, are used here for two reasons. First, the volume equivalence principle is very simple, and leads to the most straightforward formulation, regardless of the complexity of the material regions to which it is being applied. The volume formulation can be applied to inhomogeneous anisotropic regions without difficulty, whereas surface formulations generally are applicable to a smaller class of problems (although surface formulations can be applied quite easily to the homogeneous isotropic regions of primary interest here). Second, since the typical geometry used in an integrated electronics application consists of relatively wide, thin strips of material, the partitioning of the volume in the direction tangential to the background planar layering is much larger than that normal to the background layering. As such, the number of "cells" in the volume approach may not be much greater than the number of "elements" in a surface approach. The volume formulation is also amenable to judiciously chosen entire-domain basis functions [6], which may in some cases lead to increased computational efficiency.

The relevant set of coupled integral equations is formed by enforcing appropriate boundary and field conditions. Specifically, the boundary condition of vanishing total tangential electric field at the location of the conducting strips, and the field condition that the total electric field in the ridge region equals the field maintained by all currents in the system plus the incident field, leads to a coupled system of integral equations. Upon expansion of the unknown strip currents and ridge fields in a suitable set of known functions, the coupled set of IE's is converted to a matrix system

$$[M(k_z,f)][J] = [e^{inc}] \tag{1}$$

where J represents the unknown currents (conduction and polarization), e^{inc} is the incident field, and M is the matrix of all self and mutual interactions among the various fields and currents.

The desired matrix characteristic equation for the characteristic values of propagation constant k_z can be arrived at from (1) by considering several different views. One idea is that the characteristic values of k_z are such that they are associated with fields which can exist in the absence of an excitation, i.e., $\vec{e}^{\,inc} = 0$, leading to $|M| = 0$. Another method arises from the fact that these discrete modes occur at pole singularities in the complex k_z-plane. With the excitation assumed to be bounded at these values of k_z, the

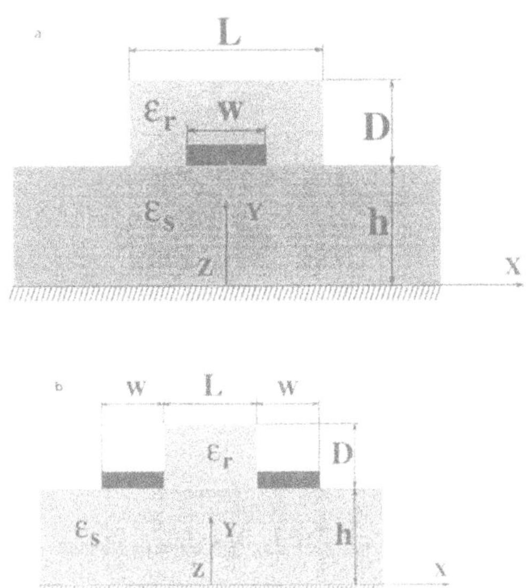

Fig. 1. Geometry of transmission line structures to be considered, single (a) and coupled (b) microstrip lines.

characteristic equation $|M|=0$ is again obtained. Furthermore, one can consider the general eigenvalue problem associated with (1), $[M][J]=\lambda[J]$, and determine the characteristic values of k_z which are associated with the eigenvalue(s) $\lambda=0$.

The desired problem is then reduced to the study of the properties of the operator-function $H(k_z,f,\epsilon_r,L,D)=|M(k_z,f,\epsilon_r,L,D)|$. The explicit dependance of the system matrix on frequency and on the electrical and geometric parameters of the dielectric ridge is retained here, since these will be important in the suppression problem to be considered. The system determinant is also dependent on many other parameters which characterize the background environment and the conducting strips, but these will be assumed fixed for some structure of interest, with the ridge added to dynamically suppress leakage from the transmission line over an appropriate frequency range.

For the transmission lines considered in Fig. 1 (without the dielectric ridge), and for many other types of transmission line structures, it has been found that the onset of mode leakage occurs in a region called the "spectral gap." The spectral gap region encompasses the intersection of three curves. In a small region surrounding the intersection point, two of the curves resemble branches of a parabola, and the third curve resembles a straight line. Each curve of the parabola corresponds to a real mode (improper or proper), and the straight line corresponds to an improper complex mode. It has been observed [7] that the intersection point within the spectral gap region is a fold or turning point from bifurcation theory [8], and obeys the set of equations

$$H(u,v)|_{(u_f,v_f)}=H'_u(u,v)|_{(u_f,v_f)}=0$$
$$H''_{uu}(u,v)H'_v(u,v)|_{(u_f,v_f)}\neq 0 \tag{2}$$

for a given (u,v) plane, with (u_f,v_f) denoting the fold point. The normal form associated with this bifurcation analytically predicts the characteristic spectral gap behavior. While

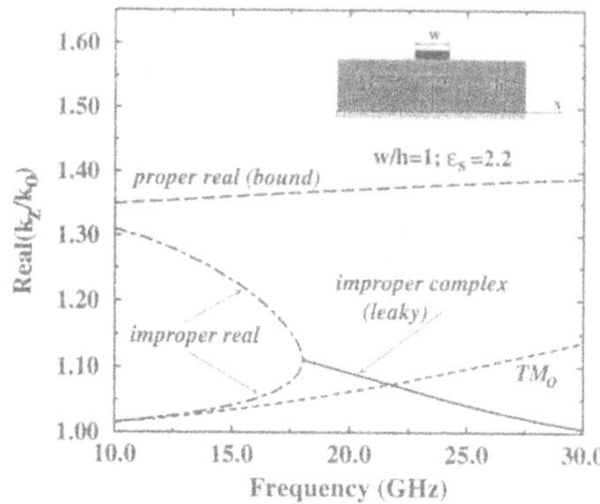

Fig. 2. Dispersion characteristics of single microstrip transmission line, h=0.15 cm.

(2) was used in [7] to provide an alternative explanation of the leakage phenomena from the standpoint of bifurcation theory, it may also be useful as an analytic or numerical tool to study methods of leakage suppression. That is, conditions which prohibit the satisfaction of (2) within some range of electrical and geometric parameters could be used to guide the development of leakage suppression techniques, although this subject is not further developed here.

In the numerical results to follow, a finite-width dielectric ridge as depicted in Fig. 1 will be used to modify the appropriate modal behavior in such a way as to suppress leakage from occurring over some useful parameter range. Viewed another way, the addition of a dielectric ridge can be used to change the position of the fold point (2), or even to suppress the occurrence of a fold point altogether, for some specific mode.

NUMERICAL RESULTS

Suppression of mode leakage on the two transmission line structures shown in Fig. 1 will be demonstrated. Leakage suppression will be discussed first for the single microstrip transmission line structure depicted in Fig. 1(a), followed by the coupled lines shown in Fig. 1(b). Although leakage on the two structures is substantially different, the same technique can be applied to suppress mode leakage on both structures. The general idea is to add a dielectric superstrate, which affects (slows) the appropriate transmission line mode, thereby pushing the location of the fold point to higher frequencies. This technique was utilized in [1] for the single microstrip case using an infinite superstrate, and has also been examined for leakage suppression in other transmission line structures [3].

The use of a finite-width dielectric superstrate has some advantages compared to the infinite superstrate technique. First, the ridge superstrate does not affect the background surface-wave mode. While the infinite superstrate primarily slows the transmission line mode, it also slows the background mode. Both curves are pushed up in the k_z-f plane, and so leakage suppression can be achieved, but perhaps not as readily as for the case when only the transmission line mode is affected. Also, in the case of the coupled transmission lines with the ridge placed between the lines, the even mode is almost unaffected, while the odd mode is greatly affected. Since the odd mode leads to leakage,

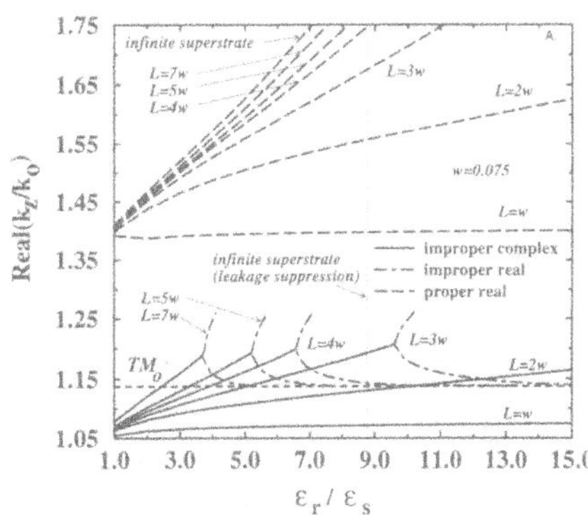
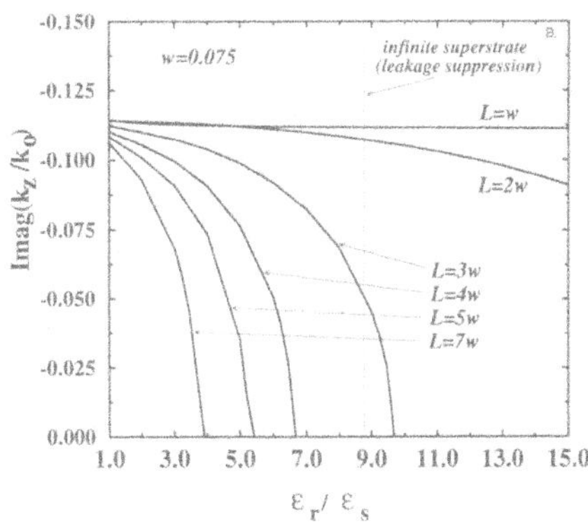

Fig. 3. **Propagation characteristics of single microstrip line with ridge superstrate versus ridge/substrate permittivity ratio. h=0.15 cm, D=0.1h, f=30 GHz, ϵ_s=2.2.**

precise control of the odd mode is desirable. This is achieved using the ridge positioned as shown in Fig. 1(b). Furthermore, the ridge only affects the transmission line under consideration, and not other lines or devices which may be integrated on the same substrate. One obvious drawback is the greater fabrication difficultly in producing a non-planar layer.

Consider first the dispersion curves for the single microstrip transmission line structure as shown in Fig. 2. The proper mode is always bound, but it has recently been discovered that an improper leaky dominant mode also exists [1]. Since the leaky mode is dominant-like, it may be strongly excited by a traditional microstrip feed. Suppression of this mode was also examined in [1], using an infinite superstrate. Suppression using a dielectric ridge is shown in Fig. 3, for different values of ridge width L. It can be seen that for a wider ridge, leakage suppression occurs at a smaller value of ridge permittivity. In fact, for the leaky mode, the curves do not asymptotically approach the infinite superstrate case as ridge width becomes larger. This is because the leaky mode always "sees" the termination of the superstrate. The proper bound mode curves approach the infinite superstrate case as $L\to\infty$, since this mode is only sensitive to the dielectric media immediately surrounding the line. For the leaky mode though, the behavior is obviously quite different. While leakage is suppressed more quickly as ridge permittivity is increased compared to the infinite superstrate case, the entire ridge/conducting strip composite geometry must be considered as the waveguiding structure. When leakage is suppressed, fields outside of the ridge/strip structure should be small and decay in transverse directions, but away from the strip, yet still under the ridge, the fields may be large.

For the coupled strips depicted in Fig. 1(b), a dispersion curve for the case without a superstrate is shown in Fig. 4(a) [7],[9]. The even mode is always bound, whereas the proper odd mode is converted into an improper complex (leaky) mode at the fold point, which occurs at the intersection of the odd mode dispersion curve with that of the TM_0 surface-wave mode. This intersection occurs at approximately 8 GHz. In Fig. 4(b), an infinite superstrate having $\epsilon_r/\epsilon_s=1$ is shown to move the fold point to 14.75 GHz, suppressing leakage for f<14.75 GHz by slowing the odd mode. Figures 5(a) and 5(b) show a similar situation, although in this case a finite width ridge is placed between the

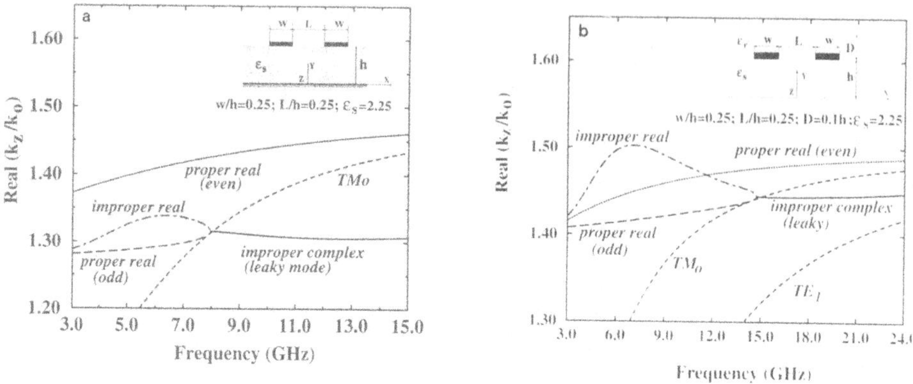

Fig. 4. Dispersion curves for coupled microstrip lines, (a) Uncovered and (b) infinite superstrate, h=1 cm.

strips. The odd mode is greatly affected by the finite ridge, while the even mode is almost unaffected. This is due to the concentration of the odd mode fields between the strips, and of the even mode fields under the strips. For the ratio $\epsilon_r/\epsilon_s=1$, the fold point is pushed to approximately 9.8 GHz. Fig. 5(b) shows the results for a geometry similar to the one analyzed in Fig. 5(a), but for the case $\epsilon_r/\epsilon_s=3$. It can be seen that leakage is totally suppressed, since the odd mode dispersion curve is well above the TM_0 surface wave mode for all frequencies. It can be seen that good control can be achieved over the odd mode characteristics with the ridge between the strips. While the application is for leakage suppression, this can also be used to equalize even and odd mode phase velocity, etc.. In Fig. 6, leakage suppression versus ridge/substrate permittivity ratio at f=10 GHz is shown for different values of ridge height D. As the ridge height increases, the fold point occurs at lower values of permittivity ratio.

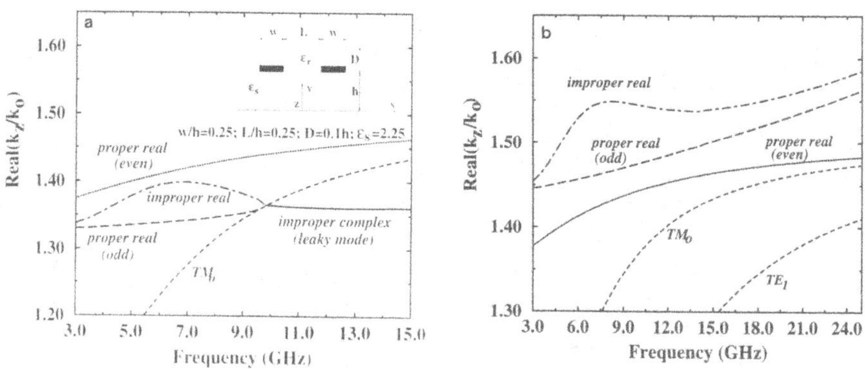

Fig. 5. Dispersion curves for coupled microstrip lines with finite ridge between lines for (a) $\epsilon_r/\epsilon_s=1$ and (b) $\epsilon_r/\epsilon_s=3$, h=1 cm.

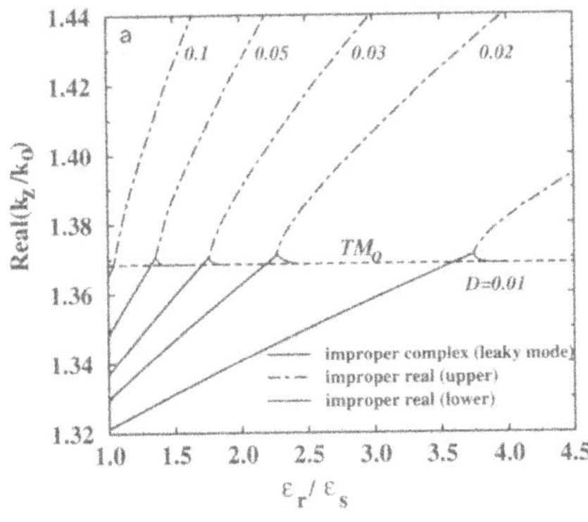

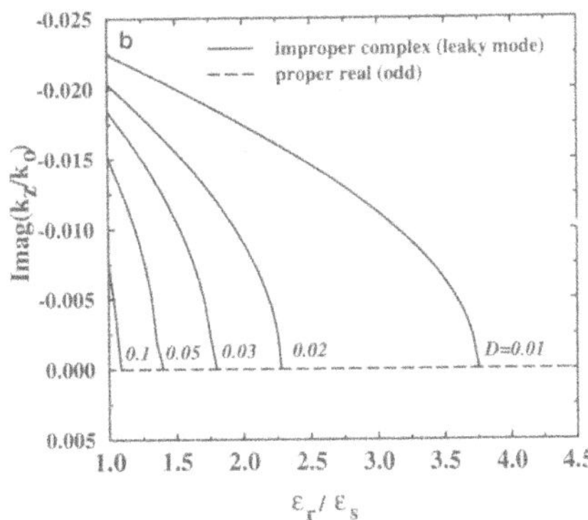

Fig. 6. Dispersion curves for coupled microstrip lines with a finite ridge between the lines versus ϵ_r/ϵ_s, for various values of ridge thickness D. Geometrical parameters are the same as in Fig. 5, except for ϵ_r, which varies as shown.

CONCLUSION

Propagation characteristics of single and coupled microstrip transmission lines having finite-width dielectric superstrates have been examined. The considered microstrip configurations exhibit dominant or dominant-like leaky modes in the absence of a superstrate. It was found that the addition of an appropriately placed dielectric ridge, forming a finite-width superstrate, can lead to leakage suppression or elimination. Although an infinite superstrate can be used in a similar manner for leakage suppression, the finite-width superstrate may have some advantages. One considerable advantage is that a ridge superstrate does not contribute to enhanced surface-wave propagation, which occurs for the infinite superstrate case, especially at higher frequencies and when the permittivity of the superstrate is large. The ridge superstrate is confined to the immediate vicinity of the conducting strips, and does not significantly interact with other devices/circuits.

REFERENCES

1. D. Nghiem, J.T. Williams, D.R. Jackson, A.A. Oliner, "Suppression of leakage on stripline and microstrip structures," 1994 IEEE/MTT-S Int. Microwave Symp. Dig., San Diego, CA, 1994, pp. 145-148.
2. N.K. Das, "Methods of Suppression or Avoidance of Parallel-Plate Power Leakage from Conductor-Backed Transmission Lines," IEEE Trans. Microwave Theory
3. Y. Liu, K. Cha, T. Itoh, Non-Leaky Coplanar (NLC) Waveguides with Coplanar Backing," *IEEE Trans. Microwave Theory Tech.*, vol. 43, pp. 1067-1072, May. 1995.
4. G.W. Hanson, "Propagation Characteristics of a Microstrip Transmission Line on an Anisotropic Material Ridge," *IEEE Trans. Microwave Theory Tech.*, vol. 43, pp. 2608-2613, Nov. 1995.
5. G.W. Hanson, "Complex Media Microstrip Ridge Structures: Formulation and Basic Characteristics of Ferrite Structures," *IEEE Trans. Microwave Theory Tech.*, vol. 44, Sept. 1996.

6. G. Athanasoulias and N.K. Uzunoglu, "An Accurate and Efficient Entire-Domain Basis Galerkin's Method for the Integral Equation Analysis of Integrated Rectangular Dielectric Waveguide, " *IEEE Trans. Microwave Theory Tech.*, vol. 43, pp. 2794-2804, Dec. 1995.
7. A.B. Yakovlev and G.W. Hanson, "On the Nature of Critical Points in Leakage Regimes of a Conductor-Backed Coplanar Strip Line," to appear, *IEEE Trans. Microwave Theory Tech.*
8. R. Seydel, *Practical Bifurcation and Stability Analysis*, 2nd Ed., Springer-Verlag, New York, 1994.
9. H. Shigesawa, M. Tsuji, and A.A. Oliner, "Simultaneous Propagation of Bound and Leaky Dominant Modes on Printed Circuit Lines: A New General Effect," *IEEE Trans. Microwave Theory Tech.*, vol. 43, pp. 3007-3019, Dec. 1995.

ELECTROMAGNETIC SIMULATION OF HIGH-SPEED AND RF MULTI-CHIP MODULES -- AN OVERVIEW

Madhu S. Gupta

Radar and Communications Systems
Hughes Aircraft Company, El Segundo, CA 90245
and
Department of Electrical Engineering
Florida State University, Tallahassee, FL 32310

INTRODUCTION

Electromagnetic (EM) simulators have already become established in the design of a variety of electromagnetic devices, such as MMICs (monolithic microwave integrated circuits), antennas, shields, electromechanical machines, and magnetic circuits. Such broad applicability results from the fact that the EM simulators are software packages for numerically solving the Maxwell's fundamental electromagnetic field equations; for the same reason, and unlike some of the other simulators, they are independent of technology, models, foundry, and other such ephemeral considerations. Although designers have long carried out electromagnetic analysis of RF components and packages where necessary, the availability of commercial EM simulators[1,2] since 1989 has made this task much more common, rapid, and easier. This development parallels the concurrent development of multi-chip module (MCM) technology for high-speed digital systems, as well as the front-ends of radio-frequency (RF) systems, such as receivers, transponders, transceivers, synthesizers, and transmit-receive modules. MCMs are a modern method of electronic packaging and integration that has the advantages of low-cost, compact size, and reliability. At low speeds/frequencies, the analysis, design, simulation, layout, and verification, of MCMs requires a number of software tools for computer-aided design (CAD), including process simulators, device simulators, circuit simulators, thermal simulators, and system simulators. For high-speed and RF MCMs, the design requirements make an additional CAD tool, namely EM simulator, indispensable. Numerous papers have appeared on numerical algorithms for EM simulation, modeling of individual planar and non-planar components with EM simulation, and case studies of its successful application. The purpose of this paper is to assess the current status of EM simulation technology as applied to the design of MCMs, and to suggest some areas in need of further development.

Directions for the Next Generation of MMIC Devices and Systems
Edited by N. K. Das and H. L. Bertoni, Plenum Press, New York, 1997

THE GENERIC PROBLEM

The class of problems of interest in MCM design, that is solved by EM simulation, is as follows. EM simulation is used to predict the performance of a high-speed or RF MCM, given its constituent materials (described in terms of their electromagnetic parameters) and its structure (i.e., spatial distribution of those materials). From the point of view of EM simulation, a typical high-frequency/high-speed MCM consists of multiple planar parallel layers of dielectrics, possibly housed in a metallic enclosure with ports (in the form of different types of transmission lines) for signal access, and with cavities for chip placement. Within the dielectric layers (on the surface and/or in the cavities) are embedded a number of chips and components that are typically rectangular, have different dielectric constants, and are metallized over parts of their surface. These chips and components are electrically interconnected by conducting ribbons placed parallel to, and at the interfaces between, the dielectric layers. Typically, there are some metallic posts (the "vias"), oriented perpendicular to the dielectric layers, to provide a conducting paths through one or more of the dielectric layers. There may also be some other non-planar conducting paths in the structure, such as metallic bond-wires and air-bridges on the surface, wrap-around straps or connections at the periphery, and possibly some connectors (bumps, elastomerics, fuzz-buttons) with associated dielectric beads or separators.

The electrical stimuli applied at the ports include RF and microwave (i.e., high-frequency analog) signals, high-speed digital signals, low-frequency control signals, DC biasing voltages, and ground. Since these excitations may originate outside the volume of space included in EM simulation, and come from different types of sources (e.g., the DC signals typically are supplied by low-impedance sources, RF signals by impedance-matched sources, and a digital signal may come from a high-impedance source), the excitations are incorporated by appropriate "effective sources" of fields at ports.

The purpose of EM simulation is to deduce a variety of signal characteristics, usually referred to the ports of the MCM, including RF scattering parameters, impedance levels and reflections, step and pulse response, other types of transient response for estimating delay and distortion, coupling coefficients and isolation, signal attenuation due to dielectric and conductor losses, field strengths, radiation level and characteristics, electrical functionality, DC potentials, and some global performance measures of the module that can be derived therefrom, like interference level and noise margin.

JUSTIFICATION FOR EM SIMULATION

Since many of these parameters can also be (and at lower frequencies, are) calculated by circuit simulation tools, a comparison of the EM simulator with circuit simulators is inevitable. A circuit simulator conceptually subdivides a module spatially into sections whose electromagnetic behavior can be represented by an equivalent circuit model, or at least described as a black-box in circuit terms (e.g., in terms of port parameters). A nodal description of the interconnection of these sections serves as a circuit representation of the module, and requires only a circuit analysis for performance prediction. Such a simulation of MCMs is, however, not sufficient for all design needs.

The reasons why a circuit simulation of a high-speed or RF MCM is inadequate can be grouped into four categories :

(a) Unavailability of a circuit model for a custom geometry. Analytically or empirically deduced circuit models are known only for some commonly occurring structures, and all circuit simulators have a limited library of models. Novel structures cannot be analyzed until a circuit model is first established for them.
(b) Overidealization of the structural features. Even when models are available, they have been idealized in order to make them tractable. The seemingly minor structural features, such as discontinuities in an otherwise uniform transmission line, or non-idealities of the ground plane, may contribute significant parasitics that manifest themselves as resonances or frequency shifts.
(c) Neglect of coupling with neighboring structures (traces, components). Even if a circuit model faithfully captures the features of a component, it assumes that the component is electromagnetically independent, (i.e., interacts with the remainder of the universe through exchange of electromagnetic energy only at its ports, typically defined where a conductor intersects an imaginary closed surface surrounding the component). Such electromagnetic isolation becomes an increasingly poor approximation as MCMs become more compact, are operated at higher speeds/frequencies, and employ materials with a wider range of dielectric constants.
(d) Disregard of radiative fields. Even if coupling effects between neighboring elements are adequately incorporated through lumped and distributed circuit models, there may be radiated fields present within the module, produced intentionally (e.g., when an antenna is integrated into the MCM by design), or unintentionally. The radiated fields, and the coupling they introduce between non-neighboring components, can be properly accounted for only by a full-wave EM treatment.

Electromagnetic simulation of an electronic package or module is a more demanding activity than a circuit simulation of the same object, both in terms of manpower (designers' time, skill, and careful attention) and in terms of computational resources (machine size and computation time). Therefore the use of EM simulation requires a justification. The technical rationales for EM simulation include higher accuracy; design of more complex circuits, macrocells, and modules; design integration (i.e., concurrent design of circuit, package, antenna, etc.); more adequate accounting of coupling and cross-talk; and a more aggressive design approach (meaning one with smaller safety margins) due to the confidence resulting from the higher reliability of simulation results. The business rationales for EM simulation are two. The first is the reduction in cost and cycle time of product design and development; it results from fewer prototype fabrications (hopefully just one !); smaller size (and therefore cost); and more robust designs which result in higher yields, or alternatively require less stringent manufacturing tolerances. The second is the prospect of gaining a competitive advantage through more aggressive designs and the introduction of new product features (such as the replacement of a protruding antenna in a product by a conformal one) for compact size, greater functionality, or user convenience.

AVAILABLE EM SIMULATION TOOLS

The available electromagnetic simulators, both custom developed and commercial[3,4,5,6], differ from each other in a variety of ways, such as the dimensionality (2, 2.5, or 3) of the region of analysis; its extent (open structure or closed box); type of gridding used for discretizing that region; quasi-static or full-wave formulation of field equations; programming languages, operating systems, and computational resources required; and whether the software stands-alone, or is a part of an integrated suite of design tools. For the present purposes, perhaps the most useful discriminator among them is the method used for the solution of Maxwell's field equations, because it determines the limitations, and the domain of applicability, of the EM simulator. Excluding the techniques (such as mode matching) that do not apply to MCMs due to their specialized boundary conditions, four principal methods are presently available for EM simulation of MCMs :

(a) Finite Element (FE) Methods[7], that are the longest established, general in their applicability, computationally intensive, and take no particular advantage of the layered structure of the MCM for reducing the computational load.

(b) Spectral Domain Methods, such as the method of moments[8] (MoM) that are also well developed and used in many commercial EM simulation packages, and are the basis of all general-purpose simulation tool that are fast enough at present to be used for an iterative design of MCMs in real time.

(c) Finite-Difference Time-Domain (FDTD) Methods[9,10], that also have a large computational load, are limited to finite (closed box) geometries, but are useful if time-domain information is needed (for example where the signals are very broadband, or nonlinear devices are embedded).

(d) Spatial Domain Methods, such as the multi-pole methods, that promise efficient computation, but need further development to be available as a general purpose tool.

The first two of the above methods account for nearly all commercially available EM simulation software at the present time, and for all known cases of EM simulation applied to MCM design. The other two methods have been applied to other domains of problems (e.g., antenna design; scattering cross-section calculation), and are being intensely investigated in the academic community.

USES OF ELECTROMAGNETIC SIMULATION

The designers of high-speed digital or RF MCMs may sometimes narrowly view EM simulation as "merely" an extension of circuit simulation, whose purpose is to increase the accuracy of circuit models and circuit simulation. More generally, the following are the principal applications of EM simulation in MCM design :

Development of Simulation-Based Component Models[11]

The circuit models of interconnect lines, distributed circuit elements, and other building blocks of MCMs, that are required in the circuit-level analysis and simulation of MCMs, have in the past been derived from analytical, quasi-static, asymptotic, or empirical approximations. For unusual components, or unconventional geometries, such standardized models may not exist at all. When they do exist, such models necessarily have a limited range of validity, and can become increasingly inaccurate, and therefore less useful, as the operating frequencies increase (so that the parasitics are not insignificant), or as the packaging becomes more compact and three-dimensional (with stacked components and interconnects, so that the possibilities for cross-talk are higher). EM simulation is a more powerful method for establishing circuit models or terminal descriptions of black-boxes, because it can be applied to realistic geometries, and is not limited to restricted ranges of values of parameters like frequency and dielectric constant.

Simulation-Based Design and Optimization[12]

A second application of EM simulation is in the design and optimization of individual components of MCMs by iterative calculation of their performance as the design variables are changed. The ability to make any desired change in the structure, and reanalyze it, is a tremendous convenience (particularly appreciated by those who have designed in an earlier era), and is helpful not only in carrying out a design to meet required goals, but also for negotiating a compromise between opposing preferences, such as for higher isolation and greater compaction.

Simulation of Layout-Dependent Interactions and Performance Verification[13]

A third application for EM simulation is in performance verification and virtual prototyping, with the objective of reducing the number of design iterations, and hence the design cost and cycle time. Compared to a single-layer circuit with similar physical dimensions, an MCM has a higher level of complexity, both in its electrical functionality and its spatial structure. As a result, there are more opportunities for interaction among different parts of the circuit that are functionally distant, or that carry signals with vastly different power levels. Many of the problems typically encountered in MCM design, such as instabilities, oscillations, undesired modulation, or inadequate isolation, can be traced back to these interactions. Although an accurate simulation of such effects by circuit simulators is possible in principle, the effort involved in constructing a custom model that includes all potential sources of coupling, is usually prohibitive. An EM simulator is the only practical method for simulating such layout-dependent coupling effects.

Field Visualization

One of the unique advantages of EM simulators is their ability to provide a visualization of electromagnetic fields. The EM simulator can present this information in a variety of ways, including vector plots, equipotential contours, graphical or color-coded field intensity plots, and surface current plots. These are helpful in the identification of hot spots, placement of other structural elements or components so as to ensure or minimize coupling, and in robustness and reliability considerations.

PRESENT LIMITATIONS OF EM SIMULATION TOOLS

Although as in any other software tool, numerous shortcomings can be pointed out in EM simulators, perhaps the most significant one from the perspective of a MCM designer, in the short-term, is the limitation on the size of simulation region. At the present stage of development, the commercially available EM simulation tools are used primarily for the analysis of small subsets of fabricated objects : for example, a section of a MMIC chip, or of a MIC substrate with a few components on it, rather than an entire MMIC chip or multi-chip module. This is partly due to the speed limitations of the EM simulators, which may take several hours to yield a single-frequency [S] parameter matrix for a structure, and partly due to the inability of the simulators to converge on a solution for a structure of realistic complexity. The ability to simulate larger, denser, more complex MCMs has a strong economic incentive, and is related to the issues of design accuracy and optimality.

The simulation of only a partial module is a disadvantage for several reasons :
(a) Partitioning Uncertainty. When a single component or section is analyzed, it is tacitly assumed that there is no electromagnetic coupling between any part of it and the remainder of the circuit, except at the defined "ports". The validity of this assumption is not known *a priori*, and there are no guidelines available to the designers for partitioning a circuit into such "isolated" islands that can be analyzed one at a time. Moreover, for cost effectiveness and ease of design, it is desirable to transfer the geometrical and structural description of the module from a layout package directly to the EM simulator. However, if the simulator is unable to handle the complexity of the structure as designed, manual intervention becomes necessary to arbitrarily subdivide and isolate a section of the structure for analysis.
(b) Design errors. Measurements on fabricated prototypes often show unexpected resonances and couplings, and frequency shifts relative to the designed response, that can cause a module to fail the specifications and be rejected. Experience shows that the RF performance of a module is influenced by such features as a module housing or a lid, which may have been ignored in piece-meal EM simulation. These resonances, and accurate values of port isolation, cannot be predicted if only a part of the electromagnetic structure is analyzed, because the assumed absence of electromagnetic coupling introduces errors in the simulated results, and unless EM simulation is applied to an entire module, it may entirely miss phenomenon such as oscillations in high gain circuits, resulting from the feedback caused by the ignored coupling.
(c) Suboptimal Designing. If the EM simulator cannot analyze an entire module, it cannot be used in conjunction with an optimizer to optimize the entire module. Optimization of a piece of the module is neither convenient (for lack of a direct criterion of goodness) nor reassuring (since a local optimum may not be optimal for the overall module).

Future generations of EM simulators will resolve these problems through greater robustness of algorithms, higher speed of simulation, and lower demands of machine capability; although seemingly universal for all CAD software, these three needs are critical in the context of MCMs.
(a) Algorithmic robustness is a problem in existing commercial EM software that arises in the form of a failure of the simulator to converge on a solution. Unfortunately, even the conditions under which convergence is certain cannot be precisely stated. One general observation, based on experience, is that the convergence is governed by the ratio of (i) the size of the smallest significant feature in the structure, to (ii) the largest dimension (typically the size of the structure) being simulated, and is worsened with decreasing values of this ratio. However, convergence is not determined by the structure alone, but also by the formulation of the problem; e.g., for a given structure it can change with the use of a different gridding.
(b) Simulation times of the order of hours to days (for calculating the performance at one frequency) discourage use of EM simulators, and tend to limit its application for a "final check" or diagnostic purposes only. Speed-up can be achieved in a variety of ways, including more efficient gridding, eliminating extensive computation by table look-up or reuse of previously computed and stored data; exploitation of structural symmetries; isolation of summary parameters (such as pole-zero data) that obviate the need for repeated computation for multiple parameter values; and of course, by devising new algorithms for solving the field equations.
(c) Machine capability requirements can also be a deterrent to users. While the research community discusses redesigning algorithms to take advantage of supercomputers, it is the EM simulators introduced for the low-end workstations and personal computers, that are

presently enjoying the most widespread use in industry. By extension, it appears likely that the most useful EM simulators designed for parallel processors and other distributed computer architectures, will be those that are able to run on groups of desktop machines typically available in the industrial environment.

FUTURE DEVELOPMENTS

EM simulation is far from being a mature technology, as evidenced by the fact that the new developments are proceeding at several fronts : not only are the commercially available EM simulation tools being constantly updated and improved, but entirely new tools are being introduced; and new algorithms are being researched for solving field equations that will likely become the basis still more new tools yet to be developed. As a result, the state-of-the-art in EM simulation is rapidly advancing, and many of the advancements are motivated by applications other than MCMs. A wish list for future developments particularly relevant to MCM design includes the following.

Tool Integration

The integration of EM simulators with other MCM design tools is desirable for a variety of reasons, including savings in design time, designer convenience, reduced likelihood of design errors, more globally optimized designs, and concurrent design. Present EM simulators do permit importing layouts from drafting tools, exporting results to circuit simulators, and sharing a common database with other design tools. The MCM design process may benefit from still other forms of integration, and may be able to take advantage of some common steps such as gridding for discretizing the MCM structure that occurs not only in EM simulators, but also in thermal and mechanical simulators. Similarly, embedding of optimization tools in the simulator will allow a more systematic optimization than is now practiced.

True Synthesis Tools

At the circuit level description, synthesis tools are available both for electrical and for physical synthesis. Thus the network synthesis tools automate the electrical design of a circuit (e.g., a 4-th order Butterworth filter) to meet a desired specification, while automatic layout tools can create the physical layout of a component given a desired component value (e.g., a spiral inductor from a given inductance value). At the electromagnetic level of description, the problem is more complex, and no direct synthesis tools are available. Perhaps the first step in this direction will be design advisories that will offer the designer some alternatives for meeting the frequently-occurring needs, such as reduction of coupling, loss, or delay time.

Coarse-Grained Analyses

An EM simulation can only be carried out if the structure to be analyzed is specified in complete detail, both with respect to the geometry and the material parameters, so that the problem is well-posed. As a result, EM simulation cannot be carried out during the early phases of design, when the structure has not been sufficiently defined. This is in contrast to the modern concurrent design philosophy, in which different aspects of module performance are considered at every stage of design, starting with the conceptual stage. EM simulation information could indeed be useful in the early stages of design, e.g., in system partitioning, material selection, and tradeoff analysis. Therefore the need is to develop a fuzzy or coarse-grained description of the structure (e.g., one in which parameters are specified by ranges rather than values) that nevertheless permits EM simulation and provides guidelines for the development of a detailed design.

Simulation-Based Design Rules

Design rules of a technology or foundry, used for implementing a module, not only govern the detailed design, but also establish the ultimate limits on the achievable figures of merit of a module, such as its compactness, or the degradation in the performance of the module relative to that of the bare chips in respect of some critical parameter. The design rules are often empirical, and originate from processing limitations. Empirical design rules lead to cautious designing, and hence sparse circuit layouts, whether created manually or by automatic layout routines. As the degree of control on the processing increases, the limitations to the design rules will increasingly arise from electromagnetic considerations, and will have a rigorous theoretical basis. Indeed, one can foresee the replacement of a "hard" design rule (one that leaves no choice to the designer) by "soft" design rules that can be bent with an associated cost. Future EM simulators can provide the means to get there.

REFERENCES

1. Z. J. Cendes, "EM simulators = CAE tools," *IEEE Spectrum*, vol. 27, no. 11, pp. 73-77, 93, November 1990.
2. D. G. Swanson, Jr., "Simulating EM fields," *IEEE Spectrum*, vol. 28, no. 11, pp. 34-37, November 1991.
3. J. X. Zheng, "A 3-D electromagnetic simulator for high frequency applications," *IEEE Trans. Components, Packaging, and Manufacturing Technology*, Pt. B, vol. 18, no. 1, pp. 112-118, February 1995.
4. MacNeal-Schwendler Corp., "High frequency design using three-dimensional finite element analysis," *Microwave Jour.*, vol. 38, no. 9, pp. 174-175, September 1995.
5. Ansoft Corp., "A multilayer planar electromagnetic field solver," *Microwave Jour.*, vol. 38, no. 11, November 1995.
6. *High-Speed Digital Interconnect Modeling*, Product note 85240-1, Hewlett Packard Co., January 1996.
7. B. E. MacNeal, J. R. Brauer, and R. N. Coppolino, "A general finite element vector potential formulation of electromagnetics using a time-integrated electric scalar potential," *IEEE Trans. Magnetics*, vol. 26, no. 5, pp. 1768-1770, September 1990.
8. T. Becks and I. Wolff, "Analysis of 3-D metallization structures by a full-wave spectral domain technique," *IEEE Trans. Microwave Theory and Techniques*, vol. 40, no. 12, 2219-2227, December 1992.
9. R. Mittra, W. D. Becker, and P. H. Harms, "A general purpose Maxwell solver for the extraction of equivalent circuits of electronic package components for circuit simulation," *IEEE Trans. Circuits and Systems - I : Fundamental Theory and Applications*, vol. 39, no. 11, pp. 964-973, November 1992.
10. W. D. Becker, S. Chebolu, and R. Mittra, "Finite difference time domain modeling of high speed electronic packages," *Microwave Jour.*, vol. 38, no. 12, pp. 62-74, December 1995.
11. R. W. Jackson, "Circuit topology for microwave modeling of plastic surface mount packages," *IEEE Trans. Microwave Theory and Techniques*, vol. 44, no. 7, Pt. 1, pp. 1140-1146, July 1996.
12. L. Zu, et al., "High Q-factor inductors integrated on MCM Si substrates," *IEEE Trans. Components, Packaging, and Manufacturing Technology*, vol. 19, no. 3, pp. 635-643, August 1996.
13. F. Nusseibeh, et al., "Three dimensional modeling of MQUAD packages at 100 MHz," *Proc. 1995 International Electronics Packaging Conference* (San Diego, CA), pp. 385-389, September 1995.

MICROSTRIP ANTENNA ANALYSIS USING A SUITE OF MODELING TECHNIQUES --A REVIEW

Raj Mittra

Applied Research Laboratory
The Pennsylvania State University
P.O. Box 30
State College, PA 16804

1. INTRODUCTION

In this work we present a brief examination of the problem of modeling microstrip antennas using a variety of numerical techniques. One of the simplest among these is the cavity model approach,[1] which is useful for developing an intuitive understanding of the performance of the microstrip antenna. However, it and similar approximate methods (see for instance[2]) are limited in their accuracy, and are only useful either for the design of simple structures, or for providing good initial estimates of the performance characteristics of the design.

The three numerically-rigorous microstrip antenna modeling techniques are the Method of Moments (MoM); the Finite Element Method (FEM); and, the Finite Difference Time Domain (FDTD) approach. The moment method has been extensively applied to the problem of microstrip antenna analysis both in the spectral and spatial domains,[3-8] because it is numerically the most efficient among the three techniques mentioned above. Initially, it suffered from somewhat of a disadvantage that it required the evaluation of Sommerfeld-type integrals for the construction of the Green's function for layered dielectric media backed by a ground plane. However, considerable progress has been made recently toward circumventing this difficulty, and accurate closed form expressions have been derived that totally bypass the evaluation of the time-consuming integrals. Much has been written on this topic in recent literature and the reader is referred to[9-13] for further details. Additionally, Alatan *et al.*[14] have shown how the matrix elements themselves can be evaluated in closed form, further enhancing the efficiency of the closed form Green's function technique.

The Moment method does have the drawback, however, that it is unable to handle, efficiently, inhomogeneous structures or antennas on a finite ground plane. This prompts us to examine finite methods, e.g., the Finite Element Method (FEM), or the Finite Difference

Directions for the Next Generation of MMIC Devices and Systems
Edited by N. K. Das and H. L. Bertoni, Plenum Press, New York, 1997

Time Domain (FDTD) techniques as tools for analyzing general microstrip antennas that may be embedded in a complex structure. These methods have the advantage that they do not require the use of a Green's function, which may not be known explicitly for a given problem domain. In addition, they are inherently capable of modeling material inhomogeneties inside the dielectric regions of the microstrip patch antenna without additional computational cost.

The FEM approach to microstrip antenna modeling has recently been detailed in a series of works by Volakis and his co-workers,[15,16] who have used either the boundary element method or the absorbing boundary conditions (ABCs) approach to truncating the Finite element mesh. In the FEM/ABC approach,[15] a second-order conformal ABC of vectorial nature is employed in the FEM mesh truncation of a problem domain that comprises a cavity-backed microstrip patch antenna located on a cylindrical platform. In contrast to a more rigorous and accurate Finite Element Boundary Integral (FEBI) formulation[16] which results in a partially full system matrix, the imposition of the approximate local ABC at the truncation surface yields a fully-sparse system matrix, this being a characteristic feature of the FEM technique when applied to interior domain problems. However, this approach may not be sufficiently versatile for some applications, and this prompts us to discuss the hybrid FEM/MoM techniques[17,18] that are more general in nature.

In such hybrid techniques, the Finite element method is combined with the method of moments (MoM) that are applied in complementary regions. The well-known computational disadvantage of FEM in the form of a rapid increase in the CPU time and memory requirements for the case of large problem domains that extend from the patch antenna to the truncation surface, is a compelling reason to combine the strengths of the MoM and FEM approaches to obtain a hybrid technique in which the disadvantages of both approaches can be overcome.

The hybrid technique for microstrip antennas embedded in conducting structures, described by Pekel and Mittra,[18] will now be summarized below.

2. A HYBRID FEM/MOM TECHNIQUES

The hybrid technique we describe in this section utilizes the MoM approach in the domain exterior to the microstrip patch antenna, obviating the need for mesh truncation surfaces and any type of absorbing boundary condition (ABC), and employs the FEM approach in the interior of the dielectric regions to account for the possible presence of material inhomogeneties. The exterior and interior domain solutions are coupled on the corresponding interfaces by imposing the continuity condition on the tangential magnetic field components. The continuity of the tangential electric field is directly enforced through the use of magnetic currents that are defined on the interfaces by invoking the Equivalence Principle. Two different formulations can be considered for the FEM analysis in the interior domains, viz., the total field formulation[17] and the scattered field formulation.[18] The latter may be preferable in terms of minimizing error propagation, and effectively accounting for the presence of fine features, such as thin wires and coaxial cable feeds, in the interior domain by including their "incident field contributions" directly in the FEM formulation. Thus, fine features are also analyzed via the MoM/FEM hybridization in the latter approach; consequently, the use of an unnecessarily fine mesh in the vicinity of these features is avoided. In addition, unlike the MoM, the hybrid technique does not require the use of the Green's function for an electromagnetically complex problem domain, but utilizes the well-known form of the free-space Green's function in the exterior domain analysis. In addition, the hybrid approach leads to fully-sparse system matrices in the FEM, as opposed to the partially-full matrices that are generated when the more rigorous and accurate Finite Element Boundary Integral (FEBI) technique is applied.

In the hybrid approach, the FEM analysis is carried out in the interior regions of the patch antenna where inhomogeneous material properties may be encountered. The sparsity and symmetry properties of the FEM matrix are retained due to the fact that the interior domain FEM solution and the exterior domain MoM solution are decoupled at the corresponding interfaces. In the hybrid approach, the Equivalence Principle is invoked to define magnetic surface currents at the interfaces, which are expanded, similar to the ordinary surface electric currents on the PEC surfaces, in terms of a suitable set of vectorial expansion functions. In this particular application, the expansion functions have been selected to be triangular surface patch vector functions.[19]

An FEM total field formulation is employed to determine the magnetic field in the interior domain, with the magnetic surface currents at the interfaces appearing in the right-hand-side of the FEM system equations. In the interior domain FEM analysis, the unknown magnetic field is computed by using tetrahedral edge elements, subject to the constraint that the triangular bases of the tetrahedral elements located at the interface overlap exactly with the triangular surface patches covering the surface in the exterior domain MoM analysis. This constraint implies that the mesh that covers the problem domain has to be generated in a careful manner. We also note that in this particular application, an iterative bi-conjugate gradient method (Bi-CGM) solver has been utilized to be able to efficiently handle systems with a large number of unknowns.

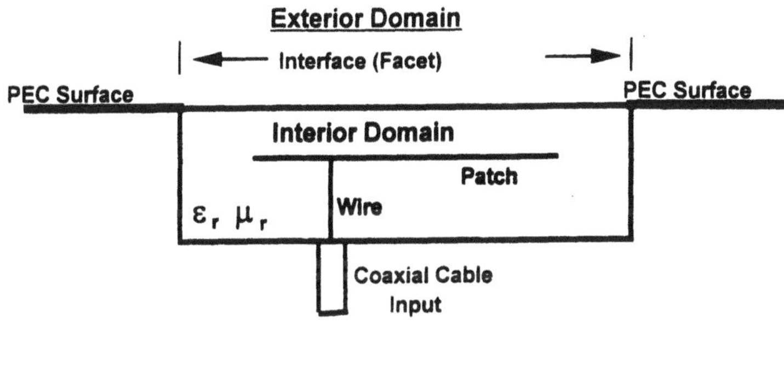

Figure 1. The cavity-backed microstrip patch antenna model embedded in a PEC body.

Next, we present a few representative results that have been obtained with the above-outlined hybrid approach for a coaxial-line-fed cavity-backed patch antenna. For the sample problem geometry depicted in Figure 1, with a patch size of 27.78 mm x 27.78 mm, a cavity size of 32.52 mm x 32.52 mm x 3.0 mm with no material properties inside, a dielectric cover of thickness 0.508 mm and $\varepsilon_r = 2.2$, a coaxial line with inner and outer radii of 1.037 mm and 3.620 mm and $\varepsilon_r = 2.25$, respectively, and a feed point that has been shifted by an amount of 5.7 mm from the patch center, this version of the hybrid approach has been applied to determine the reflection coefficient and input impedance of the patch antenna. The resonant frequency was found to occur at about 4.19 GHz and the calculated results were found to be in good agreement with reference results obtained with the FDTD technique.

We have also developed a second version of the hybrid MoM/FEM approach based on the scattered field formulation, and have applied it to a representative geometry depicted in Figure 2. The cavity-backed patch antenna in question has been designed with the aim of achieving a circularly polarized radiation pattern at around 1.55 GHz, which is cited to be the main reason for the off-center location of the coaxial cable feed with respect to the conducting patch. The input impedance variation that was found for the patch antenna with the above-outlined hybrid technique over a frequency band from 1.4 to 1.75 GHz is plotted in Figure 3 using a Smith Chart format. It is observed that the "circular polarization" feature manifests itself with a "cusp" in the input impedance plot at around 1.55 GHz, where a VSWR value of 2.11 is obtained on the 50 Ω coaxial cable. The input impedance attains a value of approximately 50 Ω at around 1.72 GHz with a virtually negligible reflection coefficient.

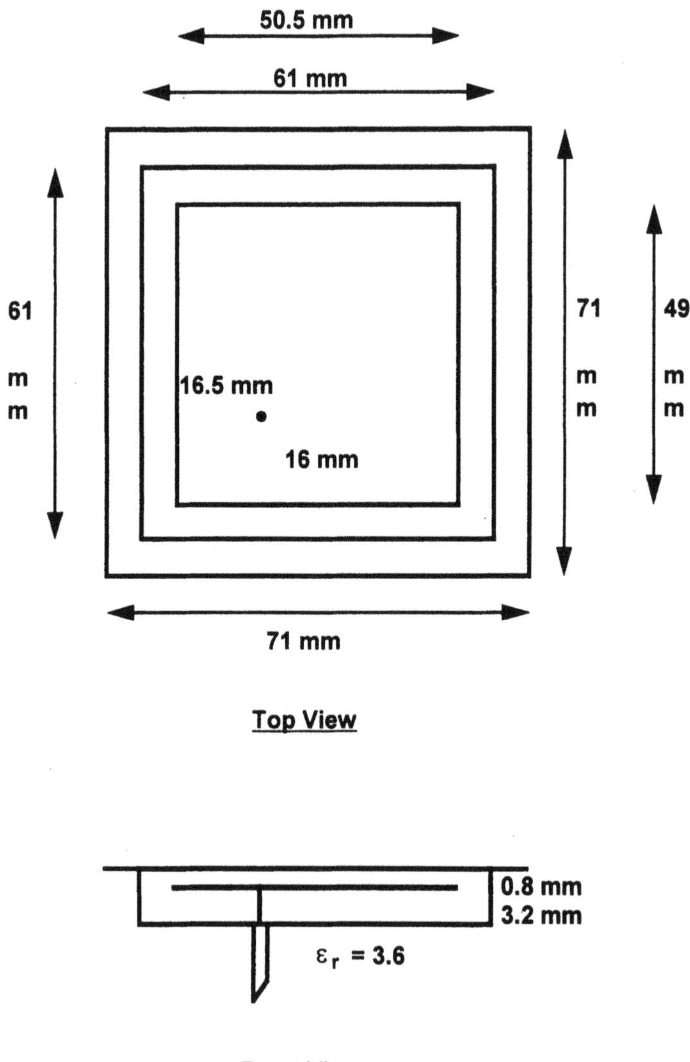

Figure 2. Geometry of the cavity-backed patch antenna for Circular polarization.

3. THE FDTD METHOD FOR ANTENNA MODELING

The FDTD technique is also well-suited for handling complex microstrip antenna configurations, as it can conveniently model the numerous inhomogeneties encountered in these structures. Furthermore, it has the distinct advantage over the frequency domain methods that it can generate the characteristics of the patch over a broad band of frequencies with a single simulation. The matrix-free nature of the algorithm that enables the FDTD method to routinely handle upward of 10^6 unknowns on conventional workstations.

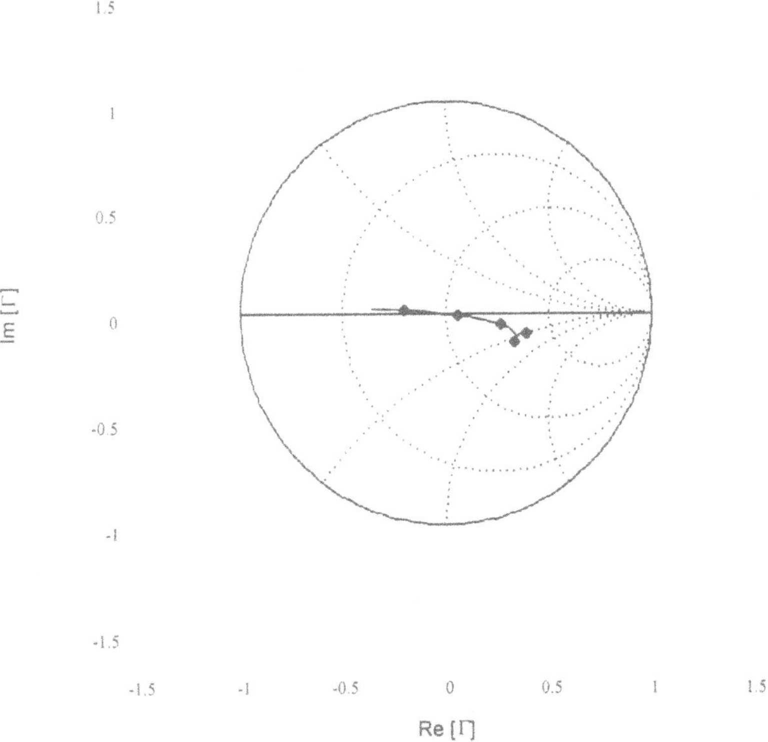

Figure 3. Input impedance variation of the patch antenna from 1.4 to 1.75 GHz, plotted in a Smith Chart format.

Simple microstrip antenna structures were first modeled using the FDTD method by Reineix and Jecko[20] in 1989. Since then, many different configurations such as parasitically coupled patches[21], active antennas[22], and microstrip antennas mounted on curved surface, etc., have been successfully analyzed with this approach. The papers by Wu *et al.*[23,24] considerably improved the modeling technique that enabled it to accurately characterize multi-layer patch antennas with various feed structures such as microstrip, coaxial and aperture coupled feeds. In this work, we describe some recent advancements with the FDTD algorithm that improve the computational efficiency of the conventional FDTD method, and thereby extend its scope for analyzing complex microstrip antenna configurations.

The FDTD method is sufficiently well established and, hence, the details of the conventional FDTD approach will be omitted here. However, we will discuss some of the difficulties encountered in the conventional method and suggests ways by which they can be overcome. In the conventional form, FDTD can be highly computer-intensive in terms of *cpu* time and memory. For example, accurate characterization of a typical dual-layer microstrip patch via the conventional FDTD algorithm could require as much as 400 MB of RAM and the simulation could run for several days on a workstation. Recently, a number of techniques have been developed to improve the computational efficiency of the conventional FDTD method, applied to the problem of modeling microstrip antennas. These efficient techniques typically involve a trade-off between accuracy and speed of computation. We will now briefly review some of these techniques.

Spatial Discretization

A problem that is commonly encountered when modeling microstrip antenna geometries with a uniform orthogonal discretization is that it becomes necessary to use a small cell size, typically on the order of $\lambda_0/40$ - $\lambda_0/100$, to accurately represent the fine features of the antenna such as coaxial connectors, shorting pins and substrate thickness. This value of the spatial increment is much smaller than the nominal value of $\lambda_0/10$, which is required to maintain the accuracy of the conventional FDTD algorithm. Similarly, since the Courant stability condition necessitates the time step to be proportionally small, the time signature is highly oversampled. These two requirements can result in a FDTD simulation that has to deal with an excessively large number of cells, and the computation time can go up dramatically. Thus, the conventional FDTD method can be an inefficient approach to modeling microstrip antennas with fine features.

Over the last decade, several approaches have been developed to model fine features of an object in an efficient manner. We will now describe a few of these techniques.

Subcell Gridding and Expansion Techniques

In the subcell gridding and expansion techniques[25,26], the geometry is modeled with a uniform fine grid in certain regions enclosing discontinuities and fine features, while a uniform coarse grid is employed elsewhere. In the subgridding approach,[25] spatial and temporal interpolations are used to update the fields on the interface between the fine and coarse grids. It has been found that the accuracy of this model is not good unless the domain of the fine feature region is sufficiently large. The problems under consideration involve fine structural features that may be distributed over the entire volume of the structure and their dimensions do not necessarily extend over large spatial domains.

Subcell Modeling

This technique is commonly employed to model thin sheets, slots and wires.[27] In microstrip antenna analysis, the thin-sheet approximation is valid even in complex environments, including one where the sheet is in the proximity of several conductors, and is used, where applicable, to reduce the computational domain.

On the other hand, the thin wire approximation breaks down when there are several conductors in its vicinity as, for instance, in a coaxial cable. The thin wire model is based on the assumption that the electric fields are normal to the surface of a thin wire and the tangential magnetic fields exhibit a $1/r$ dependence in the vicinity of the wire, where r is the radial distance from the center of the wire. In such situations, one edge of a Yee cell can be used to model the thin wire and the field update equations have to be modified accordingly. Unfortunately, for the structures of interest here, the original assumption regarding the field behavior is not valid; hence, this model is also not useful for the problems being considered.

Non-Orthogonal, Curvilinear And Globally Unstructured Grids

To obtain an accurate field solution in an arbitrary structure by using the FDTD algorithm, it becomes necessary to choose a suitable grid discretization such that it accurately models both the geometry and the field variations in the structure, and one approach to doing this is to use a general, finite-element type of unstructured grid.[28] The generalization of the FDTD algorithm to include the surface-curve integral form and the volume surface integral form of Maxwell's equations has led to the development of the hybrid FDTD/FVTD technique[29] which can solve for the fields on a globally unstructured volume grid. However, this approach is not always very robust, as it is not suitable for handling different structures with a minimum amount of preprocessing, which is one of the principal attractive features of the FDTD method.

An alternative strategy is to use the curvilinear FDTD approach,[30,31] which employs a structured non-orthogonal grid, and is well-suited for modeling patch antennas with curved surfaces.[18] However, for the class of structures analyzed in this work, the curvilinear method turns out to be prohibitively expensive in terms of memory requirements and CPU time, owing to the fact that it is necessary, in the aforementioned approach, to store the coordinate transformation metrics, and because the time required to convert the covariant and contravariant field components back and forth is computationally expensive. In predominantly rectangular geometries, curved objects such as shorting pins and circular coaxial feeds can be simply modeled with square geometries. Since the dimensions of these fine features are very small, this approximation yields good results and obviates the use of the curvilinear FDTD.

Non-Uniform Orthogonal Grids

From the earlier discussion, we have seen that an accurate representation of the antenna geometry using a non-orthogonal grid leads to a significant increase in the solution time and modeling the fine features using sub-cell approximations can lead to erroneous results. In this respect, the use of a non-uniform orthogonal grid[32] is a good compromise to increasing the modeling accuracy while preserving the computational speed and accuracy of the FDTD technique.

In this method, the simplicity of the FDTD update equations is retained by using an orthogonal grid. Therefore, the computational speed of this method is the same as the conventional Yee algorithm.

By employing non-uniform spatial increments, this technique allows us to model the geometry more accurately than is possible in the uniform approach, which requires the dimensions of a structure to be integral multiples of the chosen discretizations in the x, y, and z directions. This flexibility in varying the mesh dimensions is especially advantageous for modeling a circulary polarized (CP) patch antenna accurately, since its characteristics, such as the axial-ratio and the input impedance, are very sensitive to variations in the dimensions of the patch and the location of the probe. For instance, in an ordinary CP patch, the ratio of the two sides (a/b ratio) is of the order of 1.01 to 1.05. Hence, it becomes very difficult to accurately describe these dimensions as well as the feed, and its position on the patch, using a uniform mesh with a fixed spatial discretization. On the other hand, a non-uniform grid can easily model the structural dimensions precisely with a few cells.

With this technique, a fine discretization can be used to model the regions where there is a rapid variation in the fields, and a coarse mesh can be employed in regions where the field is well behaved. This process can lead to a significant savings in the memory requirements for the simulations. Also, since microstrip antennas are open radiating structures, the ability to vary the mesh resolution enables us to move the absorbing boundaries of the computational domain farther away from the radiating structure, without an undue increase in the number of cells. This can be quite advantageous when modeling large, complex structures, because the presence of spurious reflections from an imperfect ABC can contaminate the FDTD solution when these boundaries are placed too close to the antenna being modeled.

As a rule of thumb, the growth factor of the mesh (which is the ratio of the spatial steps of two adjacent cells) should be kept below 1.2 ~ 1.3 to prevent artificial discontinuities introduced by the abrupt changes in the cell size. However, larger growth rates are acceptable as long as the cell size is very small compared to the wavelength.[33] Also, it should be noted that the non-uniform grid no longer preserves the second-order accuracy that is obtained with the use of the uniform grid. Nevertheless, if the mesh spacing changes slowly, the error can be close to that of a second-order method. It is possible to achieve about 40% ~ 80% reduction in the problem size and a corresponding decrease in the computation time by using this modeling procedure without an undue sacrifice of the accuracy.

Considering all of these advantages, the non-uniform orthogonal FDTD method appears to be the preferred approach for modeling predominantly rectangular microstrip antenna geometries. For analyzing circular patches, it is perhaps more advantageous to use a curvilinear mesh. Recent advances in the contour path and conformal FDTD technique[34,35] may enable one to use a regular grid with the option of locally deforming the mesh to conform to the curvature of the geometry.

Note that although a significant reduction in the computational domain has been achieved through non-uniform spatial discretization, the time step is still determined by the smallest cell size. This issue can be addressed through signal processing techniques.[36-38]

Finally, one of the most recent developments in the Finite methods is the introduction of a new concept for mesh truncation called the Perfectly matched layer (PML)[39] approach. The use of this approach enables one to reduce the computational domain significantly by bringing the boundary closer to the object than is possible with the conventional absorbing boundary conditions, without unduly sacrificing the accuracy of the computation.

ACKNOWLEDGMENT

This review paper is based on the works of many individuals to whom the author is grateful. They include Drs. M. I. Aksun of Bilkent University, S. Chebolu of Celwave, J. Svilegj of Texas Instruments, D. Becker of IBM, S. Dey of Penn State University, and U. Pekel, formerly with the University of Illinois. The paper could not have been written without the valuable assistance of these colleagues. The author also acknowledges the support of Applied Research Laboratory of the Penn State University, and especially the help of Ms. Shelby Sharp, during the preparation of the manuscript.

REFERENCES

1. W. F. Richards, Microstrip antennas, in: *Chapter 10, Antenna Handbook*, Y. T. Lo and S. W. Lee, eds., Van Nostrand Rheinhold, New York (1988).
2. A. Bennela and K. C. Gupta, Multiport network model and transmission characteristics of two-port rectangular microstrip antennas, *IEEE Trans. Antennas & Propagat.*, 36(10):1337 (Oct. 1988).
3. E. H. Newman and P. Tulyahan, Analysis of microstrip antennas using moment methods, *IEEE Trans. Antennas & Propagat.*, 29(1):47 (Jan. 1981).
4. R. C. Hall and J. R. Mosig, The analysis of coaxially fed microstrip antennas with electrically thick substrates, *Electromagnetics*, 9:367 (1989).
5. J. R. Mosig and F. E. Gardiol, General integral equation formulation for microstrip antennas and scatterers, *Proc. Inst. Elec. Eng.*, 132(H):424 (Dec. 1985).
6. J. T. Aberle and D. M. Pozar, Accurate and versatile solutions of probe-fed microstrip patch antennas and arrays, *Electromagnetics*, 11(1):1 (1991).
7. W. Chen, K. F. Lee and R. Q. Lee, Spectral Domain full wave analysis of the input impedance of coaxially-fed rectangular microstrip antennas, *J. Electromagnetic Waves and Applications*, 8(2):248 (1994).
8. J. P. Damiano and A. Papiernik, Survey of analytical and numerical models for probe-fed microstrip antennas, *IEE Proc. Microw. Antennas Propag.*, 141(1) (Feb. 1994).
9. Y. L. Chow, J. J. Yang, D. F. Fang and G. E. Howard, A closed form spatial Green's function for the thick microstrip substrate, *IEEE Trans. on Microwave Theory & Tech.*, MTT-39:588 (Mar. 1991).
10. M. I. Aksun and R. Mittra, Derivation of closed form Green's functions for general microstrip geometries, *IEEE Trans. on Microwave Theory & Tech.*, MTT-40:2055 (Nov. 1992).
11. G. Dural and M. I. Aksun, Closed-form Green's functions for general sources and stratified media, *IEEE Trans. on Microwave Theory & Tech.*, MTT-43:1545 (Jul. 1995).
12. M. I. Aksun, A robust approach for the derivation of closed form Green's function, *IEEE Trans. on Microwave Theory & Tech.*, MTT-44:651 (May 1996).
13. M. I. Aksun and R. Mittra, Estimation of spurious radiation from microstrip etches using closed form Green's functions, *IEEE Trans. on Microwave Theory & Tech.*, MTT-40:2063 (Nov. 1992).
14. L. Alatan, M. I. Aksun, K. Mahadevan and M. T. Birand, Analytic evaluation of the MoM matrix elements, *IEEE Trans. on Microwave Theory & Tech.*, MTT-44:519 (Apr. 1996).
15. J. L. Volakis, J. Gong and T. Ozdemir, Application of the FEM to conformal antennas, in: *Chap. 13, Finite Element Software for Microwave Engineering*, T. Itoh, G. Pelosi and P. P. Silvestor, eds., John Wiley and Sons (1996).
16. J. L. Volakis, J. Gong and A. Alexanian, A finite element boundary integral method for antenna RCS analysis, *Electromagnetics*, 14(1):63 (1994).
17. J. C. Cheng, N. I. Dib, and L. P. B. Katehi, Theoretical modeling of cavity-backed patch antennas using a hybrid technique, *IEEE Trans. Antennas Propagat.*, 43(9):1003 (Sept. 1995).
18. U. Pekel and R. Mittra, A hybrid MoM/FEM technique for the analysis of cavity-backed patch antennas embedded in large conducting surfaces, *Microwave and Optical Technology Letters*, 12(5):255 (August 1996).

19. S. M. Rao, D. R. Wilton and A. W. Glisson, Electromagnetic Scattering by surfaces of arbitrary shape, *IEEE Trans. Antennas & Propagat.*, 30(3):409 (May 1982).
20. A. Reineix and B. Jecko, Analysis of microstrip patch antennas using the finite difference time domain method, *IEEE Trans. Antennas & Propagat.*, 37(11):1361 (Nov. 1989).
21. G. S. Hilton, C. J. Railton and M. A. Beach, Modeling parasitically-coupled patch antennas using the finite-difference time-domain technique, *IEE Eighth Int. Conf. Antennas Propag.*, 1:186 (1993).
22. B. Toland, J. Lin, B. Houshmand, and T. Itoh, FDTD analysis of an active antenna, *IEEE Microwave and Guided Wave Letters,* 3(11) (Nov. 1993).
23. C. Wu, K. L. Wu, Z. Bi and J. Litva, Modeling of coaxial-fed microstrip patch antenna by finite difference time domain method, *Electron. Lett.*, 27(19):1691 (Sept. 1991).
24. C. Wu, K. L. Wu, Z. Q. Bi, and J. Litva, Accurate characterization of planar printed antennas using the finite-difference time-domain method, *IEEE Trans. Antennas Propagat.*, 40(5) (May 1992).
25. S. S. Zivanovic, K. S. Yee and K. K. Mei, A subgridding method for the time-domain finite-difference method to solve Maxwell's equations, *IEEE Trans. Microwave Theory Tech.*, 39(3) (March 1991).
26. K. S. Kunz and R. L. Luebbers, A technique for increasing the resolution of finite-difference solution of the Maxwell equations, *IEEE Trans. Electromag. Compat.*, 23:1320 (Nov. 1981).
27. A. Taflove, *Computational Electromagnetics: The Finite-Difference Time-Domain Method,* Artech House, Boston (1995).
28. K. Mahadevan, R. Mittra and P. M. Vaidya, Use of Whitney's edge and face elements for efficient finite element time domain solution of Maxwell's equations, *J. Electromagnetic Wave and Applications*, 8(9/10):1173 (1994).
29. K. S. Yee and J. S. Chen, Conformal hybrid finite difference time domain and finite volume time domain technique, *IEEE Trans. Antennas Propagat.*, 42(10) (Oct. 1994).
30. R. Holland, Finite difference solution of Maxwell's equations in generalized nonorthogonal coordinates, *IEEE Trans. Nuc. Sci.*, 30(6):4589 (Dec. 1983).
31. J. F. Lee, R. Palandech and R. Mittra, Modeling three-dimensional discontinuities in waveguides using nonorthogonal FDTD algorithm, *IEEE Trans. Microwave Theory Tech.*, 40(2):346 (Feb. 1992).
32. J. Svigelj, Efficient solution of Maxwell's equations using the nonuniform orthogonal finite difference time domain method, Ph.D. dissertation, *Univ. of Illinois at Urbana-Champaign* (1995).
33. J. Svigelj and R. Mittra, Grid dispersion error using the nonuniform orthogonal finite difference time domain method, *Microwave and Opt. Tech. Lett.*, 10(4):199 (Nov. 1995).
34. C. J. Railton, I. J. Craddock, and J. B. Schneider, Improved locally distorted CPFDTD algorithm with provable stability, *Electronics Lett.*, 31:1585 (August 1995).
35. S. Dey, R. Mittra and S. Chebolu, A technique for implementing the FDTD algorithm on a Non-orthogonal grid, to appear.
36. W. L. Ko and R. Mittra, A combination of FD-TD and Prony's methods for analyzing microwave integrated circuits, *IEEE Trans. Microwave Theory Tech.*, MTT-39(12):2176 (Dec. 1991).

37. J. Litva, C. Wu, K. L. Wu and J. Chen, Some considerations for using the finite difference time domain technique to analyze microwave integrated circuits, *IEEE Microwave and Guided Wave Letters*, 438 (Dec. 1993).
38. V. Jandhyala, E. Michielssen, and R. Mittra, FDTD signal extrapolation using the forward-backward autoregressive model, *IEEE Microwave and Guided Wave Lett.*, 4(6):163 (Jun. 1994).
39. J. P. Berenger, A perfectly matched layer for the absorption of electromagnetic waves, *J. Computational Physics*, 11:185 (1994).

A NEW APPROACH TO THE DESIGN OF MICROWAVE AMPLIFIERS

E. Kerherve,[1] P. Jarry,[1] and C. Tronche[2]

[1]Microelectronic Laboratory IXL
URA 846 CNRS
Bordeaux University - FRANCE
[2]ALCATEL ESPACE - Toulouse - FRANCE

INTRODUCTION

This paper presents a new method to simulate and optimize microwave amplifiers more quickly. We present the results obtained for a low-noise microwave amplifier designed according to this method. At present, in order to design a microwave amplifier it is first necessary to choose the correct type of transistor in function of the applications required. After this, classical simulation and optimization software is used to calculate the elements of the equalizers placed before and after the transistor (Figure 1), taking into account the specifications of the amplifier.

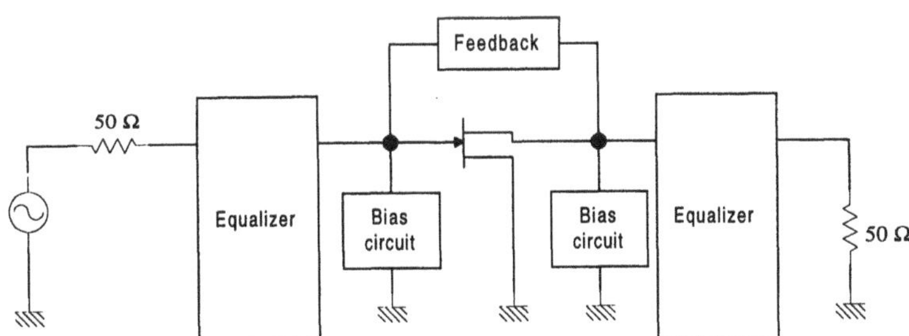

Figure 1. Microwave amplifier

However, there is no indication a priori of the values of the constituent elements of these equalizers. Therefore, optimization can prove long, tedious and without any guarantee as regards the convergence of the final result (this optimization is even longer, the greater

Directions for the Next Generation of MMIC Devices and Systems
Edited by N. K. Das and H. L. Bertoni, Plenum Press, New York, 1997

the number of parameters to be optimized and the longer the bandwidth length). For this reason we suggest a new approach to the design of microwave amplifiers firstly using software developed in our laboratory and based on the method of real frequencies. One of the particularities of this method is that it does not require any equivalent transistor model; we directly use the S parameters file obtained under point during practical measurements, or the file provided by the maker. Moreover, when this method is associated to an optimization algorithm and to a classic synthesis method with distributed or lumped elements, it gives a very rapid and accurate set of values of the equalizer's constituent elements. Then, using « Academy » (or an other classical software) it is simply necessary to introduce the dielectric losses associated with the use of the substrate.

In order to obtain a finer result, a second slight optimization may be carried out. This method has been applied to the design of a low-noise amplifier over the 5,925-6,425 GHz bandwidth.

GENERAL OVERVIEW

A numerical approach called the « real frequency » or « line-segment technique was introduced[1] in 1977 to overcome the limitations of the analytical methods, when applied to the single-matching problem. Using only measured 2-port active device data, the Carlin method consists in generating a positive real (PR) impedance, $Z_q = R_q(\omega) + jX_q(\omega)$, looking into a resistively terminated lossless matching network (Figure 2). This impedance is assumed to be a minimum reactance function so as to be able to determine $X_q(\omega)$ uniquely from $R_q(\omega)$ by a Hilbert transformation.

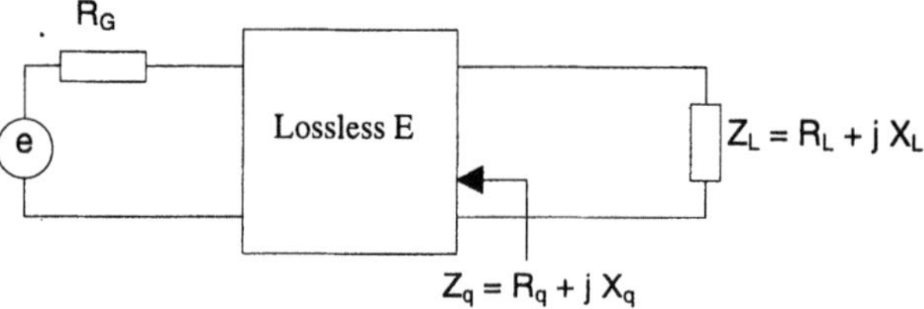

Figure 2. Real frequency technique for single-matching problems

In this manner, the transducer power gain function of Z_q and Z_L, has only one unknown R_q, which is computed by using a set of line segments to approximate the desired transducer power gain bandpass response. $Z_q(\omega)$ is approximated by a realizable rational function (describing a ladder network for example) which fits the computed data. Finally, Z_q is synthesized using the Darlington procedure as a lossless 2-port with a resistive termination. Despite several attempts, it has not proved convenient to apply this method to the double-matching problem.

The potential power of the real frequency technique led to the development of a new numerical synthesis procedure by Yarman and Carlin[2] in 1982, which has all the merits of the line-segment technique. In double-matching, the final result of the new procedure is an optimized, physically realizable, unit-normalized reflection coefficient $e_{11}(p)\,(p = \sigma + j\omega)$, which describes the equalizer alone. The equalizer is placed between a complex source Γ_G and complex load Γ_L (Figure 3).

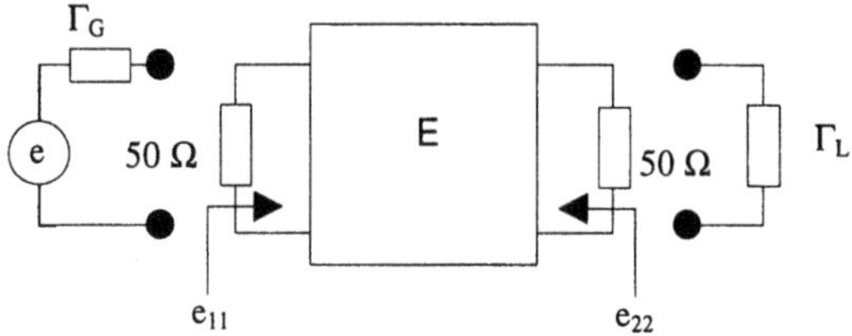

Figure 3. Real frequency technique for double-matching problems

If $e_{11}(p)$ is appropriately determined, then the equalizer E may be synthesized using the Darlington theorem, which states that any bounded real (BR) reflection coefficient $e_{11}(p)$ is realizable as a lossless reciprocal 2-port terminated in a pure resistance[2] and a ladder type network may be extracted.

This method has the further advantages of generality, being applicable to all matching problems, and universality, as it involves neither equalizer values nor a predefined equalizer topology. The simple formalism of this technique allows us, without complex calculations, to optimize many performance parameters of single and multistage microwave amplifiers.

THE REAL FREQUENCY METHOD[3]

Formalism

In the case of the double matching problem, it has been shown[4] that the scattering parameters of an equalizer E, can be completely determined from the numerator polynomial h(p) of the input reflection $e_{11}(p)$. E is assumed to be a ladder network, thus the scattering parameters are given as follows (Belevitch representation) :

$$e_{11}(p) = \frac{h(p)}{g(p)} = \frac{h_0 + h_1 p + \ldots + h_n p^n}{g_0 + g_1 p + \ldots + g_n p^n} \tag{1}$$

$$e_{12}(p) = e_{21}(p) = \frac{\pm p^r}{g(p)} \tag{2}$$

$$e_{22}(p) = -\frac{(-1)^r h(-p)}{g(p)} \tag{3}$$

where r is an integer and specifies the order of the transmission zeros. The polynomial $h(p)$ is chosen as the unknown and assumes the equalizer to be without loss.

The polynomial $g(p)$ is generated from the Hurwitz factorization of

$$\left|e_{11}(p)\right|^2 + \left|e_{12}(p)\right|^2 = 1 \tag{4}$$

or

$$g(p)\,g(-p) = h(p)\,h(-p) + (-1)^r\,p^{2r} \tag{5}$$

To obtain the scattering parameters of E, it is therefore sufficient to generate the Hurwitz denominator polynomial $g(p)$ from $h(p)$.

The optimization is performed simultaneously on the transducer power gain, on the noise figure and on the input and output VSWR.

Transducer power gain, noise figure and VSWR

Transducer power gain

Referring to Figure 4 for the first k cascaded amplifier stages, the transducer power gain is :

$$G_k = G_{k-1} \frac{\left|e_{21_k}\right|^2 \left|S_{21_k}\right|^2}{\left|1 - e_{11_k}\,\Gamma_{g_k}\right|^2 \left|1 - \Gamma_{s_k}\,S_{11_k}\right|^2} \tag{6}$$

i.e.,

$$G_k = G_{k-1} \cdot E_k \tag{7}$$

where G_{k-1} is the gain of the first $(k-1)$ stages with normalized resistive terminations, $(e_{ij})_k$ are the scattering parameters of the kth equalizer E_k, $(S_{ij})_k$ are the scattering parameters of the kth transistor with bias networks.

The overall gain $G(\omega)$ is defined after the final equalizer E_{k+1} has been added:

$$G = \left(G_1 \cdot G_2 \cdots G_k\right) E_{k+1} \tag{8}$$

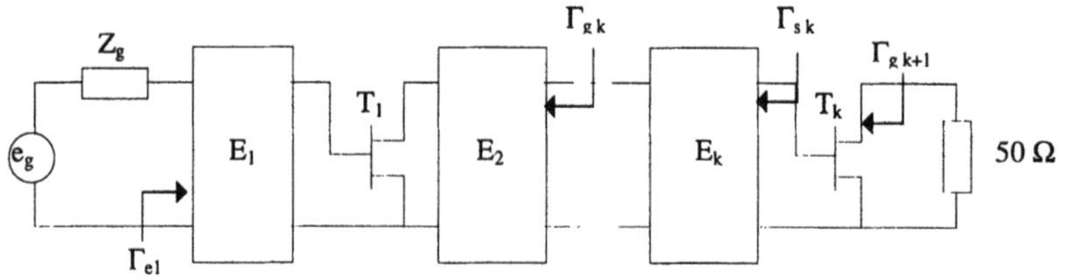

Figure 4. Multistage microwave amplifier

Input and output matching

Using the same method, we are able to define the input and output VSWR of the multistage microwave amplifier. For the first k cascade stages, the input VSWR is given by:

$$VSWR_{in_k} = \left.\frac{1 + \left|\Gamma_{e_1}\right|}{1 - \left|\Gamma_{e_1}\right|}\right|_{k\ \text{stages}} \tag{9}$$

The overall output VSWR is only defined after the final equalizer E_{k+1} has been added (Figure 5)

$$\mathrm{VSWR}_{out} = \frac{1+\left|\Gamma_{s_{k+1}}\right|}{1-\left|\Gamma_{s_{k+1}}\right|} \tag{10}$$

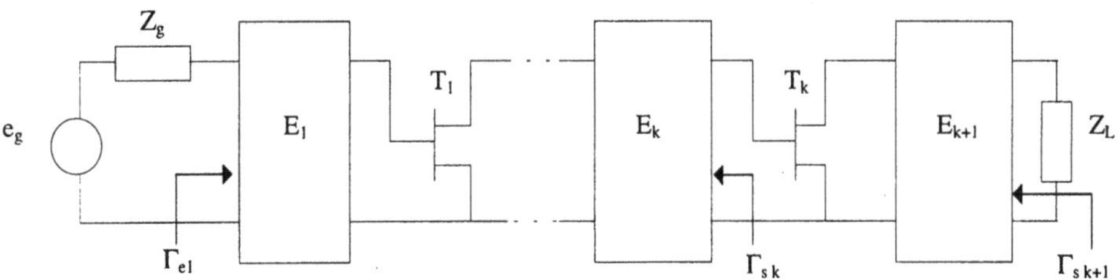

Figure 5. Multistage microwave amplifier with the final equalizer E_{k+1}

Noise figure

The noise figure is given by:

$$F = F_k = F_{k-1} + \frac{1}{G_{D_{k-1}}}\left(F_{min_k} + 4\frac{R_{n_k}}{R_0}\frac{\left|\Gamma_{s_k} - \Gamma_{Gmin_k}\right|^2}{\left|1+\Gamma_{Gmin_k}\right|^2 (1-\left|\Gamma_{s_k}\right|^2)} - 1 \right) \tag{11}$$

where

$$G_{D_k} = G_k \frac{1}{1-\left|G_{g_{k+1}}\right|^2} \tag{12}$$

F_{min} and F_{min_k} respectively represent the minimum transistor noise figure in general, and at the *k*th stage. R_n and R_{n_k} are respectively the noise resistance in general, and at the *k*th stage. $\Gamma_{G_{min_k}}$ is the optimum noise reflection coefficient at the *k*th stage. F_{k-1} represents the overall noise figure at the input to the $(k-1)$th stage.

Multi-objective Optimization

In the CAD program called « RFCAD », that we have developed, optimization is performed simultaneously upon the gain, the noise figure the input and output VSWR.
The multi-objective function at the *k*th equalizer is described using a least squares criterion and appropriate weighting functions as follows :

$$U_k = \sum_{j=1}^{m} W_1\left(\frac{G_k(\omega_j)}{G_{0_k}} - 1\right)^2 + W_2\left(\frac{R_{in\,k}(\omega_j)}{R_{in0\,k}} - 1\right)^2 + W_3\left(\frac{F_k(\omega_j)}{F_{0\,k}} - 1\right)^2 \tag{13}$$

G_{0_k}, R_{in0_k}, R_{out0_k} and F_{0_k} are the specified gain, input VSWR, output VSWR and noise figure. m is the number of bandpass sampling frequencies. W_i are the appropriate weighting functions.

With the $(k+1)$th equalizer, the multi-objective function is described as follows :

$$U_{k+1} = \sum_{j=1}^{m} W_1\left(\frac{G_{k+1}(\omega_j)}{G_{0_{k+1}}} - 1\right)^2 + W_2\left(\frac{R_{in\,k+1}(\omega_j)}{R_{in0_{k+1}}} - 1\right)^2 + W_4\left(\frac{R_{out\,k+1}(\omega_j)}{R_{out0_{k+1}}} - 1\right)^2 \tag{14}$$

At each iteration, coefficients $h_i + \Delta h_i$ are corrected to minimize the objective function using the Levenberg-Marquardt algorithm. Vector Δh is given as[5]

$$\Delta h = -\left[J^T J + \alpha\, D^T D\right]^{-1} J^{-1} e_0 \tag{15}$$

e_0 is the initial error vector, J is the Jacobian matrix of e (where the elements are $\partial e_j/\partial h_i$; $j = 1 \cdots m$; $i = 1 \cdots n$), D is a diagonal matrix and α is the Levenberg-Marquardt parameter. J. J. More[6] introduced relationships between J, D and α which permit rapid convergence.

Transmission line circuit topology

The extraction of lumped element topologies of the ladder form using the Darlington procedure is well known[7], the real frequency technique can also be applied using the Richards transformation to distributed commensurate transmission line extraction[8]. This extraction approach has been applied to the following three-stage low-noise amplifier.

HYBRID REALIZATION

This design method has been applied to the design of a low-noise amplifier over the 5,925-6,425 GHz bandwidth. This amplifier is realized with distributed elements and uses NE242 transistors from the NEC catalog.

Specifications

The specifications (Table 1) are provided by our industrial partner ALCATEL ESPACE. This amplifier is in the C band (Figure 6).

Table 1. Specifications of the C band amplifier

Parameter	Value
Gain	≥ 33 dB
VSWRs	≤ -10 dB
Noise Figure	≤ 1,4 dB
Flatness	± 0,2 dB

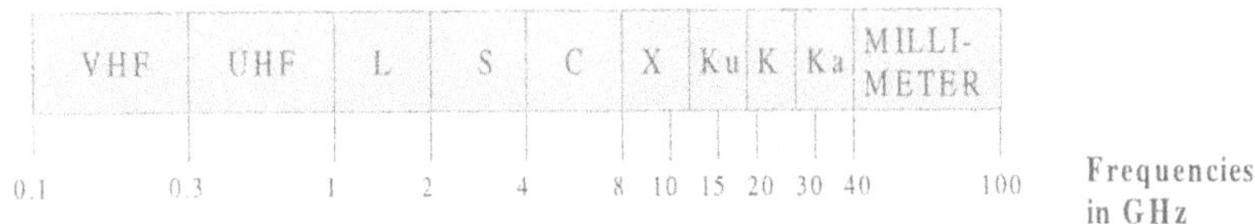

Figure 6. Designation of frequencies band

Simulation

The following figures (Figure 7, 8 and 9) present the results of the simulation by the program associated to the real frequency method.

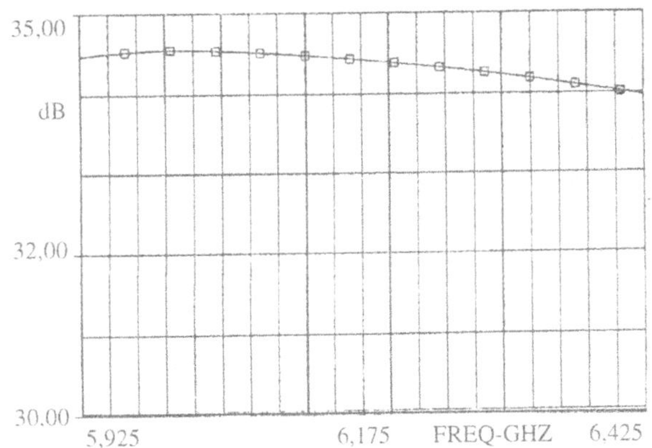

Figure 7. Simulated gain of the C band amplifier.

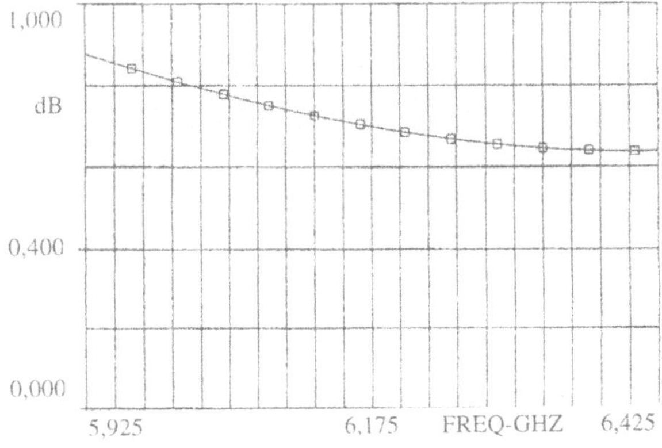

Figure 8. Simulated noise figure of the C band amplifier.

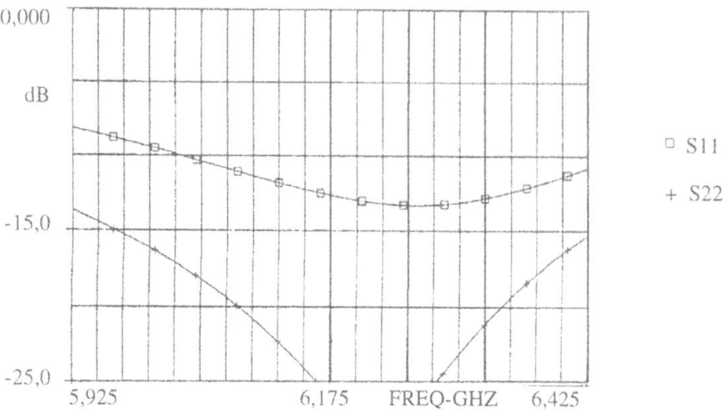

Figure 9. Simulated VSWRs of the C band amplifier.

Therefore, we totally satisfy the specifications. The simulation software permits to the optimal solution to be found quickly.

Realization

Figure 10 shows the hybrid implementation of the three-stage amplifier fabricated on a 0,635mm thick alumina substrate.

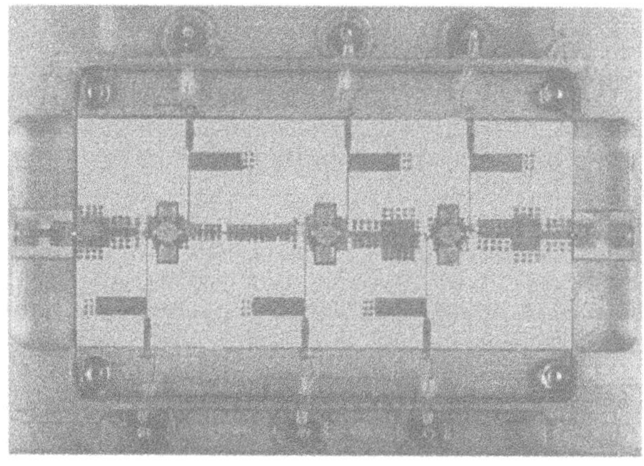

Figure 10. Photography of the C band amplifier.

Measure

The results of the measures were obtained after a few adjustments.

The signal gain is 32 dB with a very good flatness (±0,16 dB) on the 5,925-6,425 GHz band (Figure 11).

A noise figure less than 1,5 dB (Figure 12) was obtained.

Input and output VSWRs are less than -12 dB (Figure 13 and Figure 14).

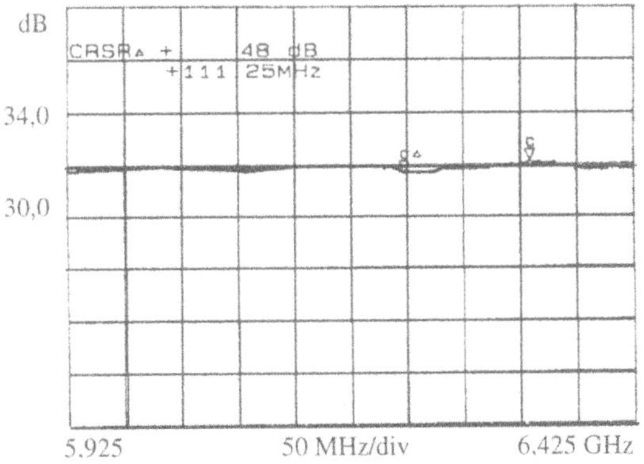

Figure 11. Measured gain of the C band amplifier.

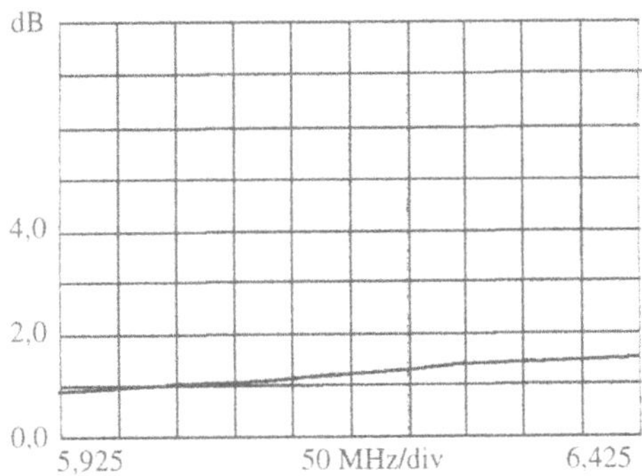

Figure 12. Measured noise figure of the C band amplifier.

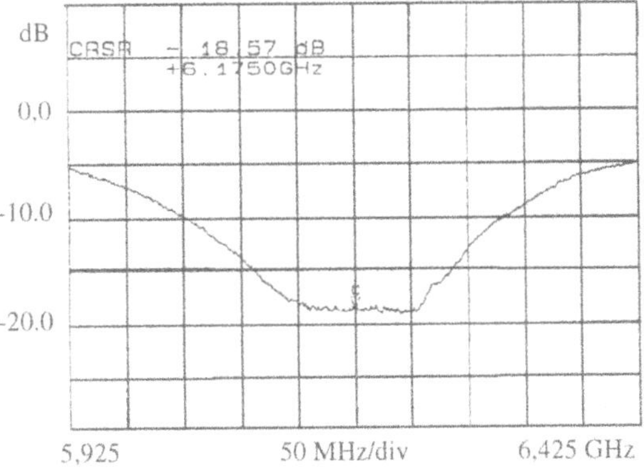

Figure 13. Measured input VSWR of the C band amplifier.

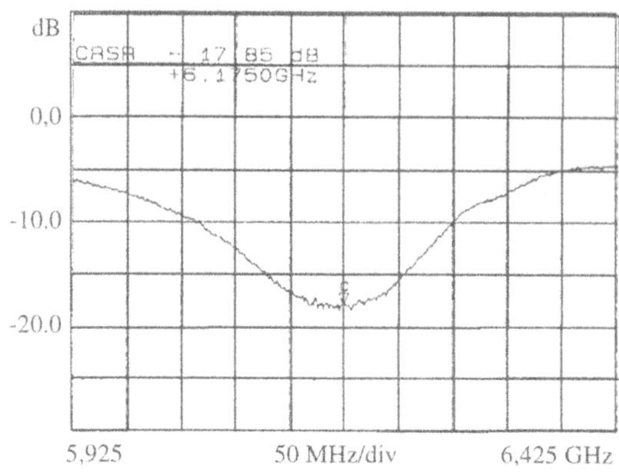

Figure 14. Measured output VSWR of the C band amplifier.

CONCLUSION

The measures obtained with the low-noise amplifier are similar to those obtained from the software simulation, which is based on the real frequency method.

Table 2 is a summary table of desired, simulated and measured results. It shows that the matching equalizers obtained by the real frequency technique are realizable.

Table 2. Summary of results for 5,925-6,425 GHz amplifier

Parameter	Desired	Simulated	Measured
Gain	≥ 33 dB	34,2 dB	32 dB
Flatness	± 0,2 dB	± 0,2 dB	± 0,16 dB
Noise Figure	≤ 1,4 dB	≤ 0,85 dB	≤ 1,5 dB
Input VSWR	≤ -10 dB	≤ -10 dB	≤ -15 dB
Output VSWR	≤ -10 dB	≤ -15 dB	≤ -12 dB

REFERENCES

1. H. J. Carlin and J. J. Komiak. "A New Method of Broadband Equalization Applied to Microwave Amplifiers", *IEEE Trans. Microwave Theory Tech.*, Vol. MTT-27, pp. 93-99, (Feb.1979).
2. B. S. Yarman and H. J. Carlin. "A Simplified Real Frequency Technique Applied to Broadband Multistage Microwave Amplifiers", *IEEE Trans. Microwave Theory Tech.*, Vol. MTT-30, pp. 2216-2222, (Dec.1982).
3. P. Jarry, E. Kerhervé, H. El Hendaoui, H. Tertuliano. "CAD of Microwave Integrated Active Filters by the Real Frequency Computer Program FREELCD", *Annals of Telecommunications*, pp. 191-205, (May-June 1996).
4. V. Belevitch. "Elementary Application of the Scattering Formalism to Network Design", *IRE Trans. Circuit theory*, CT-Vol. 3, (June 1956).
5. R. Soares, A. Pérennec, A. Olomo Ngongo, P. Jarry. "Application of a Simplified Real-Frequency Synthesis Method to Distributed-Element Amplifier Design", *Int. Journal of MIMICAE*, Vol.1, No. 4, pp. 365-378, (1991).
6. J.J. More. "The Levenberg-Marquardt Algorithm : Implementation and Theory", *Lecture Notes in Mathematics 630*, Springer Verlag, (1978).
7. S. Darlington. "Theory of Reactance 4-Poles which Procude Prescribed Insertion Loss Characteristics, *Journal Math. Phys.*, Vol. 18, pp. 257-353, (1939).
8. P. I. Richards. "Resistor transmission line circuits", Proc. IRE, Vol. 36, pp. 217-220, (Feb. 1948).

BROADBAND CHARACTERIZATION OF REPRESENTATIVE CIRCUIT-TO-CIRCUIT INTERCONNECTIONS

Alina Deutsch

IBM Research Division, Thomas J. Watson Research Center,
P. O. Box 218, Yorktown Heights, NY 10598
Phone: (914) 945-2858; FAX: (914) 945-2141
email: deutsch@watson.ibm.com

Abstract - This paper reviews the major class of chip-to-chip interconnections used in high-performance computers and communication systems. The critical electrical characteristics are highlighted and the most beneficial type of measurement techniques are exemplified for cables, printed-circuit-boards, ceramic, thin-film, and on-chip wiring.

I. INTRODUCTION

Continued advances being made in circuit density and speed, both at the chip and package level, are placing increasing demands on the performance of interconnection technologies. Designers are reducing the wiring cross sections and trying to pack the lines closer together, while at the same time the propagated signals switch with faster rise times. New insulator materials with lower dielectric loss and dielectric constants, and conductors with lower resistivity are extending operation to higher speeds. Improvements in integration efficiency need to be balanced against cost effectiveness at the system level. Reducing wiring dimensions results in appreciable resistive lines. Transient analysis of coupled lossy transmission lines having frequency-dependent parameters becomes a necessity. Interconnection performance is limited by the ability to control reflections caused by discontinuities in the signal paths, such as vias, crossing lines, wire bonds, connectors; it is also limited by the higher level of crosstalk and switching-induced noise due to the packaging density and speed increase.

This paper addresses the need to understand the increase in signal propagation delay, rise-time degradation, attenuation and coupled noise caused by loss mechanisms such as skin-effect and dielectric dispersion. High-speed signal propagation and the use of lossy

Directions for the Next Generation of MMIC Devices and Systems
Edited by N. K. Das and H. L. Bertoni, Plenum Press, New York, 1997

transmission lines are narrowing the gap between digital and microwave circuit designers. The circuit-to-circuit interconnection in today's high-performance computers and communication systems has many varieties, from cables, printed-circuit-boards, ceramic carriers, to thin-film and on-chip wiring. Signals on these carriers have rise times of t_r = 100 to 1000 ps, resistance of 0.05 - 500 Ω/cm, and maximum usable lengths of 1 cm - 5 m.

We will review the electrical characteristics of representative package interconnections from each category and highlight the performance limiting factors. A novel broadband technique for completely characterizing the frequency-dependent electrical properties of resistive transmission lines by short-pulse propagation is described. The technique is illustrated with measurements on cables, card wiring, thin-film and on-chip interconnections. Broadband dielectric loss extraction is shown for new low loss printed-circuit-board material. A simple method is shown for determining the dielectric anisotropy of polyimide insulators used in thin-film multilayer structures (such as MCM-D carriers). Signal propagation, coupled noise, and eye-diagram measurement results on representative test vehicles are compared with simulations based on full-wave, electromagnetic modelling.

II. PACKAGING TECHNOLOGIES

The packaging technology for high-performance digital computers and communication systems has to provide the connectivity, power distribution and cooling for chips that could have over nine million transistors and over 1,000 signal I/O's. **Figure 1** shows a generic configuration where the circuits on one printed-circuit-card have to communicate through a backplane board to the circuits on another card.

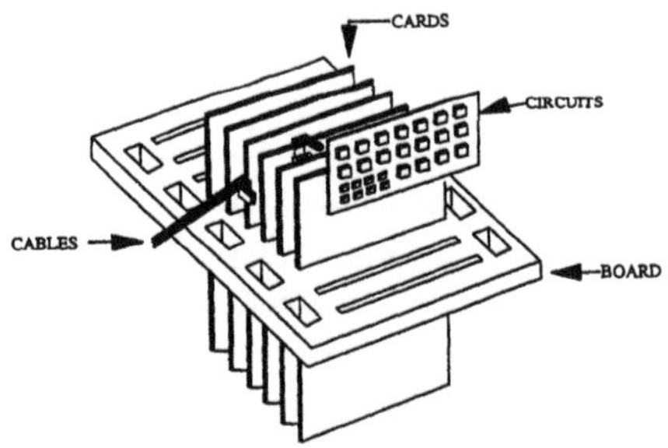

Figure 1. Schematic representation of card-on-board packaging.

The driver and receiver circuits could be packaged on single-chip, or multi-chip carriers, and the card-to-card path might involve either a mother board or many cables. The interconnection technology can be broken down into five categories as shown in **Table I**, namely shielded cables, printed-circuit-board, ceramic, thin-film, or on-chip wiring. All these transmission lines have non-uniform cross sections caused by discontinuities such as connectors, vias, wire bonds, flip-chip solder balls, redistribution leads, orthogonal lines, sparsely placed power buses, insulators with anisotropic dielectric constant, lossy dielectrics, stacks of many dielectrics. Signal integrity and system performance has to be balanced against fabrication limitations, cost and system size.

Table I. Interconnection Technologies

Interconnection Type	Line Width (um)	Line Thickness (um)	Line Resistance (ohm/cm)	Rl $_{MAX}$	Maximum Length (cm)
On-Chip	0.5 - 2	0.7 - 2	100 - 1000	> 2Zo	0.3 - 1.5
Thin-film	10 - 25	5 - 8	1.25 - 4	0.5Zo - 2Zo	20 - 45
Ceramic	75 - 100	16 - 25	0.4 - 0.7	< 0.2Zo	20 - 50
Printed-circuit-board	60 - 100	30 - 50	0.06 - 0.08	< 0.1Zo	40 - 70
Shielded Cables	100 - 450	35 - 450	0.0013 - 0.033	< 0.1Zo	150 - 500

Modelling of a typical signal path of Fig. 1 such as depicted in **Figure 2** is extremely challenging due to the variety of structures.

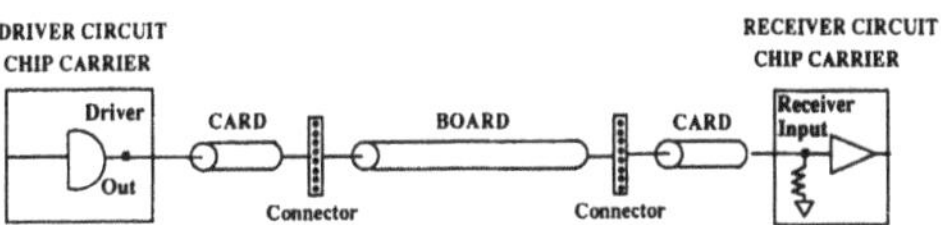

Figure 2. Representative signal propagation path between driver and receiver circuit for the type of package shown in Fig. 1.

It is the task of the modelling engineer to subdivide the problem into tractable components that can be analyzed with available electromagnetic tools and to concatenate or superimpose various effects when CPU run times or storage limitations restrict such studies. It is also imperative to verify the validity of such simplifications experimentally or to supplement the modelling with results obtained through testing of relevant test vehicles. Relevant measurements involve signal propagation integrity, crosstalk, reflections, dielectric constant and dielectric loss, attenuation, and eye-diagrams. Time-domain-measurement of signal propagation is mostly relevant for digital applications but will be affected by distortions, noise sources and loss mechanisms that are frequency dependent. Measurement techniques have to be adapted for the special needs of packaging structures. The following sections will highlight relevant characterization results for the various packaging interconnection technologies.

III. SHORT-PULSE PROPAGATION TECHNIQUE

In most of the package interconnections of Table I, the transmission line cross section and extent of non-uniformities are smaller than the wavelength for the frequency range of interest. The solutions to Maxwell's equations for electric and magnetic field can be considered quasi-TEM waves. The line voltage and current can be expressed as[1]

$$V = V_A\, e^{-\Gamma x} + V_B\, e^{\Gamma x} \tag{1}$$

$$I = I_A\, e^{-\Gamma x} + I_B\, e^{\Gamma x} \tag{2}$$

where the propagation constant is defined as $\Gamma = \sqrt{ZY} = \sqrt{(R + j\omega L)(G + j\omega C)}$ and the characteristic impedance is

$$Z_0 = \sqrt{\frac{Z}{Y}} = \sqrt{\frac{R + j\omega L}{G + j\omega C}} \tag{3}$$

V_A, V_B, I_A, and I_B are constants that can be determined by boundary conditions at the two ends of the line, and R, L, C and G are the line resistance, inductance, capacitance and dielectric conductance per unit length and are in general frequency-dependent. The propagation constant can also be written as $\Gamma = \alpha + j\beta$, where $\alpha = \mathrm{Re}\Gamma$ is the attenuation constant and $\beta = \mathrm{Im}\Gamma$ is the phase constant. Consequently, the key elements that need to be determined for characterizing any transmission line system then, are α, β, and Z_0. In practical digital logic applications, however, the key parameters of interest are propagation delay, characteristic impedance, risetime degradation and crosstalk, all of which are determined in the time domain. This is why most of the measurement and simulations of package interconnections are being made using TDR (time-domain reflectometry) and TDT (time-domain transmission) techniques[1]. Skin-effect and dielectric losses are best handled in the frequency domain by most circuit simulators, and the results are Fourier or Laplace transformed into time-domain waveform modelling. The accuracy of these measurements is limited by the finite risetime of the step source, the bandwidth of the commercially available sampling oscilloscopes and of the probing system connecting the source signal to the test pads on the sample being measured. The driving-end impedance Z_0 measurement is impeded by unwanted distortions caused by the probes and the large separations between signal and ground pads. Moreover, in the case of lossy transmission lines, the impedance increases with loss[1], and the TDR trace is not a flat reflected step that can easily be related to a constant reflection coefficient. The incident-impedance, however, is extremely important in package design because it affects the transient current the driver circuits have to supply and therefore, the simultaneous-switching capability of the carrier, due to ΔI-noise limitation. The propagation delay of lossy interconnections (measured at the 50 % level of the waveforms for digital applications) increases in proportion with line length and cannot yield a direct measurement of the dielectric constant of the insulating material being used.

Based on dimensional and material data collected during fabrication, the cross section is modelled and the frequency-dependent attenuation α, phase-constant β, and characteristic impedance Z_0 are calculated. The accuracy of the modelling is gated by the availability and accurate determination of structure dimensions and characteristics. The results of these models are then used in simulating pulse propagation with transient circuit analysis programs. By using such a procedure, there is no direct verification of the transmission line models. Instead, their correctness is inferred by comparing simulated with measured waveforms.

An alternative frequency-domain technique is often used[2]. S-parameter measurements are made using a network analyzer from which the transmission line characteristics are calculated. The accuracy of such a measurement depends on the ability to de-embed the frequency-dependent parasitics introduced by the probing system and of the interface between the test pads and the actual wiring which could be buried deep into the substrate. As the lines start having appreciable series loss, lossless line standards used for de-embedding become useless. In general, the entire calibration procedure is extremely cumbersome and the frequency-domain results give no direct physical interpretation of reflections caused by structure inhomogeneities or crosstalk waveshapes.

A new technique has been developed for completely characterizing resistive transmission lines. A short electrical excitation is launched onto two identical transmission lines of different lengths. The digitized waveforms are then numerically Fourier transformed. Time windowing is used to eliminate any unwanted reflections, and therefore the Fourier spectra contain information about the forward-travelling wave only (the forward term in (1) is given by $V = V_A e^{-(\alpha+j\beta)l}$ where l is the line length). The ratio of the complex spectra then yields the propagation constant.

$$\alpha(f) + j\,\beta(f) = -\frac{1}{(l_1 - l_2)}\ln\frac{A_1(f)}{A_2(f)} + j\,\frac{\Phi_1(f) - \Phi_2(f)}{(l_1 - l_2)} \tag{4}$$

where $\alpha(f)$ and $\beta(f)$ are the frequency dependent attenuation coefficient and propagation constant, respectively. $A_i(f)$ and $\Phi_i(f)$ (i = 1, 2) are the amplitude and phase of the transforms corresponding to lines of lengths l_1 and l_2, respectively, with $l_1 > l_2$. The effect of interface discontinuities, which are the same for both lines, linearly cancel out making it unnecessary to do any de-embedding. In the special case of insulators with small losses $G \ll \omega C$ (which is usually the case for most practical insulators in packaging structures) equation (3) reduces to

$$Z_0(f) \cong \sqrt{\frac{R + j\omega L}{j\omega C}} = \frac{\Gamma(f)}{j\omega C} = \frac{\beta(f)}{\omega C} - j\frac{\alpha(f)}{\omega C} \tag{5}$$

The frequency dependence of the capacitance is usually dominated by the dispersion of the dielectric constant. For low loss dielectric this is very small. Thus, the measured low-frequency capacitance together with the experimentally determined phase constant and loss coefficient from (4) can be used to determine the complex impedance with (5).

IV. PRINTED-CIRCUIT-BOARD WIRING

a) Dielectric Loss

In the case when $Rl_{max} \ll Z_0$ ($Rl_{max} \leq 0.1\ Z_0$) the interconnection transmission lines are considered lossless and will transmit signals with negligible distortion. Such is the case with wide (75 - 100 μm) and thick (15 - 50 μm) printed-circuit-board wiring that have negligible resistance of about R_{dc} = 0.06 Ω/cm. Even for l_{max} as large as 70 cm, Rl_{max} = 4.2 Ω which is less than one tenth of Z_0 (Z_0 = 50 - 80 Ω). These lines are generally terminated to avoid reflections and the amplitude is only reduced by the dc drop. The onset of skin effect can occur for risetimes as slow as 7 ns[1]. The attenuation caused by skin-effect losses (tan δ = 0.0), however, is quite small, 0.029 dB/cm at 0.5 GHz, and even for 70-cm-long lines, the loss is only 2 dB. The generally used FR-4 insulator which is a composite of epoxy and fiber-glass cloth has a very high dielectric loss tangent tan δ of 0.025. Due to the low resistive losses, dielectric loss will dominate and substantially increase signal attenuation and dispersion. Even at 0.5 GHz, attenuation increases to 0.053 dB/cm. Dielectric loss increases in proportion to frequency since $G = \omega C \tan\delta$ (attenuation is proportional to G). At 2 GHz, the total loss on a 70-cm long signal line increases to 10.8 dB which is substantial. Such large losses inhibit the propagation of fast signal transitions on long printed-circuit-board wiring. The most typical range of interest for digital applications is f = 0.5 - 2 GHz, which corresponds to risetimes of 300 - 1,000 ps. Lower dielectric loss thus increases the bandwidth of board wiring. In addition, the use of low dielectric constant ε_r materials (lower than $\varepsilon_r = 4.3$ for FR-4) results in lowering of signal propagation delay τl in proportion to $\sqrt{\varepsilon_r}$. It also reduces the wiring capacitance in proportion to ε_r which results in lower power dissipation since lower amplitude currents are needed to charge the signal lines.

The short-pulse propagation technique was used to characterize specially built test vehicles having either FR-4 dielectric or a newly developed material, a thermoplastic toughned cyanate ester[3]. The attenuation for 102 x 31 μm, 5-cm and 10-cm long lines was measured for the 0.8 - 8 GHz range. The frequency-dependent attenuation $\alpha(f)$ was also calculated and the calculation was repeated for differing values of tanδ until a good agreement was obtained between calculated and measured attenuation as shown in **Figure 3**.

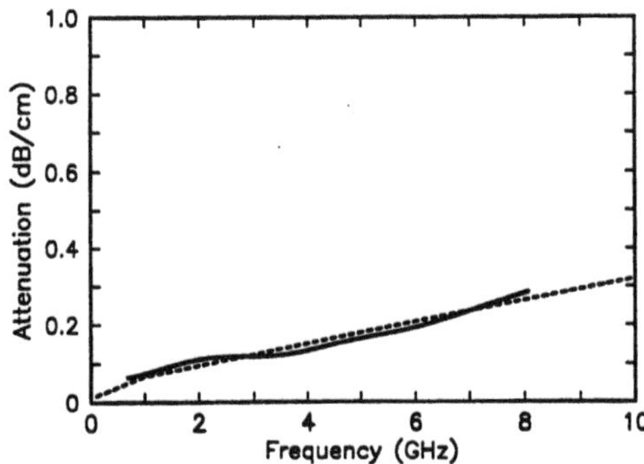

Figure 3. Measured (solid trace) and calculated (dashed trace) attenuation with tanδ = 0.009. Copyright 1996 by IEEE Transactions on Components, Packaging, and Manufacturing Technology - Part B.

It was found that the new material has tanδ = 0.0095 ± 0.0025 which is lower than for FR-4.

Large, circular, 1.524-cm diameter, parallel-plate capacitors were built in the test vehicle. Measurements of capacitance at low frequency (1 MHz) allowed the extraction of ε_r to be $\varepsilon_r = 3.64 \pm 2$ %. The large diameter-to-dielectric-height aspect ratio of the plates allowed the capacitance calculation to be performed without the inclusion of any fringe components.

Simulations were performed of a representative signal path as the one shown in Fig. 2 which is part of a serial communication link that has 500 Mb/s data rate which require the propagation of 2-ns wide data pulses. The requirement at the receiver input is to have adequate "eye-opening" or waveform area.

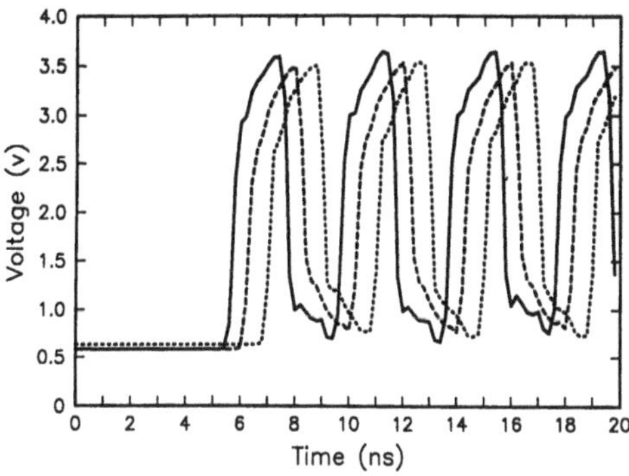

Figure 4. Simulated waveforms for card-board-card path with FR-4 and CYTUF insulators. Responses at the receiver circuit input are shown for 70 cm CYTUF-LC total wiring (solid trace), 70 cm FR-4 total wiring (dashed trace), and 90 cm CYTUF-LC total wiring (dotted trace). The 70 cm wiring is divided into 20, 40 and 10 cm segments, while the 90 cm wiring has 30, 40 and 20 cm parts. Copyright 1996 by IEEE Transactions on Components, Packaging, and Manufacturing Technology - Part B.

These waveforms are shown in **Figure 4** for a total length of 70 cm with either the new insulator or FR-4, and for 90-cm long wiring with the new material. The less lossy material results in a card wiring length improvement of 20 cm over FR-4 which is significant. Propagation delay improvement is 8 %.

b) Card-to-Board Connectors

Typically 300 - 500 signal connections and soon even 1,000, between daughter cards and mother board are required, and distortionless signal propagation of 300 - 1,000 ps rise-time transitions has to be assured with connector crosstalk contribution in the order of 5 - 10 %. Pinned connectors are limited in density since printed-circuit-boards cannot be fabricated with through-via-holes on grids smaller than 1.25 mm using conventional techniques. 5 - 6 rows of contacts are generally used but a large number of contacts have to be allocated to ground current returns. It is important to evaluate the capacitive and inductive discontinuities introduced by connectors. If the card-to-board interface delay is close to $t_r/2$, the waveform is distorted and displays an increased rise time. Inductive and capacitive discontinuities cause reflections which have amplitudes proportional to $[L/(2Z_0 t_r)] \cdot V_{in}$ or $(CZ_0/2t_r) \cdot V_{in}$, respectively (where V_{in} is the input source signal)[4]. Such reflections can become significant if they are large enough to cause logic failure. TDR measurements are generally performed using step excitations from sampling oscilloscopes such as the HP 54120 with 35-ps risetime and 50-Ω-impedance source. **Figure 5** shows typical response for a 5-row pinned-connector having 1:1 signal-to-ground contacts. The TDR waveforms are shown for two locations namely, the shortest lead and the longest lead. The shortest lead has an equivalent impedance Z_C = 55.5 - 61.5 Ω, while the longest lead has Z_C = 56.0 - 67.4 Ω.

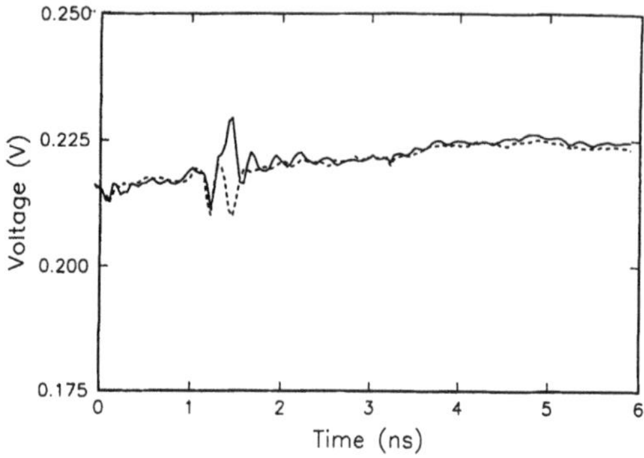

Figure 5. TDR measurements using a 35-ps-risetime, 200-mV source at 10-cm distance from the 5-row pinned connector. Measurements are shown for connections made to the fifth row (longest lead in the receptacle, solid trace) and the first row (dashed trace).

A very important performance determining criterion is the amount of noise generated in the connector leads. Crosstalk is generated by the transfer of energy through capacitive and inductive coupling. The capacitive K_C $(-C_{12}/C_{11})$, and inductive K_L (L_{12}/L_{22}) couplings between adjacent lines add at the driving end and subtract at the far end of the line[1]. Crosstalk is generally measured with two types of configurations, for near-end-noise, NEN, at the end close to the source, and far-end-noise, FEN, at the far end of the quiet line. **Figure 6** shows the good agreement between measured and simulated results with the pinned connector in a 70-cm-long path on a specially designed test vehicle. The connector contribution is riding on top of the large card wiring crosstalk and is best measured by considering two adjacent pin-pairs at a time. The connector leads were modelled using a three-dimensional finite-element algorithm to extract the equivalent R, L, and C matrices[5].

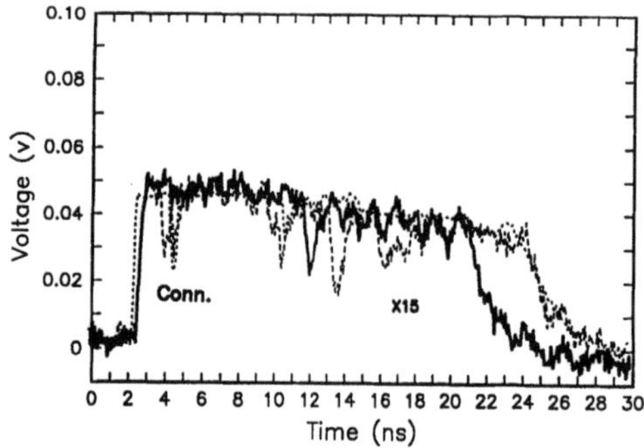

Figure 6. Measured and simulated (dotted) near-end-crosstalk, NEN, for the 70-cm-long card path of Fig. 2 with a 289-ps-risetime, 200-mV swing, step source with (dashed) and without (solid) the two 5-row pinned connectors. The connector contribution is riding on top of the card crosstalk.

Simulations included the probe tip discontinuities and sample test pad models as well. The equivalent L and C for the probe tips were modelled using the three-dimensional tools described in Ref. 6 and 7 and were found to be L_p = 0.1625 nH and C_p = 0.03 pF which represent very small discontinuities. The probes were characterized using a 70-GHz bandwidth sampling oscilloscope[8] and it was found that the 5.5-ps fall-time of the instrument output only degraded to 14.5 ps at the output of the probe. This implies a 3-dB bandwidth of 24 GHz and acceptable behavior across a wide frequency range. This last point is crucial for time-domain measurements, since poor phase performance will result in distorted signals even when amplitude bandwidth is broad. The custom-built probe tips are shown in **Figure 7** and are accomplishing the 24-GHz bandwidth with only one ground contact which is mostly the case for packaging applications.

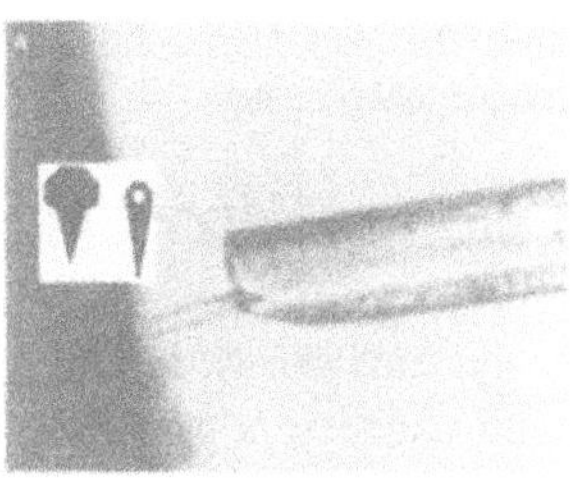

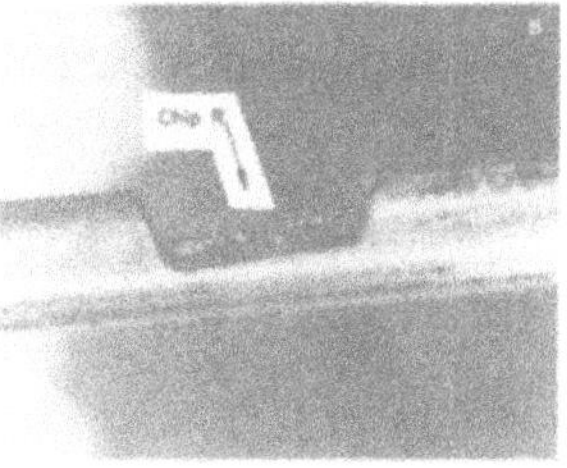

Figure 7. a) Bird-beak probe tips mounted on semi-rigid coaxial cable. The actual source. The actual ground (left) and signal (right) probe plates are shown in inset. b) Side view of opening in the semi-rigid cable where the chip resistor is mounted in the high-impedance probe.

All the broad-band co-planar probes described by others[9] achieve their highest bandwidth with ground-signal-ground probe configurations. The low-cost, bird-beak design allows reliable contact for a large range of planar and non-planar geometries such as encountered in representative packaging interconnections.

Crosstalk measurement was also used to assess the need for large number of ground contact assignment in a surface-mount connector case[10]. **Figure 8** compares the NEN noise for 1:1 and 5:1 signal-to-ground cases. The connector crosstalk has to be contained in the overall noise budget of the receiver circuit which has to also accommodate other noise sources such as simultaneously-switching-driver-circuits-induced noise (ΔI-noise).

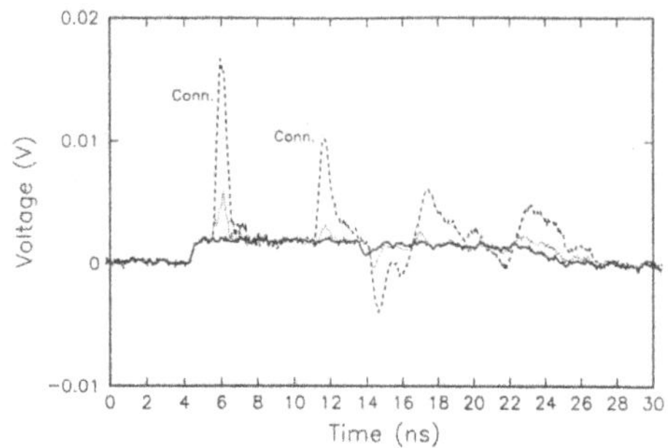

Figure 8. Measured NEN waveforms for a path with 70-cm card wiring using a 289-ps-risetime, 200-mV swing, step source. Path is tested without the connectors (solid) and with two pad-on-pad connectors having either 5:1 (dashed) or 1:1 (dotted) contact assignments.

V. HIGH-PERFORMANCE HIGH-DENSITY CABLES

The cabling required for the systems shown in Fig. 1 must be 1 - 5 m in length inside a single electronic enclosure with connectors providing contact to associated card wiring at a 10-20-per-inch linear density. Up to several hundred cables might have to be attached to each card. In such cases, shielding is mandatory in order to avoid loss of energy through radiation, minimize intra-cable crosstalk, and maintain controlled impedance.

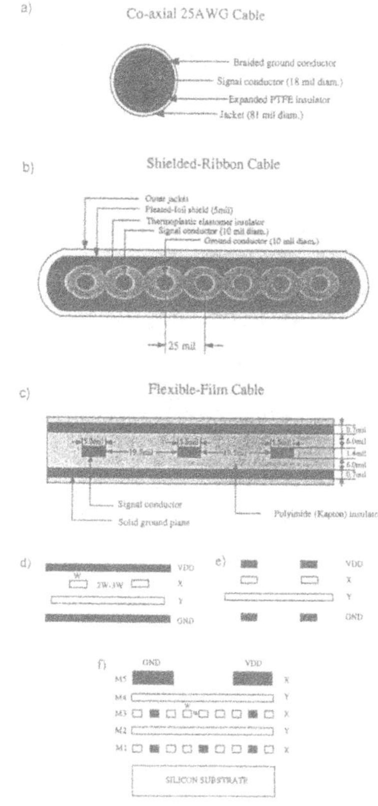

Figure 9. Schematic cross sections of cables: a) coaxial, b) shielded ribbon, c) flexible-film, and X and Y wiring with d) solid reference planes, e) mesh planes, and f) five-layer on-chip wiring. Copyright 1994, and 1995 by International Business Machines Corporation.

Three types of cross sections are shown in **Figure 9**, namely coaxial, shielded-ribbon, and flexible-film structures.

The short-pulse propagation technique was used to analyze their attenuation and characteristic impedance over a broad frequency range. Measurements were performed with 2- and 3-m-long cables. **Figure 10** shows the attenuation for the 25AWG and 30AWG coaxial cables, the ribbon cable and the flexible-film design.

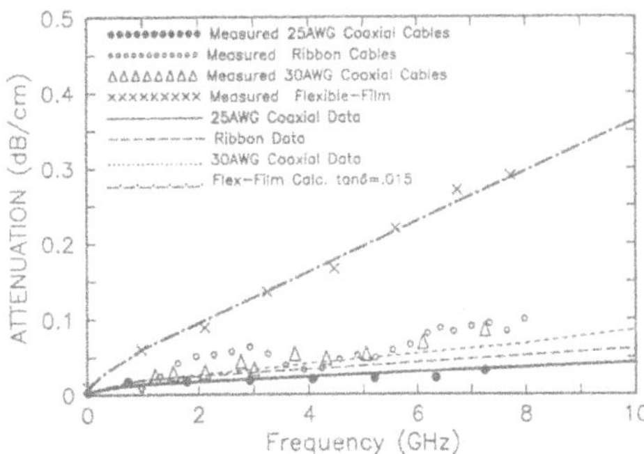

Figure 10. Measured (dots) and manufacturers' data (lines) for coaxial, shielded-ribbon, and flexible-film cable attenuations as a function of frequency.

The vendor-supplied data is generally available only up to 400 MHz. Extrapolation based on skin-effect losses was made in order to compare with our measured results up to 8 GHz. Such measurements were extremely beneficial in highlighting the short-comings of some of the insulators used such as foamed polypropylene, thermoplastic elastomer (used in the ribbon cable) or Kapton (used in the flexible-film cable). The flexible-film with 5.5 x 1.4-mil cross section and measured $\tan\delta = 0.015$ had the highest attenuation. It is not obvious from manufacturers' specifications that such losses exist and that they limit extendibility to GHz data rates while the short-pulse propagation technique can provide the means to identify such salient effects as dielectric loss. Once again, dielectric loss is dominant over resistive loss which are small.

Similarly to card-board-card paths, the card-cable-card paths have to be analyzed for signal distortion and crosstalk in order to determine the maximum useful length and bandwidth. TDR and TDT measurements are generally performed to determine the cable connector capacitive and inductive discontinuities and crosstalk[11]. In addition, signal degradation caused by dc drop, risetime dispersion and reflections is evaluated from eye-diagram data streams. Pseudo-random data patterns with variable data rates are used as input source. The random pattern simulates typical data streams found in communication links. The variable data pattern exacerbates the signal distortion because each pulse has a variable starting steady-state level, depending on the number of consecutive low or high states in the data. A 40 - 50 % loss of the "eye-opening" or the area of the waveform is associated with a reasonable signal at the receiver-circuit input to overcome all the anticipated noise sources. **Figure 11** shows the results obtained with a 2-m-long coaxial 25AWG cable at 500-Mb/s, 1-Gb/s and 1.6-Gb/s rates. The path included two 12.7-cm lengths of card wiring at each end of the cable. The acceptable bit-rate limit was found to be 1.3 Gb/s.

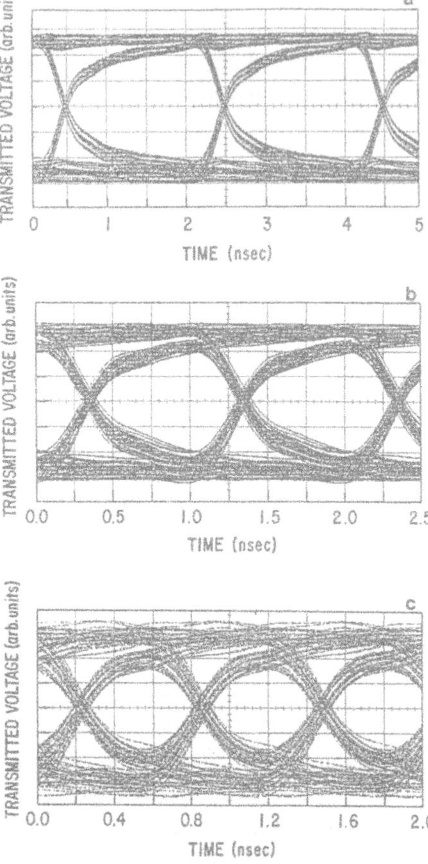

Figure 11. Eye diagrams obtained with 2-m-long coaxial 25 AWG cable, at a) 500-Mb/s, b) 1-Gb/s, and c) 1.6 Gb/s data rates. The path traversed by the signal included two 12.7-cm lengths of card wiring and connectors at each end of the cable. Copyright 1994 by International Business Machines Corporation.

VI. CERAMIC CARRIERS

Thick-fim ceramic carriers have lines, whether terminated or unterminated, with finite series loss since $Rl_{max} \leq 0.2\ Z_0$, and some rise-time distortion will be encountered. Typical cross section is shown in Fig. 12e with mesh ground planes dictated by fabrication limitations. Line widths are in the range of 75 - 100 μm and center-to-center spacing of 350 - 500 μm. Large wiring stacks of 30 - 60 layers and 64 - 150 mm size carriers are common[12]. A typical multi-chip module is shown in **Figure 12.** This type of wiring uses fairly thick lines (16 - 25 μm) and skin-effect will occur with signals of around 1-ns rise time. Typical transitions are in the order of 200 - 1,000 ps. Distortion, however, will be minimal because of the low R_{dc} (R_{dc} = 0.4 - 0.7 Ω/cm). Dielectric loss can again play a strong role, however, most ceramics have been shown to have very low tanδ. For example, alumina can have tanδ = 0.0005, and mullite type ceramics have tanδ = 0.0015 to 0.0035, and therefore dielectric dispersion can also be neglected for most ceramic packages. Most of the distortion is introduced by the orthogonal wire capacitive loading and the very long via loading which will cause most of the risetime distortion. Modeling of such complex multi-chip carriers as shown in Fig. 12 becomes a very challenging task and is very much needed due to the stringent electrical requirements and high fabrication cost[13].

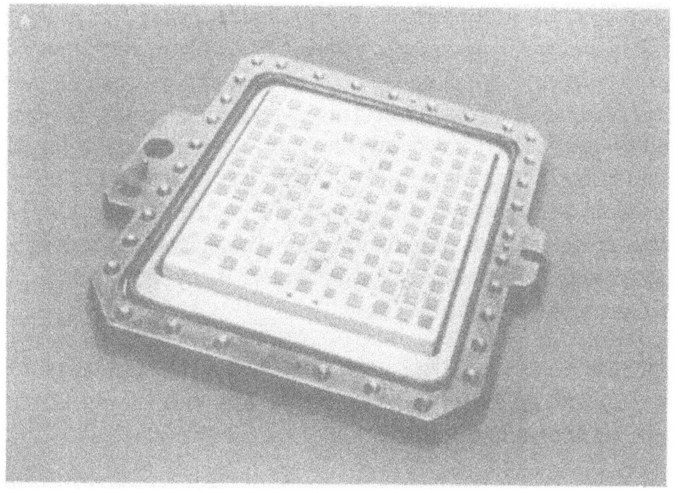

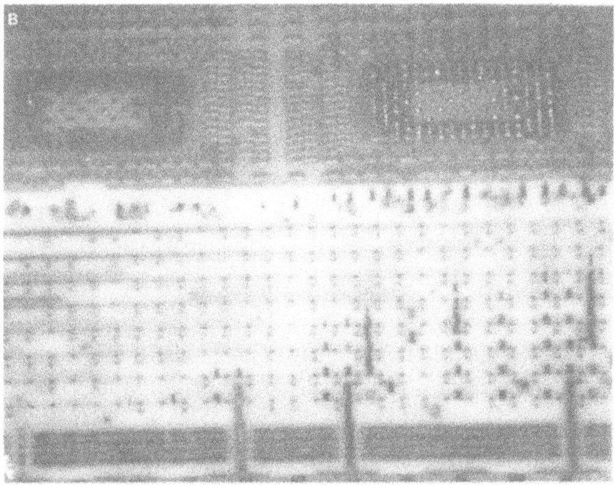

Figure 12. a) Top view and b) cross section of a 100-mm multi-chip alumina/mollibdenum module with up to 11 x 12 chips and eight plane-pairs of mesh wiring of the type shown in Fig. 9e.

VII. THIN-FIM WIRING

In the case of chip-to-chip wiring on thin-film carriers, the lines have substantial series loss since Rl_{max} can be in the rangeof $0.5Z_0$ to $2Z_0$. For 100-mm carrier, l_{max} could approach 20 cm, and if the resistance is in the range R = 1.25 to 4 Ω/cm, unterminated connections need to be used as was recommended in Reference 14. A design guideline has been to use such lines for $l_{max} \leq 2\ Z_0\ /\ R$, such that the attenuation $e^{-Rl/2Z_0}$ will be less than or equal to 36.8 %. A controlled amount of attenuation can be tolerated because of the voltage doubling at the open end. Rise-time dispersion will occur, even for frequency-independent resistive loss. It was shown in Ref. 15 that since the resistive loss is very small for short lines (but long enough to have propagation delay greater than the rise time), the line behaves like an LC line, and sustained reflections from both ends generate unwanted over-and under-shoots. An active clamp network, such as one using Schottky-barrier-diodes, can be used to suppress the oscillations and maintain a fixed steady-state level at one tenth the current needed for a terminating resistor. The conductors generally used are fairly thin (5 - 8 μm), so that skin-effect in these cases could become significant at frequencies greater than 0.7 to 2 GHz or rise times t_r less than 200 ps (where $\delta \leq t$). Practical switching speeds t_r of current circuits are in the range of 200 to 1,000 ps; therefore, skin-effect-induced dispersion will degrade the signal rise times but total circuit-to-circuit interconnection delays are not substantially increased. Due to the high R_{dc}, skin effect will dominate and dielectric loss can generally be ignored. Since such interconnections can propagate fast signals, the switching speeds will be limited by the noise margin of digital receiver circuits. For typical systems, crosstalk and power-supply noise generated by simultaneous switching of many drivers and crosstalk will limit the transitions to not less than 200 ps.

TDT measurements and simulations were performed on representative thin-film wiring of a four-layer structure of the type shown in Fig. 12d with R_{dc} = 4 Ω/cm and l = 5.06 cm. The results are shown in **Figure 13.**

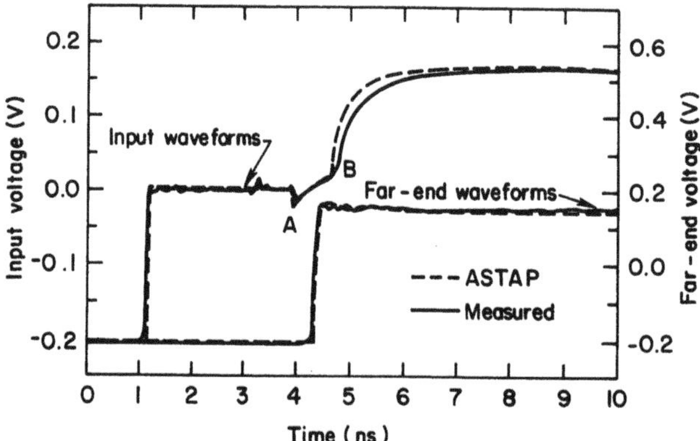

Figure 13. Measured (dashed curves) and simulated (solid curves) waveforms for line with R_{dc} = 4 Ω/cm and length l = 5.06 cm, using 10:1 coaxial probes. Input monitored at sending oscilloscope channel, output measured at far end of line. Copyright 1990 by International Business Machines Corporation.

A 500-Ω coaxial, custom-built probe shown in Fig. 7b was used which then presents a quasi-open termination to the transmission line with Z_0 = 40 Ω. This simulates the un-terminated environment used in typical thin-film wiring applications. A 450-Ω thin-film chip resistor in-line with the semi-rigid coaxial cable is used as shown in Fig. 10b. The 500-Ω probe was characterized as before with the same 70-GHz oscilloscope and its bandwidth was determined to be 19 GHz. The simulation includes the discontinuities associated with the chip resistor and wire bonds needed for assembly as seen in **Figure 14.** In this case the transmission line model calculation included frequency-dependent R and L due to skin-effect.

Most thin-film wiring use organic polymer insulators due to their low dielectric constant, and ease of processing. The low dielectric constant of the polymer assures fast signal propagation and low crosstalk, and therefore high-wiring density. The polymers that have excellent mechanical properties could exhibit large dielectric anisotropy. Special test structures were built and characterized to extract such properties. The test vehicle was built using BPDA-PDA polyimide as the dielectric and copper as the conducting material. The three-layer structure was fabricated on a large (166 x 166 mm) pyrex glass substrate having 90 mm x 93 mm active area representative of actual use. The minimum number of layers were built that still allowed high-speed characterization

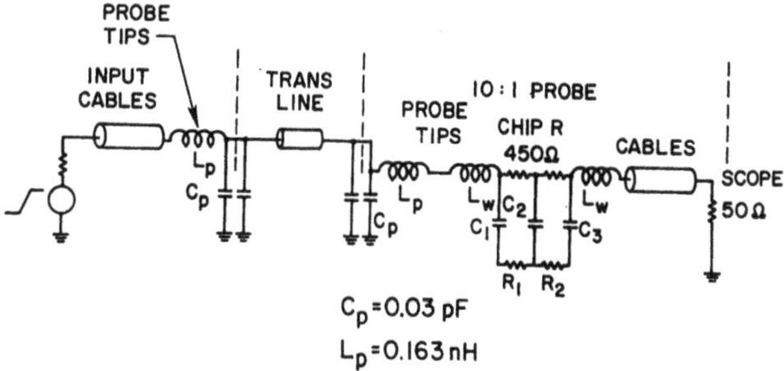

Figure 14. Equivalent circuit representation of 19-GHz-bandwidth high-impedance test system for measuring propagation on single line. Copyright 1990 by International Business Machines Corporation.

and at the same time underwent all the relevant processing steps used to fabricate chip-to-chip interconnection structures.

Moreover, relevant dimensions were used throughout. The cross section of the test vehicle is shown in **Figure 15a**. It consists of two metallic and two polymer layers and has a microstrip transmission line design. The out-of-plane dielectric constant ε_Z (in the direction of the film thickness) was obtained from capacitance measurements on 600 μm diameter parallel-plate capacitors as shown in Fig. 15b to be $\varepsilon_Z \pm = 0.1$. The in-plane dielectric constant was obtained through a combination of experimental and theoretical techniques. Specially designed comb structures, as shown in Fig. 15c were used. 10 inner and 11 outer, 3 cm long conductors are interleaved in order to amplify the mutual capacitance, C_{12}, between any two lines by a factor of 20. This enhances measurement accuracy since C_{12} is usually a very small value in typical designs. The following technique was used to extract the in-plane dielectric constant, ε_{XY}. The out-of-plane dielectric constant was fixed at $\varepsilon_Z = 3.22$. Modelling of the cross section was iterated until the best agreement between measured and calculated mutual capacitance values was obtained. It was concluded then that BPDA-PDA exhibits an in-plane dielectric constant of $\varepsilon_{XY} = 3.8 - 4.0 \pm 0.1$ which is significantly higher than the out-of-plane value of $\varepsilon_Z = 3.22$. It is explained in Ref. 1 that both FEN and NEN depend directly on the capacitive coupling coefficient $K_C = C_{12} / C_{11}$, the accuracy of which is determined by the ε_{XY} and ε_Z that was extracted. This effect is strongest for the far-end crosstalk which is proportional to the difference between the capacitive and inductive coupling, $K_C - K_L$ (where $K_L = L_{12} / L_{11}$). Small differences show up as large noise spikes and the inaccuracy in the FEN prediction increases with longer line lengths since FEN is proportional to the coupled length[1]. **Table II** summarizes the effect of dielectric anisotropy on the key transmission line electrical parameters, The self, C_{11}, and mutual, C_{12}, capacitance values increased by 4.6 - 9.6 % and 21.4 - 25.4 %, respectively. The far-end coupled noise, FEN, for the 3 cm line length, is overestimated by 4 - 5.3 %. The experimental characterization performed on representative structures was essential for determining the dielectric properties as was shown in detail in Ref. 16.

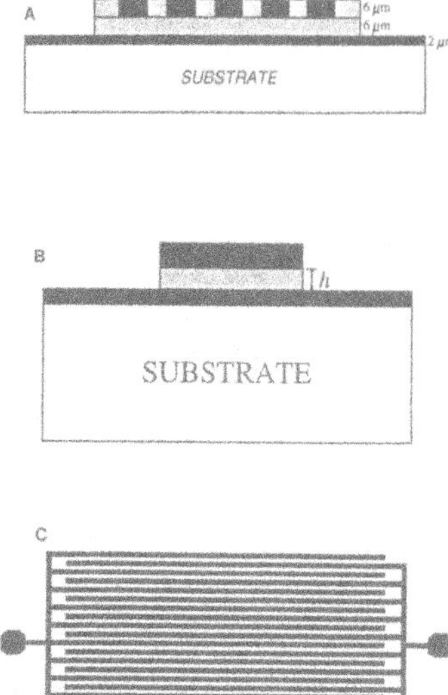

Figure 15. a) Three-layer test vehicle cross section, b) parallel-plate circular capacitor test site, and c) comb structure for extracting in-plane dielectric constant ε_{XY} by measuring the amplified mutual capacitance 20 x C_{12}. Copyright 1994 by IEEE Transactions on Components, Packaging, and Manufacturing Technology - Part B.

Table II. Effect of anisotropy on capacitance, impedance, propagation delay, and far-end coupled noise.

ELECTRICAL CHARACTERISTICS		Pitch (μm) 24	31	40
ΔC (%)	ΔC_{11}	9.6	6	4.6
	ΔC_{12}	21.4	22.8	25.4
Z_0 (Ω)	Measured	42	42.4	44
	Calculated w. Anisotrophy	36	39.5	40
	Calculated w/o Anisotrophy	38	40.7	41
τ (ps / cm)	Measured	61.7	61	60.8
	Calculated w. Anisotrophy	59.7	57	56.7
	Calculated w/o Anisotrophy	57	55.6	55.4
FEN @ 3cm (%)	Measured	Neg.	5	11
	Calculated w. Anisotrophy	Neg.	5.5	9.5
	Calculated w/o Anisotrophy	5.3	11.3	13.5

VIII. ON-CHIP WIRING

On-chip interconnections have very high resistance R, ranging from 35 to 500 Ω/cm and beyond, and $Rl_{max} \geq 2Z_0$. In the case of large microprocessor chips approaching 20 mm on a side, critical global wiring can be 1 - 2 cm in length. Device scaling to sub-quarter-micron channel length results in 50 - 100-ps-risetime initial signal transition. The propagation delay is comparable to the signal risetime and the conventional lumped-circuit RC-circuit representation is no longer adequate. It results in significant under-estimation of both delay and crosstalk. The combination of fast risetimes and long lengths require treatment as lossy coupled configurations[4]. A typical cross section is shown in Fig. 9f with five-metal layers. Most on-chip power distribution relies on wide power buses placed sparsely on the topmost layer. This layer has thick, non-planarized metallization with narrower buses on the lower levels distributing power to the individual logic cells. Such a nonuniform reference mesh results in increased inductance and resistance, which in turn increase the propagation delay and line-to-line crosstalk.

Special test sites were designed to analyze the different type of possible transmission line configurations in the various layers of the five-layer stack. Each site was designed to include all the representative details of orthogonal wiring, vias and wide and narrow reference buses. One such site is shown in **Figure 16** with two 2.7-μm-wide lines, with 2.7-μm-separation on the fourth layer and having 110.0-μm-wide orthogonal power buses on 450.0-μm pitch on the fifth layer.

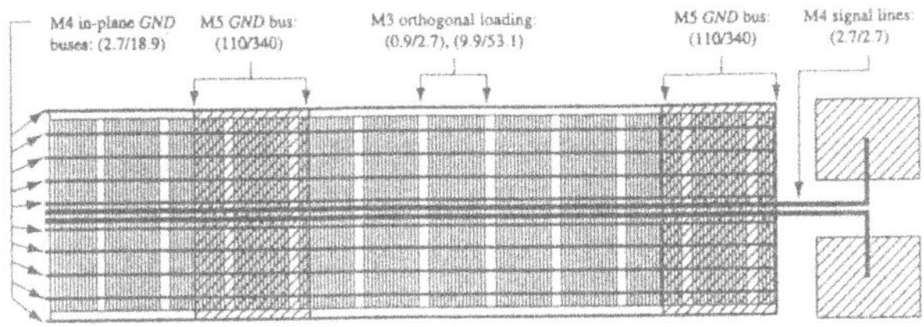

Figure 16. Layout of site with coupled 2.7-μm-wide signal lines on M4 having orthogonal 110.0-μm-wide ground buses on the fifth layer. (Numbers in parantheses are the line width and line spacing, respectively, both in μm). Copyright 1995 by International Business Machines Corporation.

Such complex structures are difficult to model accurately using conventional techniques. Time-domain measurements were performed using again the 35-ps risetime step excitation provided by a 20-GHz sampling oscilloscope. Signal propagation measurements were repeated on short structures that were identical with the long lines. The difference in the measured delay on the long and short lines yielded the propagation delay per unit length, without the error introduced by pads and probes. Signal propagation and crosstalk were measured on 1.6-cm-long lines, 2.7-μm-wide in the third layer having R = 166 Ω/cm. The measurements were repeated at -160 °C, 22 °C, and 100 °C. There was a 2.9x reduction in rise time at -160 °C compared to room temperature (22 °C). As a consequence, the delay measured at the 50 % level of the signals is lowered by 1.5x. At 100 °C, delay and rise time increase by 15 % and 16 %, respectively. This confirms the resistance-dominated performance due to the very large R ($Rl \cong 6.4Z_0$). At -160 °C, $Rl \cong 1.6Z_0$, which means that the line has a fast LC-line type of behavior explained in Ref. 1. The lower signal attenuation and dispersion, however, results in much higher crosstalk at -160 °C as seen **Figure 17.**

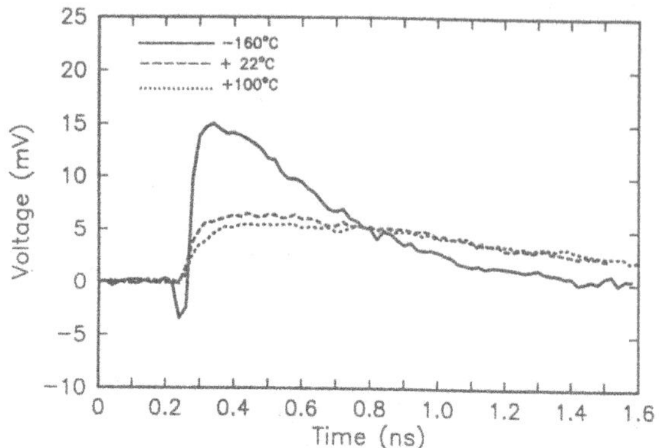

Figure 17. Measured crosstalk at the far-end of a quite line (FEN) for two adjacent 1.6-cm-long, 2.7-μm-wide lines, with 2.7-μm separation, on the third layer, with solid ground on the fifth layer Input is a 35-ps-transition, 200-mV swing source, and FEN waveforms are shown for - 160 °C, + 22 °C, and +100 °C operation.

Figure 18 shows measured and simulated waveforms on 1.56-cm-long, 4.8-μm-wide line on the fifth layer with R = 36.0 Ω/cm. The simulated waveforms are shown both with (dotted trace) and without (dashed trace) taking frequency-dependent losses into account and with a 50-section distributed RC-circuit representation (dot-dashed trace). The delay is under-predicted by 54 % with the RC-circuit. The frequency-indepent model is within 25 % of measurement while the frequency-dependent model improves agreement to within 11 % compared to measurement. These results highlight the need for using distributed RLC models for the accurate prediction of performance for long on-chip wiring. The modelling and simulation technique used throughout is explained in great detail in Ref. 17.

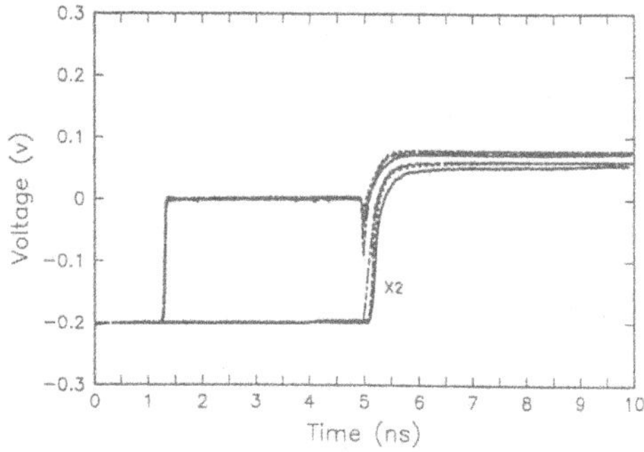

Figure 18. Measured (solid curves) and simulated waveforms using frequency-independent (R and L parameters (dashed curves), frequency-dependent R(f) and L(f) (dotted curves), and 50-section distributed RC-circuit representation, for 1.56-cm-long, 4.8-μm-wide line on the fifth layer, with parallel ground reference bus on the same layer. Input is monitored at sending oscilloscope channel; output measured at far end of the line and terminated with 50 Ω.

IX. CONCLUSIONS

A large range of packaging interconnections were reviewed and the different type of characterization techniques were highlighted. The experimental analysis was shown essential in all these complex configurations which are too difficult to model or have processing-induced effects too hard to predict. New approaches that rely on simple structures and instrumentation need to be pursued to complement, verify or substitute for the shortcomings of modelling and simulation tools.

REFERENCES

1. A. Deutsch., G. V. Kopcsay, et al., "High-speed signal propagation on lossy transmission lines", *IBM J. Res. Develop.*, 34:601 (1990).
2. S. B. Goldberg, M. B. Steer, P. D. Franzon, J. S. Kasten, "Experimental electrical characterization of high speed interconnections", *Proc. 41st Electr. Comp. & Tech. Conf.*, ECTC'91:85 (1991).
3. A. Deutsch, et al., "Broadband characterization of low dielectric constant and low dielectric loss CYTUF cyanate ester printed circuit board material", *IEEE Trans. Comp., Pack. and Manuf. Technology-Part B*, 19:331 (1996).
4. R. E. Matick, *Transmission Lines for Digital and Communication Networks*, McGraw-Hill Book Co., Inc., New York, (1969).
5. Z. J. Cendes, J. F. Lee, D. N. Shenton, B. Shyamkumar, E. Smetak and D. K. Sun, "Integrated microwave field simulation using three dimensional finite elements", *1990 IEEE MTT-S Int. Microwave Symp. Dig.*, 2:721 (1990).
6. A. E Ruehli, "Inductance calculation in a complex integrated circuit environment", *IBM J. Res. develop.*, 16:470, (1972).
7. A. E. Ruehli and P. A. Brennan, "Efficient capacitance calculations for three dimensional multiconductor systems", *IEEE Trans. Microwave Theory Tech.*, MTT-21:76, (1973).
8. E. R. Hanson, G. K. G. Hohenwarter, S. R. Whiteley, and S. M. Faris, "Ultra-wide bandwidth instrument product based on Josephson junctions", *Proc. Int. Conf. on Computer Design, ICCD'87*, 472, (1987).
9. E. W. Strid, "26 GHz, wafer probing for MMIC development and manufacture", *Microwave J.*, (1985).
10. J. Campbell, J. Holton, A. Knight, "A high density edge connector", *Proceed. of the 45th Electr. Comp. & Tech. Conf., ECTC'95*, IEEE/EIA, 140, (1995).
11. A. Deutsch, G. Arjavalingam, et al., "Performance limits of electrical cables for intrasystem communication", *IBM J. Res. Develop.*, 38:659, (1994).
12. R. R. Tummala and E. J. Rymaszewski, *Microelectronics Packaging Handbook*, Van Nostrand Reinhold, New York, (1989).
13. B. J. Rubin, "An electromagnetic approach for modeling high-performance computer packages", *IBM J. Res. Develop.*, 34:585, (1990).
14. C. W. Ho, D. A. Chance, C. H. Bajorek, and R. E. Acosta, "The thin film module as a high-performance semiconductor package", *IBM J. Res. Develop.*, 26:286, (1982).
15. A. Deutsch and C. W. Ho, "Triplate structure design for thin film lossy, unterminated lines", *Proceed. of the 1981 IEEE Int. Symp. on Circuits and Systems*,1, (1981).
16. A. Deutsch, M. Swaminathan, et al., "Measurement of dielectric anisotropy of BPDA-PDA polyimide in multilayer thin-film packages", *IEEE Trans. Comp., Pack. and Manuf. Technology-Part B*, 17:486, (1994).
17. A. Deutsch, G. V. Kopcsay, et al., "Modeling and characterization of long on-chip interconnections for high-performance microprocessors", *IBM J. Res. Develop.*, 39:547, (1995).

SPECULATIONS ON THE NEXT TEN YEARS IN ELECTROMAGNETIC FIELD SIMULATION

John M. Dunn

University of Colorado
Department of Electrical and Computer Engineering
Boulder, CO 80309-0425

ABSTRACT

The author speculates on the future uses of electromagnetic field, simulation tools for MMIC designer in the next decade. First, a brief overview is given as to the current types of the most common classes of numerical, field algorithms, and their typical uses. Arguments are then made that simulation problem size will increase by a factor of 100 to 1000 from present day. This will allow field simulation tools to be used in circuit and system level design, and optimization loops. Finally, the features of future software programs are discussed.

INTRODUCTION

Electromagnetic field simulation codes have become a commonplace tool for rf and microwave designers. These tools have been commercially available for only a short time, about ten years, yet have become a relatively commonplace fixture in the computer aided design (CAD) suite of the engineer. Undoubtedly, these tools will continue to be important in the next decade. Indeed, they will be a vital component of future, high-speed design, CAD software. However, it is also probable that their usage will change somewhat, as the power of computers, software, and numerical algorithms increase.

The purpose of this paper is to give the author's opinions on the future usage of field simulation tools. In particular, I will speculate on the effective power of these tools, on their usage by the practicing engineer, and on needed areas of improvement if their potential is to be reached. At the outset, the word **speculative** is emphasized.

The paper is broken up into a number of sections. First, I discuss the present role of field simulation tools in microwave, monolithic, integrated circuit (MMIC) CAD design. As part of the discussion, a basic taxonomy of the types of field simulation tools available is given. An understanding of this is important, as a number of the speed estimations discussed later in the paper depend on the specifics of the methods being used. The second section of the paper estimates the size of simulations possible based on Moore's law, and projections of simulation times. A discussion of the impact of parallelization on field simulation algorithms is given. I also raise the possibility of new mathematical algorithms to speed up the computation times of large simulations. The section concludes with the prediction of simulation tools will be increased in power by 100 to 1000 times in the next decade. In the third section of the paper, I speculate as to how field simulation tools will be used in the future; in particular, what uses are expected to continue to be used, and what applications will change. Examples of field based optimization, system level parasitic calculations, and automatic model generation will be examined. Nonlinear, time based applications will also be seen. The fourth and final section of the paper will briefly speculate on the field program of the future. References are given to appropriate literature, although they are meant to be a starting point, not exhaustive.

Directions for the Next Generation of MMIC Devices and Systems
Edited by N. K. Das and H. L. Bertoni, Plenum Press, New York, 1997

FIELD SIMULATION METHODS AND PRESENT DAY APPLICATIONS

First, a tentative definition of a field simulation tool needs to be given. In particular, it is important to understand the difference between field simulation software, and circuit simulation and system simulation software. Each of these three types of simulation tools has an important place in modern day CAD design. Each is different. Most design today is carried out at the circuit and system levels. The circuit simulation level works by solving standard circuit equations, in which circuit models of components are linked together by transmission line structures. System simulation solves a set of system equations, in which abstract models are interconnected by a set of transfer equations. System simulation has the advantage of being able to simulate more complex systems; it has the disadvantage of being more abstract and therefore possibly neglects important physical phenomena. For example, an "on-off" model of a gate will neglect all analog effects of the underlying transistors. Field simulators today are used as tools to aid in developing new models. They solve the underlying equations of electromagnetics, Maxwell's equations, using some form of numerical algorithm. The data generated from these simulators are then interpreted to predict the S parameters of the structure. From these S parameters, a model is developed. In the case of circuit simulators, a lumped or distributed circuit model is extracted, although this is not strictly necessary. In the case of a system simulator, an abstract response model is generated.

A number of important details of this scenario need to be emphasized. The field simulation tool is typically not directly used for the design of the circuit. Rather, its purpose is to generate data for models of components in the circuit. (Of course exceptions to this statement undoubtedly exist; but, my experience is that they are the exception rather than the rule.) My belief is that this trend will continue in the future. Electromagnetic simulators will be used predominantly as analysis tools, rather than design tools. MMIC design will be carried out more at the system level a decade from now. This is essential as MMIC circuits and systems increase in complexity. The future of field solvers is to fit within this framework, and to be as unobtrusive as possible in doing so.

The second point to emphasize is that field simulation tools allow the designer to generate data faster than by relying completely on experiments. Also, the designer can see how the performance of the system is changed when various parameters change, the values of which may be difficult or even impossible to carry out experimentally. However, the purpose of field simulation tools is not to avoid experiments. The purpose of field simulation tools is to gain insight to a system's performance quickly, and to reduce the number of experiments needed for the building of databases for models.

The final point to be made is that intelligent use of field simulation software requires an experienced designer. Field simulation tools are based on solving Maxwell's equations numerically. As such, they have the important advantage that they can incorporate any electromagnetic effect, at least in principle. The word in principle must be emphasized in that the simulator will give results which can disagree with reality for two reasons. First, there is always numerical error, and computer limitations in memory, and speed. Therefore, a feature in a system's performance might be missed. For example, the response of a high Q resonator may be incorrectly predicted if care is not taken. Second, it is always necessary to omit unimportant details in the geometry of the physical problem in order to get a reasonably sized numerical problem. Sometimes, the designer can neglect an important effect without realizing it. The exception to this case will be for field simulation tools that are incorporated into higher level simulators automatically. This will be discussed in more detail later.

I now discuss the types of field simulation tools used today. This discussion is not intended to be exhaustive. Rather, only the generic types of methods are mentioned, in order to provide a framework for the next section's discussion of numerical methods. No commercial products are listed here. I do not discuss quasistatic tools, which are not as useful to MMIC designers; normally, they are interested in full-wave simulation methods, that is methods that solve the full set of Maxwell's equations in either time or frequency domain. The frequency domain methods can be broken down into planar type solvers, and full 3D type solvers. The time domain methods usually are 3D solvers. Please note that the classifications used in this paper are not universally accepted. Each of the basic categories is now discussed in more detail. For a more detailed discussion, see for example References 1 and 2.

Frequency Domain Methods

Moment Methods: Moment methods are arguably the most popular type of method used today. They rely on solving an integral equation version of Maxwell's equations for unknown currents on conductors. Moment methods rely on the availability of a Green's function. Unfortunately, not all geometries have known Green's functions. Therefore, moment methods techniques are commonly limited to certain types of structures. Fortunately, Green's functions are known for planar structures; that is, structures where there are planar layers of dielectric and air, with infinitely thin ground planes and signal lines sandwiched between them. This is precisely the geometry that is of importance to the MMIC designer. When moment method techniques can be used, they compete favorably against other methods, as the simulator only solves for the unknown currents. In contrast, other methods solve for the fields all over the 3D space.

Moment methods reduce the integral equation to a matrix equation by approximating the current as a sum of basis functions. The unknown amplitudes of the basis functions are then calculated. There are two computationally intensive steps in carrying out this process. The first is the filling of the matrix. If the unknown current is broken up into N unknowns, the filling of the matrix takes order N^2 time. The solution of this matrix using direct solve techniques takes N^3 time. Therefore, if the number of unknowns is doubled, the simulation time increases by a factor of 4 to 8. The break-even point between fill time and solve time on current computers is about 1500 unknowns. For larger problems, the solve time will dominate. Today's computers can solve problems on the order of a 1000 unknowns in about one hour per frequency point. This size problem would typically be a medium level simulation, for example a meander line, or a interdigitated filter.

Moment methods will continue to be popular. They will be incorporated in circuit simulation programs for on the fly model generation and simulation. They are a good candidate for a field solving engine for circuit layout optimization routines. They will remain the leading method for planar antenna simulation. Their future use as standalone codes will depend on how many of their features are already incorporated in circuit simulation packages.

Finite Element Methods: These methods have become popular in the electromagnetic field simulation community because of their ability to solve a much larger class of problems than moment method techniques. Instead of solving for the currents on conductors, the electromagnetic fields are solved all over space. The method breaks all spatial regions up into small tetrahedra, and approximates the electric field variation in each cell in a simple manner. Typically, a linear or quadratic variation is chosen. The unknown amplitudes of the fields in the cells are determined by solving the matrix equation which results after a variational principle is employed. The resulting matrix is much larger than the matrix resulting from the moment method technique. However, the matrix is sparse; that is, most of the elements are zero. The moment method matrix is dense; no elements are zero. The solution time for a sparse matrix of this type is order N^1.6, where N is the number of unknowns. Typically, finite elements are slower than moment methods, but work on a larger variety of problems. For example, finite element methods have been used with success in waveguide problems, novel stripline problems, 3D discontinuity problems, and problems where the thickness of lines cannot be neglected.

Finite element methods will continue to be widely used. It is expected that they will be seen in nonlinear types of problems, where moment method type methods cannot be applied. A problem with a few thousand tetrahedra can be simulated on today's computers at about an hour a frequency point. Typical examples that could be examined are vias, airbridges, waveguide discontinuities, and package interconnects.

Boundary Element Methods: Boundary element methods work by taking each homogeneous, spatial region in the problem and surrounding it by fictitious electric and magnetic currents. The fictitious currents are then related to each other by coupled integral equations, which are then solved. One way of looking at the method is that it attempts to generalize moment methods to arbitrary 3D structures. However, it must be remembered that one must solve for all fictitious magnetic and electric currents on all boundaries, not just the real currents on the conductors. Therefore, much more surface must be gridded than in the moment method. The solve time for the method is the same as for the moment method: order N^2 to fill the matrix, and order N^3 to solve it. Boundary element methods have not been as popular in the electromagnetics community as the first two mentioned methods. In this author's opinion, this is primarily for nontechnical reasons. They should be competitive with finite element methods on linear problems.

Time Domain Methods

Finite Difference Time Domain: Maxwell's equations in the time domain are solved by gridding up all of space at a given time, and marching forward in time. The method has the advantage that transitory behavior can be observed. It has the disadvantage that it is computationally extremely intensive. Typically, high end workstations or supercomputers are needed to solve interesting problems. Because of this drawback, and the fact that in the MMIC community, frequency domain design is emphasized, the method is not in as widespread use as the previously mentioned methods. The method shows promise for problems where nonlinear, transitory behavior is to be studied. Currently, most problems using field simulators have been linear. Future applications in nonlinear optics, and semiconductor device modeling, may provide an essential role for this method.

Transmission Line Matrix (TLM) Method: Another type of finite difference, time domain method. The formulation differs from the FDTD method in that the nodes in space are connected by fictitious transmission lines, and generalized voltages and currents are examined. It can be shown that the resulting equations are equivalent to the FDTD equations to first order, after transformations between the fields and the voltages are currents are made. The method has the same general drawbacks and strengths as the FDTD method.

Finally, I mention that the methods will start to merge together. It is well known how to incorporate the various methods into each other.[3] Any fields problem can be broken up into various regions, and different methods applied within those regions. The regions are interconnected by fictitious currents at the boundaries. Future field simulation packages will allow the designer to break a problem into subproblems, and specify the method being used for each subproblem. For example, a boundary element method is chosen to look at the package, which is coupled to a moment method solution of the coupling of the lines in the circuit, which is coupled with a finite element solution of some localized regions of complicated 3D structures, maybe an airbridge to a via. Obviously, sophisticated graphical front end capabilities will have to be developed for the designer to efficiently implement these features.

ESTIMATION OF FIELD SIMULATION SPEEDS

Obviously, the use of field simulators will be driven by the size of the simulation possible. This is determined by three major factors: the computer hardware being used; the software being used - for example, the operating system, and compiler; and the numerical algorithms being used. I now look at each of these three areas in detail.

Inherent Computer Speed

The power of computers has increased dramatically over time due to increased circuit size and density. We use a variant of Moore's law which states that the processing power of computers will increase by a factor of 2 every 18 months. (Moore stated the number of processing elements on a chip will increase by a factor of two every two years. We use the more useful criterion of processing power.) This results in an increase of computer power of 100 times in the next ten years. If we go back in time 10 years, 1986, and medium sized simulations of commercially based workstations typically involved about 100 unknowns, requiring memory sizes of a few Mbytes. Today, medium level simulations are a few thousand unknowns requiring order 25 Mbytes. Therefore, we have seen an increase of several tens in power. It is probable that Moore's law will continue to be upheld for the next 10 years.

Please note the important point that simulation size does not increase linearly with number of unknowns. The actual relationship depends on the specific algorithm employed to solve the matrix equation. The simulation time increases faster than the increase in number of unknowns. For example, moment method codes result in a dense matrix which is most typically solved in order N^3 time, where N is the number of unknowns. Therefore we expect to see field simulations of about 10 times larger based on Moore's law. Finite element and finite difference methods generate sparse matrices. Typically, the equations can be solved in about $N^{1.6}$ time. Therefore, they will be able to look at problems about 20 times bigger in physical size. Remember that these methods typically grid a 3D region, as opposed to a 2D surface. The important point is to expect increase of problem size by about 10 times, at most in the next ten years, based on pure increase in computer performance, and using straightforward algorithms.

Software Enhancements

Compilers and operating systems will increase in their efficiency. The most obvious room for improvement in speed is in the area of parallelization. There is no doubt that multiple processors will be used on the same computer, or between networked computers. Significant speed enhancements will be made, but it is difficult to predict how much it will improve performance, as it is outside the author's area of expertise.

Numerical, Field-Based Algorithms

The third way to increase simulation speeds to improve the algorithms that are used in solving the problem. As mentioned before, moment method based codes solve a dense matrix equation in order N^3 time; finite element and difference schemes in order $N^{1.6}$ time, where N is the number of unknowns. Most commercial software available today uses standard, direct solve techniques, for example variants of Gaussian elimination for dense matrices. Fortunately, there are a number of new techniques that have been developed in the mathematics community, which are starting to be employed by developers of field simulation software. These techniques have the advantage of reducing the solve time to nearly order N. Memory usage is reduced accordingly. The discussion can be broken down into two groups: methods which rely on new mathematical techniques, and methods which take advantage of the physical characteristics underlying the problem being solved.

The first category takes advantage of a number of new methods developed by mathematicians. A few of the more promising possibilities are listed. Wavelets[4,5] are becoming popular for solving integral equation problems. A wavelet is a new type of special function that has a number of desirable features. Chief among these for the present discussion is their localized extent (compact support) and relative smoothness. This second characteristic implies that a number of their moments are zero, which in turn means they lead to a sparse representation of the matrix being examined. Test cases have shown the ability to transform a matrix into a sparse representation in a wavelet basis, which then can be solved in almost order N time. A problem that still needs to be solved effectively is that wavelets do not make full wave matrices as sparse as quasistatic ones.[6] Therefore, their effectiveness can be reduced for electrically large simulations. A second promising development is in the recent generalization of the fast Fourier transform (FFT) to nonuniform grids. The FFT has been used with success to speed up the matrix fill portion of moment method problems,[7] but has the disadvantage of requiring uniform gridding. The new developments promise to provide a way around this bottleneck.[8] A third interesting technique has to do with predicting simulation results over a large range of frequencies based on relatively few simulations.[9,10] The methods work by approximating the system response as a relatively low order polynomial. Sample results in the literature have been encouraging. In some cases, filter behavior over a resonance region has been well approximated with only a few frequency points.

The second category of methods relies on taking advantage of the underlying physics of the problem. Specifically, it is well known that the interaction between elements decays as 1/R where R is the separation. For planar circuit problems, the decay goes approximately as $1/R^2$. This means that elements in the matrix representing far away interactions are small compared with closer elements. New methods take advantage of this fact when the matrix is solved. Essentially, a sparse matrix is created by successive approximations, which can then be solved quickly.[11,12] A related development is the breaking of a problem up into subregions, each of which can be independently solved, owing to relatively weak interaction between the various regions. In all these methods, the direct solve of a dense matrix is replaced by an iterative solve of a sparse matrix. Problems will need to be overcome if these methods are to attain common usage. First, the numbering of the cells is critical. It is not obvious what the optimal numbering scheme is for realistic problems. A poor numbering scheme can render the new techniques useless. Second, the integral equations used in electromagnetics are poorly conditioned. This is not the optimum situation for these methods.

Based on the discussion above, it is my opinion that the solve time for field simulation will be reduced to close to order N. This will therefore allow simulation sizes to increase by 100 to 1000 in the next decade.

THE FUTURE USES OF FIELD SIMULATION SOFTWARE

It was mentioned earlier in this paper that field simulation tools for MMICs are primarily used today for generation of data, which in turn is used to develop a model for insertion into circuit and system simulation programs. This use of field simulation tools is expected to continue into the next decade. Undoubtedly the size and complexity of the simulations possible will allow more sophisticated, accurate models to be developed. It is my opinion that field simulation software will not be used as the primary design software in most cases. Rather, designers will work more and more at the system level. This will allow for more complicated circuits and systems to be built in an efficient manner. Circuit simulators will be used when a more accurate simulation of a part of a system is needed. Finally, the field simulator will be used as a last result to obtain accurate models. I base these predictions on the parallel development of CAD tools in the digital circuit community. The present level of chip complexity has been achieved by designing at a high, functional level description of the circuitry. Indeed, this trend is starting to be observed in the MMIC community. Various manufacturers are beginning to develop standard FET cells that can perform a number of system level functions, for example: power dividers, or isolators. The possible reduction in performance may well be offset with the ease and sophistication of design possible.

I would now like to look at a number of new uses of field simulation software in the next decade. The list, of course, is speculative and driven by the author's interests and opinions.

Automatic Model Generation

As mentioned above, the predominant use of field simulators will be to generate data, from which models can be developed. The model can be a circuit model, or more abstract in form. In some cases, simply inserting the data into the simulator as a "black box" is adequate. Many of the types of models needed today in MMIC design are relatively standard. An important category is planar, discontinuity models, for example: microstrip bends, tees, and crosses. Many models already exist, and are standard components in circuit programs. Often existing models are not sufficiently accurate for a specific design. The reason for this is that the CAD developer is forced to supply models for large ranges of parameters. These models are typically lumped circuit approximations, which are taken from the literature. Field simulation software has the possibility of improving this situation dramatically. Essentially, field simulation software will be incorporated into circuit simulation software automatically. Most discontinuity simulations need only a few hundred unknowns. With the type of speed increases mentioned in the last section, it should be possible to generate the data for the discontinuity when needed, and still have a reasonably fast simulation. The circuit program will be intelligent enough to call the simulator, generate the model, and save it for future use if required. Since most designers only work with a limited set of process parameters, the circuit program would quickly "train" itself as it is used. The engineer will not be aware that field simulation tools are being used. The process will be completely transparent.

Circuit and System Level Parasitic Calculations

Circuits and systems will be more complicated in the next ten years. As was mentioned above, this will occur partially because of the designer's increased ability to work at the system CAD level, only going down to the circuit simulation and field simulation levels when absolutely necessary. The increased size and density of these future systems will make them susceptible to parasitic interactions between various parts of the circuit. The possibility of degradation in performance will be enhanced because of the inherently high frequencies of operation of these circuits. System and circuit simulators of today do not include parasitic analysis in their models, unless the engineer specifically includes a model for a given interaction. For example, transmission lines in a MMIC could be coupled in some way, without the designer ever knowing. The problem with carrying a full field simulation out on a circuit is obvious; it is much too large a problem to be effective, especially when the effect looked for is often insignificant.

The answer to this problem is to develop new algorithms specifically geared toward calculating parasitics. These types of algorithms are designed to be very fast, and give the designer an idea if parasitics are a problem. Speed will be accomplished by making the algorithms relatively inaccurate when compared to a full field simulation. Parasitic analysis does not require great accuracy; it is not a design tool. Rather, it is intended to be a quick estimate of order of magnitude effects. As such, these tools are very different in philosophy than traditional field simulation tools. The next decade will see the inclusion of parasitics checkers in system and circuit level software, which will automatically calculate parasitics accurately enough to show the designer if there is a problem. If there is, redesign and more accurate analysis need to be carried out on the trouble spots in the system.

The algorithms will be of a variety of forms, the specifics of which will depend on the parasitic coupling being examined. Obvious candidates include coupling of transmission lines,[13] which will be calculated by fast moment methods; coupling between specific small structures, for example between packaged circuit components, which can be modeled as dipole antennas; and between elements and the package, which can be modeled as some kind of abstract transfer function at the system level.[14]

Field Simulations in Optimization Loops

One of the more exiting possibilities is the use of field based simulators in optimization loops. The problem is, of course, that field simulators are slow; it is not realistic to expect an optimizer to call a field simulator several thousand times. An interesting way around this is to perform the optimization at the circuit or system level, and correct for the errors in the models being used at the circuit level by using a field simulator. In this manner, only a relatively few calls to the field simulator need be made. Notice that this approach requires a circuit or system level model. An interesting example of this approach is Bandler's space domain mapping method.[15]

Automatic Layout and Layout Compaction Algorithms

The physical layout of a circuit is carried out by hand at the present time; i.e., the engineer tells the computer where to put the various parts of the circuit. This process must become automated if circuit and system performance levels are to increase. Therefore, automated layout routines are necessary. Field simulation tools must be coupled into these routines somehow, if the layout is to keep the parasitics below a desired level. Another obvious step will be to include optimization of the layout so that circuit compaction can take place.

Nonlinear, Transient Applications

Another new area for full wave simulation methods will be in nonlinear phenomena. So far, field simulators developed assume linear, homogeneous materials are being used. Certain areas of applications are inherently nonlinear. Magnetic machinery applications are an obvious example - although quasistatic simulators are adequate. Integrated optics promises to be a new and existing area for field simulators. For example, researchers are starting to examine integrated laser structures, and nonlinear waveguides. Another interesting area starting to emerge is in field based device modeling. Traditionally, this area has relied on semiconductor equations being coupled with a quasistatic version of Maxwell's equations. More recently, inherently full wave structures have been proposed. I expect to see new field simulation development in these areas as the commercial markets mature. New techniques will be developed using time domain techniques, coupled with nonlinear circuit models; possibly analogies to harmonic balance methods will also make sense.

THE SOFTWARE PROGRAM OF THE FUTURE

The general trend in commercial products will be for increased ease of use, and increased power. Many field simulation tools will be included in circuit and system level software, in a transparent way to the engineer. When a problem must be solved by specifically using a field simulator, the designer will be able to break the problem up into sub-problems, and use the most appropriate method for each of the regions. This requires the various tools being used to have a standardized format for geometry and field descriptions. There will be increased pressures for new field simulation tools to interact smoothly with higher level software packages.

REFERENCES

1. T. Itoh, ed.. Numerical Techniques for Microwave and Millimeter-wave Passive Structures, John Wiley & Sons, Inc., New York (1989).
2. Numerical Methods for Passive Microwave and Millimeter-wave Structures, IEEE Press, New York (1989).
3. D. Jankovic, M. LaBelle, D. Change, J.M. Dunn, and R.C. Booton, A hybrid method for the solution of scattering from inhomogeneous dielectric cylinders of arbitrary shape, IEEE Trans. on Ant. and Prop., 42:9 (1994).
4. B. Alpert, G. Beylkin, R. Coifman, and V. Rohklin, Wavelet-like bases for the fast solution of second-kind integral equations, SIAM J. Sci. Comp. 14:1 (1983).
5. G. Wang, and G. Pan, Full wave analysis of microstrip floating line structures by wavelet expansion method, IEEE Trans. on MTT, 43:1 (1995).
6. D. Gines, and J. Dunn, The quasi-static limit of wavelet based algorithms, submitted to IEEE Trans. on Ant. and Prop. (1996).
7. M. Catedra, R. Torres, J. Basterrechea, and E. Gago, The CG-FET Method: Application of Signal Processing Techniques to Electromagnetics, Artech House, Boston (1995).
8. G. Beylkin, On the fast Fourier transform of functions with singularities, submitted to App. and Comp. Harmonic Anal. (1996).
9. T. Dhaene, J. Ureel, N. Fache, and D. De Zutter, Adaptive frequency sampling algorithm for fast and accurate S-parameter modeling of general planar structures, IEEE MTT Symp. Dig. (1995).
10. M. Celik, and A. Cangellaris, Simulation of multiconductor transmission lines via Pade approximation and the fast Lanczos process, IEEE MTT Symp. Dig. (1996).
11. R. Coifman, V. Rohklin, and S. Wandzura, The fast multipole method for the wave equation: a pedestrian approach, IEEE Ant. and Prog. Mag., 35:3 (1993).
12. E. Michielssen, and A. Borg, A multilevel matrix decomposition algorithm for analyzing scattering from large structures, IEEE Trans. on Ant. and Prop., 44:8 (1996).
13. J.M. Dunn, L.C. Howard, and K. Larson, An efficient algorithm for the calculation of parasitic coupling between lines in MIC's, IEEE Trans. on MTT, 41:8 (1993).
14. R. Perry, J.M. Dunn, and R.C. Booton, A computationally efficient method for modeling circuits housed in non-rectangular packages, IEEE MTT Symp. Dig. (1994).
15. J. Bandler, R. Biernacki, and S. Chen, Fully automatic space mapping optimization of 3D structures, IEEE MTT Symp. Dig. (1996).

ANN AND KNOWLEDGE-BASED APPROACHES FOR MICROWAVE DESIGN

K.C. Gupta

University of Colorado at Boulder
Department of Electrical and Computer Engineering
Boulder, CO 80309-0425

ABSTRACT

Two approaches that are likely play significant roles in the future developments of RF, microwave and millimeter-wave CAD are: (i) modeling and optimization based on artificial neural network (ANN) computing, and (ii) use of knowledge-based techniques for development of initial design and also for design training/instruction activities. This article reviews these two areas as relevant to RF and higher frequency CAD.

INTRODUCTION

Computer-aided techniques for RF and microwave design have undergone revolutionary changes in the last decade. The significant advances in this area include: accurate modeling of passive components[1] and active devices,[2,3] electromagnetic simulation software,[4] and non-linear analysis tools[5] using harmonic balance and allied approaches. While these developments have brought microwave CAD to a level of maturity, we are on the threshold of a new era in the development of microwave design methodology and tools. The two key aspects of these forthcoming developments are:

(i) use of artificial neural network (ANN) computing[6] for microwave design, and

(ii) development of knowledge-based approaches[7] for initial steps in design process and for design instruction/training.

This article discusses applications of these two approaches to the next generation of microwave CAD. There have been a few reported applications of ANN to microwave design in the last four years. These are reviewed in the next section, ANN Computing in Microwave Design, which includes a discussion of potential applications also. The third section, Knowledge-Based Design Tools, describes the need of a knowledge-based approach for the initial design stages in microwave CAD as well as for design instruction. The article ends with Concluding Remarks.

Directions for the Next Generation of MMIC Devices and Systems
Edited by N. K. Das and H. L. Bertoni, Plenum Press, New York, 1997

ANN COMPUTING IN MICROWAVE DESIGN

Recently Reported Applications

A few researchers have tried ANN computing for: (i) component modeling, (ii) circuit design, and (iii) optimization. Also, application of ANNs to real time circuit adjustment has been proposed. These research reports are summarized in this section.

ANNs are neuroscience-inspired, massively parallel, distributed computer systems. It has been shown that multilayer ANNs are capable of approximating virtually any function to any desired accuracy.[8] In addition, ANNs have the ability to learn from data and to generalize patterns in the data, giving these networks excellent predictive capabilities.[9] The ANN learns relationships among sets of input/output data which are characteristic of the component under consideration. The most common method for training ANNs is the error back-propagation training algorithm.[9]

In the past, ANNs have been used for many complex tasks. Applications have been reported in areas such as control,[10] telecommunciations,[11] biomedical,[12] remote sensing,[13] pattern recognition,[14] and manufacturing,[15] just to name a few. However, ANNs have been used only to a very limited extent in the area of microwave design.

Component Modeling Using ANNs. ANNs have been used for modeling selected passive and active microwave components. The developed models are then used like any other model in available CAD programs in order to develop the complete circuits. Applications to component modeling follow.

Active Device Modeling: In Reference 16, an ANN has been used to model the input/output characteristics of a MOS transistor. Inputs to the network are the drain-to-source voltage, v_{ds}, and the gate-to-source voltage, v_{gs}. The drain current, I_d, is the calculated output. The ANN is trained using data from a circuit simulator, covering all regions of device operation. Another example of active device modeling is given in Reference 17. In this case, an ANN is used to model the small-signal and noise parameters of a MESFET over a range of 1 to 30 GHz. Inputs to the ANN include frequency and an additional frequency dependent bias for fast convergence. Outputs are the small-signal and noise parameters of the device at a particular DC bias. The developed model is valid only at a single bias point for the particular FET modeled. Modeling of heterojunction bipolar transistors (HBTs) by using ANNs has also been reported.[18] In this work, ANNs are used in conjunction with simulated annealing to extract equivalent circuit parameters of the device from measured data. This model can then be used in available CAD programs. None of these reports take temperature effects into account. ANNs are capable of more general modeling combining electrical and thermal effects.

Passive Component Modeling: Passive microwave components have also been successfully modeled using ANNs. In Reference 19, an ANN has been trained to predict the S-parameters of an X-band spiral inductor. Inputs to the network consist of physical dimensions and frequency, and outputs are the S-parameters corresponding to the inputs. Training data is produced by EM simulation of various structures. The developed model, however, is valid only from 7 to 11 GHz. This frequency range limits its use to linear simulations (as it may not be valid for harmonics). ANNs have also been used to construct a broadband model of GaAs microstrip grounding vias.[20] Inputs to the network include geometrical dimensions of the vias and frequency. Outputs consist of S-parameters. EM simulation was used to provide the training data. The developed via model has been implemented in a commercial microwave simulator. It is shown to give the accuracy of the EM simulator used for training the ANN at a fraction of the simulation time. Since the developed model is valid from 5 to 55 GHz. It is useful for both linear and nonlinear designs where harmonic frequency components are generated.

Microwave Circuit Design Using ANNs. The authors of Reference 21 have used ANNs to analyze a microstrip corporate feed for antenna design. The inputs to the ANN are the geometrical parameters for the feed structure. The network outputs are the magnitude and phase of current ratios formed by the currents of the various feed arms. EM simulation is used to obtain the training data. Once generated, the current ratios are used to calculate the array pattern.

Another example of using ANNs for microwave circuit design is given in References 22 and 23. Here, an ANN has been trained to represent a normalized impedance and admittance

Smith Chart. The trained ANN is then used to develop appropriate impedance matching and stabilizing networks.

ANNs have also been proposed for waveguide filter alignment.[24] Inputs to the network are the desired center frequency, bandwidth, and insertion loss. Outputs are the positions of a set of tuning screws which bring the filter response to within the prescribed specifications.

Microwave Circuit or Component Optimization. The authors of References 25 and 26 have used ANN models for microwave circuit optimization and statistical design. Trained ANN models represent either a nonlinear device, or an entire circuit. On the device level, and ANN is used to model a nonlinear MESFET, using simulated data from an existing physics-based model for training. The developed ANN model is then used in the optimization and yield analysis of large signal amplifiers. For given amplifier design specifications, it is shown that using the ANN model instead of the existing physics-based model results in a speed-up ratio of 6 (7.13 min. : 1.10 min.) for optimization.

Optimization of high speed digital circuits has been presented in Reference 27. Here, an ANN model is used to simulate propagation delays in high speed VLSI interconnect networks. Tremendous speed-ups were obtained by using the developed ANN model in place of existing simulation techniques, such as Numerical Inversion of Laplace Transforms (34.43 hr. : 6.67 min.) and Asymptotic Waveform Evaluation (9.56 hr. : 6.67 min.).

Real Time Circuit Tuning/Adjustment. In Reference 18, an ANN has been developed for automatic impedance matching using a double-stub tuning network. The developed ANN model is used to control the adjustment of stub locations and lengths. The developed model can be used as a CAD tool for synthesizing stub-tuning networks or it can be developed into an in-circuit, real-time adjustable impedance matching network. Suggested implementation for hybrid circuits uses a moveable block of silicone rubber with a small conducting strip at its bottom to change the length of the tuning microstrip stubs. The output of the neural network is used to mechanically slide the block on the microstrip. For MMICs, the microstrip stub lengths can be adjusted by using the output of the neural network to turn on (or off) PIN diodes, switching in (or out) microstrip sections. This suggestion has not been implemented so far.

Potential Application of ANNs in Microwave Design

ANN computing is likely to have much more wide spread applications in microwave CAD. These potential areas correspond to what are the weak points in spectrum of microwave CAD developments today. Some of these are listed below.

Modeling of Nonlinear Devices. Efficient and accurate models for nonlinear behavior of active microwave devices like HBTs, MESFETs, and HEMTs including thermal effects could be developed using the ANN modeling approach. This could be achieved by starting with the existing models and modifying these models by developing an ANN configuration to model the differences among the available approximate models and the results based on experimental characterization. This approach (called the hybrid ΔS model in Reference 6) has been used successfully earlier[6] in development of EM-ANN models for microstrip vias and vertical interconnects[6] and also for chamfered 90° bends in CPWs.[29]

Modeling of CPW Components and Discontinuities. CAD tools for CPW circuits have not yet been developed adequately because of the lack of accurate and efficient models. Compared to microstrip circuit modeling, we have at least one additional parameter (gap width) that makes the conventional model fitting techniques more difficult. Also, the need to locate air-bridges near discontinuties calls for the effects of air-bridges to be accounted for in the model development process. The EM-ANN modeling approach appears to be well suited for this purpose.[29]

Modeling of Multilayered Circuits. Design of multilayered microwave circuits required two new classes of components to be characterized and modeled: (i) multilayered multiconductor transmission line components, and (ii) inter-layer interconnects using either vertical vias or electromagnetic coupling through apertures in the inter-layer ground planes. Lack of CAD models for these components is the current bottleneck in the design of multilayer microwave and millimeter-wave circuits. Again, the EM-ANN modeling approach will help in providing a design solution.

Design of Integrated Circuit-Antenna Modules. Design of integrated circuit-antenna modules (or active antennas[30] as they are more popularly known as) is another emerging area that could benefit by the ANN modeling approach. Appropriate ANN models for microstrip patches and other printed radiating elements could be linked to currently available powerful microwave circuit simulators, thus allowing designers to handle active antenna design in a convenient manner.

Design of Millimeter-Wave Circuits. CAD packages currently available for microwave circuit design become less accurate at millimeter-wave frequencies, because of degradation in the accuracy of models used for various components. The ANN modeling approach could be used to overcome this difficulty.

Circuit Optimization. Apart from modeling, ANN computing could play a key role in circuit (and component) optimization. Here the repeated analysis performed by linear/nonlinear/electromagnetic simulator(s) can be replaced by an ANN configuration yielding the circuit performance as a function of various designable parameters (whose optimum values we are trying to choose). Efficiency resulting from such an optimization process needs to be investigated for different classes of microwave/mm-wave circuits.

KNOWLEDGE-BASED DESIGN TOOLS

In addition to computer-aided design (CAD) methodology, which is used currently for microwave and several other design demands, a methodology knows as "knowledge-based design" (KAD) has been proposed.[7] KAD may be defined as a system that enhances design by having computers make knowledge available to the designers. In order to appreciate the role of KAD in design, we need to discuss in detail the various steps involved in the design process.

Anatomy of the Design Process

Various steps in a typical design process[31] are illustrated in Fig. 1. Problem identification phase is concerned with determining the need for a product. A problem is identified, resources allocated, and end-users are targeted. The next step is drawing up the product design specification (PDS), which describes the requirements and performance specifications of the product. Preliminary design decisions are made at the concept generation stage, with satisfying a few key constraints as the goal. Several alternatives will normally be considered. Decisions taken at this stage determine the general configuration of the product, and thus have enormous implications for the remainder of the design process. Upwards arrows on the left-hand sides of the blocks in Fig. 1 denote feedback to earlier stages and reworking of the previous steps if needed. The analysis and evaluation of the conceptual design lead to concept refinement, for example by placing values on numerical attributes. The performance of the conceptual design is tested for its response to external inputs and its consistency with the design specifications. These steps lead to an initial design.

The step from initial design to the final detailed design involves modeling, computer-aided analysis and optimization. CAD tools available to us for microwave design today address primarily this step only.

Role of Knowledge Aids

The design process outlined above can be considered to consist of two segments. Initial steps starting from he product identification to the initial design may be termed as "design-in-the-large".[32] The second segment that leads from an initial design to the detailed design has been called "design-in-the-small". It is for this second segment that most of the microwave CAD tools have been developed.

It is in the "design-in-the-large" segment that important, and expensive, design decisions are made. Here a "knowledge-based system" is the most likely candidate technology that could help designers. Understanding this part of the design process is a prerequisite for developing knowledge aids for design. An extensive discussion on these and related topics is available in a three-volume treatise on artificial intelligence in engineering design.[33]

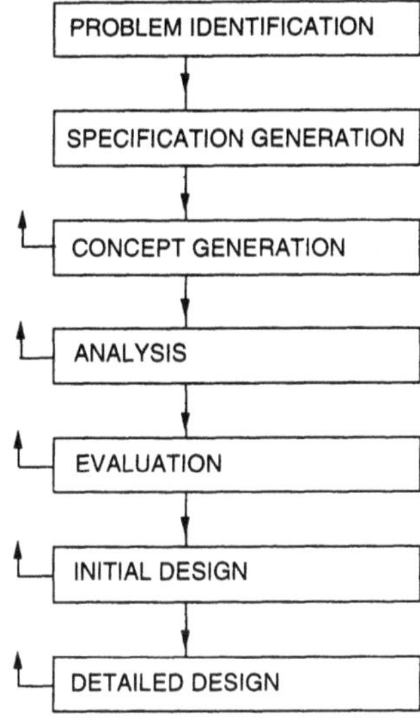

Fig. 1. Various Steps in a Design Process.

Knowledge Aids for Design

Development of knowledge aids may be based on developing a task structure[34] for the design process. A generic task oriented-methodology involves: (i) a description of the tasks, (ii) proposed methods for it, (iii) decomposition of the task into subtasks, (iv) methods available for various subtasks, (v) knowledge required for implementing various methods, and (vi) any control strategies for the methods.

A method for accomplishing a generic design task is known as Propose-Critique-Modify (PCM)[35] approach. This approach consists of the following: (i) a proposal of partial or complete design solutions, (ii) verification of proposed solutions, (iii) critiquing the proposal by identifying causes of failure, if any; and (iv) modification of proposals to satisfy design goals. A flow chart for this approach is shown in Fig. 2.

Applications to Microwave Design

Use of knowledge-based approaches to the initial stages of microwave design is an area that needs to be explored. Currently we depend heavily upon the accumulated experience of senior designers for executing these design steps. Recognizing the significant contribution of these steps to the final design, efforts in developing technology aids for this purpose would be worthwhile.

Knowledge-based systems developed for initial design of microwave circuits would also be very helpful for instruction or training of desing engineers. For example, a system that can present all the relevant options for (say) designs of digital phase-shifters at microwave frequencies, could educate the designer about the relative merits of various phase-shifter configurations as well as lead to a design for meeting a particular set of specifications.

CONCLUDING REMARKS

Two approaches that could play significant roles in the future developmetn of microwave-mm-wave CAD have been discussed. Modeling based on ANN computing has been used in recent years for some microwave applications. Much wider applications of this approach for design of CPW circuits, multilayered designs, and integrated circuit-antenna modules are expected to emerge. Applications of knowledge-based methods for microwave/mm-wave design have not been explored so far. This approach could play a significant role in initial stages of design and also for training of design engineers.

ACKNOWLEDGEMENTS

Ideas expressed in this article were developed in the MIMICAD Center's exploratory project on "Intelligent Software for MMIC Design" and a current CAMPmode project on "EM-ANN Models for MMIC Components." Contributions of Paul Watson and several helpful discussions with Professor Roop L. Mahajan are gratefully acknowledged.

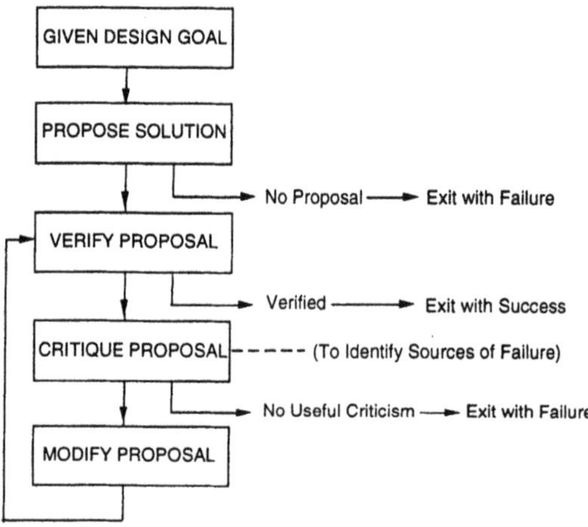

Fig. 2. Propose-Critique-Modify (PCM) Approach for Task Design.

REFERENCES

1. D.I. Wu et al., Accurate numerical modeling of microstrip junctions and discontinuties, *Intl. J. Microwave Mm-Wave CAE*, 1:48 (1991).
2. C.M. Snowden, Microwave and millimeter-wave device and circuit design based on physical modeling, *Intl. J. Microwave Mm-Wave CAE*, 1:4 (1991).
3. R.J. Trew, MESFET models for microwave CAD appliations, *Intl. J. Microwave Mm-Wave CAE*, 1:143 (1991).
4. F. Gardiol, Electromagnetic simulation of planar microwave and millimeter-wave circuits, Special Issue of *Intl. J. Microwave Mm-Wave CAE*, 2:221 (1992).
5. R.J. Gilmore and M.B. Steer, Nonlinear circuit analysis using the method of harmonic balance - a review of the art, Part I, Introductory Concepts, and Part II, Advanced Concepts, *Intl. J. Microwave Mm-Wave CAE*, 1:22 and 159 (1991).
6. P.M. Watson and K.C. Gupta, EM-ANN models for microstrip vias and interconnects in multilayer circuits, *IEEE Trans. MTT*, 43:scheduled for publication (1996).

6. P.M. Watson and K.C. Gupta, EM-ANN models for microstrip vias and interconnects in multilayer circuits, *IEEE Trans. MTT*, 43:scheduled for publication (1996).
7. M. Green, *Knowledge Aided Design*, Academic Press, San Diego, CA (1992).
8. K. Hornik, M. Stinchcombe, and H. White, Multi-layered feed-forward neural networks are universal approximations, *Neural Networks*, 2:259 (1990).
9. J.M. Zurada, *Introduction to Artificial Neural Systems*, West Publishing Company, St. Paul, MN (1992).
10. S.N. Balakrishnam and R.D. Weil, Neurocontrol: a literature survey, *Math. & Computer Modeling*, 23:101 (1996).
11. B.S. Cooper, Selected applications of neural networks in telecommunication systems, *Australian Telecom. Research*, 28:9 (1994).
12. T. Alvager, et al., The use of artificial neural networks in biomedical technologies: an introduction, *Biomedical Inst. & Tech.*, 28:315 (1994).
13. K. Goita, et al., Literature review of artificial neural networks and knowledge based systems for image analysis and interpretation of data in remote sensing,*Canadian J. Electrical & Computer Engr.*, 19:53 (1994).
14. Yu.G. Smetanin, Neural networks as systems for pattern recognition: a review, *Pattern Recognition & Image Analysis*, 5:254 (1995).
15. J.F. Nunmaker Jr. and R.H. Sprague Jr., Applications of neural networks in manufacturing, *Proc. 29th Hawaii Intl. Conference on Sys. Sci.*, 2:447 (1996).
16. V.B. Litovski, et al., MOS transistor modeling using neural netowrk, *Electronics Ltrs.*, 28:1766 (1992).
17. F. Gunes, F. Gurgen, and H. Torpi, Signal-noise neural network model for active microwave devices, *IEE Proc.-Circuits, Devices, Sys.*, 143:1 (1996).
18. M. Vai and Z. Xu, Representing knowledge by neural networks for qualitative analysis and reasoning, *IEEE Trans. on Knowledge & Data Engr.*, 7:683 (1995).
19. G.L. Creech, et al., Artificial neural networks for accurate microwave CAD applications, *MTT-S Intl. Microwave Symp. Dig.* p.733 (1996).
20. P. Watson and K.C. Gupta, EM-ANN models for via interconnects in microstrip circuits, *MTT-S Intl. Microwave Symp. Dig.* p.1819 (1996).
21. T. Horng, C. Wang, and N.G. Alexopoulos, Microstrip circuit design using neural networks, *MTT-S Intl. Microwave Symp. Dig.* p.413 (1993).
22. M. Vai, S. Prasad, and H. Wang, A Smith chart represented by a neural network and its applications, *MTT-S Intl. Microwave Symp. Dig.* p.1565 (1992).
23. M. Vai and S. Prasad, Microwave circuit analysis and design by a massively distributed computing network, *IEEE Trans. MTT*, 43:1087 (1995).
24. N.S. Sengor and T. Sengor, Using neural networks for waveguide filter alignment, *Proc. PIERS* p.285 (1995).
25. A.H. Zaabab, Q.J. Zhang, and M. Nakhla, Analysis and optimization of Microwave circuits and devices using neural network models, *IEEE Trans. MTT* p.393 (1994).
26. A.H. Zaabab, Q.J. Zhang, and M. Nakhla, A neural network modeling approach to circuit optimization and statistical design, *IEEE Trans. MTT* 26:1349(1995).
27. A. Veluswami, et al., A neural network model for propagation delays in systems with high speed VLSI interconnect networks, *Proc. IEEE Custom Integrated Circuits Conf.* (1995).
28. M. Vai and S. Prasad, Automatic impedance matching with a neural network, *IEEE Microwave & Guided Wave Ltrs.*, 3:353 (1993).
29. P.M. Watson and K.C. Gupta, *Optimal Chamfering and EM-ANN Modeling of 90° CPW Bends*, University of Colorado at Boulder, CAMPmode Technical Report (1996).
30. J. Lin and T. Itoh, Active integrated antennas, *IEEE Trans. MTT*, 42:2186 (1994).
31. A.A. Hopgood, Systems for design and selection, Chapter 8 in *Knowledge-Based Systems for Engineers and Scientists*, CRC Press, Inc., Boca Raton, FL (1993).
32. M. Green, Conceptions and misconceptions of knowledge aided design, Chapter 1 in *Knowledge Aided Design*, Academic Press, San Diego, CA (1992).
33. C. Tong and D. Sriram, *Artificial Intelligence in Engineering Design*, Academic Press, San Diego, CA (1992).
34. B. Chandrasekaran, Generic tasks in knowledge-based reasoning - high-level building blocks for expert system design, *IEEE Expert* 1:23 (1986).
35. B. Chandrasekaran, Design problem solving: a task analysis, Chapter 2 in *Knowledge Aided Design*, Academic Press, San Diego, CA (1992).

CAD Needs for Flip Chip Coplanar Waveguide Monolithic Microwave Integrated Circuit Technology

C. P. Wen, Michael Cole, C. K. Pao, and R. F. Wang

Hughes Aircraft Company
P. O. Box 2999
24120 Garnier Street
Torrance, California 90509

Rapid expansion of microwave systems in phased array radar and commercial applications in the emerging wireless communication system areas requires affordable, compact, lightweight, reliable, solid state, integrated microwave transmit/receive (T/R) modules. Traditional microwave modules fabricated using microstrip-based hybrid/ monolithic integrated circuitry are incompatible with low cost, high volume, robotics manufacturing techniques for multi-chip microwave modules, because of the fragile nature of current 4 mil-thick gallium arsenide monolithic microwave integrated circuits (MMICs). Backside processing, including wafer thinning and via-hole ground electrode access, are also significant yield limiting factors for MMIC manufacturing. Hughes has developed a novel, robust, flip-chip approach for microwave T/R modules using self-aligned, solder reflow, multichip assembly techniques. The rugged (25 mil thick), passivated MMIC chips with coplanar waveguide circuitry, silicon-like, on-chip multi-layer metal interconnects, with off-chip plated metal electrical/thermal paths offers major advantages over conventional MMICs (Table 1). Elimination of wafer thinning will be critical for processing and handling of large-diameter wafers. The use of plated metal thermal bumps directly on top of active circuit elements leads to 30% thermal property improvement over conventional GaAs MMICs.

We have recently demonstrated the operation of various flip-chip MMICs, including a 4-watt, X-band power amplifier. However, we are severely limited by our ability to optimize the design of flip-chip MMICs. Lack of accurate active device and passive circuit element models and computer aided design tools is a major challenge in developing coplanar waveguide based MMIC designs. Models for transistors with thermal bumps, and circuit discontinuities such as bends, T-junctions, crosses do not exist in widely used design tools. Neither is the effect of metal thickness and dielectric thickness on circuit element impedance readily available. There is no commercial CAD circuit optimization software to link circuit layout to RF performance simulation.

Uncertainty of CPW circuit element models generated by using EM simulation tools remains to be a major issue. Four different EM simulation tools give four different

Directions for the Next Generation of MMIC Devices and Systems
Edited by N. K. Das and H. L. Bertoni, Plenum Press, New York, 1997

models (Fig. 1) for a simple CPW T-junction. The corresponding computation time varies from 14 minutes to 7 hours.

We are taking an empirical approach to address the flip-chip MMIC/assembly modeling issues. Discrete circuit elements, including transistors, transmission line segments, spiral inductors, capacitors, bends, T-junctions, and crosses of various line width to gap ratios, are fabricated and measured. These basic building blocks, stored in our cell library, are used for MMIC circuit design. The next goal is to establish analytic expressions for interpolation between measured samples.

In our presentation, we will describe the status of our flip chip MMIC technology and a summary of our assessment of the commercially available flip-chip MMIC design/layout/ EM simulation tools.

TABLE 1 ADVANTAGES OF DIELECTRIC COATED FLIP-CHIP MMICS

	Advantage of flip chip MMIC module technology	Conventional microstrip T/R module
Mechanical strength	**Rugged, thick** (625 mm) **chips,** compatible with proven robotics manufacturing methods, **critical for processing large diameter wafers.**	Fragile, thin (100 mm) chips, difficult for both chip handling and wafer processing.
Reliability improvement	**Dielectric coated circuitry, silicon-like, multilayer metal interconnects, no air bridge, no via hole, no back side process.**	Exposed metal, un-supported air-bridges, via hole ground interconnect, and multiple wire bonds are potential reliability problems.
Thermal property improvement	Improved power handling capability with **plated metal thermal bumps** on active circuit elements.	Fragile, thin chips required to reduce thermal resistance.
T/R module producibility improvement	**Precision, self-aligned, single step**, solder reflow multi-chip module assembly.	Labor intensive sequential multi-chip, multi-wire-bond module assembly.

EM Tool	L1 (pH)	L2 (pH)	L3 (pH)	C (pH)	Computing Time (hr.)
Tool A	74	74	13	141	NA
Tool B	28	28	29	133	0.7
Tool C	16	16	16	210	7
Tool D	36	36	42	135	0.25

Fig. 1 Four EM simulation tools five four different models for the same CPW T-junction

COMPUTER AIDED DESIGN TOOLS FOR MICROSTRIP CIRCUITRIES - AN APPLICATION TO MICROSTRIP PATCH ANTENNAS OF CIRCULAR GEOMETRY

Hoton How[1] and Carmine Vittoria[2]

[1] ElectroMagnetic Applications, Inc.
Boston, MA 02109
[2] ECE Department, Northeastern University
Boston, MA 02115

INTRODUCTION

In the past we have considered the radiation problem of a circular microstrip patch antenna using the Green's function approach.[1,2] The traditional Green's functions were modified to include dielectric and conductor losses. That is, we have used the complex dielectric constant for the dielectric substrate material, and have considered the finite conductivity for the metal patch and the ground plane. Conductor loss is added to the Green's function by introducing finite surface impedance to the metal patch and Wheeler's incremental impedance to the ground plane.[1,2] As such, surface-pole singularities are pushed off the real axis and the Sommerfeld-type integrals can be processed as proper integrals. We have also shown that up to second order in skin depth, the induced current on the metal patch can be expressed in terms of irrotational currents which can be derived from current potentials.[2] The use of current potentials can not only reserve the symmetry of the antenna geometry, but also it simplifies the resultant Galerkin elements to involve only the one-fold Sommerfeld-type integrals. This can largely simplify numerical calculation, and many important engineering data on antenna performance can now be readily calculated.

To accurately evaluate Sommerfeld-type integrals we have encountered three kinds of difficulties. The first difficulty is associated with the quasi-resonant modes of the cavity which appear as zeros in the denominators of the integrands. Since these modes also appear as zeros of the same order in the numerators of the integrands, this kind of difficulty does not physically cause problem. However, in order to avoid large truncation errors, Taylor expansion of the integrands near these quasi-singular points needs to be applied. The second difficulty arises from the surface modes which are identified as simple poles near the real axis of integration. In the vicinity of these quasi-singularities the integrands cancel sharply in narrow regions such that the conventional fixed-stepsize integration routines fail to provide a specified accuracy in general. Instead, we have used variable stepsize algorithm which provides adaptive control over integration. The last difficulty is due to oscillation of Bessel functions at infinity. Since the oscillation period is not strictly 2π, numerical integration involving Bessel function at infinity is by no means trivial. We have circumvented this difficulty by invoking asymptotical expansion of Bessel function for large arguments. The resultant series expansions are then analytically evaluated using either the sine and cosine integrals and their derivatives, or the error functions and their derivatives. As such, to be equipped by the aforementioned three numerical techniques, we have calculated Sommerfeld-type integrals up to six significant digits.

In this paper we consider the coupling between multiple circular patch antennas deposited on the same piece of dielectric substrate. The formulation considers M circular patches to be fed by N feeder lines. The

Directions for the Next Generation of MMIC Devices and Systems
Edited by N. K. Das and H. L. Bertoni, Plenum Press, New York, 1997

patches can possess different size and the feeders can be either coax lines or microstrips. The coupling effect is retained in the Green's function which clearly indicates the nature of the coupling. That is, the coupling is through exchange of surface waves between microstrip patches, and the coupling strength diminishes in a rate proportional to the inverse of the square root of the distance separating the patch antennas. Calculations have been applied to two cases consisting of either an isolated antenna or two identical closely coupled antennas. Our calculations compared very well with experiments. The effects on dielectric thickness, feeding postion of the feeder lines, separation distance of the patches, and radiation patterns will be published elsewhere.[3,4]

FORMULATION

We consider the metal patches to be deposited on top of a dielectric substrate whose thickness is d and dielectric constant ϵ_r. The lateral dimension of the microstrip structure is assumed to be infinitely wide so that the previously derived Green's functions can be used.[1,2] We consider the antenna to be constructed from lossy materials such that the metal patches and the ground plane possess the same value of finite conductivity, σ, and the loss tangent of the substrate is denoted as $\tan\delta$. We assume copper is used as the metal part of the microstrip, whose thickness is negligible.

As derived previously,[1,2] the electric field at position ($\underline{\rho}$, z) produced by a unit current source located at position ($\underline{\rho}'$, d) is termed as the dyadic Green's function $\underline{\underline{G}}(\underline{k}; z)$. Therefore, for a given current distribution $\underline{j}_s(\underline{\rho})$ on the metal patches above the dielectric substrate, the resultant electric field is

$$\underline{E}_s(\underline{\rho}; z) = \frac{iZ_o}{4\pi^2 k_o}\iint d^2\underline{k}\, d^2\underline{\rho}' \exp[i\underline{k}\cdot(\underline{\rho} - \underline{\rho}')]\underline{\underline{G}}(\underline{k}; z)\cdot\underline{j}_s(\underline{\rho}'), \tag{1}$$

where $k_o = \omega(\epsilon_o\mu_o)^{1/2}$, $Z_o = (\mu_o/\epsilon_o)^{1/2}$, and $\underline{\underline{G}}(\underline{k}; z)$ is given by

$$\underline{\underline{G}}(\underline{k}; z) = [G_o(k)\,\underline{\underline{I}} - G_2(k)\,\underline{k}\,\underline{k}]\,\sin\beta_1 z/\sin\beta_1 d + i\,G_1(k)\,\underline{e}_z\,\underline{k}\cos\beta_1 z/\cos\beta_1 d. \tag{2}$$

Here

$$G_o = k_o^2 D_e', \tag{3}$$

$$G_1 = \beta D_m', \tag{4}$$

$$G_2 = (\beta + \beta_1\tan\beta_1 d)\, D_e' D_m, \tag{5}$$

$$D_m = (\epsilon\beta + \beta_1\tan\beta_1 d)^{-1}, \tag{6}$$

$$D_e = (\beta + \beta_1\coth\beta_1 d)^{-1}, \tag{7}$$

$$D_e' = D_e - (1 - i)\delta_s(1 - \coth\beta d)/(1 + \coth\beta d), \tag{8}$$

$$D_m' = D_m - (1 - i)\delta_s(1 - \tanh\beta d)/(1 + \tanh\beta d), \tag{9}$$

and β and β_1 are given by

$$\beta = (k^2 - k_o^2)^{1/2}, \tag{10}$$

$$\beta_1 = (k^2 - \epsilon k_o)^{1/2}. \tag{11}$$

We denote δ_s as the rf penetration depth of the ground plane given by

$$\delta_s = (2/\omega\mu_o\sigma)^{1/2} \tag{12}$$

and the surface impedance of the metal patches is

$$Z_s = (1 - i)\sqrt{\frac{\omega\mu_o}{2\sigma}} = \frac{1 - i}{2}(k_o\delta_s)\,Z_o. \tag{13}$$

Note that in Eqs.(6) and (11) we have used the complex dielectric constant ϵ given by

$$\epsilon = \epsilon_r\,(1 + i\tan\delta), \tag{14}$$

which characterizes the loss property of the dielectric substrate. The finite conductivity of the metal patches gives rise to finite surface impedance Z_s in Eq.(12) and finite conductivity of the ground plane results in shifting of the TM surface-wave poles in the integrand of Eq.(1) shown in Eqs.(8) and (9).[1,2]

Let M circular patches be located at position $\underline{r}_\mu$ with radius R_μ, $1 \le \mu \le M$. We assume these metal patches are well separated in space on top of the substrate such that they do not overlap each other. Up to second order in δ, we omit the solenoidal current and express the patch current in terms of irrotational current specified by the basis of $\{\underline{j}_{mn}^{(\mu)}(\underline{\rho})$: $\mu = 1, \cdots, M\}$:[2]

$$\underline{j}_s(\underline{\rho}) = \sum_{\mu=1}^{M} \sum_{mn} a_{\mu mn} \underline{j}_{mn}^{(\mu)}(\underline{\rho}), \tag{15}$$

where $a_{\mu mn}$'s are the unknown coefficients subject to determination. In Eq.(15) $\underline{j}_{mn}^{(\mu)}(\underline{\rho})$ denotes the irrotational multipole current local to the μ-th patch given by the following equation

$$\underline{j}_{mn}^{(\mu)}(\underline{\rho}) = c_{\mu mn}^{-1} \underline{\nabla}_t^{(\mu)} [J_m(\beta_{\mu mn}\rho) \exp(im\phi)],$$
$$= 0, \quad \text{if } \rho > R_\mu. \tag{16}$$

where $\underline{\nabla}_t^{(\mu)}$ denotes the two dimensional gradient operator operating on position (ρ, ϕ) whose coordinate origin is located at the center of the μ-th metal patch. In Eq.(16) $\beta_{\mu mn}$ is given by

$$J_m'(\beta_{\mu mn}R_\mu) = 0, \tag{17}$$

where $J_m(x)$ and $J_m'(x)$ denote the Bessel function of the first kind of order m and its derivative with respect to x, respectively. Note that the current basis of Eq.(18) has been normalized such that

$$\int d^2\rho \underline{j}_{mn}^{(\mu)}(\underline{\rho})^* \cdot \underline{j}_{m'n'}^{(\mu')}(\underline{\rho}) = \delta_{\mu\mu'}\delta_{mm'}\delta_{nn'}, \quad 1 \le \mu, \mu' \le M. \tag{18}$$

This determines the normalization constants $c_{\mu mn}$'s as

$$c_{\mu mn} = d_{\mu mn} R_\mu J_m(\beta_{\mu mn}R_\mu), \text{ and } d_{\mu mn} = \sqrt{\pi}\,(1 - \frac{m^2}{\beta_{\mu mn}^2 R_\mu^2})^{1/2} \beta_{\mu mn}. \tag{19}$$

For the above employed current basis, m denotes the angular mode and n the radial mode.

Under external driving condition one obtains

$$\underline{E}_s - Z_s \underline{j}_s = \underline{E}_e, \tag{20}$$

where $\underline{E}_e$ is the electric field generated by the excitation current $\underline{j}_e(\underline{\rho}\ z)$, and $Z_s \underline{j}_s$ denotes the conduction current due to the finite conductivity of the metal patches. $\underline{E}_s$ is given by Eq.(1) which involves four-fold integration in $\underline{k}$ and $\underline{\rho}'$. The Galerkin's method further requires both sides of Eq.(20) to be multiplied by $\underline{j}_{mn}^{(\mu)}(\underline{\rho})$ followed by integration over $\underline{\rho}$ and z. Integration over z can be conveniently carried out, since it involves δ-function distribution of surface current on microstrip patches. There still remains six-fold integration. However, five-fold integration can be analytically carried out if one expands the patch current $\underline{j}_s$ in terms of the multipole currents, Eq.(15), and utilizes Eqs.(18) and (19) and the following identities:

$$\int_0^{2\pi} d\theta\, e^{im\phi} e^{\pm i\underline{k}\cdot\underline{\rho}} = 2\pi i^{\pm m} J_m(k\rho) e^{im\phi}, \tag{21}$$

$$\int_0^{R_\mu} \rho d\rho J_m(k\rho) J_{m'}(\beta_{\mu'm'n'}\rho) = \delta_{\mu\mu'}\delta_{mm'} \frac{kR_\mu}{\beta_{\mu mn}^2 - k^2} J_m(\beta_{\mu mn}R_\mu) J_m'(kR_\mu), \tag{22}$$

$$J_n(k\rho) e^{in\phi} = \sum_{m=-\infty}^{\infty} [e^{-im\phi_1} J_m(k\rho_1)]\, [e^{i(m+n)\phi_2} J_{m+n}(k\rho_2)]. \tag{23}$$

In Eq.(23) we have used the following notations:

$$\underline{\rho}_1 = (\rho_1\cos\phi_1, \rho_1\sin\phi_1), \quad \underline{\rho}_2 = (\rho_2\cos\phi_2, \rho_2\sin\phi_2),$$

and

$$\underline{\rho} = \underline{\rho}_1 - \underline{\rho}_2 = (\rho\cos\phi, \rho\sin\phi). \tag{24}$$

The resultant Gelerkin elements only involve one-fold integration over k whose functional dependence is given in Eq.(26) below.

The Galerkin equation is therefore written in the following matrix equation

$$\sum_{\mu'm'n'} [B_{\mu\mu'mm'nn'}]\,[a_{\mu'm'n'}] = [b_{\mu mn}], \tag{25}$$

and the kernel matrix, $[B_{\mu\mu'mm'nn'}]$, is given as

$$B_{\mu\mu'mm'nn'} = -Z_s \delta_{\mu\mu'} \delta_{mm'} \delta_{nn'} + \frac{2\pi i Z_o}{k_o} d_{\mu mn}^{-1} d_{\mu'm'n'}^{-1} e^{i(m'-m)\phi_{\mu\mu'}} \cdot$$

$$\cdot \int_0^\infty kdk J_{m-m'}(kR_{\mu\mu'}) \left[G_o(k,d) \frac{mm' J_m(kR_\mu) J_{m'}(R_{\mu'})}{k^2 R_\mu R_{\mu'}} + \right.$$

$$\left. + \frac{[G_o(k,d) - k^2 G_2(k,d)] J_m'(kR_\mu) J_{m'}'(kR_{\mu'})}{(1 - k^2/\beta_{\mu mn}^2)(1 - k^2/\beta_{\mu'm'n'}^2)} \right], \tag{26}$$

where $R_{\mu\mu'}$ and $\phi_{\mu\mu'}$ define the relative position of patch μ to patch μ' given by

$$\underline{\mathbf{R}}_{\mu\mu'} = \underline{\mathbf{R}}_\mu - \underline{\mathbf{R}}_{\mu'} = (R_{\mu\mu'} \cos\phi_{\mu\mu'}, R_{\mu\mu'} \sin\phi_{\mu\mu'}). \tag{27}$$

For an isolated circular patch the partial waves associated with different angular modes, $m \neq m'$, do not couple to each other due to the symmetry of the assumed patch geometry, as it can be checked from Eq.(26) by setting $\mu = \mu' = 1$. For multiple patches, this circular symmetry is broken, and all of the partial waves admix together as shown in Eq.(26) for $\mu \neq \mu'$. We note also that the coupling between patches and between angular modes are through the term $J_{m-m'}(kR_{\mu\mu'})$ in the kernel matrix $[B_{\mu\mu'mm'nn'}]$, Eq.(26). For large patch separation, $kR_{\mu\mu'} >> 1$ and $kR_{\mu\mu'} >> |m - m'|$, $J_{m-m'}(kR_{\mu\mu'})$ approaches zero in a rate proportional to the inverse of the square root of $R_{\mu\mu'}$. This implies that the coupling between microstrip patches is long range in nature. The is indeed true if one recognizes the fact that the interaction is actually taking place through the exchange of surface waves.[2]

The inhomogeneous term in Eq.(25) represents the source excitation. We assume the M patches are excited through N feeders utilizing either coax lines or microstrips. Under the i-th port excitation, $1 \leq i \leq N$, we have

$$b_{\mu mn}^{(i)} = \int_0^d dz \int d^2\underline{\rho}\, \underline{j}_{mn}^{(\mu)}(\underline{\rho})^* \cdot \underline{E}_e^{(i)}(\underline{\rho}, z) = \int_0^d dz \int d^2\underline{\rho}\, \underline{E}_{mn}^{(\mu)}(\underline{\rho}, z)^* \cdot \underline{j}_e^{(i)}(\underline{\rho}, z), \tag{28}$$

where $\underline{\mathbf{E}}_e^{(i)}$ denotes the electric field generated by excitation current $\underline{\mathbf{j}}_e^{(i)}$, and $\underline{\mathbf{E}}_{mn}^{(\mu)}$ is the field generated by $\underline{\mathbf{j}}_{mn}^{(\mu)}$, which can be derived by using Eq.(1) with $\underline{\mathbf{j}}_s$ being replaced by $\underline{\mathbf{j}}_{mn}^{(\mu)}$. In deriving Eq.(28) we have used the reciprocity theorem.[5] We consider the i-th port to be located at $\underline{\rho}_e^{(i)} = (\rho_e^{(i)}\cos\phi_e^{(i)}, \rho_e^{(i)}\sin\phi_e^{(i)})$ which creates the following excitation current at the junction of the feeder to the microstrip patch:

$$\underline{\mathbf{j}}_e^{(i)}(\underline{\rho}, z) = \underline{\mathbf{e}}_z\, \delta(\rho - \rho_e^{(i)})\, \delta(\phi - \phi_e^{(i)})\, \delta(z - d)/\rho, \text{ if coax line excited,} \tag{29}$$

and

$$\underline{\mathbf{j}}_e^{(i)}(\underline{\rho}, z) = \underline{\mathbf{e}}_\rho\, \delta(\rho - \rho_e^{(i)})\, S(\phi)\, \delta(z - d)/\rho, \text{ if microstrip excited.} \tag{30}$$

In Eq.(29) we have modeled the coax line to possess an inner filament of zero diameter ending in a point charge at the junction with the patch. For microstrip line the feeder is located at the edge of the patch with uniform current distribution in an angular width of $2\phi_o^{(i)}$. That is, the angular function $S(\phi)$ in Eq.(30) is

$$S(\phi) = 1/(2\phi_o^{(i)}), \text{ if } |\phi - \phi_e^{(i)}| < \phi_o^{(i)},$$
$$= 0, \text{ otherwise.} \tag{31}$$

We note here that both the excitation currents of Eqs.(29) and (30) are of unit magnitude:

$$\int_{\Sigma_i} d^2s\, |\underline{j}_e^{(i)}(\underline{\rho}, d)| = 1, \tag{32}$$

where Σ_i denotes the cross-sectional area of the port at the junction with the patch.

After substituting Eqs.(29), (30), and (31) into Eq.(28), we obtain

$$b_{\mu mn}^{(i)*} = \frac{-2\pi i Z_o}{k_o} d_{\mu mn}^{-1} \sum_{m',n'}^{\{i\in\mu'\}} R_{\mu'}^{-1} d_{\mu' m' n'}^{-2} F_e(\rho_e^{(i)}) \int_0^\infty \frac{k^3 dk}{\beta_1} G_1(k,d) \tanh\beta_1 d \cdot$$

$$\cdot \frac{J_m'(kR_\mu) J_{m'}'(kR_{\mu'})}{(1 - k^2/\beta_{\mu mn}^2)(1 - k^2/\beta_{\mu' m' n'}^2)}, \tag{33}$$

where $F_e(\rho_e^{(i)})$ is given as

$$F_e(\rho_e^{(i)}) = \exp(im'\phi_e^{(i)}) \frac{J_{m'}(\beta_{\mu' m' n'} \rho_e^{(i)})}{J_{m'}(\beta_{\mu' m' n'} R_{\mu'})}, \quad \text{if coax excited,} \tag{34}$$

$$F_e(\rho_e^{(i)}) = \exp(im'\phi_e^{(i)}) \frac{\sin m'\phi_o^{(i)}}{m'\phi_o^{(i)}}, \quad \text{if microstrip excited.} \tag{35}$$

In Eq.(29) we have used the notation $\{i \in \mu'\}$, which denotes that the patch μ' contains the junction imposed by the i-th port feeder.

Interport impedance can be calculated from[6]

$$Z_{ij} = \int_0^d dz \int d^2\rho \, \underline{j}_s^{(i)}(\underline{\rho})^* \cdot \underline{E}_e^{(j)}(\underline{\rho}, z) / [\int_{\Sigma_i} d^2 s \, |\underline{j}_e^{(i)}(\underline{\rho}, d)|][\int_{\Sigma_j} d^2 s \, |\underline{j}_e^{(j)}(\underline{\rho}, d)|], \tag{36}$$

where $\underline{j}^{(i)}(\underline{\rho})$ is the induced patch current through excitation of the i-th port, and $E_e^{(i)}(\underline{\rho}, z)$ is the generated electric field associated with excitation of the j-th port. By using the reciprocity theorem [5] together with Eqs. (25), (28), and (32), eq. (36) can be written as

$$[Z_{ij}]_{N\times N} = - \sum_{\mu\mu' mm' nn'} [b_{\mu mn}^{(i)}]^* [B_{\mu\mu' mm' nn'}]^{-1} [b_{\mu' m' n'}^{(j)}]. \tag{37}$$

From Eq.(37) we note that $Z_{ij} = Z_{ji}$, since $[B_{\mu\mu' mm' nn'}]$ is a symmetric matrix, as it can be explicitly checked from Eq.(26). Let the source voltage at these N-port feeders be expressed by a column vector $[V_j]_{N\times 1}$. The input currents are therefore

$$[I_i] = \sum_{j=1}^{N} [Z_{ij}]^{-1} [V_j], \tag{38}$$

and the input impedance at the i-th port is

$$Z_{IN}^{(i)} = V_i / I_i, \quad 1 \le i \le N. \tag{39}$$

We have ignored the feeder self-impedance in Eq.(37). If self-impedance is not negligible, one needs to add to Eq.(37) a diagonalized matrix $[Z_L]_{N\times N}$ where

$$(Z_L)_{ij} = \delta_{ij} Z_i \tag{40}$$

and Z_i denotes the self-impedance of the i-th port.[7]

RESULTS

There exist three kinds of difficulties in performing numerical integration for the matrix elements of $[B_{\mu\mu' mm' nn'}]$, Eq.(26), and $[b_{\mu mn}]$, Eq.(33), known as Sommerfeld-type integrals.[8] The first difficulty is associated with the quasi-resonant modes of the metal disks occurring at $k = \beta_{\mu mn}$. However, since these singularities always appear in the following form

$$\frac{[J_m'(kR_\mu)]^2}{(1 - k^2/\beta_{\mu mn}^2)(1 - k^2/\beta_{\mu' m' n'}^2)}, \tag{41}$$

they also occur as zeros to the numerator of Eq.(41). Eq.(41) remains finite even when $\mu = \mu'$, $m = m'$, and $n = n'$. Therefore, these singularities do not physically cause problem. However, they cannot be accessed directly. In order to avoid large truncation error, Eq.(41) needs to be evaluated using its Taylor expansion in the vicinity of these quasi-resonant points.

The second difficulty arises from the simple poles in the Green's function of Eq.(2), identified as surface TM and TE waves described by D_m and D_e in Eqs.(6) and (7), respectively.[7] However, by taking into account of dielectric and conductor imperfections, Eqs.(6) and (7) have been modified into Eqs.(8) and (9) with the corresponding surface poles being pushed off from the real axis toward the lower half plane (of the k-variable). As such, the sense of integration along real axis is proper. However, due to sharp cancellation of the integrands near these quasi-singular points, accurate numerical integration is not so easy. We have employed the fifth order Runge-Kutta method to evaluate these Sommerfeld-type integrals near surface-pole singularities. This adopted method allows for adaptive adjustment of its integration steps such that the local error can be monitored and the global error is well controlled.[9] Accuracy in one part per million has thus been achieved for the Sommerfeld-type integrals in the interval from k_o to $\epsilon_r k_o$ to which the surface poles will occur.

The third difficulty is associated with the improper integral of Bessel functions at infinity. Due to mild oscillation of Bessel functions at infinity, integrals can hardly be evaluated accurately even when extrapolation scheme is employed. We have circumvented this difficulty by exploiting asymptotic expansion of Bessel function at infinity. For large arguments, $x \gg m$, and $x \gg 1$, Bessel functions can be expanded as[10]

$$J_m(x) = \sqrt{\frac{2}{\pi x}}\left[\cos(x - \frac{\pi x}{2} + \frac{\pi}{4})\sum_{k=0}^{\infty} \frac{(-1)^k}{2^{2k}} \frac{\Gamma(m+2k+1/2)}{(2k)!\,\Gamma(m-2k+1/2)} x^{2k} - \right.$$

$$\left. - \sin(x - \frac{m\pi}{2} + \pi/4)\sum_{k=0}^{\infty} \frac{(-1)^k}{2^{2k+1}} \frac{\Gamma(m+2k+3/2)}{(2k+1)!\,\Gamma(m-2k-1/2)} x^{2k+1}\right], \tag{42}$$

where Γ's denote gamma functions. As such, integration of Bessel functions at infinity depicted in Eqs.(26) and (33) can be converted into the following form

$$\int_{cutoff}^{\infty} \frac{A_p \sin(\alpha x + \beta)}{x^p}\, dx, \tag{43}$$

where p can be an integer, when $\mu = \mu'$, or a half-integer, when $\mu \neq \mu'$. Eq.(43) can be explicitly calculated by using sine and cosine integrals and their derivatives for integer values of p, or using error functions and their derivatives for half-integer values of p. We have verified that by keeping 8 terms in Eq.(42) the associated improper Bessel integrals can be evaluated up to 12 significant digits for a moderate choice of the cutoff value used in Eq.(43).

The remaining task is to expand the surface-wave kernels, G_o, G_1, G_2, at infinity in terms of powers in 1/k. That is, we choose cutoff such that exp(-2kd) can be ignored for k > *cutoff.* The surface-wave kernels, described by Eqs.(3) to (5), can be then derived whose explicit forms can be found elsewhere.[3,4] We note here that although it is most tedious to perform integration around infinity for Bessel functions, the most time-consuming procedure occurs to the computation associated with surface-wave pole contributions. Sommerfeld-type integrals, Eqs.(26) and (33), have thus been calculated up to six significant digits.

In the following calculations we have considered the Galerkin method to consist of the following matrix elements in Eqs.(25), (26), (33): $\mu = 1$, $m = \pm 1$, $1 \le n \le 20$, for a single isolated patch, and $1 \le \mu \le 2$, $-9 \le m \le 9$, $1 \le n \le 20$, for coupled two patches. The size of the kernel matrix $[B_{\mu\mu' mm' nn'}]$ is, therefore, 40×40 and 760×760 for a single patch and for two coupled patches, respectively. By ignoring the other multipole terms the resultant truncation error has been verified to be within a few parts in one thousand in the frequency range near the fundamental mode excitation. In this paper we have considered two calculations. The first calculation applied to the case previously reported in Ref.11 consisting of a single antenna possessing the following parameters: $R_1 = 6.75$ cm, $d = 0.1588$ cm, $\epsilon_r = 2.62$, and $\tan\delta = 0.001$. The antenna is fed by a microstrip line with a half-suspension angle $\theta_1 = 0.03254$ rad, which results in 50 Ω line in the present geometry. The resonance frequency is defined as the frequency at which the input impedance loci intersects with the real axis. The calculated resonance frequency of the fundamental mode is 0.7936 GHz, which compares exactly to the measured value of 0.794 GHz. This is contrasted to the cavity model which predicts a resonant frequency of 0.805 GHz.[11] We therefore conclude that at resonance surface waves and fringing fields have effectively increased the radius of the metal patch by a fraction of 1.37%. The calculated input impedance of the antenna is shown in Fig.1. It is seen in Fig.1 that the calculated impedance loci in Smith chart compare very well with the measured values which are shown as small circles cited from Ref.11. It is also seen in Fig.1 that slightly larger discrepancy occur near the two ends of the impedance loci, indicating that multipoles of higher orders need to be considered when frequency drifts away from resonance. The effects on dielectric thickness and on the feeder line position, and the resultant radiation patterns have also been calculated, which will be published elsewhere.[3]

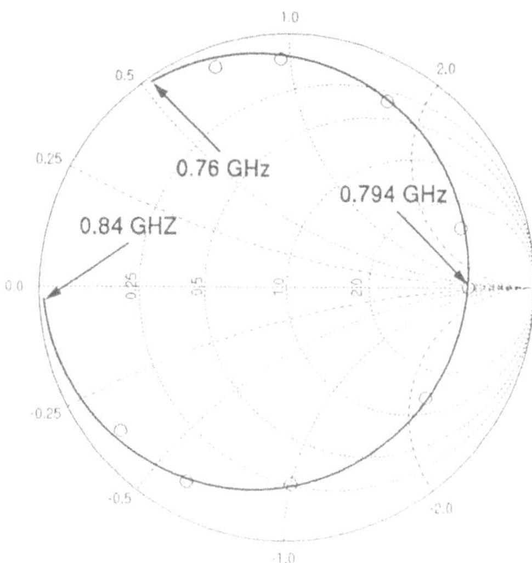

Figure 1 Calculated and measured input impedance loci of a single isolated patch antenna reported in Ref.11.

The second calculation applies to two identical circular patche antennas of radius $R_1 = R_2 = 1$ cm, either touching each other, $R_{12} = 2$ cm, or separated by large distance, $R_{12} = 5$ cm. These microstrip patches are fed by microstrip lines arranged parallel to each other, but orthogonal to the line joining the centers of the patches. The dielectric substrate considered is duroid material with dielectric constant $\epsilon_r = 2.2$, loss tangent $\tan\delta = 0.001$, and thickness $d = 0.01$ cm (RT/duroid 5880, Rogers Co., Chandler, AZ, USA). The microstrip feeders are of width 0.0312 cm such that their characteristic impedance is 50 Ω. We consider the two circular microstrip patches to be excited near their respective fundamental modes. The excitation of the patches is assumed to be totally symmetric and the input voltages at the two feeder ports are therefore of equal phase and magnitude: $V_1 = V_2$, Eq.(38). Fig.2 shows the calculated input impedance loci in Smith chart for the coupled case, $R_{12} = 2$ cm, and the uncoupled case, $R_{12} = 5$ cm. We see from Fig.2 that not only the impedance values of the two patches, $Z_{in}^{(1)} = Z_{in}^{(2)}$, Eq.(39), have been changed as a result of patch coupling, but also their resonant frequencies have been changed. Calculations have been compared with experiments where the patch antennas were fabricated using RT/duroid 5880 material. Measurements are shown in Fig.2 as crosses and small circles for the closely coupled and uncoupled patches, respectively. The measured resonant frequencies were 5.514 and 5.561 GHz for the coupled and uncoupled case, which correspond almost exactly to their respective calculated values of 5.5137 and 5.5642 GHz, see Fig.2. The calculated impedance loci also comply very well with measurements, except when frequencies are drifted away from the resonant points where it seems that more multipole terms should be included in the calculation for the Galerkin elements, Eqs.(26) and (33). We have also calculated the resonant frequencies and input impedances, as well as their compound radiation patterns, for intermediate R_{12} values. These calculations, together with measurements, will be published elsewhere.[4]

ACKNOWLEDGMENT

We acknowledge the support of this research from US Navy Air-Warfare Center under contract number N62269-96-C-0013.

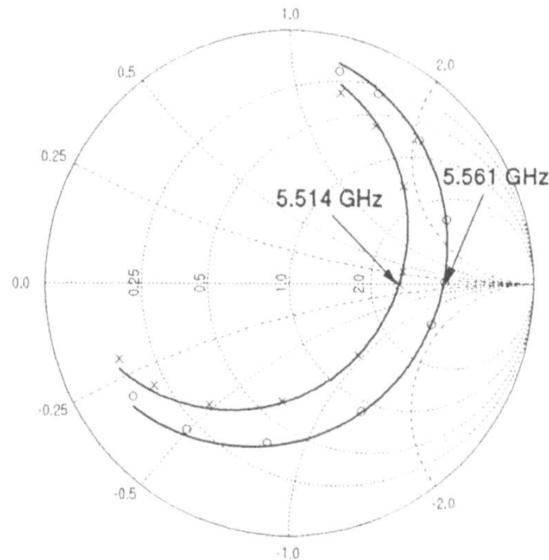

O: Uncoupled case (R = 5 cm) x: Coupled case (R = 2 cm)

Figure 2 Calculated and measured input impedance loci of two coupled patch antennas with the following parameters: $R_1 = R_2 = 1$ cm, $R_{12} = 2$ cm and $R_{12} = 5$ cm.

REFERENCES

1. H. How and C. Vittoria, "New Formulation of Dyadic Green's Function: Applied to a Microstrip Line," IEEE Trans. Microwave Theory Tech., Vol. MTT-42, pp.1580, 1994.
2. H. How and C. Vittoria, "Radiation Modes in Dielectric Circular Patch Antennas," IEEE Trans. Microwave Theory Tech, Vol. MTT-42, pp.1939, 1994.
3. H. How and T. -M. Fang, "Green's Function Calculation on Circular Microstrip Patch Antennas," to be published.
4. H. How and T. -M. Fang, "The Coupling Between Microstrip Patch Antennas of Circular Geometry," to be published.
5. J. A. Stratton, "Electromagnetic Theory," McGraw-Hill, New York, 1941.
6. R. F. Harrington, "Time Harmonic Electromagnetic Fields," McGraw-Hill,, New York, 1961.
7. J. R. James and P. S. Hall, Ed., "Handbook of Microstrip Antennas," Chapter 8, "Numerical Analysis of Microstrip Patch Antennas," by J. R. Mosig, R. C. Hall, and F. E. Gardiol, London, U.K., P. Peregrinus on Behalf of the Institute IEE Electromagnetic Waves Series, 1989
8. A. Sommerfeld, "Partial Differential Equations," Academic Press, New York, 1962.
9. G. E. Forsythe, M. A. Malcolm, C. B. Moler, "Computer Methods for Mathematical Computations," Prentice-Hall, Englewood Cliffs, NJ, 1977.
10. M. Abramowitz and I. A. Stegun, Ed., "Handbook of Mathematical Functions with Formulas, Graphs, and Mathematical Tables," National Bureau of Standards, Applied Mathematics Series, 55, Washington, DC, 1964, pp.363.
11. Y. T. Lo, D. Solomon, and W. F. Richards, "Theory and Experiment on Microstrip Antennas," IEEE Trans. Antenna Propagat., Vol.AP-27, pp.137, 1979.

Matching and Compensation Network Synthesis for MMICs

Stephen E. Sussman-Fort

Department of Electrical Engineering
State University of New York
Stony Brook, New York 11794-2350 USA
E-mail: sussman@sbee.sunysb.edu

ABSTRACT

Matching network synthesis continues to be of importance in microwave amplifier design. Such networks are required between consecutive transistor stages typically to manage the transfer of power between the stages, to compensate for imperfections of the devices as manifested by gain roll-off and complex terminating impedances, and to minimize noise figure. It is often necessary to obtain the best compromise between potentially conflicting requirements. Synthesis techniques have evolved from the classical analytic gain-bandwidth theory to the more practical numerically-based real-frequency method, and to algorithms which allow the specification of arbitrary circuit structures. As of yet, there have been no means of identifying the best circuit structure (or even the best algorithm) for a given set of requirements, and some of the newer techniques systematically evaluate all topologies of a given degree.

A closely related synthesis problem occurs in certain types of MMIC active filter design. In this case, a passive MMIC one-port, comprised generally of lossy elements, is designed to compensate for the imperfections of both the active and passive elements throughout the circuit to yield a desired filter response over a finite bandwidth. Two-port matching network design techniques can be adapted to serve here, but again we lack a unified theory to identify the best circuit structures and design approaches.

APPLICATIONS

Matching Networks

Matching networks are employed in microwave amplifiers to manage the transfer of power between successive stages, between the source and the input stage, and between the output stage and the load. Such matching networks, or equalizers, as they are often

Directions for the Next Generation of MMIC Devices and Systems
Edited by N. K. Das and H. L. Bertoni, Plenum Press, New York, 1997

called, attempt to provide maximum transfer of power through the amplifier, subject to the potentially conflicting constraints of amplitude equalization and noise-figure minimization. The most effective equalizer design methods employ a numerical, computer-based approach, not the least reason being that MMIC transistors themselves are described by tables of S-parameters. This being the case, the *numerical broadband matching problem* can conveniently be stated as follows:

Given

- n frequency points: $f_1, f_2, ..., f_n$

- n complex source impedances Z_{si}:
  $(R_{s1} + jX_{s1}),\ \ (R_{s2} + jX_{s2}),\ ...,\ (R_{sn} + jX_{sn})$

- n complex load impedances Z_{Li}:
  $(R_{L1} + jX_{L1}),\ (R_{L2} + jX_{L2}),\ ...,\ (R_{Ln} + jX_{Ln})$

- n transducer-gain function magnitudes:
  $S_{21}(1),\ S_{21}(2),\ ...,\ S_{21}(n)$

Design a Network that Provides

$S_{21}(i)$ at f_i *with* source $(R_{si} + jX_{si})$, load $(R_{Li} + jX_{Li})$ *for* $i = 1, 2, ..., n.$

We will briefly review various matching design methods in the next section; however, the reader is referred to [1] for a more detailed overview of some of the earlier techniques.

Compensation Networks

A new synthesis application in MMIC circuit design arises from the development of active filter building blocks. In audio frequency active filter design, one can easily build negative impedance converters (NICs), gyrators, and frequency-dependent negative resistances (FDNRs) using op amps, resistors, and capacitors. At microwave frequencies, we endeavor to realize these same circuits using imperfect MMIC transistors, resistors, capacitors, and inductors. Note that in the microwave regime, unlike at low frequencies, inductors can be realized as integrated circuit elements. However, both the active and passive MMIC elements possess substantial parasitics that must be accounted for to achieve satisfactory circuit performance.

The strategy for designing a microwave negative resistance [2] can be understood by referring to Fig. 1. In the ideal case, using an infinite-transconductance model for the transistors and terminating port 2 with a resistance R_1, the input impedance Z_{in} is seen to be negative and real. With real microwave components, this will certainly not be the case. To achieve a purely negative-real Z_{in} under these conditions, we must terminate port 2 instead with a special impedance Z_L whose specifications can be easily derived. In particular, the chain matrix T of a nonideal NIC possesses elements A, B, C, and D which are functions of the Laplace-transformed frequency variable ***s***. From Fig. 1, for our nonideal NIC, Z_{in} at port 1 will be bilinear in the port-2 terminating impedance Z_L, where Z_L has replaced the resistance R_1 originally employed in the ideal case. We can solve for that special Z_L that will produce a $Z_{in} = -R_{in}$ at port 1, as shown at the bottom of Fig. 1. In [2], we discuss the physical realizability of Z_L as well as a numerical method for synthesizing this impedance. Referring to Fig. 2, we briefly summarize the process. First, a *lossless* matching network is designed between the port-1 termination Z_L^* and an arbitrary resistive port-2 termination R so that $|S_{21}|=1$ over a prescribed passband. The input impedance to the network at port 1 will then be the desired Z_L when port 2 is terminated in R. The next step is to invoke realistic models for the MMIC elements and optimize the overall circuit to still achieve Z_L despite the component parasitics. Finally, the MMIC realization for Z_L is inserted at port-2 of the NIC to produce a negative resistance at port-1 over a finite passband.

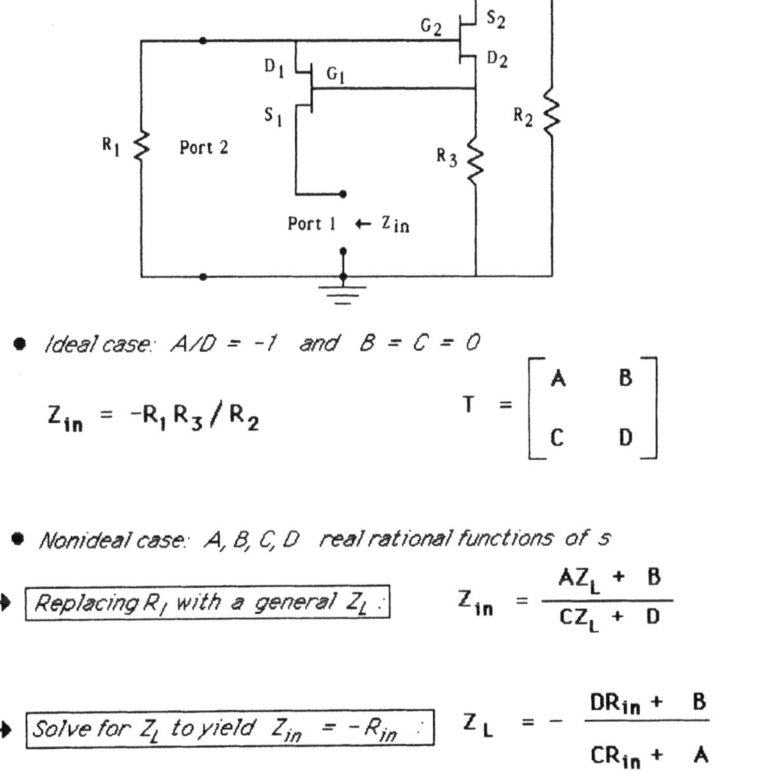

Fig. 1: Negative Impedance Converter (NIC) (terminated in R_1 at port 2) - ideal and nonideal cases

Similar approaches have been employed to realize negative capacitance [3] and simulated inductance [4]. Further research might include (1) optimal topology selection and (2) numerical methods to directly realize the (lossy) compensating one-ports required in these circuits, rather than having to start with a lossless matching network design.

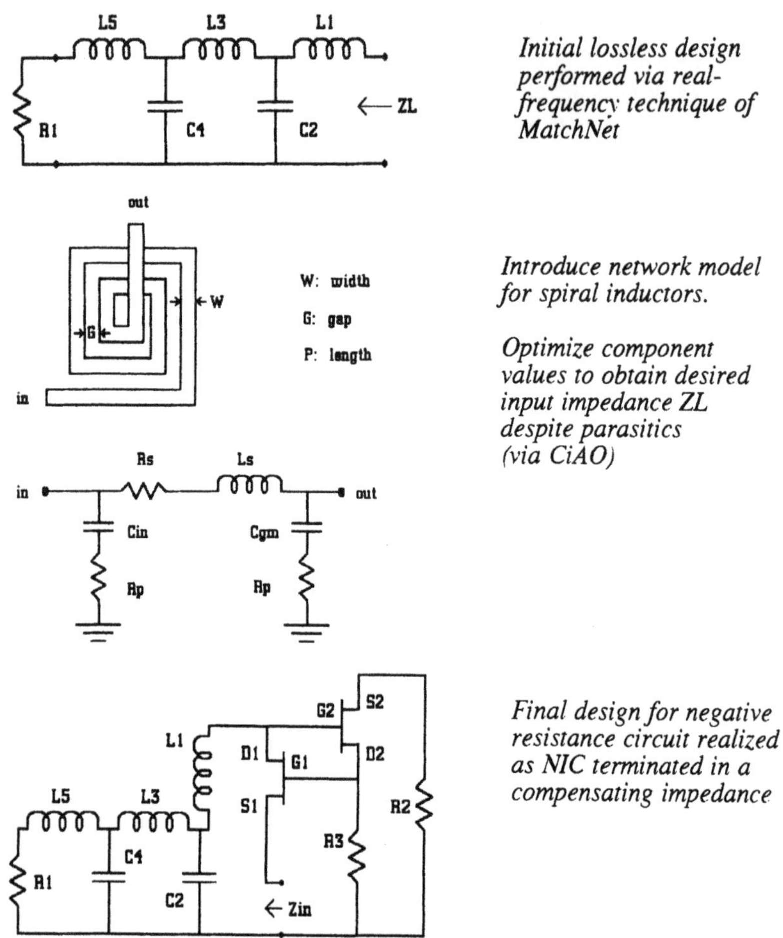

Fig. 2: One - Port Synthesis and Negative Resistance from a Compensated NIC

ALGORITHMS

Matching network design for amplifiers and compensation network design for active filters both require synthesis algorithms which realize two-ports with a prescribed $|S_{21}|$ between complex terminations. The evolution of this methodology has been discussed in detail in [1], however we shall outline the salient points here and discuss the newest approach in somewhat more detail.

The earliest synthesis methods arose from the analytic gain-bandwidth theory of Fano and Youla [5 , 6], and implementations of expanded versions of this have been given by Mellor and Linvill [7]. In these approaches, it is necessary to construct an equivalent circuit for the source and load, and the overall circuit produces a Chebyshev equiripple response, which is optimal only when the gain-bandwidth constraints allow a 0 dB $|S_{21}|$ passband gain. A brilliant departure from conventional practice was first developed by Carlin [8] with his "real-frequency technique (RFT)" which allows a numerical description of the source and load, and realizes optimal networks, in the maxi-min sense, i.e. maximizing the minimum value of gain across the passband, even when a flat loss must be accepted because of gain-bandwidth constraints. However, many questions have remained since the discovery of the RFT; for example: (1) since the network structure is a result of the design process in RFT and earlier methods, could there be other topologies that can yield better circuit responses, and (2) what algorithms should be employed to find these other solutions.

The most fruitful area of research in matching network design --with clear application to one-port synthesis as well-- has been the development of *iterated analysis* methods. In this technique, we select a network topology first, and then consider the transducer-gain function $|S_{21}(\text{calc.})|$ calculated from an initial set of component values and the desired values $|S_{21}(\text{req.})|$. We form an error function as, for example, the sum of the squared-differences between $|S_{21}(\text{calc.})|$ and $|S_{21}(\text{req.})|$ over a number of discrete frequency points in the passband, and then attempt to minimize this function. Orchard [9] was the first to apply this idea to filter design, and the result was an elegant and simple algorithm that had the same or better accuracy as previous methods but with much greater computational speed. Orchard [10] then expanded his efforts to matching networks, but met with somewhat less success, because in filter design one could employ the characteristic function K(s), rather than $H(s)=1/S_{21}(s)$, and also because the classical Gauss-Newton Least Squares minimization employed sometimes yielded convergence to local minima.

Our own efforts [11 , 12 , 13] in using iterated analysis for matching network design include using a direct-search method of minimization and a special error function, which has yielded excellent results. Dedieu [14] has developed a gradient-based iterated analysis method (Recursive Stochastic Equalization: *RSE*) which achieves excellent convergence through the use of a stochastic Gauss-Newton least-squares algorithm.

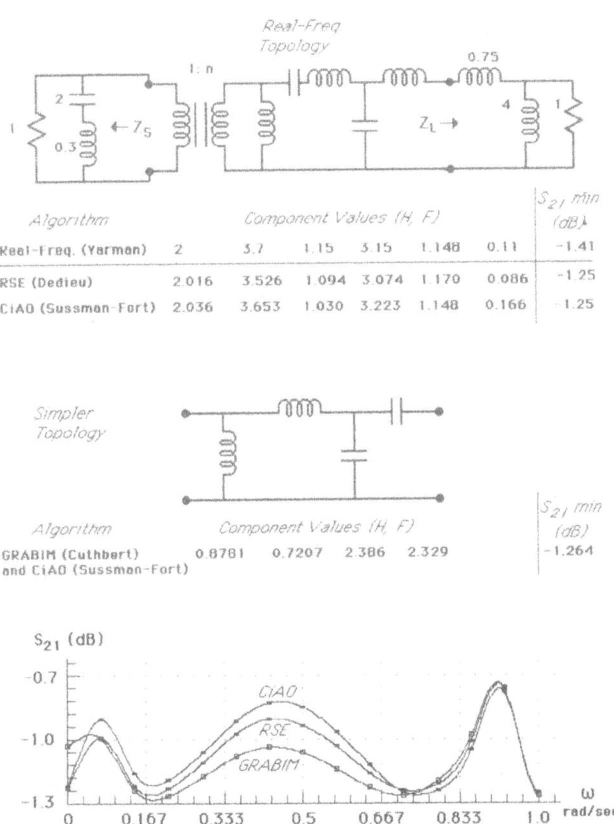

Fig. 3: Matching Network Design Example - Comparison of Real Frequency, RSE and CiAO Techniques

Cuthbert's [15] approach (*GRABIM*) to iterated analysis is to perform a global minimax search over several network topologies first, and then to apply a deterministic Gauss-Newton optimizer to achieve final convergence. In many examples, these three approaches appear to be roughly equal in their synthesis capability. However, our program *CiAO* [12] has the greatest flexibility, being able (1) to design network structures of truly arbitrary topology, and (2) to employ, if desired, mixed lumped-distributed lossy elements.

We have compared the performance of the real-frequency, *RSE, CiAO* and *GRABIM* methods in the design of a matching network between complex loads as shown in Fig. 3. The first topology shown is that which arises from an implementation of the real-frequency method. *RSE* and *CiAO* do even better, providing a higher gain floor, with each yielding a different set of component values. *GRABIM* and *CiAO* achieve the same solution and almost as good a response with a simpler topology. Plots of the response for the higher degree network achieved with *CiAO* and *RSE* are compared to that achieved with *GRABIM* (and duplicated by *CiAO*).

CONCLUSIONS

MMIC applications in matching network and one-port synthesis will continue to be of importance. While the iterated analysis method of design appears to be the best method to date, different implementations achieve different solutions, so the question of just what is optimal comes into play. Furthermore, other than exhaustively searching the available topologies, there seems to be no way as yet to predict which circuit structures will yield the best results in any given application.

REFERENCES

1 . S. E. Sussman-Fort, "The Computer-Aided Design of Microwave Matching Networks," *Int. J. MIMICAE,* Vol. 1, No. 3, Sept. 1991, pp. 288-305.

2 . S. E. Sussman-Fort, "An NIC-Based Negative Resistance Circuit for Microwave Active Filters," *Int. J. MIMICAE,* Vol. 4, No. 2, April 1994, pp. 130-139.

3 . S. E. Sussman-Fort and L. Billonnet, "An NIC-Based Negative Capacitance Circuit for Microwave Active Filters," *Int. J. MIMICAE,* Vol. 5, No. 4, July 1995, pp. 271-277.

4 . S. E. Sussman-Fort and L. Billonnet, "MMIC Simulated Inductors Using Compensated Gyrators," *Int. J. MIMICAE,* to appear.

5 . R. M. Fano, "Theoretical Limitations on the Broadband Matching of Arbitrary Impedances," *J. Franklin Inst.*, 249, Jan. 1950, pp. 52-83; 249, Feb. 1950, pp. 129-155.

6 . D. C. Youla, "A New Theory of Broadband Matching," *IEEE Trans. Circuit Theory*, CT-11, Mar. 1964, pp. 30-50.

7 . D. J. Mellor and J. G. Linvill, "Synthesis of Interstage Networks of Prescribed Gain versus Frequency Slopes," *IEEE Trans. Microwave Theory and Tech.*, MTT-23, Dec. 1975, pp. 1013-1020.

8 . H. J. Carlin, "A New Approach to Gain-Bandwidth Problems," *IEEE Trans. Circ. and Syst.*, CAS-24, April 1977, pp. 170-175.

9 . H. J. Orchard, "Filter Design by Iterated Analysis," *IEEE Trans. Circ. and Syst.*, CAS-32, Nov. 1985, pp. 1089-1096.

10 . H. J. Orchard, Dept. Of Electrical Engineering, University of California at Los Angeles, private communication.

11 . S. E. Sussman-Fort, "Synthesizing Broadband Matching Networks of Arbitrary Structure," *Microwave Systems News,* Jan. 1988, pp. 98-102.

12 . S. E. Sussman-Fort, "Approximate Direct-Search Minimax Circuit Optimization, " *Int. J. for Numerical Meth. in Engrg.*, 28, Feb. 1989, pp. 359-368.

13 . S. E. Sussman-Fort, *CiAO: A Program for the Analysis, Optimization,, and Synthesis of Microwave and RF Networks,* SPEFCO Software, Stony Brook, NY 1994.

14 . H. Dedieu et. al., "A New Method for Solving Broadband Matching Problems, " *IEEE Trans. Circ. and Syst.-- I: Fund. Theory and Appl.*, Vol. 41, No. 9, Sept. 1994, pp. 561-571.

15 . T. R. Cuthbert, Jr., "Broadband Impedance Matching -- Fast and Simple," *RF Design*, Nov. 1994, pp. 38-50.

INDEX

www.ingramcontent.com/pod-product-compliance
Ingram Content Group UK Ltd.
Pitfield, Milton Keynes, MK11 3LW, UK
UKHW051131260726
13967UKWH00010B/2976

* 9 7 8 1 4 8 9 9 1 4 8 1 1 *